Communications
in Computer and Information Science   488

T0183284

Aboul Ella Hassanien   Mohamed F. Tolba
Ahmad Taher Azar (Eds.)

Advanced Machine Learning Technologies and Applications

Second International Conference, AMLTA 2014
Cairo, Egypt, November 28-30, 2014
Proceedings

 Springer

Volume Editors

Aboul Ella Hassanien
Cairo University, Egypt
E-mail: aboitcairo@gmail.com

Mohamed F. Tolba
Ain Shams University, Cairo, Egypt
E-mail: tolba@asunet.shams.eun.eg

Ahmad Taher Azar
Benha University, Benha, Egypt
E-mail: ahmad_t_azar@ieee.org

ISSN 1865-0929   e-ISSN 1865-0937
ISBN 978-3-319-13460-4   e-ISBN 978-3-319-13461-1
DOI 10.1007/978-3-319-13461-1
Springer Cham Heidelberg New York Dordrecht London

Library of Congress Control Number: Applied for

Typesetting: Camera-ready by author, data conversion by Scientific Publishing Services, Chennai, India

Printed on acid-free paper

Springer is part of Springer Science+Business Media (www.springer.com)

Preface

This book constitutes the refereed proceedings of the Second International Conference on Advanced Machine Learning Technologies and Applications, AMLTA 2014, held in Cairo, Egypt, in November 2014. The papers included in the proceedings volume went through a thorough review process by at least three highly qualified peer reviewers. Comments and suggestion from them have helped improve the quality of the papers but also the division of the volume into parts, and assignment of the papers to the best suited parts.

The papers are organized in topical sections on machine learning in Arabic text recognition and assistive technology, Web-based application and case-based reasoning construction, social networks and big data sets, recommendation systems for cloud services, machine learning in watermarking /authentication and virtual machine, fuzzy multicriteria decision making, features extraction and classification, and rough/ fuzzy sets and applications

Thanks are due to many people and parties involved. First, in the early stage of the preparation of the conference's general perspective, scope, topics and coverage, we received invaluable help from the members of the international Program Committee.

At the stage of the running of the conference, many thanks are due to the members of the Organizing Committee, chaired by Prof. Nashwa El-Bendary and supported by their numerous collaborators. And last but not least, we wish to thank Dr. Alfred Hofmann, Leonie Kunz, Aliaksandr Birukou, and Christine Reiss for their dedication and help in implementing and completing this publication project on time while maintaining the highest publication standards.

October 2014

Aboul Ella Hassanien
Mohamed F. Tolba
Ahmad Taher Azar

Organization

Honorary Chairs

Janusz Kacprzyk, Poland

General Co-chairs

Aboul Ella Hassanien, Egypt
M. Fahmy Tolba, Egypt

International Advisory Board

Dominik Slezak, Canada
Hiroshi Sakai, Japan
Mohamed Medhat Gaber, UK
Qiangfu Zhao, Japan

Václav Snášel, Czech Republic
Zbigniew Suraj, Poland
Nashwa El Bendary, Egypt

Program Co-chairs

Ahmed Taher Azar, Egypt
Azizah BTE Abdul Manaf, Malaysya

Publicity Chairs

Abder-Rahman Ali, France
Ammar Adl, Egypt

Mohamed T. Azab, German

International Program Committee

Tomasz Smolinski, USA
Soumya Banerjee, India
Hiroshi Sakai, Japan
Nikhil R. Pal, India
Pawan Lingras, Canada
Qiangfu Zhao, Japan
Václav Snášel, Czech Republic
Zbigniew Suraj, Poland
Zied Elouedi, Tunisia

Khaled Shaalan, Cairo University,
    Egypt
Adel M. Alimi, Tunisa
Siby Abraham, India
Azizah Abd Manaf, Malysia
Barna Iantovics, Romania
Emilio Corchado, Spain
Guoyin Wang, China
Hala Own, Kuwait

Local Organizers

Nashwa El Bendary, Egypt
Mohamed Mostafa, Egypt
Mostafa Salama, Egypt

Hassan Aboul Ella, Egypt
Tarek Gaber, Egypt

Table of Contents

XII   Table of Contents

Machine Learning in Watermarking/Authentication and Virtual Machine

Features Extraction and Classification

Rough/Fuzzy Sets and Applications

Fuzzy Multi Criteria Decision Making

Web-Based Application and Case Based Reasoning Construction

Social Networks and Big Data Sets

Part I
Machine Learning in Arabic Text Recognition and Assistive Technology

Arabic Text Recognition Based
on Neuro-Genetic Feature Selection Approach

Marwa Amara[1] and Kamel Zidi[2]

[1] University of Tunis, SOIE-Management Higher Institute,
Le Bardo 2000, Tunisie
amaralmarwa@gmail.com
[2] University of Tabuk, Community College,
Saudi Arabia
kamel_zidi@yahoo.fr

Abstract. The recognition of a character begins with analyzing its form and extracting the features that will be exploited for the identification. Primitives can be described as a tool to distinguish an object of one class from another object of another class. It is necessary to define the significant primitives. The size of vector primitives can be large if a large number of primitives are extracted including redundant and irrelevant features. As a result, the performance of the recognition system becomes poor, and as the number of features increases, so does the computing time. Feature selection, therefore, is required to ensure the selection of a subset of features that gives accurate recognition. In our work we propose a feature selection approach based genetic algorithm to improve the discrimination capacity of the Multilayer Perceptron Neural Networks (MLP).

Keywords: Arabic character recognition, Feature selection, Perceptron multilayer, Genetic algorithm.

1   Introduction

In recent years, the recognition of Arabic scripts has received increasing attention. Many approaches for the recognition of Arabic characters have been proposed. However, no high recognition rate has been achieved from existing recognition systems [1]. The main reason for getting low accuracy is accounted for by the particularity of the Arabic script. Unlike other languages, the Arabic script has morphological characteristics that are the cause of the failure of treatment. As stated earlier, in order to recognize a given character, every character needs to be analyzed in terms of its form, and its features need to be extracted for characters' identification purposes.

Since the selection of the best discriminant features is considered as a crucial step in the recognition system, many researches have stressed that the inclusion of additional features leads to a worse rather than better performance. Additionally, the choice of features to represent the characters under treatment affects several aspects of the recognition system (such as the recognition rate, the learning time, and the

A.E. Hassanien et al. (Eds.): AMLTA 2014, CCIS 488, pp. 3–10, 2014.
© Springer International Publishing Switzerland 2014

necessary number of samples). The preliminary works on feature selection started in the 1960s [2].

Our paper is organized as follows: In Section 2, we provide an overview of feature selection methods and Genetic algorithm. Section 3 thoroughly exposes the details of the proposed method. Section 4 will be devoted to the experimentation and evaluation. To conclude, we discuss the results in Section 5.

2   Feature Extraction and Genetic Algorithms

In this section we present a brief introduction about feature selection and genetic algorithm.

2.1   Feature Selection

Feature selection is typically a search problem for finding an optimal or suboptimal subset of m features out of original M features [2]. Feature selection is important in many pattern recognition problems as it excludes irrelevant and redundant features. It allows reducing system complexity and processing time and often improves the recognition accuracy. For large number of features, exhaustive search for best subset out of 2M possible subsets is infeasible. Therefore, many feature subset selection algorithms have been proposed. The feature selection method is commonly used to select best feature subsets from the extracted features. Then, these feature subsets are evaluated to select a small subset of features that provides high recognition accuracy. Finally, we analyze the recognition accuracy using our classifier. There are two essential steps in the feature selection Research procedures and Evaluation functions [3, 4]:

1. Research procedures: If the all the features contains N primitives defined initially then the total number of candidates generated opportunities is 2^N. This number is very high. The goal is to find a solution among the number of probabilities. To solve this problem, the search process consists of three approaches: complete, heuristic and random.
2. Evaluation functions: These functions measure all candidate solutions generated by the research procedures. These values are compared with previous values. The best value will be kept

[5, 6, 7] have studied the different optimization techniques and found that GAs are suitable when the number of primitives to be selected is huge. The GAs has a high probability of finding the best solutions in comparison with other optimization techniques. They recommend applying GAs several times with different parameters.

2.2   Genetic Algorithms

GA has been developed by Holland in the 60s. An approach to genetic type is designed, in the literature to adapt the partition of the parameter space with respect to all the tools available. The GA is an example of how exploration that uses random choice as a tool to guide exploration in the parameter space coded. This technique guarantees not to reach the global optimum of the function, but to reach this optimum. A GA aims to evolve a set of candidate solutions to a problem to the optimal solution. This development is done on the basis of transformations inspired by genetics, providing the generations, exploring the solution space.

3   The Proposed Algorithm

Despite the extensive work on the recognition of Arabic script, several problems exist [1]. Feature selection can be considered as a problem of global combinatorial optimization. Our contribution is to make a selection of relevant primitives using GAs. These primitives are represented in vector of size N. To do so, we must therefore find M primitives such as M<N. Feature extraction aims at not only reducing the dimension of the space of representation, but also improving the classification in order to recognize shapes [23]. We select some features to encode character; these primitives are described below; Horizontal and Vertical Density Derivatives Signatures, Existence of Holes, Descending and Ascending Characters and Character Width. The construction steps of GA used are:

Chromosomes Coding: We have chosen binary encoding. The presence of bit 1 means that the primitive is selected and the presence of bit 0 mean that the primitive is not selected.

Fitness function: The fitness function is the weighted sum of objective functions. The first objective is to minimize the number of primitives. The second is to minimize the error rate generated during the classification. So, the fitness function is a combination of two functions to be minimized. This function is described by the formula Equ.1. Each chromosome of the population is evaluated by this function:

$$\text{Fitness} = \text{Minimize} (\alpha F_1 + \beta F_2) \tag{1}$$

Such that F_1: Error rate generate during the classification; and F_2: Ratio between the numbers of features selected by the feature selection process and the total number of primitives;

The two parameters α and β are fixed arbitrarily to 1 in 1000 based to work of numbers recognition [8]. The proposed technique has the advantage of selecting the most representative primitives that improve recognition quality. Then, we performed several experiments to determine such parameters like population size and probabilities of mutation and crossover. These experiments are described below.

Crossover Probability P_{cross}: The crossover method is assumed to be one point crossover. To determine the probability of crossover (Fig.1), we have varied P_{cross} between 0.6 and 0.7 {0.6, 0.65, and 0.7}.

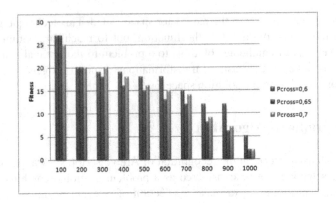

Fig. 1. Crossover Probability

An observation can be considered for P_{cross} = 0.6; although the optimal solution is not reached with the fitness function, it is attained in the case of P_{cross} = 0.7 and P_{cross} = 0.65. We can notice that the convergence is slower when the probability of crossover is equal to 0.7. The optimal solution is obtained when a value which tends to zero of fitness function is reached; and this is observed in P_{cross} = 0.65. In the next section, the crossover probability is fixed to P_{cross} = 0.65.

Mutation Probability P_{mut}: We have used the technique of bit string mutation to build our AG. The mutation of bit strings is ensued through bit flips at random positions. To determine the mutation probability P_{mut} (Fig.2), we have varied the values between 0.02 and 0.04 {0.02, 0.03, and 0.04}.

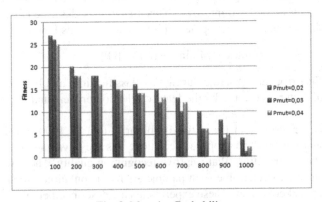

Fig. 2. Mutation Probability

Another remarkable observation can be made regarding the fitness function which reaches the optimal solution in the case of P_{mut} = 0.03. After some experiments, we have found that in order to choose the probability of crossover and mutation, the values chosen are equal respectively to P_{cross} = 0.65 and P_{mut} = 0.3, because the rate of convergence is faster with those values.

Initial population Size: In the same context, we do some experiments to observe the influence of population size with one point crossover P_{cross} = 0.65 and Bit string mutation P_{mut} = 0.03. Our results are based on an initial population of 100 individuals; we conducted experiments with a larger population of 200 individuals. Fig. 3 shows the average rating of the Fitness functions with our population and a wider population

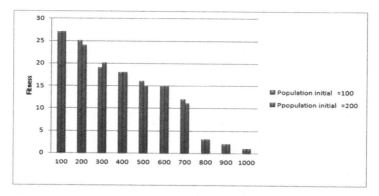

Fig. 3. Comparison of the convergence of Fitness for two populations of different sizes

We observed that the convergence rate of the two populations is almost the same. The optimal solution is reached at the 800th generation. So, it is better to use a population size equal to 100 people instead of 200 because the computation time for a large population is slower.

Finally, these values (Tab.1) will be used for primitives selection.

Table 1. Genetic algorithm Parameters

GA Parameters	Values
Chromosome Coding	Binary
Chromosome Size	27
Population Size	100
Number of generations	1000
Selection type	Roulette wheel
Crossover type	one point
Crossover rate	0.65
Mutation type	Bit string
Mutation rate	0.03
Insertion	Elitism
Stopping criteria	Maximum number of generation

4   Experimental Results and Discussion

We used the relevant feature subset with a PML classifier and examined the recognition accuracy of every letter form. PML is most common neural network model, called supervised network because it requires a desired output in order to learn. The goal of this type of network is to create a model that correctly maps the feature vector of input to the class [8]. An overall rate of 87.94% was recorded. Variation rate character recognition is shown in Fig.4.

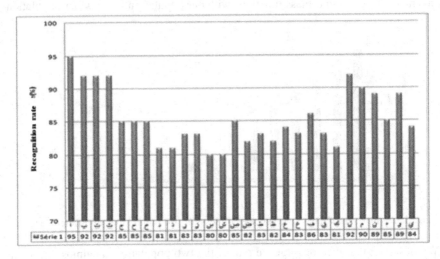

Fig. 4. Recognition Rate

We found that the classification error is 10% for easy letter forms such as (ﺍ), (ﻝ), (ﺩ) and (ﺕ). For the hardest letter forms the classification error achieves 40%. The following table shows the letter forms that have the worst classification errors.

Table 2. Character misclassification

Target Position	Target Classification	Wrong Classification
Initial	ﺻ	ﻫ
Medial	ﺨ	ﻘ
Medial	ﻬ	ﻤ
Medial	ﻗ	ﻔ,ﻗ
Fianl	ﺢ	ﻊ
Final	ﻭ	ﻕ
Isolated	ﻥ	ﻑ
Isolated	ﺭ	ﻥ

We propose to compare our results with other work. In fact, OLIVERI and his colleagues propose in their 2002 paper, a neuro-genetic hybridization approach for the recognition of handwritten digits [9] with an accuracy of 97.1% and in 2007, KROUCHI and DJEBBAR propose an extension in the domain of OCR is to perform a hybridization of a neural network of type PMC and AG with an accuracy of 87.54% [10]. The recognition rate of our contribution (87.94%) seems low in relation to that of [9]. This gap is justified by the fact that [10] uses a database of numbers from 0 to 9, as far as our system is applied on Arabic script which is characterized by a wide vocabulary. Although the phase of post-treatment can improve the recognition rate to 10%, we did not achieve this step which may have a tremendous impact on the recognition rate of AOCR system.

Table 3. Comparison of results

Authors	Accuracy (%)	Classifiers
Our contribution	87.94	PML-AG
[9]	97.1	PML-AG
[10]	87.54	PML-AG

It is difficult to make an exhaustive comparison between the performance of our system and those of other published systems. Indeed, the test sets are different and the performance measures are not clearly defined. In addition, it is not possible to determine if learning samples and tests are mutually exclusive. The effort required for the preparation of a set of tests is actually a significant obstacle for valid experimentation from a statistical point of view. It is therefore necessary to focus efforts in developing a standard basis of the Arabic script that could be used in the evaluation of various Arabic recognition systems.

5 Conclusion and Future Works

In this paper, we have presented a feature selection method based on genetic algorithms. The experimental results prove that feature selection using the GA reduces from 25 to 30% of features during the feature selection process. This reduction improves the recognition accuracy. The results obtained in this research have been very promising and identification accuracy has reached 87%.

The evaluation of our proposed recognition system is employed to improve the performance of the algorithm, its efficiency as well as its weaknesses. Moreover, this evaluation may give clues about points to be improved later. The problem remains open and further efforts are made to improve current approaches to Arabic writing recognition. As a future work, the author plans to improve the recognition rate of AOCR system using SVM classifiers. The performance criteria would include recognition rate and time complexity.

References

1. Parvez, M.T., Mahmoud, S.A.: Offline arabic handwritten text recognition: a survey. ACM Computing Surveys (CSUR) 45(2), 23 (2013)
2. Marwa, A., Kamel, Z.: Feature Selection Using a Neuro-Genetic Approach For Arabic Text Recognition. In: Metaheuristics and Nature Inspired Computing (2012)
3. Nunes, C.M., Britto Jr., A.S., Kaestner, C.A., Sabourin, R.: An optimized hill climbing algorithm for feature subset selection: Evaluation on handwritten character recognition. In: Ninth International Workshop on Frontiers in Handwriting Recognition, IWFHR-9 2004, pp. 365–370. IEEE (2004)
4. Dash, M., Liu, H.: Feature selection for classification. Intelligent Data Analysis 1(3), 131–156 (1997)
5. Alaei, A., Pal, U., Nagabhushan, P.: A comparative study of persian/arabic handwritten character recognition. In: International Conference on Frontiers in Handwriting Recognition (ICFHR), pp. 123–128. IEEE (2012)
6. Yang, J., Honavar, V.: Feature subset selection using a genetic algorithm. In: Feature Extraction, Construction and Selection, pp. 117–136. Springer US (1998)
7. Kudo, M., Sklansky, J.: Comparison of algorithms that select features for pattern classifiers. Pattern Recognition 33(1), 25–41 (2000)
8. El-Bendary, N., Zawbaa, H.M., Daoud, M.S., Nakamatsu, K.: ArSLAT: Arabic Sign Language Alphabets Translator. In: 2010 International Conference on Computer Information Systems and Industrial Management Applications (CISIM), pp. 590–595. IEEE (October 2010)
9. Oliveira, L.S., Sabourin, R., Bortolozzi, F., Suen, C.Y.: A methodology for feature selection using multiobjective genetic algorithms for handwritten digit string recognition. International Journal of Pattern Recognition and Artificial Intelligence 17(06), 903–929 (2003)
10. Ghizlaine, K., Bachir, D.: Reconnaissance hors ligne des chiffres manuscrits isolés par l'approche Neuro-Génétique. RIST 17(1) (2007)

Isolated Printed Arabic Character Recognition Using KNN and Random Forest Tree Classifiers

Marwa Rashad and Noura A. Semary

Faculty of Computers and Information, Menofia University, Egypt
{marwa.hassan,noura.samri}@ci.menofia.edu.eg

Abstract. Classification step is one of the most important tasks in any recognition system. This step depends greatly on the quality and efficiency of the extracted features, which in turn determines the efficient and appropriate classifier for each system. This study is an investigation of using both K- Nearest Neighbor (KNN) and Random Forest Tree (RFT) classifiers with previously tested statistical features. These features are independent of the fonts and size of the characters. First, a binarization procedure has been performed on the input characters images, and then the main features have been extracted. The features used in this paper are statistical features calculated on the shapes of characters. A comparison between KNN and RFT classifiers has been evaluated. RFT found to be better than KNN by more than 11 % recognition rate. The effect of different parameters of these classifiers has also been tested, as well as the effect of noisy characters.

Keywords: Random forest tree, KNN, Classification, Arabic character recognition, Statistical features.

1   Introduction

Optical Character Recognition (OCR) is a process of converting images of written text to machine editable text. A lot of work has been done on English OCR, but Arabic OCR is still a challenge [1, 2]. Arabic text has some special characteristics that make Arabic (OCR) considered to be a difficult task [3]. These characteristics can be summarized as follows

- Arabic text (printed or handwritten) is cursive and written from right to left. Arabic letters are normally connected to each other on the baseline.
- Arabic text uses letters (which consists of 28 basic forms), ten Hindi numerals, punctuation marks, as well as spaces and special symbols.
- Arabic letters might have up to four different forms (beginning, middle, end and isolated) depending on their position in the word for example (ـغ , ـغـ , غـ ,غ respectively). This feature increases the number of the classes from 28 to 100. Actually, a new character (ﻻ) is created when Alif (ﺍ) is written immediately after Lam (ﻝ), this kind of new characters increase the number of classes to 120.

A.E. Hassanien et al. (Eds.): AMLTA 2014, CCIS 488, pp. 11–17, 2014.
© Springer International Publishing Switzerland 2014

- Several Arabic letters have exactly the same primary part. However, they are distinguished from each other by the addition of dots in different locations.
- Arabic characters do not have a fixed size. The width and height of the character vary according to its position in the word.
- Some Arabic letters include a loop that can have different shapes, the shape of the loop changes with respect to the font [4]. (See Figure.1).

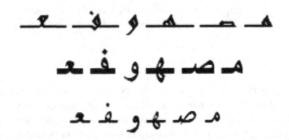

Fig. 1. Different shapes of holes

Arabic OCR has attracted many researches because it contributes to the advancement of the automation process and it improves the interaction between man and machine in many applications, including office automation, check verification and a large variety of banking, business and data entry applications [5]. Arabic character recognition falls into either online or offline. In offline recognition the image is acquired using an optical scanner. The image then is read by the system and is analyzed for recognition. In online recognition system input is usually acquired from a tablet computer or pen based device as cell phone and sign pad [6].

Our research is designed for offline printed individual Arabic characters. The purpose of the proposed work is to build a high accurate font and size independent Arabic OCR using statistical features used in [6] and RFT classifier.

The rest of the paper is organized as follows: section 2 gives an overview of proposed system methodology. Section 3 explains experimental results and conclusion is mentioned in section 4.

2   Methodology

2.1   Image Binarization

Binarization plays a key role in processing degraded images. In general, binarization is either global or local. In a global approach, threshold selection leads to a single threshold value for the entire image often based on an estimation of the background level from the intensity histogram of the image. Unlike global approaches, local binarization use different values for each pixel according to the local area information [7, 8]. In this paper, global binarization has been used [9].

2.2   Feature Extraction

One of the basic steps of pattern recognition is features selection. Features should distinguish between classes, be invariant to input variability, and also be limited in number to compute discriminate functions efficiently and to limit the amount of needed training data [10]. Statistical features look for a typical spatial distribution of the pixel values that define each character. There are 14 statistical features extracted from each character [6], four of them are for the whole image as listed below:

1. Height / Width.
2. Number of black pixels / number of white pixels.
3. Number of horizontal transitions.
4. Number of vertical transitions.

The horizontal and vertical transitions are a technique used to detect the curvature of each character and found to be effective for this purpose [6]. The procedure runs a horizontal scanning through the character box and finds the number of times that the pixel value changes state from 0 to 1 or from 1 to 0 as shown in figure 2. The total number of times that the pixel status changes, is its horizontal transition value. Similar process is used to find the vertical transition value [6].

The other 10 features are extracted after dividing the image of the character into four regions to get the following ratios as shown in figure 3:

1. Black Pixels in Region 1/ White Pixels in Region 1
2. Black Pixels in Region 2/ White Pixels in Region 2
3. Black Pixels in Region 3/ White Pixels in Region 3
4. Black Pixels in Region 4/ White Pixels in Region 4
5. Black Pixels in Region 1/ Black Pixels in Region 2
6. Black Pixels in Region 3/ Black Pixels in Region 4
7. Black Pixels in Region 1/ Black Pixels in Region 3
8. Black Pixels in Region 2/ Black Pixels in Region 4
9. Black Pixels in Region 1/ Black Pixels in Region 4
10. Black Pixels in Region 2/ Black Pixels in Region 3

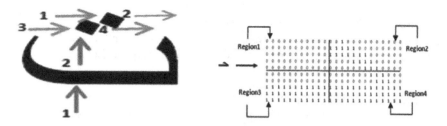

Fig. 2. Horizontal and vertical transitions   **Fig. 3.** Dividing the image and extract the features

2.3   K-Nearest Neighbor Algorithm

KNN is a type of instance-based learning, or lazy learning where the function is only approximated locally and all computation is deferred until classification. The k-nearest neighbor algorithm is amongst the simplest of all machine learning algorithms: an object is classified by a majority vote of its neighbors, with the object being assigned to the class most common amongst its k nearest neighbors (k is a positive integer, typically small). If k = 1, then the object is simply assigned to the class of its nearest neighbor. The training phase for KNN consists of simply storing all known instances and their class labels. If we want to tune the value of K, n-fold cross-validation can be used on the training dataset. The testing phase for a new instance "T", in a given known set "I" is as follows:

1. Compute the distance between T and each instance in I
2. Sort the distances in increasing numerical order and pick the first K elements
3. Compute and return the most frequent class in the K nearest neighbors

KNN learning algorithm has been tested because it needs few parameters to tune and has been frequently used in recent AOCR research [2], [11].

2.4   Random Forest Tree (RFT)

Random forest is an ensemble classifier that consists of many decision trees and outputs the class that is the mode of the classes output by individual trees. The algorithm for RFT was developed by L. Breiman [12] . RFT classifier has been tested in this study because: It is one of the most accurate learning algorithms available as it produces a highly accurate classifier for many data sets and , it also runs efficiently on large databases and it gives estimates of what variables are important in the classification [13] [14]. RFT is composed of some number of decision trees. Each tree is built as follow:

— Let the number of training objects be D, and the number of features in features vector be F.
— Training set for each tree is built by choosing D times with replacement from all D available training objects.
— Number $f <<F$ is an amount of features on which to base the decision at that node. These features are randomly chosen for each node.
— Each tree is built to the largest extent possible.
— Each tree gives a classification, which is called voting for that class. The forest chooses the class having the most votes (over all the trees in the forest) [15].

3   Experimental Results

3.1   Dataset Generation

Our Dataset has been created from 1160 sample characters of sizes 18, 24, 32 and 64 from 6 different Arabic fonts shown in figure 4. The process starts by acquiring the text image using 300 dots per inch (dpi) scanner then the image of the text is saved as

bit map file format. The experiments were executed using 1.38 GHz CPU, 2 GB
RAM PC and windows Vista platform. Generating the dataset has been done using
MATLAB 7.0.1 software.

To test the efficiency of the tested features and classifiers, it was necessary to study
the impact of the noise using a set of noisy characters. Different types of background
noise were tested, these types are: Gaussian, Poisson and salt & pepper. It was also
necessary to test other types of noise that affect the shape of the character itself. Photoshop has been used to simulate these noises. Some examples of different noisy
characters are shown in figure 5.

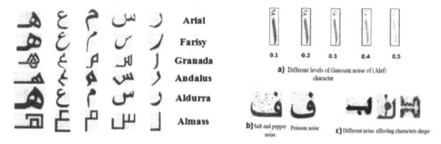

Fig. 4. Six fonts used in training and testing   **Fig. 5.** Different tested noise

3.2   KNN Experimental Results

Weka package has been used for training and testing purposes by both KNN and RFT
classifiers. The performance of KNN algorithm basically depends on two factors:
Suitable similarity function and appropriate value for the parameter k [16].

After several experiments using different values of parameter K and distance functions, we found that very acceptable results have been obtained by setting K to 1 and
Euclidean distance function. Nearest neighbor search algorithm was also tested as it is
found to be an influential factor that affects training and testing time. Tested search
algorithms were: Linear nearest neighbor search, Ball Tree and KD Tree.

We found that changing search algorithms has no effect on recognition rates as
they give the same recognition rate, but at the same time it has a great effect on training and testing time. The shortest time was obtained using KD tree search algorithm
as shown in Table 1. Using the same parameters, noisy data were tested and showed
that KNN is very sensitive to noise as the obtained recognition rate was 33.5%.

3.3   RFT Experimental Results

Several experiments have been done taking into account changing the number of trees
and features to test their effect on recognition rate, training and testing time. The results obtained are shown in Table2.

Table 1. Results of KNN

Recognition Rate	87 %
Search Algorithm	KD Tree
Training Time	0.02 sec
Testing Time	0.01 sec

Table 2. Results of RFT

Number of trees	10	20	50
Recognition Rate	94%	96.5 %	98 %
Number of features	3	2	3
Training Time	0.77 sec	1.06 sec	3.90 sec
Testing Time	0.01 sec	0.02 sec	0.08 sec

It has been noticed from the experiments that increasing number of trees often increases the recognition rate but to a certain extent as when increasing the number of trees from 50 to 70 and to 100 there was no improvement in recognition rate. According to number of features selected to split tree nodes, while it has no specific relationship with recognition rate it has a remarkable effect on training and testing time, as it is clear from presents results that increasing number of features increases training and testing time dramatically.

Testing noisy data obtained about 43.5 % which shows that random forest tree is also sensitive to noise, as the rates have been decreased by about 55% compared to the results of testing data without noise.

4   Conclusion and Future Works

This paper presented an isolated printed Arabic character recognition system using a set of effective and invariant statistical features. The proposed system was independent of the character font and size. Although many researchers have found that KNN algorithm achieves very good performance for character recognition in their experiments on different data sets [12], we found that RFT performs better in our research. Using the same training and testing set RFT achieves 98% recognition rate compared to 87 % using KNN. On the other hand KNN is very fast in training and testing the datasets compared to RFT. Experimental results also showed that both KNN and RFT are very sensitive to noisy data. In our future works, rough sets-based feature extraction, rule generation and classification will provide more challenging and may allow us to refine our learning algorithms and/or approaches to the Arabic pattern recognition [17].

References

[1] Abd, M.A., Paschos, G.: Effective arabic character recognition using support vector machines. In: Innovations and Advanced Techniques in Computer and Information Sciences and Engineering, pp. 7–11. Springer (2007)

[2] Al-Jamimi, H.A., Mahmoud, S.A.: Arabic character recognition using gabor filters. In: Innovations and Advances in Computer Sciences and Engineering, pp. 113–118. Springer (2010)

[3] Thein, Y., Yee, S.S.S.: High accuracy Myanmar handwritten character recognition using hybrid approach through MICR and neural network. International Journal of Computer Science Issues (IJCSI) 7, 22–27 (2010)

[4] Hassin, A.H., Tang, X.-L., Liu, J.-F., Zhao, W.: Printed Arabic character recognition using HMM. Journal of Computer Science and Technology 19, 538–543 (2004)

[5] Amin, A.: Offline Arabic character recognition: a survey. In: Proceedings of the Fourth International Conference on Document Analysis and Recognition, vol. 2, pp. 596–599 (1997)

[6] Almohri, H., Gray, J., Alnajjar, H.: A Real-time DSP-Based Optical Character Recognition System for Isolated Arabic characters using the TI TMS320C6416T. Doctoral dissertation, University of Hartford (2007)

[7] Arica, N., Yarman-Vural, F.T.: An overview of character recognition focused on off-line handwriting. IEEE Transactions on Systems, Man, and Cybernetics Part C: Applications and Reviews 31(2), 216–233 (2001)

[8] Gatos, B., Pratikakis, I., Perantonis, S.J.: Adaptive degraded document image binarization. Pattern Recognition 39(3), 317–327 (2006)

[9] Dhingra, K.D., Sanyal, S., Sharma, P.K.: A robust OCR for degraded documents. In: Huang, X., Chen, Y.-S., Ao, S.-I. (eds.) Advances in Communication Systems and Electrical Engineering. LNEE, vol. 4, pp. 497–509. Springer, Heidelberg (2008)

[10] Jensen, J., Kendall, W.S.: Networks and chaos-statistical and probabilistic aspects, 50. CRC Press (1993)

[11] Mahmoud, S.: Recognition of Arabic (Indian) check digits using spatial gabor filters. In: Proceeding of 5th IEEE GCC Conference & Exhibition, Kuwait, pp. 1–5 (2009)

[12] Briman, L.: Random Forests. Machine Learning 45(1), 5–32 (2001)

[13] Caruana, R., Karampatziakis, N., Yessenalina, A.: An empirical evaluation of supervised learning in high dimensions. In: Proceedings of the 25th International Conference on Machine Learning, pp. 96–103 (2008)

[14] http://www.stat.berkeley.edu

[15] Homenda, W., Lesinski, W.: Features Selection in Character Recognition with Random Forest Classifier. In: Jędrzejowicz, P., Nguyen, N.T., Hoang, K. (eds.) ICCCI 2011, Part I. LNCS, vol. 6922, pp. 93–102. Springer, Heidelberg (2011)

[16] Elglaly, Y., Quek, F.: Isolated Handwritten Arabic Characters Recognition using Multilayer Perceptrons and K Nearest Neighbor Classifiers (2012)

[17] Hassanien, A.E., Suraj, Z., Slezak, D., Lingras, P.: Rough computing: Theories, technologies and applications. IGI Publishing Hershey, PA (2008)

Arabic Character Recognition Based M-SVM: Review

Marwa Amara[1], Kamel Zidi[2], Salah Zidi[3], and Khaled Ghedira[1]

[1] University of Tunis, SOIE-Management Higher Institute,
Le Bardo 2000, Tunisie
amaralmarwa@gmail.com,
khaled.ghedira@isg.rnu.tn
[2] University of Tabuk, Community College,
Saudi Arabia
kamel_zidi@yahoo.fr
[3] Archimed 49 Boulevard Strasbourg
Lille Cedex 59042, France
sazidi@archimed.fr

Abstract. Optical Character Recognition Systems (OCR) provide human-machine interaction and are widely used in many applications. Classification is the most important step in an OCR system. Support Vectors Machines (SVM) is among the tool of classification that appears these days. This tool proves its ability to discriminate between the forms and gives encouraging result. In this paper, we present an overview of the Arabic optical character recognition (AOCR) work done using SVM classifiers.

Keywords: AOCR, multi-class SVM, SVM survey.

1 Introduction

Characters recognition within the field of pattern recognition interests in characters shapes. Since the 40s, the character recognition has been the subject of extensive research. Many researches can be done to improve OCR system, particularly in case of Arabic letters recognition. Characters recognition is an active research area in pattern recognition and has many practical applications. Some of these applications include postal address and zip code recognition, forms processing, automatic processing of bank cheques, etc. Recognition requires the definition of the knowledge which we have about forms to be treated based on learning. Learning can acquire the knowledge and organize reference model. It is impossible to learn a great number of samples and different forms because of their variability. We should replace this number by better quality of characteristics [1]. Learning procedures are different depending on whether to recognize the printed or manuscript, mono-font or multi-font ...The classification approaches are divided into three main categories; statistics, structural and stochastic. The problem of recognition of characters stills an active area of research and is challenging and interesting task in the field of pattern recognition. It is the process of

A.E. Hassanien et al. (Eds.): AMLTA 2014, CCIS 488, pp. 18–25, 2014.
© Springer International Publishing Switzerland 2014

classifying Arabic characters into appropriate classes based on appropriate classification method. Many classification methods have been applied in the field of recognition. However, few AOCR systems have applied SVMs compared to other traditional techniques know in literature. This has recently begun to change. The improvement of the recognition rate in AOCR system is the reason to introduce new tools of classification such as SVM.

This paper is organized into 4 sections. Section 2 presents properties of SVM. We discuss different methodologies in AOCR development using SVM Section 3 explains experimental results and conclusion is mentioned in section 4.

2   SVM for Classification

SVM tool is based on statistical learning theory proposed, in 1979, by Vapnick [2]. The principal motivation of this technique is the probabilistic search of limits that minimize errors and "empirical risk" while maximizing the margin of separation [3]. SVM tool constitutes a separation of regions with optimal hyper planes in a multidimensional data space which ensures the learning systems convergence. Applications of SVM concern, essentially, pattern recognition and statistical analysis.

2.1   Binary SVM

The idea of SVMs is to find a hyper plane that separates classes the best. If such hyper plane exists, in other words, if the data is linearly separable, we talk about a SVM Hard Margin.

1. *Hard Margin*

The hyper plane $W^T X + b = 0$ represents a separating hyper plane of two classes, and the distance between hyper plane and nearest example is called **margin**. The region located between; $W^T X + b = +1$ and $W^T X + b = -1$ is called region of generalization. The maximization of this region is the objective of training phase. It consists to search hyper plane that maximizes margin. This hyper plane is called "Optimal Separating Hyper plane". The determination of this hyper plane passes through the determination of minimum Euclidean distance between hyper plane and nearest example of two classes called margin.

To find separating hyper plane that maximizes margin, we must determine **w** vector which has the minimum Euclidean norm. The optimal separator can be obtained by solving following equation:

$$\begin{cases} \min imize \dfrac{1}{2}\|w\| \\ y_i(w^t x_i + b) \geq 1 \end{cases} \tag{1}$$

The problem of the equation (1) is a quadratic programming problem with linear constraints. In such cases problem is converted into a dual problem without constraints equivalent to equation that introduced Lagrange multipliers. We obtain, then, following dual problem to maximize:

$$
\begin{cases}
\text{maximize} \ \sum_{i=1}^{n} \alpha_i - \frac{1}{2} \sum_{i=1}^{n} \sum_{j=1}^{n} \alpha_i \alpha_j y_i y_j \mathbf{x}_i^T \mathbf{x}_j \\
\alpha_i \geq 0 \ and \ \sum_{i=1}^{n} \alpha_i y_i = 0
\end{cases}
\tag{2}
$$

If the classification problem is linearly separable, an optimal solution for α_i exists.

2. Soft Margin

A separating hyper plane does not always exist. And if it exists, it doesn't represent, usually, best solution for classification. In addition, an error class in training data can affects the choice of hyper plane.

In the case where data are not linearly separable or incorrectly labeled constraints of equation (2) cannot be verified, and there is need to relax. This can be done by assuming a certain error of classification which is called Soft Margin. We introduce variables ξi called of relaxation; it can be added to allow mis-classification of difficult or noisy data. In this case, instead of searching only a separating hyper plane that maximizes margin, we look for a hyper plane which also minimizes sum of allowed errors. Equation becomes:

$$
\begin{cases}
\text{minimize} \ \frac{1}{2} \|w\| + C \sum_{i=1}^{n} \zeta_i \\
y_i (w^t x_i + b) \geq 1 - \zeta_i \ and \ \zeta_i \geq 0
\end{cases}
\tag{3}
$$

C is a positive parameter which represents a balance between margin and allowed errors. In other words, C is the balance between the maximization of margin and the minimization of classification error.

3. Kernel

Instead of a straight, ideal representation of decision function is a representation that fits the data best training. The determination of such non-linear function is very difficult. To do, data is brought in a space where this function becomes linear. This space of transformation is often made with a function called "Mapping function". The new space is called "Features space ".

A kernel function is defined as a function that corresponds to a dot product of two feature vectors in some expanded feature space. Some examples of kernel functions proposed by Vapnik allowing a good approximation of properties of decision space are presented in [2].

2.2   Multi-Class SVM

Problems of real world are in most cases multiclass, the simplest example is OCR. In such cases, we don't want to assign a new instance to one of two classes but one among many. The decision is not binary and one hyper plane is not enough. Multi class SVMs reduce this problem to a composition of several hyper planes bi classes allow drawing the boundaries of decision between classes [3]. SVMs are originally designed for binary pattern classification. Multi-class pattern recognition problems are commonly solved using a combination of binary SVMs and a decision strategy to decide the class of input pattern.

Others works use MSVM considering multi class at once. We will discuss, following, some methods of most known MSVM.

1. One-Against-all strategy (OAA): In this approach, k binary SVM models are constructed. An SVM is constructed for each class by discriminating that classes against the (k- 1) others classes. To classify a test pattern using OAA method, it' is assigned to the class with the maximum value of the discriminant function f(x) using the winner-takes-all decision strategy.
2. One-Against-one strategy (OAO): The OAO method is introduced as pairwise SVM. An SVM is constructed for each pair of classes by training it to discriminate the two classes. Thus, the number of SVMs used is K (K- 1)/2.The maxwins strategy is generally used to determine the class of pattern x using   a majority voting. The class with maximum number of votes is assigned to pattern.
3. Directed Acyclic Graph Support Vector Machines strategy (DAGSVM): DAGSVM is a new learning architecture which is used to combine two classifiers into one. Its works as OAO method solving k (k - 1)/2 binary SVMs, in the training step. However, it uses a rooted binary DAG which has k (k - 1)/2 internal nodes and k leaves, in the testing step. Given a new pattern x, starting at the root node, a pairwise SVM decision is made. Then it moves to either left or right depending on the result, and continues until reaching to one of leaves which indicates the predicted class.

3   SVM for Optical Caracter Recognition

In this paper, we are interested in AOCR systems based SVM classifiers. There are many reasons to substantiate the performance of SVM. The essential difference with other similar approaches such as neuronal networks is on the fact that the SVM method finds the global minimum whether there are [4].   Like all parametric methods, we should make appropriate choices notably in the choice of Kernel to ensure a better delimitation of classes. The MSVM technology presents a good generalization of the method of the binary SVM initially dedicated to the dichotomous classification. A tutorial on SVM for Pattern Recognition can be finding in [5] [6]. Following, we will illustrate few examples of application of classifiers MSVM.

3.1 Studies on Arabic Character Recognition

In 2007, Abdullah and Paschos [7] proposed an Arabic character recognition system focuses on employing SVM. The feature vector feed to the SVM classifier is composed by: Invariants Moment, number of dots and their position and number of holes. The classification phase takes place using SVMs, by applying one-against-all technique with Gaussian RFB kernel. Ten-fold cross validation is used to determine the best RFB kernel parameters.

Samples are obtained from 58 characters/shapes. Recognition rates obtained are in the range of 98-99% correct classification depending on the type and number of classes used. LIBSVM [8] is used for tests. This system can be applied to different recognition area such as finger prints, iris and multi-fonts characters for many languages.

Another work in the fields of handwritten Arabic script recognition based on SVM classifier was proposed in 2008 [9]. Feature used are; Moment and Fourier descriptor of profile projection and centroïde distance. Principal component Analysis (PCA) was applied to features vector in order to reduce dimension. This system tested to classify only principal shape of the character without dots. This system uses the multiclass one against one SVM, with Gaussian RBF kernel. To validate the developed methods, database composed of 1000 Arabic characters is used. 800 are used for learning and 200 for the test. The recognition rate obtained without PCA is 94% and with PCA 96 %.

Faouzi et al [10] proposed a new segmentation approach applied to Arabic handwriting with application of MSVM classifier, in 2010. After features extracting, these vectors are used as input for the MSVM. SVM classifier should find a hyper plan that separates better these vectors. This hyper plan must be found for each character. Each new character will be exposed to each of possible classes in order to find the appropriate class using one against all MSVM strategy. The goal of SVM is to search for a hyper plan that separates the examples in the learning stage and takes a decision in the classification phase. This work uses Libsvm [8], with a Gaussian RFB kernel. To validate the proposed model, samples used are scanned using a scanner EPSON CX3400. Resolution varied between 150 and 300 dpi. The used characters, and the recognition rate, are presented in the following table:

Table 1. Recognition rate[10]

Characters	ك	ء	.	د	ع	ى	و	ح
Accuracy (%)	100	91.6	100	95	100	78.72	97.29	90.32

In 2012, Al-Hamadi et al [11] propose a MSVM classifier for handwritten Arabic text recognition. Features are extracted based Gabor-filter. For training and testing, IESK-ar Database was used [12]. IESK-ar is an Arabic off-line handwritten database, developed in the Institute for Electronics freely available. In this work, a class for each letter shape was created. One-vs-all SVM Classifiers with RBF kernel was used. In experiments step, libsvm [8] library is used. 5436 letters were used in training of SVM classifier. Recognition rates varied among 93% and 43%, with a median of 77% and an average of 74.3%.

In 2012, Alalshekmubarak et al proposed a novel approach to classify Off-Line handwritten Arabic word using SVMs with normalized poly kernel [13]. SMO is used to train SVMs. SMO divides the quadratic programming optimization problem, which is required to train SVMs, into the smallest possible series. The choice of normalized poly kerned is proved by the following result:

Table 2. Recognition rate using different kernels[13]

Kernel	Accuracy
Normalized	95.27%
Polynomial	93.47%
Linear	93%

To evaluate the proposed system IFN/ENIT-database is used [14]. The recognition rate achieved 95.27% in a subset of 24 classes and 7,971 entries. A recognition rate of 92.34% was obtained with 56 classes and 12,217 entries.

3.2   Comparative Analysis

Researchers listed below present encouraging recognition rate on most occasions. However, those works can be intended in order to compare the different SVM kernel, and choose the best learning parameters. The reported results were obtained using only one SVM methods in experiments. The choice of kernel function is generally arbitrary. This leaves a question that how these kernel function and MSVM-methods will perform when applied to Arabic scripts. Therefore, it is important to look into the discriminative power of each SVM-method and function kernel proposed in the literature before using it. The hybridization of SVM with other tools such as KNN and MLP presents an encouragement recognition rate when it's applied to Latin character [15]. Experiments show that SVM achieves better generalization ability than MLP classifier and Naive Bayes [13].

Table 3. Script recognition methods based SVM

Research's	Script Recognition Methods			
	Script type	Method	kernel	Accuracy (%)
2007[7]	Printed	OAO	Guassian RFB	In the range of 98-99
2008[9]	Handwritten	OAO	Gaussian RFB	94 without PCA / 96 with PCA
2010[10]	Handwritten	OAA	Guassian RFB	TABLE 4.
2012[11]	Handwritten	OAA	Gaussian RFB	Between 93 and74.3
2012[13]	Handwritten	SMO	Normalized	95.27
			Polynomial	93.47
			Linear	93

4   Conclusion and Future Works

In this paper, we introduce a survey of AOCR system based SVM classifier. Firstly, we have illustrated the basic concept of SVM in the binary case and the multi-class case. After that we presented the different works on Arabic script. Finally, we discussed those works and we expose the recognition rate obtained, the MSVM method and kernel function for each research. AOCR is one of the most attractive and difficult areas of pattern recognition. The uses Of MSVM prove encouragement result compared to other classification tools. An open problem is still; precision, consistency and efficiency of the algorithm. Future work will be focused on the development of an AOCR system based SVM, rough sets classifiers or hybridization of SVM and rough sets [15]. Moreover, rough sets-based feature extraction, rule generation and classification will provide more challenging and may allow us to refine the learning algorithms and/or approaches to the Arabic character recognition.

Acknowledgment: Researches are conducted in an MOBIDOC thesis financed by the EU within PASRI program.

References

1. Amara, M., Zidi, K.: Feature Selection Using a Neuro-Genetic Approach For Arabic Text Recognition. In: Proceedings of the International Conference on Metaheuristics and Nature Inspired Computing, Tunisia (2012)
2. Vladimir Vapnik, N.: Statistical Learning Theory. John Wiley & Sons Inc., New York (1998)
3. Zhou, W., Zhang, L., Jiao, L.: Linear programming support vector machines. Pattern Recognition Society (2002)
4. Gay Thomé, A.C.: SVM Classifiers – Concepts and Applications to Character Recognition, book edited by Xiaoqing Ding (2012)
5. Ghosh, D., Dube, T., Shivaprasad, A.P.: Script Recognition – A Review. IEEE Transactions on Pattern Analysis and Machine Intelligence (2009)
6. Borovikov, E.A.: An Evaluation of Support Vector Machines as a Pattern Recognition Tool. University of Maryland at College Park (1999)
7. Abdulla, M., Paschos, G.: Effective Arabic Character Recognition using Support Vector Machines. In: Innovations and Advanced Techniques in Computer and Information Sciences and Engineering, pp. 7–11 (2007)
8. Chang, C.-C., Lin, C.-J.: LIBSVM – A Library for Support Vector Machines, http://www.csie.ntu.edu.tw/~cjlin/libsvm
9. Boucharebet, F., Hamdi, R., Bedda, M.: Handwritten Arabic character recognition based on SVM Classifier. In: Information and Communication Technologies: From Theory to Applications, Algeria (2008)
10. Faouzi, Z., Abdelhamid, D., Mohamed, B.: An Approach Based on Structural Segmentation for the Recognition of Arabic Handwriting. In: Advances in Information Sciences and Service Sciences, vol. 2 (2010)
11. Al-Hamadi, A., Saeed, A., Dings, L., Elzobi, M.: Arabic handwriting recognition using Gabor wavelet transform and SVM. In: Signal Processing (ICSP), Germany (2012)

12. Guericke, O.V.: IESK-arDB-A database for off-line handwritten Arabic words, http://www.iesk-ardb.ovgu.de/
13. Alalshekmubarak, A., Hussain, A., Wang, Q.-F.: Off-line handwritten arabic word recognition using SVMs with normalized poly kernel. In: Huang, T., Zeng, Z., Li, C., Leung, C.S. (eds.) ICONIP 2012, Part II. LNCS, vol. 7664, pp. 85–91. Springer, Heidelberg (2012)
14. Zanchettin, C., Azevedo, W.W., Bezerra, B.L.D.: A KNN-SVM hybrid model for cursive handwriting recognition. In: Proceedings of the International Conference on Neural Networks (IJCNN), Brazil, pp. 1–8 (2012)
15. Hassanien, A.E., Suraj, Z., Slezak, D., Lingras, P.: Rough computing: Theories, technologies and applications. IGI Publishing Hershey, PA (2008)

iSee: An Android Application for the Assistance of the Visually Impaired

Milad Ghantous, Michel Nahas, Maya Ghamloush, and Maya Rida

Lebanese International University
{Milad.ghantous,Michel.nahas}@liu.edu.lb

Abstract. Smart phone technology and mobile applications have become an indispensable part of our daily life. The primary use however, is targeted towards social media and photography. While some camera–based approaches provided partial solutions for the visually impaired, they still constitute a cumbersome process for the user. iSee is an Android based application that benefits from the commercially available technology to help the visually impaired people improve their day-to-day activities. A single screen tap in iSee is able to serve as a virtual eye by providing a sense of seeing to the blind person by audibly communicating the object(s) names and description. iSee employs efficient object recognition algorithms based on FAST and BRIEF. Implementation results are promising and allow iSee to constitute a basis for more advanced applications.

1 Introduction

Smart phones technology has spread in the international market faster than any other technology in human history. Worldwide Smartphone vendors shipped 237.9 million units in 2013 compared to the 156.2 million units shipped in 2012 which represent 52.3% year-over-year-growth [1]. Android Operating System is one of the smart phones operating systems that have been on the top list of best-selling where 80% of the smart phones shipped ran android operating systems, while Apple's iOS and Microsoft Windows 8 Mobile take up the remaining 20%.

Any user can benefit from the wide range of mobile applications that help in organizing our daily life, connecting with people through social networks, help in education, etc. Only few of these mobile apps have surpassed the purpose of being an added value to our daily life, to becoming a crucial tool. Applications crafted for people with disabilities have emerged lately to help them achieve more what they could do before the Smartphone era. Visually impaired and blind people find it difficult to face their day-to-day problems because of their impairment. The World Health Organization (WHO) estimated that in 2002, 2.6% of the world's total population was visually imapried. In spite of the advancement in technology, there is no complete cure for such problems and they need some assistance to complete their daily tasks. Reading glasses, walking sticks, and Braille tools are few devices which could help them to complete their daily life. However, they don't provide a sense of "seeing" for the blind person. This paper discusses the development of "iSee", an Android based application that aims at serving as a virtual eye. iSee provides the sense of sight by

A.E. Hassanien et al. (Eds.): AMLTA 2014, CCIS 488, pp. 26–35, 2014.

letting the user hold the phone and point anywhere he/she desired and tap on the screen. The application's algorithm runs in the background and then communicates audibly, via a voice message, the object type, name and description. This helps the blind person recognize what's surrounding him/her, or in some cases, find an object he/she is searching for.

The remaining part of this manuscript is organized as follows: section 2 provides a survey over some existing apps in the Google play market and the AppStore. In section 3, we present the main algorithm and implementation details. Section 4 summarized the results and the possible enhancements, while section 5 concludes the paper.

2   Related Work

Several mobile applications on multiple platforms such as iOS, Android and windows were developed with the visually impaired in mind. These applications can be divided among three different categories: currency recognition oriented apps, object detection oriented apps and phone assistant oriented apps.

"eyeNote" [2] and "LookTel" [3] applications belong to the first category. The former recognizes currency bills (US dollars only) and communicates with the user via a voice message, while the latter does the same job but supports more currencies.

The "Blind sighted" fits in the second category. Its functionality is limited to buzzing whenever the user is in 2-3 cm vicinity of an object.

"Siri" [4] and "S voice" [5], iOS's and Samsung's famous voice assistant software belong to the third category. Users can communicate with them via voice messages to inquire about weather, stocks, messages, as well as perform a phone call or a Google search. Few independent applications, like "Blind Navigator" [6] and "Georgie" [7] were also developed with the same goal in mind. The former provides the user with several functions like the phone, calculator, alarm, color identifier and GPS, while the latter adds a TextToSpeech functionality to the above.

Despite the usefulness and effectiveness of the several of the aforementioned applications, the blind people still lack the ability to identify the objects that surround them. Getting to know the type and description of the objects near you not only provides you with information about that specific object but also provides a sense of seeing and overall scene re-construction.

One of the non-mobile approaches is proposed in [8]. A multi-camera network is built by placing a camera at important locations around the user. The developed algorithm can effectively distinguish different classes of objects from test images of the same objects in a cluttered environment.

3   Main Methodology

This manuscript introduces "iSee", an Android powered application for the visually impaired. iSee aims at helping the blind people recognize objects surround them using a voice message. The overall architecture of iSee is illustrated in Fig. 1.

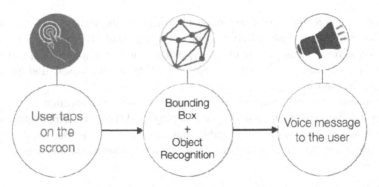

Fig. 1. Overall architecture of iSee

The user interface of iSee is crafted to be user friendly, considering the limitation of the intended users. First, the user opens the application by tapping the app icon or by using any assistant software such as S Voice (or siri for its iOS counterpart). The built-in back camera opens automatically occupying the whole screen, and the user is welcomed with an audio note notifying him/her of the application readiness. At any time, the user can point the phone anywhere he/she desires and taps anywhere on the screen using a single quick tap. An audio note is then given to the user announcing that a snapshot has been captured and ready to be analyzed. In case of a shaky result, the user is asked to re-do the process. Once the snapshot is ready, the application applies object detection and recognition algorithms. If a match is found, the name of the object is communicated to the user via a voice message. Otherwise, the user is notified with a "not found" message and an invitation to try again.

3.1 Object Detection and Recognition

The object detection algorithm constitutes the core of iSee's framework. The majority of algorithms start by finding interest points and their corresponding descriptors in both the scene and a reference image/object. A matching process is then carried between the two images to filter out the good matches. A wide variety of approaches has been proposed in literature [9-14]. Two main approaches were considered: SURF [15] and ORB [16].

SURF, or Speeded Up Robust Features, is a scale and rotation invariant detector and descriptor. SURF employs Hessian-Laplace matrix detectors to find interest points. A feature vector (or descriptor) is then computed to describe the neighborhood of the interest point, by means of local-sensitivity hashing (LSH) [17,18]. SURF relies on Euclidean distance between the matched descriptor and the most similar one. Distances below a certain pre-determined threshold are chosen as good matches. SURF has a computational reduction advantage over its predecessors (mainly SIFT [19]) due to the use of integral image in the computation of the Hessian detector [20]. This property is very attractive for real-time systems as well as power-limited battery-operated devices such as phones, tablets and phablets.

On the other hand, ORB (Oriented FAST and Rotated BRIEF) offers comparable performance to SIFT and SURF while offering two orders of magnitude in speed-up. ORB builds on the FAST detector [21] for interest point generation and on BRIEF [22] for the descriptor computation. FAST relies on the intensity threshold between the center pixel and a circular ring around it. FAST-9 employs a circular ring with a radios of 9 pixels and offers good performance. BRIEF on the other hand uses binary tests between pixels in a smoothed image patch. It is robust to lighting, blurring and distortion but sensitive to rotation. Rotated BREIF (rBRIEF) was proposed to overcome this shortcoming. Fig. 2 depicts a performance comparison between BRIEF, SURF, SIFT and rBRIEF. According to [16], ORB outperforms its competitors in terms of computation time, averaging at 15.3 ms per frame, while SIFT and SURF require 5228 and 217 ms, respectively.

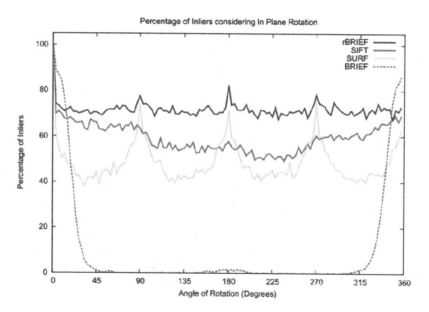

Fig. 2. Matching performance of SIFT, SURF, BRIEF with FAST, and ROB (oFAST + rBRIEF) unser synthetic rotations [16]

Due to the restrictive platform, on which iSee runs, ORB is chosen as the underlying object detection algorithm. The algorithm has 2 inputs: a training object provided by the application (or can be uploaded by the user) and the scene which is the image captured by the user using the tap described before. Interest point detection and descriptor computation take place in the next steps, followed by a nearest neighbor matching process to filter out the good matches. Fig. 3 illustrates the overall iSee object detection flowchart.

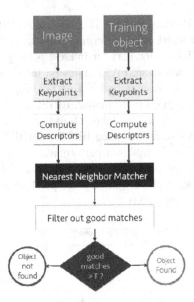

Fig. 3. iSee Matcher Flowchart

The matcher depicted in Fig. 3 is then used to recognize the object(s) found in the scene. A training set of images is provided containing a number of objects. The objects were chosen to be household objects but can be expanded to cover a wider range. Note however, that different postures of the same objects are provided too (front, back, side). Each object in the training set is matched against the scene captured by the user, by means of the aforementioned matcher. A thresholding approach based on the number of matches, is then employed to eliminate all the "bad" matches. The remaining matches are then compared to choose the maximum. At this stage, the object name/type is then communicated to the user. Implementation details are discussed in section 3.2. The overall process is depicted in Fig.4.

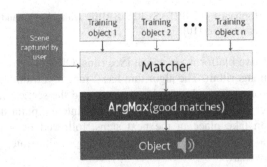

Fig. 4. iSee object recognition process

3.2   Implementation Details

The development of iSee involved several implementation tools and libraries. Android 4.3 SDK (Software developer kit) is chosen as the underlying platform. The interface of the app as well as the camera functions (initialization of the lens, screen tap, internal storage management and text to speech) are all performed using the Java libraries provided by the SDK, and developed using Eclipse IDE. On the other hand, all image processing functionalities were implemented using OpenCV 2.4.6 SDK for Android. OpenCV (Open Source Computer Vision [23]) is a library of programming functions mainly aimed at real-time computer vision, developed by Intel. Two implementation options were available for openCV: JavaCV or the native C++ alternative. Both options can be compiled and used with the Android SDK. Our choice however was set to using the native C++ due to its faster performance and wider library functions. This performance improvement came at the expense of additional implementation complexities. First, Android NDK (Native Development Kit) had to be installed and linked with Eclipse. The NDK is a toolset that allows parts of an app to be implemented using native-code languages such as C and C++. This capability helps in reusing existing code libraries written in these languages. The NDK is suitable for CPU-intensive workloads such as game engines, signal processing, physics simulation, and so on. Second, some of the libraries that are already available in openCV (e.g. SURF) are not available for its Android counterpart. To solve this problem, the libraries were compiled and added manually to the openCV SDK for Android.

The code of iSee proceeds into three main phases: (1) the welcome activity and the camera initialization, (2) the processing part in C++, and (3) the results announced to the user. During the first phase, the openCV libraries are loaded using the `system.loadLibrary` functions. The camera is then initialized and a listener is set up to watch for screen taps. Once a tap is detected on the screen, the scene image is captured and converted into a matrix, S. 'S' becomes the input for phase 2. In the second phase, matrix S and objects from the training set (O_i, $i=1,...,n$) undergo the process depicted in Fig.4 using the following functions: `OrbFeatureDetector()`, `OrbDescriptorExtractor()` and `KnnMatcher()`. Once this process is terminated, the results are passed back to Java in phase 3. During the last phase, the name of the object is converted to a voice message using the `textToSpeech()` utility. Note that all the objects in the training set are labeled with a corresponding name/description.

4   Results and Future Enhancements

Prior to the Android implementation, thorough experimentation was performed to verify the correctness as well as the accuracy of the algorithm. The algorithm was implemented in Xcode on a MacBook Pro running Intel's core i5 processor. A set of 24 objects constituted the test set, each having 3 different postures, resulting in a total of 72 images. Images were chosen to be daily use objects and categorized in 3 groups: coffee mugs, sunglasses and laptops. The algorithm's performance is depicted in Table 1. The average accuracy of all the categories was found to be 91.33%.

Table 1. iSee's performance under the desktop implementation

Category	No. of images	Correctly detected	False positives	Accuracy
Coffee mugs	24	23	1	95.8 %
Sunglasses	24	22	2	91.6 %
Laptops	24	21	3	87.5 %

A snapshot of the running algorithm is depicted in Fig. 5. The top row of the figure denotes a match success between an object and a scene with high number of good matches, while the bottom row represents a mismatch between the object and the scene with a low number of good matches.

Fig. 5. (a),(c) Training object, (b) scene containing the object (d) scene without the object

The algorithm is then implemented on the Android platform as detailed in section 3.2. The match success rate was found to be slightly less than its desktop implementation averaging at 89%. Table 2 depicts the algorithm accuracy per category. Fig. 6. Shows few snapshots of the application running on a Samsung Note 3.

Table 2. iSee's performance under Android Implementation

Category	No. of images	Correctly detected	False positives	Accuracy
Coffee mugs	24	22	1	91.6 %
Sunglasses	24	21	2	87.5 %
Laptops	24	21	3	87.5 %

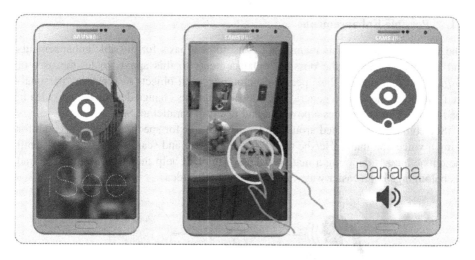

Fig. 6. Running iSee app screenshots

The processing time of iSee is finally analyzed based on the size of the training library and compared against its desktop counterpart. The results are illustrated in Fig.7. iSee was able to recognize an object within 1.8 seconds given a library size of 75 images, while its desktop counterpart outperformed it by 0.8 seconds. We strongly argue though, that with recent advancements in processor speeds employed in current-ly released smartphones, discrepancies between desktop and mobile performances will continue to decrease.

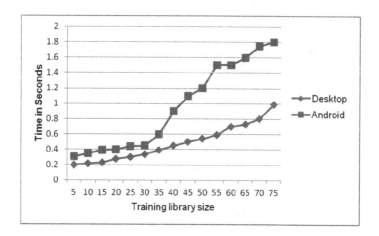

Fig. 7. Training library size versus processing time for Desktop and Android implementations

4.1 Possible Enhancements

The work presented in this manuscript constitutes a basis for possible enhancements and app ideas. One of the possible enhancements to this app is to let the user or his/her assistant upload their personal belongings and objects to create a "personalized" library rather than a generic one. Despite that this change defeats the app idea in the first place, it constitutes a powerful and fast tool parallel to iSee.

iSee could be also turned around to let the user ask for specific object to search for using a voice command. He/she then holds the phone and scans the surrounding until the phone buzzes denoting a match success. This will help the visually impaired find his belongings in an easier way. Fig. 8 illustrates this idea.

Fig. 8. Alternative use to iSee

A more advanced research can lead, not only to a better recognition rate but also to scene behavior, in which, the app can describe to the visually impaired what's happening in real time in his/her surrounding.

5 Conclusion

This paper presented the development of iSee, an Android powered application for the visually impaired. iSee lets the user taps on the screen and then communicates via a voice message the name and type of the object in the scene. iSee employs ORB as the underlying object recognition algorithm due to its computational efficiency on mobile platforms. The implementation results of iSee were promising and constituted a basis for more advanced developments.

References

1. Aaron, S.: Smartphone Ownership. Pew Research Center's Internet & American Life Project (2013), http://pewinternet.org/Reports/2013/Smartphone-Ownership-2013.aspx (accessed June 2014)
2. EyeNote App, Apple Appstore (2010), http://www.eyenote.gov/
3. Sudol, J., et al.: LookTel- A comprehensive platform for computer aided visual assistance. Paper presented at the IEEE Conference on Computer Vision and Pattern Recognition (2010)
4. Apple Siri (2011), https://www.apple.com/ios/siri/
5. S Voice (2011), http://www.samsung.com/global/galaxys3/svoice.html

6. Blind Navigator App, Google Play (2013), `https://play.google.com/store/apps/details?id=com.Blindnavigator`
7. Georgie App, Google Play (2012), `http://www.georgiephone.com/`
8. Yi, C., et al.: Finding objects for assisting blind people. Network Modeling Analysis in Health Informatics and Bioinformatics 2(2), 71–79 (2013), doi:10.1007/s13721-013-0026-x
9. Linderberg, T.: Feature detection with automatic scale selection. Computer Vision 30(2), 79–116 (1998)
10. Lowe, D.: Distinctive image features from scale-invariant keypoints, cascade filtering approach. Computer Vision 2(60), 91–110 (2004)
11. Mikolajczyk, K., Schmid, C.: An affine invariant interest point detector. In: Heyden, A., Sparr, G., Nielsen, M., Johansen, P. (eds.) ECCV 2002, Part I. LNCS, vol. 2350, pp. 128–142. Springer, Heidelberg (2002)
12. Ke, Y., Sukthankar, R.: PCA-SIFT: A more distinctive representation for local image descriptors. Paper presented at CVPR Computer Vision and Pattern Recognition (2004)
13. Tuytelaars, T., Van Gool, L.: Wide baseline stereo based on local, affinely invariant regions. Paper presented at the British Machine Vision Conference (2000)
14. Matas, J., Chum, O., Pajdla, T.: Robust wide baseline stereo from maximally stable external regions. Paper presented at the British Machine Vision Conference (2002)
15. Hebert, B., et al.: Speed-up Robust Features (SURF). Computer Vision and Image Understanding 110(3), 346–359 (2008)
16. Rublee, E., et al.: ORB: an efficient alternative to SIFT or SURF. Paper presented at the IEEE International Conference on Computer Vision (2011)
17. Qin, L., et al.: Multi- Probe LSH: Efficient Indexing for High-Dimensional Similarity Search. In: Proceedings of Very Large Database Conference, Vienna (2007)
18. Har-Peled, S., Indyk, P., Motwani, R.: Approximate Nearest Neighbors: Towards Removing the Curse of Dimensionality. Theory of Computing 8(1), 321–350 (2012)
19. Lopez-de-la-Calleja, M., et al.: Superpixel-Based Object Detection Using Local Feature Matching. In: Proceedings of the 29th Conference of the Robotics Society of Japan, Toyosu (2011)
20. Mikolajczyk, K., Schmid, C.: Indexing based on scale invariant interest points. Paper presented at the International Conference on Computer Vision (2001)
21. Rosten, E., Drummond, T.: Machine learning for highspeed corner detection. Paper presented in European Conference on Computer Vision (2006)
22. Calonder, M., Lepetit, V., Strecha, C., Fua, P.: Brief: Binary robust independent elementary features. In: Daniilidis, K., Maragos, P., Paragios, N. (eds.) ECCV 2010, Part IV. LNCS, vol. 6314, pp. 778–792. Springer, Heidelberg (2010)
23. openCV (2014), `http://opencv.org/`

Arabic Sign Language Recognition Using Spatio-Temporal Local Binary Patterns and Support Vector Machine

Saleh Aly and Safaa Mohammed

Department of Electrical Engineering, Faculty of Engineering
Aswan University, Aswan, Egypt
saleh@aswu.edu.eg, safaa.elsoghier@gmail.com

Abstract. One of the most common ways of communication in deaf community is sign language recognition. This paper focuses on the problem of recognizing Arabic sign language at word level used by the community of deaf people. The proposed system is based on the combination of Spatio-Temporal local binary pattern (STLBP) feature extraction technique and support vector machine (SVM) classifier. The system takes a sequence of sign images or a video stream as input, and localize head and hands using IHLS color space and random forest classifier. A feature vector is extracted from the segmented images using local binary pattern on three orthogonal planes (LBP-TOP) algorithm which jointly extracts the appearance and motion features of gestures. The obtained feature vector is classified using support vector machine classifier. The proposed method does not require that signers wear gloves or any other marker devices. Experimental results using Arabic sign language (ArSL) database contains 23 signs (words) recorded by 3 signers show the effectiveness of the proposed method. For signer dependent test, the proposed system based on LBP-TOP and SVM achieves an overall recognition rate reaching up to 99.5%.

1   Introduction

A gesture is a form of non-verbal communication performed with a part of body, used instead of or in combination with verbal communication [1, 5]. Similar to speech and handwriting, gestures vary from a person to another, and even for the same person between different instances. Since sign language recognition (SLR) is a kind of highly structured and largely symbolic human gesture set, SLR also serves as a good basic for the development of general gesture-based human computer interaction (HCI).

Arabic sign language (ArSL) is the natural language of hearing-impaired people in the Arabic society [11, 16]. Hearing people have difficulty to learn sign languages, also it is difficult for deaf people to learn oral languages. Sign language recognition systems can facilitate the communication between those two communities. There is a need therefore for a translation system that can

A.E. Hassanien et al. (Eds.): AMLTA 2014, CCIS 488, pp. 36–45, 2014.

convert an ArSL to written or spoken Arabic and vice versa, in order that the deaf community can communicate better with the normal people.

Appearance-based approaches are one candidate to solve SLR problem instead of traditional geometric-based approaches. Among various appearance-based features, local binary pattern (LBP) [12] features considered the most successful one due to its robustness and computation efficiency. Two variants of LBP [19] are proposed to handle dynamic textures which are: volume local binary patterns (VLBP) and local binary patterns from Three Orthogonal Planes LBP (LBP-TOP). Both methods can jointly describe motion and appearance features. These features are not only insensitive to local translation and rotation variations but also robust to monotonic gray-scale changes caused by illumination variations. Moreover, LBP-TOP features are computationally simple and easy to extend compared with VLBP. In this paper, LBP-TOP is employed to capture salient appearance and motion features efficiently and to represent each sign with a single feature vector.

In this paper, we aim at designing and implementing an automated robust signer system for Arabic sign language recognition from videos of signs at word level. The proposed method does not force the user to wear any cumbersome device or any type of gloves. The system starts with a preprocessing stage for segmenting the signers' hands and head. Hand and head segmentation is performed to focus only on the region of interest for each sign, segmentation is achieved using an effective skin detection algorithm proposed in [8]. This is followed by a feature extraction stage in which a vector of spatial and temporal features are extracted [19]. In the paper, we employ spatio-temporal local binary pattern to jointly extract spatial and temporal features of images sequences for each sign. A set of labeled signs is then used to train a multi-class support vector machine (SVM) [17] in order to classify the feature vectors resulted from previous feature extraction stage. The recognition step is achieved using the pretrained support vector machine classifier to identify each signs.

The remainder of this paper is organized as follows; a short summery of related work is presented in Section 2. The proposed system architecture is explained in Section 3 followed by Section 4 where we discuss the experimental results. Finally conclusion and future works are presented in Section 5.

2   Related Works

Hand gestures can be classified into two categories: static and dynamic. A static gesture is a particular hand shape and pose, represented by a single image. A dynamic gesture is a moving gesture, represented by a sequence of images. The proposed approach focuses on the recognition of dynamic gestures. There are two main directions for sign language recognition; glove-based methods and vision-based methods. The glove-based system [18] relies on electromechanical devices and use motion sensors to capture gesture data. Here the signer must wear some sort of wired gloves that are interfaced with many sensors. Vision-based recognition systems do not use special device such as glove, special sensor, or any additional hardware, it provides a more natural environment to capture the gesture data [5]. Vision-based recognition systems can be classified into two categories: appearance-based and geometric-based methods [13].

Although there are a lot of works have been done in sign language, less research attention have been achieved in ArSL. There have been only few research studies on alphabet ArSL. A colored glove for data collection and Adaptive Neuro-Fuzzy Inference Systems (ANFIS) [2] was used to recognize isolated signs for 28 alphabets. A recognition rate of 88% was achieved. Later, the recognition rate was increased to 93.41% using polynomial networks [4]. Spatial features and single camera were used to recognize 28 ArSL alphabets without using gloves [1] which reported a recognition rate of 93.55% using ANFIS method. Power gloves for data collection and support vector machine (SVM) learning method [9] was used to recognize isolated words. An image-based system [10] was presented for ArSL recognition. Color images where signer used a pair of colored gloves and a Gaussian skin color model was used to detect the signer's face. An automatic ArSL recognition system based on hidden Markov models is presented in [3]. A Discrete Cosine Transform (DCT) was employed to extract features from the input gestures by representing the image as a sum of sinusoids of varying magnitudes and frequencies. In the experiments, 30 isolated words from the standard ArSL database are tested. The data are collected without gloves. A recognition rate of 97.4% was achieved.

3 The Proposed Arabic Sign Language Recognition System

The proposed system is mainly consists of several stages including image capture, head and hands segmentation, feature extraction, and recognition modules. Fig. 1 illustrates the proposed system architecture; it manifests the system constituting components and the way they are connected to each other. Firstly, the signer performs word gestures in front of a camera then the gesture video is segmented into frames, the captured image sequence is then preprocessed before feature extraction. The region of interest for all signs are head and hands parts only, these regions are segmented using an efficient skin color model. The skin map is further analyzed to crop the hand and head regions, the cropped image is then rescaled to fixed size. Spatial and temporal features are extracted from the head and hands regions using LBP in three orthogonal planes (LBP-TOP). Features extracted from a set of labeled training data are used to train SVM classifier in the training phase. In recognition, the unknown sign pass through the previous steps and its label is identified using the pretrained SVM classifier.

3.1 Head and Hands Segmentation

Detection of head and hands and the segmentation of the corresponding image regions is an important step in gesture recognition systems. This segmentation is crucial because it isolates the task-relevant data from the image background

Fig. 1. Proposed Arabic sign language recognition system using LBP-TOP and SVM

Fig. 2. Head and hands segmentation using Random Forest of IHLS color model

before passing them to the subsequent feature extraction and recognition stages. The most prominent cue for detecting head and hands is skin color. Therfore, skin color detection can be used to segment head and hands area for each frame of the input image sequence.

Typically, a skin segmentation framework involves transformation of the RGB color-space to another color-space, removing the illumination component and using chromatic components of the converted color-space, finally classifying skin by an appropriate skin color modeling technique. The combination of appropriate color space and skin color modeling is important to achieve good segmentation results. Skin color segmentation has been utilized by several approaches for hand detection. For providing a model of skin color, the color space to be employed should be selected carefully. Recently, several color spaces and color modeling techniques have been tested in [8]. Six color spaces including IHLS, HSI, RGB, normalized RGB, HSV, YCbCr and CIELAB combined with nine skin color modeling approaches include tree based, neural network, and probabilistic classifiers are examined in [8]. Among all these combinations, IHLS color space combined with Random Forest classifier performed better than other methods. In this paper, we employ such combination for segmenting the face and the two hands in each frame. Fig. 2 shows examples of the detection results achieved by random forest classifiers on sample images from ArSLR dataset. Results show the robustness of the skin classification against various illumination conditions.

After detecting skin components, a connected component labeling algorithm [7] is used where subsets of connected image components are uniquely labeled. An algorithm scans the image, labeling the underlying pixels according to a predefined connectivity scheme and the relative values of their neighbors. Only large components are taken into consideration. Three components representing the two hands and the face will be detected in the normal case. However, there are some situations where fewer components are detected due to occlusion. When occlusion of two hands appears, both hands are represented by same object, also in case of head occlusion the same principle is used.

3.2 Spatio-Temporal Feature Extraction

For video segments (i.e., image sequence), feature extraction is typically done in the temporal and spatial domains in order to capture the appearance and motion information contents of the image sequence. For feature extraction, local binary patterns on three orthogonal planes (LBP-TOP) is used to capture the co-occurrence of appearance and motion features from image sequence. We give a brief background of local binary patterns (LBP) and LBP-TOP in the following:

Local Binary Patterns. Basic LBP operator was first designed for texture description [12]. This operator describes each pixel by comparing its value with neighbors; if the neighboring pixel value is higher or equal, the value is set to one, otherwise set to zero. Then the concatenation of binary patterns over the neighborhood converted into a decimal number as a unique descriptor for each pixel.

Original local binary pattern (LBP) has an extension called uniform patterns, is used in the proposed system. Uniform patterns can significantly reduce the length of the feature vector. This extension was inspired by the fact that some binary patterns occur more commonly in texture images than others. A local binary pattern is called uniform if the binary pattern contains at most two bitwise transitions from 0 to 1 or vice versa when the bit pattern is traversed circularly. Using uniform patterns in representation reduces the length of the feature vector. For example, when using 8 neighborhood pixels, there are a total of 256 patterns, 58 of which are uniform, which yields in 59 different labels. The most important property of the LBP operator in real-world applications is its robustness to monotonic gray-scale changes caused, for example, by illumination variations. Another important property is its computational simplicity, which makes it possible to analyze images in challenging real-time settings.

Spatio-Temporal Local Binary Patterns. Original local binary pattern is used mainly for static texture description but, for time variation we should use the extend of LBP for spatial and temporal domain (STLBP). STLBP is used to model the dynamic scenes using both spatial texture and temporal motion information together. Volume local binary patterns (VLBP) method [20] is an extension of LBP operator widely used in dynamic texture recognition which combine motion and appearance features. The texture features extracted in a small local neighborhood of the volume comprised space and time directions. To make VLBP computationally simple and easy to extend, LBP-TOP is considered which only compute the co-occurrences on three separated planes.

LBP-TOP considers three orthogonal planes: XY, XT and YT, and concatenates local binary pattern co-occurrence statistics in these three directions. The XY plane represents appearance information, while the XT plane gives a visual impression of one row changing in time and YT describes the motion of one column in temporal space. The LBP codes are extracted for all pixels from the XY, XT and YT planes, denoted as XY-LBP, XT-LBP and YT-LBP, and histograms from these planes are computed and concatenated into a single histogram. In such a representation, a gesture is encoded by an appearance (XY-LBP) and two spatial temporal (XT-LBP and YT-LBP) co-occurrence statistics. Feature vector from each plan is extracted simply like ordinary LBP, then they are concatenated to each other forming the final feature vector.

Spatial information is one of the most important cues to distinguish between different signs. In order to preserve the spatial information of the gesture, block based LBP-TOP is computed for all non overlapping blocks. LBP-TOP features have three advantages: 1) it is robust to monotonic gray-scale changes; 2) it is on-line and very fast to compute; 3) it can extract spatial texture and temporal motion information of a pixel. These three advantages are all very important for modeling the dynamic structure of image sequences. Fig. 3 shows the feature vector formation for one sign which includes histograms for appearance (LBP-XY), horizontal (LBP-XT) and vertical motion features (LBP-YT).

4   Experimental Results

In this section, a set of experiments have been performed to examine the performance of the proposed system. Arabic sign language (ArSL) database described in [14], [15] is employed in all experiments. A prototype has been built using hands and head segmentation, proposed feature extraction and SVM recognizer to test the effectiveness of the proposed method. In proposed system, block-based LBP-TOP is used to extract spatial and temporal feature vectors for training and testing image sequences. Features are extracted only from the region of interests which include head and hands. Features extracted using radius of 1 pixel in the 3 directions (XYT) and 8 neighborhood pixels. Extracted feature vectors (with labels) are normalized to unit length and fed to train a multi-class SVM classifier. After having the trained SVM, using it in the procedure of recognition to find the label with most probably sign.

4.1   ArSL Database

The ArSL database has been used to carry out the underlying experiments. In this database, there are 23 isolated words performed by 3 signers. The signer was videotaped without imposing any restriction on clothing or image background. The video frames are sampled at 25 frames per second and the size of the frames

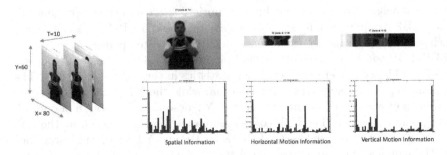

Fig. 3. Local Binary Patterns on Three orthogonal Planes (LBP-TOP) feature vector forming

is 320 × 240 pixels, the region of interests (head and hands) are segmented using skin color information. Segmented images are then cropped and rescaled to 64 × 64 pixels to reduce the time of execution without affecting the image details.

Experiment 1: The first experiment evaluates the recognition rate of the system using various kernels of SVM and LBP-TOP features. Linear, 2nd order polynomial and radial basis function (RBF) kernels are tested in this experiment. Results in Fig.4 show that linear kernel outperforms other non-linear kernels like polynomial and radial basis function. Not only the linear kernel gives better recognition rate but also it is more computationally efficient in both training and test compared with other non-linear kernels. The obtained results using linear SVM make the proposed system applicable for real time sign recognition.

Experiment 2: Because of the importance of spatial information in sign recognition, the second experiment evaluate the system using various block sizes. The volume of sign is divided in spatial domain only (XY plan) because most of the signs in the data set have small number of frames between 2 to 11 frames. For each block in the feature vector, it has (3 × 59 = 177) patterns which represent appearance, horizontal and vertical motion information. Different block sizes ranged from 1 × 1 (i.e no spatial division is performed in this case) and 16 × 16 are tested. In addition, the performance of the linear SVM classifier is compared with KNN classifier. Fig. 5 shows the accuracies of both SVM and KNN classifiers at different block sizes. As expected, spatial information is an important cue to distinguish between different signs. Therefore, the accuracy increases as the number of blocks increase. The best performance (99.5%) achieved at block size of 12 × 12, while after increasing the number of block division the accuracy decrease. In all cases, accuracies of SVM classifier are almost better than those obtained by KNN classifier.

Experiment 3: The practical significance of these results is emphasized by comparing our proposed method with the obtained results in [3]. In [3], ArSL

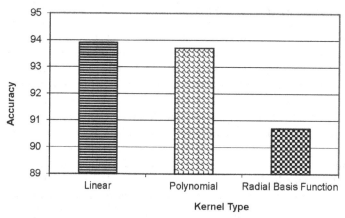

Fig. 4. Performance of the proposed system using skin detection

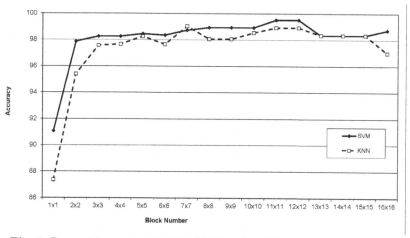

Fig. 5. Recognition rate with LBP-TOP under different number of blocks

database with different number of signers and 30 words are tested. The feature extraction method used in this work is Discrete Cosine Transform (DCT). As shown in Table 1, our system using(LBP-TOP) performs better than the system based on DCT features.

Table 1. Comparison with similar off-line, signer-dependent systems

Method	Instruments used	Mode	#of signers	Recognition rate
DCT	free hands	signer-dependent	18	97.4%
LBP-TOP	free hands	signer-dependent	3	99.5%

5   Conclusions and Future Works

This paper presents a gesture recognition system for recognizing gestures of ArSL. The system is mainly consists of several modules including skin color segmentation, feature extraction using LBP-TOP and recognition using SVM classifier. The system does not restrict user or signer to have any equipments like data/power gloves. The feature extraction module proposed for the first time for ArSL to extract jointly appearance and motion features. All experiments argue that the combination of LBP-TOP and SVM is a powerful method for sign language recognition, the recognition accuracy for the system is 99.5%. This achievement is of importance to the problem of Arabic sign language recognition, which had very limited research in its automated recognition. In future works, the proposed system will be adapted to work in signer independent scenario. Moreover, intensive experiments will be performed to examine the effect of each parameter of the LBP-TOP feature extraction method in the recognition accuracy. As a final goal, a real-time Arabic sign language translator can be implemented to help in communication with deaf and normal people. In our future works, rough sets-based feature extraction, rule generation and classification will provide more challenging and may allow us to refine our learning algorithms to the Arabic sign recognition [6].

References

1. Al-Jarrah, O., Halawani, A.: Recognition of gestures in arabic sign language using neuro-fuzzy systems. Artificial Intelligence 133(1-2), 117–138 (2001)
2. Al-Rousan, M., Hussain, M.: Automatic recognition of Arabic sign language finger spelling. International Journal of Computers and Their Applications (IJCA) 8, 80–88 (2001)
3. Al-Roussan, M., Assaleh, K., Talaa, A.: Arabic sign language recognition an image-based approach. Applied Soft Computing 9(3), 990–999 (2009)
4. Assaleh, K., Al-Rousan, M.: Recognition of Arabic sign language alphabet using polynomial classifier. EURSIP Journal on Applied Signal Processing 13, 2136–2145 (2005)
5. Dreuw, P.: Appearance-based gesture recognition. Diploma thesis, RWTH Aachen University, Aachen, Germany (2005)
6. Hassanien, A.E., Suraj, Z., Slezak, D., Lingras, P.: Rough computing: Theories, technologies and applications. IGI Publishing Hershe (2008)
7. Kang, S.K., Nam, M.Y., Rhee, P.K.: Colour based hand and finger detection technology for user interaction. In: International Conference on Convergence and Hybrid Information Technology, pp. 229–236 (2008)
8. Khan, R., Hanbury, A., Stöttinger, J., Bais, A.: Color based skin classification. Pattern Recognition Letters 33(2), 157–163 (2012)
9. Mohandes, M., Buraiky, S.A., Halawani, T., Al-Buayat, S.: Automation of the Arabic sign language recognition. In: International Conference on Information and Communication Technology (ICT 2004), pp. 117–138 (April 2004)
10. Mohandes, M., Quadri, S.I., Deriche, M.: Arabic sign language recognition an image-based approach. In: 21st International Conference on Advanced Information Networking and Applications Workshops (AINAW 2007), vol. 1, pp. 272–276 (2007)

11. Mohandes, M., Liu, J., Deriche, M.: A survey of image-based arabic sign language recognition. In: 2014 11th International Multi-Conference on Systems, Signals & Devices (SSD), pp. 1–4. IEEE (2014)
12. Ojala, T., Pietikainen, M., Maenpaa, T.: Multiresolution Gray-Scale and Rotation Invariant Texture Classification with Local Binary Patterns. IEEE Transactions on Pattern Analysis and Machine Intelligence 24(7), 971–987 (2002)
13. Ong, S.C.W., Ranganath, S.: Automatic sign language analysis: A survey and the future beyond lexical meaning. IEEE Transactions on Pattern Analysis and Machine Intelligence 27(6), 873–891 (2005)
14. Shanableh, T., Assaleh, K.: Telescopic Vector Composition and Polar Accumulated Motion Residuals for Feature Extraction in Arabic Sign Language Recognition. EURASIP Journal on Image and Video 2007(2), 9 (2007)
15. Shanableh, T., Assaleh, K., Al-Rousan, M.: Spatio-Temporal Feature-Extraction Techniques for Isolated Gesture Recognition in Arabic Sign Language. IEEE Transactions on Systems, Man, and Cybernetics, Part B: Cybernetics 37(3), 641–650 (2007)
16. Tolba, M.F., Elons, A.S.: Recent developments in sign language recognition systems. In: 2013 8th International Conference on Computer Engineering & Systems (ICCES), pp. xxxvi–xlii. IEEE (2013)
17. Vapnik, V.N., Vapnik, V.: Statistical learning theory, vol. 2. Wiley, New York (1998)
18. Zhang, X., Chen, X., Li, Y., Lantz, V., Wang, K., Yang, J.: A framework for hand gesture recognition based on accelerometer and emg sensors. IEEE Transactions on Systems, Man and Cybernetics, Part A: Systems and Humans 41(6), 1064–1076 (2011)
19. Zhao, G., Pietikainen, M.: Dynamic texture recognition using local binary patterns with an application to facial expressions. IEEE Transactions on Pattern Analysis and Machine Intelligence 29(6), 915–928 (2007)
20. Zhao, G., Pietikäinen, M.: Dynamic texture recognition using volume local binary patterns. In: Vidal, R., Heyden, A., Ma, Y. (eds.) WDV 2005/2006. LNCS, vol. 4358, pp. 165–177. Springer, Heidelberg (2007)

Improving Arabic Tokenization
and POS Tagging Using Morphological Analyzer

Michael N. Nawar

Department of Computer Engineering, Cairo University,
Cairo, Egypt
michael.nawar@eng.cu.edu.eg

Abstract. In this paper a new technique of tokenization and part-of-speech (POS) tagging for Arabic text is presented. The introduced technique uses the Arabic morphological analyzer to extract new features that will improve the stemming and the POS tagging. Applying standard evaluation metrics, the proposed tokenizer achieves an $F_{(\beta=1)}$ score of 99.99, and the POS tagger achieves an accuracy of 98.05%.

1 Introduction

Most of Natural language processing (NLP) systems such as information retrieval, text to speech, automatic translation and other use a part-of speech tagger for preprocessing. Supervised methods for part-of-speech (POS) tagging are expensive and time consuming as they depend on manually annotated data. However these methods achieve high results in NLP fields compared to unsupervised methods. Many of the Arabic words are ambiguous in their nature as tag of word can map to a noun, verb or adjective. It is believed that using a statistical approach which makes use of the morphological feature of the Arabic word would result in accurate, efficient and robust tagger that can be used in practical systems. Since both parsing and tagging Arabic words requires a stemming phase, a high accuracy in stemming phase implies a less accumulated error in further phases.

The basic idea of the proposed method is to recognize Arabic tokens and tagging them statistically using the Conditional Random Field learning approach by constructing a relevant model and feeding this model with some extra features extracted from the morphological analysis of each Arabic word. This concept is applied in the tokenization, normalization and POS tagging phase.

2 Arabic NLP and Data

There are three main categories of Arabic language; classical the language of Quran, modern standard (MSA) which is a simplified form of classical that is extracted from news and written documents, and dialectical Arabic which differs from one country to another. One variation of it is the colloquial language which is the daily used language by Egyptians.

A.E. Hassanien et al. (Eds.): AMLTA 2014, CCIS 488, pp. 46–53, 2014.

In general Arabic has a very rich morphological language where each word can include number, gender, aspect, case, mood, voice, mood, person, and state. The Arabic basic word form can be attached to a set of clitics representing object pronouns, possessive pronouns, particles and single letter conjunctions. Obviously the previous features of Arabic word increase its ambiguity. Generally Arabic stems can be attached three types of clitics orderd in their closeness to the stem according to the following formula:

$$\{[proclitic1]\{[proclitic2]\{Stem[Affix][Enclitic]\}\}\}$$

Where proclitic1 is the highest level clitics that represent conjunctions and is attached at the beginning such as the conjunction [(و, w, and), (ف, f, then)]. Proclitic2 represent particles [(ب, b, with/in), (ل, l, to/for); (ك, k, as/such)]. Enclitics represent pronominal clitics and are attached to the stem directly or to the affix such as pronoun [(ه, h , his), (هم, hm , their/them)].

The following is an example of the different morphological segments in the word that has the stem (قدر, qdr ,power), the proclitic conjunction (و, w, and) , the proclitic particle (ب, b ,with/in) , the affix (ات, At ,for plural) ,and the cliticized pronoun (ه, h , his). The set of proclitics considered in this work are the particles prepositions b, l, k, meaning by/with, to, as respectively, the conjunctions w, f, meaning and, then respectively. Arabic words may have a conjunction and a preposition and a determiner cliticizing to the beginning of a word. The set of possible enclitics comprises the pronouns and (possessive pronouns) y, nA, k, kmA, km, knA, kn, h, hA,hmA, hnA, hm, hn, respectively, my (mine), our (ours), your (yours), your (yours) [masc. dual], your (yours) [masc. pl.], your (yours) [fem. dual], your (yours) [fem.pl.], him (his), her (hers), their (theirs) [masc. dual], their (theirs) [fem. dual], their (theirs) [masc. pl], their (theirs) [fem. pl.]. An Arabic word may only have a single enclitic at the end. A token is defined as a (stem + affixes), proclitics, enclitics, or punctuation.

The data used for training and testing the stemmer and the POS tagger is the Arabic Treebank part 1 [1] which consists of 734 news articles (140kwords corresponding to 168k tokens after semi-automatic segmentation) covering various topics such as sports, politics, news, etc.

3 Related Work

A lot of the existing systems tend to target a specific application or a POS tag set that is not general enough for different applications. For example Shereen Khoja in (2001) [10] reports preliminary results on a hybrid, statistical and rule based, POS tagger, APT. APT yields 90% accuracy on a tag set of 131 tags including both POS and inflection morphology information. Diab et al. (2007) [1] perform a large-scale corpus-based evaluation of their approach. They use Yamcha SVM classifier based learner for three different tagging tasks: word tokenization, POS tagging and base phrase chunking with a collapsed tag set achieving a $F_{(\beta=1)}$ score of 99.12 on word tokenization and an accuracy of 96.6%

on POS tagging respectively. Diab (2009) [7] extended the work on Diab et al. (2007) to multiple tag set, instead of the PATB (Penn. Arabic Treebank) reduced tag set. Habash and Rambow (2005) [2] use SVM classier for individual morphological features and an ad-hoc combining scheme for choosing among competing analysis achieving an accuracy of 97.5%. Mansour (2007) [6] port an HMM Hebrew tagger to Arabic yielding to an accuracy of 96.1% for POS tagging. AlGahtani et al. (2009) [4] use transition based learning for the task of POS tagging, achieving an accuracy of 96.9%. Kulick, S. (2010) [5] performs simultaneous tokenization and POS tagging without a morphological analyzer, achieving an accuracy of 95.1% for POS tagging.

It is not a simple matter to compare results with previous work, due to differing evaluation techniques, data sets, and POS tag sets. In this paper, the results are compared with Diab et al. (2007) (SVM system) and Habash and Rambow (2005) (Majority system); because both of those papers and both of them work on the same range of data PATB (Penn. Arabic Treebank) part1, they report the results based on the PATB reduced tag set, they assume gold tokenization for evaluation of POS results, and the main concern is to report the highest accuracy unlike AlGahtani et al. (2009) and Kulick, S. (2010) where their main concern is the speedup.

4   Tokenization Phase

In this phase, the classifier takes an input of raw text, without any processing, and assigns each character the appropriate tag from the following tag set B-PRE1, B-PRE2, B-WRD, I-WRD, B-SUFF, I-SUFF. Where I denotes inside a segment, B denotes beginning of a segment, PRE1 and PRE2 are proclitic tags, SUFF is an enclitic, and WRD is the stem plus any affixes and/or the determiner Al. Two experiments have been conducted to achieve the final tokenizer: base line and binary feature experiments. The base line experiment is used to check the effect of using a CRF classifier instead of a SVM classifier in the task of tokenization. In the binary feature experiment a new feature has been proposed in addition to the features used in the base line experiment, and the effect of the binary feature in the task of tokenization is checked.

4.1   Baseline Experiment (CRF-TOK)

This experiment is based on the experiment of (Diab et al., 2007) but instead of using SVM classifier the CRF suite classifier is used. The classifier training and testing data is characterized as follows:

- Input: A sequence of transliterated Arabic characters processed from left-to-right with break markers for word boundaries.
- Context: A fixed-size window of -5/+5 characters centered at the character in focus.
- Features: All characters and previous tag decisions within the context.

4.2   Binary Feature Experiment (BF-TOK)

A new feature is proposed in this experiment and this feature is added to the feature set in the baseline experiment. BAMA-v2.0 (Buckwalter Arabic morphological analyzer version 2.0) is used to define a binary feature of length 6 where each bit in the feature is mapped to one of the 6 tags in the tokenization tag set. A bit is set if at least one analysis in the morphological analyses of the word, the character is assigned the tag corresponding to the bit.

For example the word (وحيد, wHyd) has two possible tokenization schemes: (و+حيد, w+Hyd) or (وحيد, wHyd); then (و, w) could be (B-PRE1 or B-WRD) then in the binary feature of the character there will be 2 bits set which map to B-PRE1 and B-WRD, (ح, H) could be (B-WRD or I-WRD) then in the binary feature of the character there will be 2 bits set which map to B-WRD and I-WRD, (ي, y) and (د, d) could be only (I-WRD) then in the binary feature of the characters there will be only one bit set which map to I-WRD. Table (1) shows the binary feature of each character of the word (وحيد, wHyd).

Table 1. Tokenization Binary Feature

Arabic Letter	Transliterated Letter	Binary Feature					
		B-PRE1	B-PRE2	B-WRD	I-WRD	B-SUFF	I-SUFF
و	w	1	0	1	0	0	0
ح	H	0	0	1	1	0	0
ي	y	0	0	0	1	0	0
د	d	0	0	0	1	0	0

If the word is not analyzed by the morphological analyzer (out of vocabulary); then all 7 bits of the binary feature will be set.

5   POS Tagging Phase

In this phase, the classifier takes an input of tokenized text, and it assigns each token an appropriate POS tag from the Arabic Treebank collapsed POS tags, which comprises 24 tags as follows: ABBREV, CC, CD, CONJ+NEG PART, DT, FW, IN, JJ, NN, NNP, NNPS, NNS, NO FUNC, NUMERIC_COMMA, PRP, PRP$, PUNC, RB, UH, VBD, VBN, VBP, WP, WRB}. Two experiments have been conducted to achieve the final POS tagger. The first experiment is used to check the effect of using a CRF classifier instead of a SVM classifier in the task of tokenization. In the second, the binary feature experiment a new feature has been proposed in addition to the features used in the base line experiment, and the effect of the binary feature in the task of POS tagging is checked.

5.1 Base Line Experiment (CRF-POS)

This experiment is based on the experiment of (Diab et al., 2007) but instead of using SVM classifier a CRF classifier is used. The classifier training and testing data is characterized as follows:

- Input: A sequence of transliterated Arabic tokens processed from left-to-right with break markers for word boundaries.
- Context: A window of -2/+2 tokens centered at the focus token.
- Features: Every character N-gram, N_i=4 that occurs in the focus token, the 5 tokens themselves, POS tag decisions for previous tokens within context.

5.2 Binary Feature Experiment (BF-POS)

A new feature is proposed in this experiment and this feature is added to the feature set in the baseline experiment. BAMA-v2.0 (Buckwalter Arabic morphological analyzer version 2.0) is used to define a binary feature of length 24 where each bit in the feature is mapped to one of the 24 tags in the collapsed POS tag set. A bit is set when its corresponding tag exists in the morphological analysis of a token.

For example the word (كتب, ktb) has 3 different reduced POS tags: VBD then it will mean (write), VBN then it will mean (be written), and NN then it will men (book); so there will be 3 bits set to one in the binary feature of the (كتب, ktb) word corresponding to VBD, VBN and NN. While you can find a word like (الولد, Alwld) has only one reduce POS tag which is NN and it have only one meaning the boy. In table (2), you can find the binary feature for the words of the sentence (كتب الولد الدرس, ktb Alwld Aldrs, The boy wrote the lesson).

Table 2. POS Tagging Binary Feature

Arabic Word	Transliterated Word	Binary Feature					
		VBD	VBN	NN	JJ	NNS	...
كتب	ktb	1	1	1	0	0	0
الولد	Alwld	0	0	1	0	0	0
الدرس	Aldrs	0	0	1	0	0	0

But for the word (يكتب, yktb) it has only one reduced POS tag: VBP which means (write); so there will be only one bit set in the binary feature which map to VBP. If the word is not analyzed by the morphological analyzer (out of vocabulary) like the word (الفالوجة, AlfAlwjp) which is a village in Palestine, then there will be 5 bits set in the binary feature which map to JJ, NN, NNS, NNP, and NNPS.

6   Empirical Results

For the evaluation of these experiments, k-fold algorithm was used by setting the parameter k to five so the Penn Arabic tree bank part1 is randomly partitioned into five portions of equal size. In each iteration of the k- fold algorithm four portions were used for training the model and one portion was used for testing the model. The cross-validation process is then repeated five times (the folds), with each of the k subsamples used exactly once as the testing data. The five results from the folds were averaged to produce the model evaluation. This evaluation scheme was applied for both the tokenization and POS tagging. Then the following performance measures are calculated for each experiment

$$macro\ average\ precision = \frac{1}{n} \sum_{i=1}^{n} precision(tag(i))$$

$$macro\ average\ recall = \frac{1}{n} \sum_{i=1}^{n} recall(tag(i))$$

$$macro\ average\ F_{(\beta=1)} = \frac{1}{n} \sum_{i=1}^{n} F_{(\beta=1)}(tag(i))$$

$$Accuracy = \frac{number\ of\ true\ results}{number\ of\ true\ and\ false\ results}$$

Then the proposed method is compared with the SVM based approach [1] and the Majority system [2]. The comparison between the proposed method and the SVM approach and the majority system will be in the accuracy and the $F_{(} = 1)$ of the tokenizer and in the accuracy of the POS tagger, because these are the only performance measures they have reported. The tool used for evaluation is the evaluation tool in the CRF- Suite software package.

6.1   Tokenization Phase Evaluation

Table (3) compares the different experiments applied to the Tokenization task where the row represents the experiment and the column represents the macro average performance measure.

Table 3. Tokenization Phase Evaluation Results

	Precision	Recall	$F_{\beta=1}$	Accuracy	Error
CRF-TOK	0.99835	0.99926	0.99880	99.98%	0.02%
BF-TOK	0.99998	0.99908	0.99952	99.99%	0.01%

The performance of BF-TOK is almost perfect. Comparing BF-TOK to other Arabic tokenizers like: SVM-TOK which has an accuracy of 99.77% and an F score of 99.12; and with the Majority-TOK which has an accuracy of 99.3% and an $F_{\beta=1}$ of 99.1; the improved stemmer reduces the error by about 95.65% compared to the SVM-TOK, and by 98.57% from the Majority system tokenizer.

6.2 POS Tagging Phase Evaluation

Table (4) compares the different experiments applied to the POS tagging task where the row represents the experiment and the column represents the macro average performance measures.

Table 4. POS Tagging Phase Evaluation Results

	Precision	Recall	$F_{\beta=1}$	Accuracy	Error
CRF-POS	0.83279	0.77210	0.79130	96.10%	3.9%
BF-POS	0.84872	0.81236	0.82695	98.05%	1.95%

The BF-POS is compared with other Arabic POS taggers like: SVM-POS which has an accuracy of 96.6%, and the Majority-POS which has an accuracy of 97.6%. The result was that the proposed POS tagger reduces the error by 42.65% compared to the SVM-POS tagger and by 18.75% compared to the Majority POS tagger.

7 Conclusion and Future Work

In this research, the morphological analyzer is introduced to improve stemmer and POS tagger. Using the benchmark data set improvements in both tokenization and POS stages have been reached. First the CRF classifier is used instead of SVM. This resulted in an error reduction by 91.30% in the tokenization stage. Then the new binary feature (BF) extracted from the morphological analyses of the word is added to the feature set. This binary feature is language independent and highly accurate. It resulted in an error reduction by 95.65% and 18.75% in the tokenization and POS stage, respectively.

To achieve the targeted improvement the proposed system needs extra processing for the extraction of the binary feature. This extra processing could be minimized by using caching techniques in the implementation of the task of binary feature (BF) extraction.

There are numerous ways to extend this research work. The proposed binary feature BF will be tested on other languages like English. In addition, the performance of the Arabic POS tagging system additional features will be developed to further improve the performance. Last but not least, a wider context and more data will be used for testing.

References

1. Diab, M., Hacioglu, K., Jurafsky, D.: Automated methods for processing Arabic text: From tokenization to base phrase chunking. In: van den Bosch, A., Soudi, A. (eds.) Arabic Computational Morphology: Knowledge-based and Empirical Methods. Kluwer/Springer (2007)

2. Habash, N., Rambow, O.: Arabic Tokenization, Part-of-Speech Tagging and Morphological Disambiguation in One Fell Swoop. In: Proc. of the American Association of Computational Linguistic Conference (ACL) Short Papers, Michigan, USA (2005)
3. Habash, N., Rambow, O.: Morphological analysis and generation for Arabic dialects. In: Proc. of the Workshop on Computational Approaches to Semitic Languages in the American Association of Computational Linguistic Conference (ACL), Michigan, USA (2005)
4. AlGahtani, S., Black, W., McNaught, J.: Arabic Part-of-Speech Tagging Using Transformation-Based Learning. In: Proc. of the Second International Conference on Arabic Language Resources and Tools, Cairo, Egypt (April 2009)
5. Kulick, S.: Simultaneous Tokenization and Part-of-Speech Tagging for Arabic without a Morphological Analyzer. In: Proc. of the American Association of Computational Linguistic (ACL) Conference Short Papers, Uppsala, Sweden (July 2010)
6. Mansour, S., Sima'an, K., Winter, Y.: Smoothing a Lexicon-based POS tagger for Arabic and Hebrew. In: Proc. of the American Association of Computational Linguistic Workshop on Computational Approaches to Semitic Languages: Common Issues and Resources, Prague, Czech Republic (2007)
7. Diab, M.: Second generation tools (AMIRA 2.0): Fast and robust tokenization, pos tagging, and base phrase chunking. In: Proc. of 2nd International Conference on Arabic Language Resources and Tools (MEDAR), Cairo, Egypt (April 2009)
8. Maamouri, M., Bies, A., Buckwalter, T.: The penn arabic treebank: Building a largescale annotated arabic corpus. In: Proc. of NEMLAR Conference on Arabic Language Resources and Tools, Cairo, Egypt (2004)
9. Tamah, E., Al-Shammari, J.L.: Towards an Error-Free Arabic Stemming. In: Proc. of the American Association of Computational Linguistic (ACL) Conference on Information and Knowledge Management, New York, NY, USA (2008)
10. Khoja, S., Garside, P., Knowles, G.: A tagset for the morphosynactic tagging of Arabic. In: Proc. of Corpus Linguistics. Lancaster University, Lancaster (2001)

Part II
Recommendation Systems for Cloud Services

Concept Recommendation System for Cloud Services Advertisement

Yasmine M. Afify, Ibrahim F. Moawad, Nagwa L. Badr, and Mohamed Fahmy Tolba

Faculty of Computer and Information Sciences, Ain Shams University, Cairo, Egypt
`yasmine.afify@fcis.asu.edu.eg,`
`{ibrahim_moawad,nagwabadr}@cis.asu.edu.eg, fahmytolba@gmail.com`

Abstract. Cloud computing is a major trend in Information Technology (IT), which has witnessed high adaption rate for cloud solutions. Software-as-a-Service (SaaS) providers compete to address nearly every business and IT application needs. Heterogeneous cloud service advertisements make it difficult for potential users to discover the required service offers. To overcome this problem, the cloud service registry is fundamental for both the cloud users and providers. It provides detailed information about SaaS offers from different providers in one place, which increases the services reachability. In this paper, we focus on the business perspective of the SaaS services, which has not received eligible consideration by existing literature. We propose a semantic-based system for unified SaaS service advertisements. We introduce a template for the service registration, a guided registration model, and a registration system. Moreover, we present a semantic similarity model for services metadata matchmaking. Prototypical implementation and evaluation proved the proposed system effectiveness.

Keywords: SaaS Registry, Cloud Services, Information Retrieval, Semantic Annotation, Ontology.

1 Introduction

Software-as-a-Service (SaaS) is an increasingly popular delivery model for a wide range of business applications. The major challenge to the SaaS service discovery is the lack of standard description language or naming convention for the service advertisement [1]. In particular, each cloud provider publishes his cloud services on his portal in unstructured plain text using his own terms. Traditionally, the services are searched using search engines, which is a time-consuming and error-prone process. Moreover, due to the services diversity, sometimes the user cannot settle on the service that best suits his requirements [2,3].

Recently, service directories are leveraged to list cloud services [4,5], such as CloudBook and SaaSDirectory, etc [6,7,8,9,10,11]. Most of these directories list the services according to predefined categories with a short description of the service and a link to its website. However, their search capability is limited to the service name and/or the application domain.

A.E. Hassanien et al. (Eds.): AMLTA 2014, CCIS 488, pp. 57–66, 2014.

To complement existing work, the business perspective of the SaaS services is utilized in this research in order to, firstly, eliminate the lack of standardization problem and secondly, provide proficient search capabilities based on concrete service functionalities. This research focuses on the business aspect of the SaaS service in addition to its technical aspects. We propose a business-oriented advertisement template, guided registration model, and registration system for SaaS services.

In our previous work [13,14], we introduced a system for SaaS offers publication, discovery, and selection. To extend this work, contributions of this paper include a new semantic-based SaaS service registration model, and a semantic similarity model for services matchmaking. The proposed system standardizes the advertisement process, serves as a semantic-based registry for the service offerings, and provides competent search capabilities for the user.

The remainder of this paper is organized as follows. Section 2 overviews the related work on the cloud services publication. Section 3 presents the proposed SaaS advertisement system while section 4 introduces the services matchmaking semantic similarity model. Section 5 presents the experimental evaluation results. Finally, section 6 concludes the paper.

2   Related Work

Recent research work in the area of cloud services publication is briefly discussed. Universal Description, Discovery and Integration (UDDI) [15] is a registry for web services publishing and querying. It has three main limitations: a) services are organized according to their categories not to what they really offer, b) limited syntactic service discovery, and c) lack of support for non-functional properties.

The authors of [16] proposed an ontology-based cloud computing resources catalog offered by different cloud providers. However, it only focuses on infrastructure resource capabilities and features.

The authors of [17] introduced an integrated ontology for business functions and cloud providers that matches cloud services according to their functional and non-functional requirements. Limitations of this work are the strict matching of the user query and business functions, and the query representation language, which significantly restricts the ontology use to the experienced users only.

A semantic registry of cloud services with core ontological definitions and extension mechanisms used to define ontologies for cloud services was proposed in [18]. However, this framework does not include service types for SaaS service models.

An extensible Everything-as-a-Service (XaaS) registration entry was proposed in [5]. They proposed an extensible description language for services, a registration model, a system for registration, and subsequent service discovery operations. However, no details were given on the request-service matchmaking algorithm.

Other service registries were introduced like Membrane SOA Registry [19], Service-Finder [20], and Depot [21]. However, they are dedicated to the services described using the Web Service Description Language (WSDL) files, and this is not the case for most of the SaaS service offerings. In summary, two limitations can be highlighted:

- Lack of standardization in the SaaS services offers.
- The existing registries search is limited to service name and/or application domain.

3   Proposed SaaS Advertisement System

The proposed advertisement system unifies the SaaS service offers using a common meta-model. As shown in Fig. 1, in addition to the SaaS registry, the system consists of four modules namely: service preprocessing, WordNet expansion, guided registration, and clustering.

The SaaS registry is implemented as an ontology that integrates knowledge on SaaS business service domain, service characteristics, QoS metrics, and real service offers. It is a semantic repository for the service functional capabilities and non-functional quality guarantees. The business services domain ontology comprises concepts that cover domains of four SaaS applications: Customer Relationship Management (CRM), Enterprise Resource Planning (ERP), Document Management (DM), and Collaboration. At present, the developed ontology consists of more than 700 concepts represented in the Web Ontology Language (OWL) [22]. More details about the developed ontology can be found in [13,14].

We propose a template for registering the SaaS service offers, which describes the service functionality as well as its quality information. The proposed SaaS service advertisement template consists of four sections namely: general, functional, non-functional, and quality. The general section includes the service name, cloud provider, service description, Uniform Resource Location (URL), application domain, and the price. The functional section includes a description of the service features and supported functionalities. The non-functional section includes information about the service characteristics such as: payment model, security, license type, standardization, formal agreement, user group, and cloud openness. Finally, the quality section includes Quality Of Service (QoS) values guaranteed by the cloud provider.

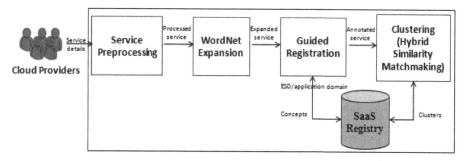

Fig. 1. SaaS Service Advertisement System

The service preprocessing module is responsible for preprocessing the service description [23]. The preprocessing process consists of tokenization, stop words removal, and stemming. The WordNet expansion module is responsible for retrieving the service description token synonyms from the WordNet [24]. The expanded service description (ESD) is then stored in the SaaS registry.

The guided registration module is responsible for semantically enriching the service with functional metadata using the business services domain ontology. In order to better characterize the new service, the cloud provider is assisted to map the service features into recommended ontology concepts. The feature-concepts mapping process is significant as it is the basis for efficient business-related cloud services searches. The recommended concepts represent service-related business functions. The SaaS service guided registration workflow algorithm is demonstrated in Fig. 2.

Algorithm: SaaS Service Registration

Input: Service S, Service Description SD, appDomain, recommendationApproach

Output: Set of Recommended Concepts RC, Expanded Service Description ESD

1.	**BEGIN**
2.	Replace any delimiter from set of delimiters D to *space*
3.	**FOR** each letter l in SD **DO**
4.	**IF** $l \in D$ **THEN**
5.	Reset l to *space*
6.	**END IF**
7.	**END FOR**
8.	Generate set of service description *tokens* by splitting SD on *space*
9.	Generate set of *relevantTokens* by removing common words from set *stopWords*
10.	**FOR** each token t in *tokens* **DO**
11.	**IF** $t \notin stopWords$ **THEN**
12.	*relevantTokens* = *relevantTokens* $\cup$ t
13.	**END IF**
14.	**END FOR**
15.	Generate ESD by finding synonyms of *relevantTokens* set from the WordNet
16.	**FOR** each token rt in *relevantTokens* **DO**
17.	$ESD = ESD \cup$ getSynonyms (rt)
18.	**END FOR**
19.	Generate set of stems *stems* of ESD using Porter Stemmer algorithm
20.	**CASE** 'recommendationApproach' **OF**
21.	'semanticAnnotation':
22.	Generate set $ontConceptStems$ by retrieving the ontology concepts stems
23.	**FOR** each stem st in *stems* **DO**
24.	**IF** $st \in ontConceptStems$ **THEN**
25.	$RC = RC \cup st$
26.	**END IF**
27.	**END FOR**
28.	'applicationDomain':
29.	Generate set DOC by retrieving relevant concepts to *appDomain*
30.	**FOR** each stem st in *stems* **DO**
31.	**IF** $st \in DOC$ **THEN**
32.	$RC = RC \cup st$
33.	**END IF**
34.	**END FOR**
35.	'hybrid':
36.	**DO** lines 22-27
37.	**DO** lines 29-34
38.	**END CASE**
39.	**Display** set of recommended concepts RC to the cloud provider
40.	**Read** selected concepts *selectedConcepts*
41.	**Return** *selectedConcepts*, ESD
42.	**END**

Fig. 2. Proposed SaaS Service Registration

We propose three concept recommendation methods: semantic annotation, application domain, and hybrid. In the recommendation via semantic annotation method, semantic annotations of the service description are retrieved. In the recommendation via an application domain method, concepts related to the application domain specified in the service information are retrieved. In the hybrid recommendation method, a combination of the two methods is retrieved. Finally, the set of selected concepts is stored in the SaaS registry.

The service clustering module clusters the service offers to functionally-similar clusters using the Agglomerative Hierarchical Clustering (AHC) [23] approach. Our previous hybrid service matchmaking algorithm is used to measure the similarity between the two services [13,14], which applies both semantic-based metadata and ontology matching. After clustering, cluster signature vectors are created and kept in the SaaS registry. To search for a service, the user enters his business function requirements, and the system returns matching SaaS services. Comprehensive details of the search process can be found in our previous work [13,14].

4   Services Matchmaking Semantic Similarity Model

In our previous work [13,14], regarding the semantic-based metadata matching, we used the Vector Space Model (VSM) to compute the similarity between two SaaS service descriptions. In this paper, we present an adaptation of the Extended Case Based Reasoning (ECBR) algorithm presented in [25]. The adapted SerECBR works as follows. First, the synonyms of the two service descriptions are retrieved (SD1 and SD2). SS1 and SS2 are synonyms of the first and second service descriptions respectively. Each service description and synonym terms are grouped into one list. T1 and T2 represent the first and second service terms respectively. Δ and Ω represent a term that occurs within T1 and T2 respectively. LT1 represents the number of terms in the first service description list. For each service description list, we associate weights for terms as follows:

$$w_\Delta = \begin{cases} 1, & if\ \Delta \in SD_1 \\ 0.5, & if\ \Delta \in SS_1 \end{cases} \quad and \quad w_\Omega = \begin{cases} 1, & if\ \Omega \in SD_2 \\ 0.5, & if\ \Omega \in SS_2 \end{cases} \quad (1)$$

To compute the semantic similarity between two SaaS services S1 and S2, we compare the lists T1 and T2 using the following cases:

Case 1: term exists in two lists with weight 1, a value of 1 is given to this term.
Case 2: term exists in two lists with weight 0.5, a value of 0.25 is given to term.
Case 3: term exists in two lists by different weights, a value of 0.5 is given to term.
Case 4: term from one list does not exist in other list, a value of 0 is given to term.

Finally, the sum of all term values is normalized by the length of the first service terms list. The semantic similarity between two services S1 and S2 is calculated using (2):

$$sim_{SerECBR}(S_1, S_2) = \frac{\sum_{\Delta \in T_1,\ \Omega \in T_2}(match(\Delta,\Omega))}{L_{T1}} \quad (2)$$

$$where\ match\ (\Delta, \Omega) = \begin{cases} 1, & if\ (\Delta = \Omega) \wedge (w_\Delta = w_\Omega = 1) \\ 0.5, & if\ (\Delta = \Omega) \wedge (w_\Delta \neq w_\Omega) \\ 0.25, & if\ (\Delta = \Omega) \wedge (w_\Delta = w_\Omega = 0.5) \\ 0, if\ (\Delta \in T_1 \wedge \Delta \notin T_2) \vee (\Omega \in T_2 \wedge \Omega \notin T_1) \end{cases} \tag{3}$$

5   Implementation and Experimental Evaluation

Real and synthetic cloud data were used to demonstrate the effectiveness of the proposed system. Experiments were conducted on an Intel Core i3 2.13 GHz processor, 5.0 GB RAM. The system was built using Java, Jena API, and WordNet API incorporated in Eclipse IDE.

The data set consists of 500 SaaS services. In particular, 40 services are live, and the remaining are pseudo services. In the following subsections, we present a case study for the guided registration process, the experimental evaluation of semantic annotation process, and the matchmaking semantic similarity model evaluation.

5.1   Guided Registration Process Case Study

In order to populate the SaaS registry, the cloud providers are expected to register their services in our system. Typically, a speedy and proficient registration process would help the proposed system to gain wide acceptance. In particular, thorough and accurate matching of the service features to domain ontology concepts is vital while considering the time factor. The aim of this case study is to demonstrate the guided registration process. A service provider registers his cloud service and uses the concept recommendation method to match domain ontology concepts to the new service features. As shown in Fig. 3, the cloud provider enters the service advertisement details for the cloud service *Box.net*, then, he matches the service features to business functions retrieved from the SaaS services domain ontology.

In this case study, the cloud provider chooses the concept recommendation via Semantic Annotation method. Tokens from the processed service description are matched to the domain ontology concepts which results in a set of recommended business functions. As shown in Fig. 3, the recommended business functions related to the service description keywords are displayed. The cloud provider selects the business functions that accurately describe the service features and registers his service. Next, this is followed by a form to allow the provider to enter the service characteristics and QoS values.

Obviously, the time taken by the registration process is significantly reduced compared to previous work in [13,14] where the cloud provider has to navigate through all domain ontology concepts to properly characterize his service features.

5.2   Semantic Annotation Process Evaluation

The objective of this experiment is twofold: to calculate the time taken by the guided registration module to semantically annotate the service description and to study the effect of the WordNet expansion of the service description on the annotation process. To achieve the first objective, we computed the processing time taken to semantically annotate the expanded service descriptions. As shown in Fig. 4, the semantic annotation process time is negligible.

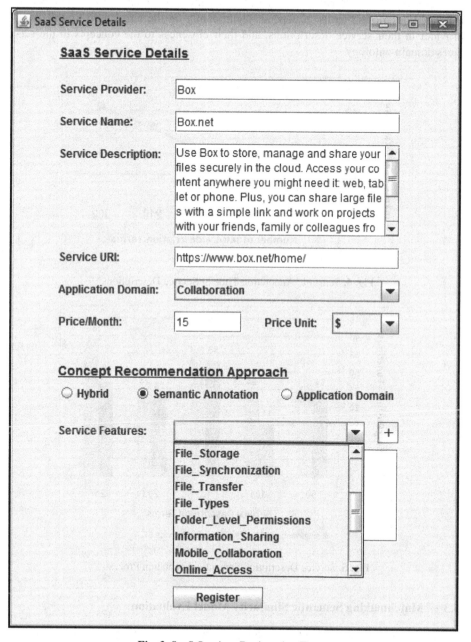

Fig. 3. SaaS Services Registration Form

To achieve the second objective, the semantic annotation process is analyzed in two cases, with and without using WordNet. The number of matching Business Functions (BF) returned by the two cases is compared. The semantic expansion of the service description generally increases the number of retrieved business functions as shown in Fig. 5. Nevertheless, the increase is neither nor relative to the number of expanded description terms. It depends significantly on the terms used by the cloud provider in their service descriptions, and their closeness to the concepts in the services domain ontology.

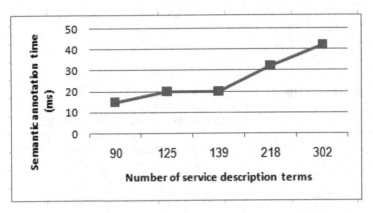

Fig. 4. Semantic Annotation Time of Service Descriptions

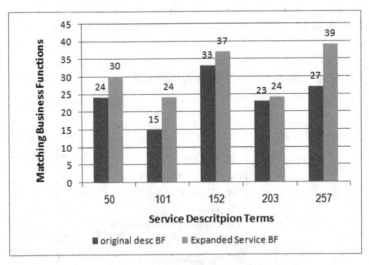

Fig. 5. Service Description Semantic Annotation Process

5.3   Matchmaking Semantic Similarity Model Evaluation

The objective of this experiment is to compare the resulting semantic similarity values of service description matchmaking of the VSM and the proposed serECBR similarity models. A sample of the results is shown in table 1. The results demonstrate that the serECBR similarity model reflects the similarity among the service descriptions to be better than the VSM model with an average increase of 49%.

Table 1. VSM vs. serECBR Service Matchmaking Semantic Similarities

Service		Service	VSM Similarity	serECBR Similarity
Oracle CRM On Demand		Intouchcrm	0.54	0.67
Box.net		Egnyte	0.65	0.37
Blue Link Elite	vs.	NetSuite	0.16	0.29
IBM Lotus Live		CubeTree	0.31	0.66
HyperOffice		GetDropBox	0.21	0.36
OrderHarmony		Plex Online	0.01	0.03
Incipi Workspace		HyperOffice	0.19	0.57

6   Conclusion

With the growth of public cloud offerings, a cloud service registry is urgently required to connect the cloud providers and users. In this paper, we proposed a SaaS services advertisement system that exploits the services business perspective by introducing functionality-related metadata to the service descriptions. The semantic metadata enrichment is accomplished in the registration process via domain ontology concept recommendation. In order to effectively group functionally-similar services, a semantic similarity model was proposed for the services metadata matchmaking. The uniformity, integrity, and comprehensive representation of SaaS offers maximize the efficiency of the services discovery and close the gap between the services supply and demand in the cloud market. The proposed system effectiveness was proved by the experimental results.

References

1. Noor, T.H., et al.: A Crawler Engine For Cloud Services Discovery on the World Wide Web. In: 20th IEEE International Conference on Web Services (ICWS), pp. 443–450. IEEE Press, USA (2013)
2. Garg, S.K., et al.: A Framework For Ranking of Cloud Computing Services. J. Future Generation Computer Systems 29(4), 1012–1023 (2013)
3. Chen, F., Bai, X., Liu, B.: Efficient Service Discovery for Cloud Computing Environments. In: Shen, G., Huang, X. (eds.) CSIE 2011, Part II. CCIS, vol. 153, pp. 443–448. Springer, Heidelberg (2011)
4. ODCA. Open Data Center Alliance[SM] Usage: SERVICE CATALOG. Open Data Center Usage Models. Open Data Center Alliance, Inc.,
   http://www.opendatacenteralliane.org/ourwork/usagemodels

5. Spillner, J., Schill, A.: A Versatile and Scalable Everything-as-a-Service Registry and Discovery. In: 3rd International Conference on Cloud Computing and Services Science (CLOSER), Germany, pp. 175–183 (2013)
6. Cloudbook. The Cloud Computing & SaaS Information Resource,
   http://www.cloudbook.net/directories/product-services/
   cloud-computing-directory?category=Applications
7. SaaS Directory, http://www.saasdirectory.com/
8. OpenCrowd Project, http://cloudtaxonomy.opencrowd.com/taxonomy/
9. SaaS Lounge, http://www.saaslounge.com/saas-directory/
10. Cloud Showplace, http://www.cloudshowplace.com/application/
11. GetApp, http://www.getapp.com/
12. ReasySaaSGo, http://www.readysaasgo.com/
13. Afify, Y.M., Moawad, I.F., Badr, N., Tolba, M.F.: A Semantic-based Software-as-a-Service (SaaS) Discovery and Selection System. In: 8th IEEE International Conference on Computer Engineering & Systems (ICCES 2013), pp. 57–63. IEEE Press, Egypt (2013)
14. Afify, Y.M., Moawad, I.F., Badr, N.L., Tolba, M.F.: Cloud Services Discovery and Selection: Survey and New Semantic-Based System. In: Hassanien, A.E., Kim, T.-H., Kacprzyk, J., Awad, A.I. (eds.) Bio-inspiring Cyber Security and Cloud Services: Trends and Innovations. ISRL, vol. 70, pp. 449–477. Springer, Heidelberg (2014), doi:10.1007/978-3-662-43616-5_17
15. UDDI Technical White Paper (2001),
    http://www.uddi.org/pubs/lru_UDDI_Technical_Paper
16. Bernstein, D., Vij, D.: Using Semantic Web Ontology for Intercloud Directories and Exchanges. In: 11th International Conference on Internet Computing, ICOMP 2010, Las Vegas, pp. 18–24 (2010)
17. Tahamtan, A., Beheshti, S.A., et al.: Cloud Repository and Discovery Framework Based on a Unified Business and Cloud Service Ontology. In: 8th IEEE World Congress on Services, pp. 203–210. IEEE Press, USA (2012)
18. Mindruta, C., Fortis, T.-F.: A Semantic Registry for Cloud Services. In: 27th International Conference on Advanced Information Networking and Applications Workshops (WAINA), pp. 1247–1252. IEEE Press, Spain (2013)
19. Membrane SOA Registry, http://www.membrane-soa.org/soa-registry/
20. Steinmetz, N., Lausen, H., Brunner, M.: Web Service Search on Large Scale. In: Baresi, L., Chi, C.-H., Suzuki, J. (eds.) ICSOC-ServiceWave 2009. LNCS, vol. 5900, pp. 437–444. Springer, Heidelberg (2009)
21. AbuJarour, M.: A Proactive Service Registry with Enriched Service Descriptions. In: 5th Ph. D. Retreat of the HPI Research School on Service-oriented Systems Engineering, Germany, pp. 191–200 (2011)
22. Web Ontology Language (OWL), http://www.w3.org/TR/owl-features/
23. Salton, G.: Introduction to Modern Information Retrieval. McGraw-Hill Computer Science Series. McGraw-Hill Companies (1983)
24. Fellbaum, C.: WordNet: An Electronic Lexical Database. MIT Press (1999)
25. Dong, H., Hussain, F.K., Chang, E.: A Service Search Engine for the Industrial Digital Ecosystems. IEEE Transactions on Industrial Electronics 58(6), 2183–2196 (2011)

Dynamic Distributed Database over Cloud Environment

Ahmed E. Abdel Raouf, Nagwa L. Badr, and Mohamed Fahmy Tolba

Faculty of Computer and Information Sciences, Ain Shams University, Cairo, Egypt
Ahmed_ezzat991@yahoo.com, nagwabadr@cis.asu.edu.eg,
fahmytolba@gmail.com

Abstract. An efficient way to improve the performance of database systems is the distributed processing. Therefore, the functionality of any distributed database system is highly dependent on its proper design in terms of adopted fragmentation, allocation, and replication methods. As a result, fragmentation, its allocation and replication is considered as a key research area in the distributed environment. The cloud computing is an emerging distributed environment that uses central remote servers and internet to maintain data and applications. In this paper, we present a dynamic distributed database system over cloud environment. The proposed system allows fragmentation, allocation, and replication decisions to be taken dynamically at run time. It also allows users to access the distributed database from anywhere. Moreover, we present an enhanced allocation and replication technique that can be applied at the initial stage of the distributed database design when no information about the query execution is available.

Keywords: Distributed database management system (DDBMS), fragmentation, replication, allocation, cloud computing.

1 Introduction

Distributed database systems typically consist of a number of distinct database fragments located at different geographic sites which can communicate through a network and they are managed by a distributed database management system (DDBMS) [1].

An efficient support is needed to databases that consist of very large amounts of data which used by applications at different physical locations. Telecom databases, scientific databases, and large distributed enterprise databases are examples of application areas [2]. The main problem of many of these applications is the delay of accessing remote databases. As a result, it is necessary to use a distributed database that employing fragmentation, allocation, and replication [2].

The design of distributed database is one of the major research issues in distributed database system area. The main challenges facing the DDBS design are: how to fragment database tables, which type of fragmentation will be used, when to replicate fragments, what is the optimal number of replica that can be taken for each fragment to enhance system performance and increase availability, how to allocate fragments to sites where they are mostly frequently accessed, do we grouping distributed database sites into disjoint clusters, and which type of clustering will be used. These issues were previously solved either by static and dynamic solution or based on a priori query analysis.

A.E. Hassanien et al. (Eds.): AMLTA 2014, CCIS 488, pp. 67–76, 2014.

Fragmentation, replication, and allocation are considered the most important design issues that lead to optimal solutions particularly in a dynamic distributed environment. They also have a great impact on the Distributed Database Systems (DDBS) performance. In distributed databases, the communication costs can be reduced by partitioning database tables into fragments. The fragments are then allocated to the sites where they are most frequently accessed, aiming at maximizing the number of local accesses compared to accesses from remote sites. The cost of the read operation can be further reduced by the replication of fragments when beneficial. Fragmentation, allocation, and replication will be referred to as FAR in the rest of the paper.

Many applications of DDBS generate very dynamic workloads with frequent changes in access patterns from different sites. Consequently, static/manual FAR may not always be optimal. As a result, FAR should be automatic and completely dynamic. Any change in access patterns should result in re-fragmentation of existing fragments or tables and reallocation of fragments to different sites, as well as creation or removal of fragment replicas [2].

In this paper, we present a dynamic DDBS over the cloud environment that allows dynamic FAR decisions to be taken dynamically over clustered distributed database sites. Dynamic FAR decisions are based on the access pattern and the load of the sites after allocation or migration of fragments or a replica to it. Moreover, we present an optimal allocation and replication technique, which can be applied at the initial stage of the distributed database design when no information about the query execution is available.

The rest of this paper is organized as follows. Section 2 reviews the related work of FAR. Section 3 introduces the proposed system architecture and its components. Section 4 presents an optimal allocation and replication technique that can be applied at the initial stage of distributed database design. Section 5 presents the experimental results. Finally the conclusion and future work are presented in Section 6.

2   Related Research Work

In [2], the authors present a decentralized approach for dynamic table FAR in DDBS. It performs FAR based on recent access history. It uses cost functions to take decision by estimating the difference in future communication costs between a given replica change and keeping it as is. However, this algorithm doesn't consider site constraints, load of the site after the allocation or migration of replica or fragment to it, and the optimal number of replicas that can be taken for each fragment to enhance system performance and increase availability. The authors of [3] present a synchronized horizontal FAR model. It adopts a new approach to perform horizontal fragmentation of database relation based on the attribute retrieval and update frequency. Both [3] and [18] perform the allocation process based on the fragment access pattern and the cost of moving the data fragments from one site to the other. The authors of [4] propose a new algorithm called region based fragment allocation (RFA). The proposed algorithm considers the frequency of fragment accessed by region as well as individual nodes to move the fragment from source node to target node. The RFA algorithm decreases the migration of fragments using knowledge of the network topology in comparison to optimal [19] and threshold [20] algorithms. In comparison to the BGBR [21] algorithm, the RFA algorithm reduces the amount of topological data required in decision making. However, this solution will not be the optimal solution in case the node in a region has a high fragment access while the other nodes in that region have low fragment access. As a result, the fragment will allocate to it. In this case, the problem is solved for only one node and is not solved for the other nodes in other regions. The authors of [5] present a model that takes site constraints into account in the process of reallocation. However, this model will be more complicated when information queries continuously change in a faster way and when the number of fragments and sites largely increase. The authors of [6] propose an algorithm named Near Neighborhood Allocation (NNA). It moves data to a neighborhood node that placed in the path to the node with the maximum access counter. The NNA could be more useful in large networks to decrease the delay of response time. The authors of [7] propose an algorithm that dynamically reallocates fragments to sites at runtime. The proposed algorithm takes into account the time constraints of database accesses, volume threshold, and the volume of data transmitted in successive time intervals in accordance with the changing access patterns. However, the volume threshold, the number of time intervals, and its duration are the most important factors that regulate the frequency of fragment reallocations. In [8] the authors perform reallocation process given the changing data access patterns, time, and sites' constraints of the DDBS. That technique reduces data transmission cost compared to the previous methods since it adopts the shortest path algorithm once data movement decision is taken. The main drawback of this algorithm is the more storage required compared to some previous algorithms. The authors of [15] solve the fragment allocation problem using the well-known Quadratic Assignment Problem solution algorithms.

The authors of [9] propose a new technique for horizontal fragmentation of the relations of distributed database. This technique can be applied at the initial stage as well as in later stages of DDBS for partitioning the relations. The authors of [10] address some important scalability issues. They provide some algorithms to ensure generality of the technique developed in [9]. In [11] the authors propose a heuristic technique to satisfy horizontal fragmentation and allocation using a cost model to minimize the total cost of distribution. Furthermore, the authors of [12] and [16] present a new framework for dynamic fragment allocation and replication. In [12], they consider the fragment correlation under a flexible network topology. However, this framework doesn't handle extra characteristics, such as bounds on the capacity of sites and constraints on the number of replicas for each fragment.

Some researches introduce a clustering approach for partitioning database sites and allocate fragments across the sites of each cluster [13]. However, it doesn't address the replication phase of database design. It also wastes a lot of time between node, LCA, LCA Validator, Resource Checker, and GCA until it reaches the required data. The authors of [17] propose a dynamic data replication strategy using historical access record and proactive deletion. The authors of [14] present a novel algorithm for grouping distributed database network sites into disjoint clusters based on communication time. The experimental outcomes confirmed that this approach can be implemented in a different DDBS environment even if the network sites are enormous. However, this work doesn't mention how to allocate fragments to cluster of sites and which sites in the cluster will hold the fragment and on which bases. Another method of clustering the sites in which low communication cost sites are grouped in one cluster[1]. Furthermore it allows the fragmentation of structured data, fragmentation of unstructured data, and it describes the allocation of fragments to the cluster of sites in order to reduce communication cost. However, this work doesn't mention which sites in the cluster will hold the fragment when the fragment is allocated to it.

3   Dynamic Distributed Database System over Cloud Environment Architecture

To overcome the limitations of existing literature highlighted by the above survey, we propose a DDBS design over cloud environment. The proposed architecture allows users to access a database from anywhere in the world without owning any technology infrastructure. It can be accessed through: a web browser, mobile application or desktop application while the database is stored on servers at remote sites. It also allows FAR decisions to be taken dynamically at run time. These decisions are based on access patterns and load of sites after the allocation and migration of fragments as well as its replicas. The proposed architecture is shown in Fig. 1. It consists of two layers: distributed database system manager and distributed database clusters layers.

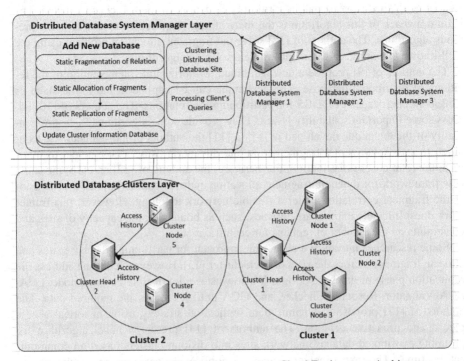

Fig. 1. Dynamic Distributed Database System over Cloud Environment Architecture

3.1   Distributed Database System Manager Layer

This layer is composed of three modules: add new database, clustering distributed database sites, and client queries processing.

Add New Database Module. This module is used to add the database tables to DDBS. Firstly, the new tables are fragmented using the static fragmentation technique that can be used at the initial stage of the DDBS design. Secondly, that fragments are allocated to the sites of each cluster. Thirdly, some fragments are replicated to different sites of each cluster in order to enhance the system performance and increase the availability. Finally, the cluster information database is updated to save the locations of each fragment and replica in the distributed database system. Cluster information database is a database that holds complete information about each fragment or replica in which site or cluster.

Clustering Distributed Database Sites Module. Distributed database system manager will use clustering technique to cluster distributed database sites into disjoint clusters. It also allocates sites to each cluster. Finally, it updates cluster information database with the recent location of each site in DDBS.

Client Queries Processing Module. When the client sends a query to the distributed database system manager firstly, the distributed database system manager uses the cluster information database to determine the closest site that holds the required data. Then, it redirects the user to that site to fetch the needed data.

3.2   Distributed Database Clusters Layer

The distributed database clusters layer consists of more than one cluster. Each cluster has one cluster head and more than one cluster node. The cluster head is a cluster node has especial and additional operations to manage the other cluster nodes. At each access to the cluster node, firstly, the cluster node checks whether it is a local access or remote. Secondly, allows the user to access local database of the node to run the query and fetch the required data [22]. Thirdly, the site access record is updated to save the information about the user access. Each cluster node sends the access history to the cluster head to take any suitable decision such as: create or delete replica, re-fragmentation or re-allocation of the fragments. The cluster head takes the decisions based on access history and load of each site after the allocation or migration of the fragments and replicas of it. The fragments and replicas will send to the site with less workload. The cluster head also collects the data that each node in its cluster holds. Afterwards, it sends it to the distributed database system manager to keep the cluster information database updated.

4   Enhanced Allocation and Replication Technique

The authors of [9] proposed a new technique of horizontal fragmentation. This technique helps in taking the fragmentation decision at the initial stage of designing distributed database. It uses knowledge gathered during the requirement analysis phase using the enhanced CRUD (Create, Read, Update, and Delete) matrix without the help of empirical data about query execution. This technique performs data allocation according to the maximum attribute locality precedence (ALP) value and the locations of sites. The attribute locality precedence (ALP) can be defined as the importance of an attribute with respect to each site. However, the allocation strategy of this technique doesn't meet the goal of the data allocation. The goal of the data allocation can be achieved by allocating the fragments to the sites that require it only. As a result the user can access data with low cost and time. The proposed module enhanced the allocation strategy of this technique by performing data allocation according to the site that has the maximum ALP single value. That methodology guarantees that no fragment duplication will happen during the allocation process. In the case we got two sites that have the maximum ALP single value, the fragment will be allocated to the site that performs more data manipulations and less read operations. Consequently, the replica is sent to the site which performs less data manipulations and more read operations. After the process of data allocation, we replicate the fragment to the site that performs more read operation than other sites. For example, if we have two sites: site1 and site2. Site1 performs CUD operations. Site2 performs R operation. The replica will be sent to site2 although site1 has the maximum ALP single value. If the replica is sent to site1, it will not be used because there is no data manipulation on the replica. The pseudo code for the enhanced allocation and replication technique is shown in Fig. 2.

5   The Experimental Results

We have implemented our technique on an HP Compaq computer with Core-two Duo 2.33 processors and 2 GB RAM using SQL Server as DBMS. We have implemented the modified technique on the MCRUD matrix of the bank account table shown in table 1. We performed the enhanced technique on account relations shown in table 2. The ALP table is generated after applying the modified technique on the MCRUD matrix. The ALP table is shown in table 3. From the ALP table, the attribute that has a maximum ALP value is the branch name, so its predicates will be used to perform horizontal fragmentation. The resulted fragments are shown in tables 4, 5, and 6 and are allocated to sites that already need it without taking the location of sites into account. Fragment1 has been allocated to site1 and fragment2 has been allocated to site2 because they have the maximum ALP single value for the fragmentation attribute. However, the last predicate of the branch name attribute has two sites which have the same maximum value. In this case, the fragment will be allocated to the site that performs more data manipulation and less read operations. In our experiments, site1 performs two read operations and site3 performs three read operations. As a result, fragment3 will be allocated to site1 and its replica will be sent to site3.

```
Input:  Number of attributes, number of predicates of each attribute [], number of sites, number of
        applications of each site [], and MCRUD matrix [total number of predicates, total number of applications]
Output: The ALP table, the sites that contain maximum ALP single value for each predicate of each
        attribute (Position of MAX [attribute, predicate]), and the sites that performs more read operations for
        each predicate of each attribute (Position of next Max [attribute, predicate]).
Foreach attribute in Number of attributes do
     Foreach predicate in number of predicates [attribute] do
          Number of read operation of max = 0
          Number of read operation of next max = 0
          Application number = 0
          Total sum = 0
          Foreeach site in number of sites do
               Application sum = 0
               Number read operation = 0
               Foreeach application in number of application [site] do
                    Application sum += Calculate MCRUD (MCRUD [Predicate number, application number])
                    Application number++     End
               Total sum += application sum
               If application sum == MAX [attribute, predicate] then
                    If Number read operation > Number of read operation of max then
                         Position of next Max [attribute, predicate] = site
                         Number of read operation of next max = Number read operation     End
                    Else if Number read operation < Number of read operation of max then
                         Position of next Max [attribute, predicate] =
                              Position of MAX [attribute, predicate]
                         Number of read operation of next max = Number of read operation of max
                         MAX [attribute, predicate] = application sum
                         Position of MAX [attribute, predicate] = site
                         Number of read operation of max = Number read operation  End End
               If application sum > MAX [attribute, predicate] and application sum > 0 then
                    If Number of read operation of max > Number of read operation of next max then
                         Position of next Max [attribute, predicate] =
                              Position of MAX [attribute, predicate]
                         Number of read operation of next max = Number of read operation of max End
                    MAX [attribute, predicate] = application sum
                    Position of MAX [attribute, predicate] = site
                    Number of read operation of max = Number read operation     End
               Else if application sum > 0 and Number read operation > 0 and
                    Number read operation > Number of read operation of next max then
                    Position of next Max [attribute, predicate] = site
                    Number of read operation of next max = Number read operation;     End End
          Predicate number++
          ALP Single [attribute, predicate] = MAX [attribute, predicate] -
               (Total sum - MAX [attribute, predicate])     End
     Foreach predicate in number of predicates [attribute] do
          ALP FINAL [attribute] += ALP Single [attribute, predicate]  End End
```

Fig. 2. Pseudo Code for the enhanced allocation and replication technique

After the allocation process, we replicate the fragments to enhance the system performance of reading only queries and increase the availability. The replicas will be allocated to the site that performs more read operations than other sites. In our scenario, replica1 of fragment1 is allocated to site3, replica2 of fragment2 is allocated to site1, and replica3 of fragment3 is allocated to site3. The final results of the allocation and the replication processes are shown in table 7.

Table 1. MCRUD Matrix

	Site1			Site2			Site3		
	AP1	AP2	AP3	AP1	AP2	AP3	AP1	AP2	AP3
Account. Account id > 20	C		RU						C
Account. Account id <= 20		R							
Account .Type= ind	CRD	RU	RUD		R				
Account .Type= cor		RU	R				CRUD	RU	R
Account .customer id > 5	C		RU						R
Account .customer id <= 5		R							
Account .open date > 1-1-2008	CRD	RU	RU		R				
Account .open date <= 1-1-2008		RU	R				CRUD	RU	R
Account .Balance < 10000	R		R			CRUD			R
Account .Balance >= 10000		CR							
Account. Branch Name = dhk	CRUD	RU	CRUD		CUD		R		
Account. Branch Name = ctg		R		CRUD	CRUD	R		R	
Account. Branch Name = khl	CRUD	CRU	U				CRUD	CRU	CR

Table 2. Account Relation

account_no	Account Type	customer id	open date	Account Balance	Account Br Name
3	ind	1	20/01/2009 ...	12500.0000	dhk
4	cor	2	20/05/2009 ...	12000.0000	dhk
7	ind	2	05/03/2009 ...	11000.0000	ctg
15	ind	3	08/05/2009 ...	11000.0000	khl
20	cor	2	08/05/2010 ...	15000.0000	ctg
21	ind	1	09/05/2012 ...	9000.0000	khl
22	cor	8	20/09/2011 ...	8000.0000	dhk
23	ind	5	06/08/2011 ...	6000.0000	khl
24	ind	9	08/09/2006 ...	15000.0000	khl
28	cor	5	07/05/2009 ...	16000.0000	ctg

Table 3. ALP TABLE

Attribute name	ALP Value
Account id	6
Type	22
customer id	6
open date	22
Balance	8
Branch Name	27

Table 4. Fragment 1

account_no	Account Type	customer id	open date	Account Balance	Account Br Name
3	ind	1	20/01/2009 ...	12500.0000	dhk
4	cor	2	20/05/2009 ...	12000.0000	dhk
22	cor	8	20/09/2011 ...	8000.0000	dhk

Table 5. Fragment 2

account_no	Account Type	customer id	open date	Account Balance	Account Br Name
7	ind	2	05/03/2009 ...	11000.0000	ctg
20	cor	2	08/05/2010 ...	15000.0000	ctg
28	cor	5	07/05/2009 ...	16000.0000	ctg

Table 6. Fragment 3

account_no	Account Type	customer id	open date	Account Balance	Account Br Name
15	ind	3	08/05/2009 ...	11000.0000	khl
21	ind	1	09/05/2012 ...	9000.0000	khl
23	ind	5	06/08/2011 ...	6000.0000	khl
24	ind	9	08/09/2006 ...	15000.0000	khl

Table 7. Final Result of Allocation and Replication

Fragment Number	Allocated to	Replicated to
Fragment 1	1	3
Fragment 2	2	1
Fragment 3	1	3

6   Conclusion

Efficient distribution of the distributed database fragments and replicas to various sites play a critical role in the function of the database in terms of performance and cost. In this paper, we present a dynamic DDBS over cloud environment. The proposed system allows FAR decisions to be taken based on access history and the load of the site. Moreover, we present enhanced allocation and replication technique, which allocates the fragments and replicas to the sites that already, requires it without taking into account the location of the sites. The enhanced technique aims at maximizing the number of local access compared to access from the remote sites, enhance the system performance, and increase the availability.

As proposed future work, we plan to use an enhanced clustering technique to cluster distributed database sites into disjoint clusters then do allocation and replication of, the fragments to sites of each cluster. Second, implement the remaining parts of the proposed architecture to efficiently allow FAR decisions to be taken automatically at run time.

References

1. Suganya, A., Science, C., Kalaiselvi, R.: Efficient Fragmentation and Allocation in Distributed Databases. Int. J. Eng. Res. Technol. 2, 1–7 (2013)
2. Hauglid, J.O., Ryeng, N.H., Nørvåg, K.: DYFRAM: dynamic fragmentation and replica management in distributed database systems. Distrib. Parallel Databases 28, 157–185 (2010)
3. Abdalla, H.I.: A synchronized design technique for efficient data distribution. Comput. Human Behav. 30, 427–435 (2014)
4. Varghese, P.P., Gulyani, T.: Region based Fragment Allocation in Non-Replicated Distributed Database System. Int. J. Adv. Comput. Theory Eng. 1, 62–70 (2012)
5. Abdalla, H.: A New Data Re-Allocation Model for Distributed Database Systems. Int. J. Database Theory 5, 45–60 (2012)
6. Gope, D.: Dynamic Data Allocation Methods in Distributed Database System. American Academic & Scholarly Research Journal 4, 1–8 (2012)
7. Mukherjee, N.: Synthesis of Non-Replicated Dynamic Fragment Allocation Algorithm in Distributed Database Systems. Int. J. on Information Technology 1, 36–41 (2011)
8. Abdallaha, H.I., Amer, A.A., Mathkour, H.: Performance optimality enhancement algorithm in DDBS (POEA). Comput. Human Behav. 30, 419–426 (2014)
9. Khan, S., Hoque, A.: A new technique for database fragmentation in distributed systems. Int. J. Comput. Appl. 5, 20–24 (2010)
10. Khan, S., Hoque, A.: Scalability and performance analysis of CRUD matrix based fragmentation technique for distributed database. In: 15th International Conference on Computer and Information Technology (ICCIT), pp. 557–562. IEEE, Chittagong (2012)
11. Abdalla, H.I., Amer, A.A.: Dynamic horizontal fragmentation, replication and allocation model in DDBSs. In: 2012 Int. Conf. Inf. Technol. e-Services, pp. 1–7. IEEE, Sousse (2012)
12. Kamali, S., Ghodsnia, P., Daudjee, K.: Dynamic data allocation with replication in distributed systems. In: 30th IEEE Int. Perform. Comput. Commun. Conf., pp. 1–8. IEEE, Orlando (2011)
13. Amalarethinam, D., Balakrishnan, C.: A Study on Performance Evaluation of Peer-to-Peer Distributed Databases. IOSR J. Eng. 2, 1168–1176 (2012)
14. Hababeh, I.: Improving network systems performance by clustering distributed database sites. J. Supercomput. 59, 249–267 (2010)
15. Tosun, U., Dokeroglu, T., Cosar, A.: Heuristic Algorithms for Fragment Allocation in a Distributed Database System. In: Gelenbe, E., Lent, R. (eds.) Computer and Information Sciences III, pp. 401–408. Springer, London (2013)
16. Amalarethinam, D., Balakrishnan, C.: oDASuANCO-Ant Colony Optimization based Data Allocation Strategy in Peer-to-Peer Distributed Databases. Int. J. Enhanc. Res. Sci. Technol. Eng. 2, 1–8 (2013)
17. Wang, Z., Li, T., Xiong, N., Pan, Y.: A novel dynamic network data replication scheme based on historical access record and proactive deletion. J. Supercomput. 62, 227–250 (2011)

18. Abdalla, H.I.: An Efficient Approach for Data Placement in Distributed Systems. In: 2011 Fifth FTRA Int. Conf. Multimed. Ubiquitous Eng., pp. 297–301. IEEE, Loutraki (2011)
19. Corcoran, L.: A Genetic Algorithm for Fragment Allocation a Distributed Database System. In: Proc. 1994 ACM Symp. Appl. Comput., SAC 1994, pp. 247–250. ACM, USA (1994)
20. Ulus, T., Uysal, M.: Heuristic Approach to Dynamic Data Allocation in Distributed Database Systems. Inf. Technol. J. 2, 231–239 (2003)
21. Bayati, A., Ghodsnia, P.: A Novel Way of Determining the Optimal Location of a Fragment in a DDBS: BGBR. In: Syst. Networks, pp. 64–69. IEEE Computer Society, Washington (2006)
22. Maghawry, E.A., Ismail, R.M., Badr, N.L., Tolba, M.F.: An Enhanced Resource Allocation Approach for Optimizing Sub Query on Cloud. In: Hassanien, A.E., Salem, A.-B.M., Ramadan, R., Kim, T.-h. (eds.) AMLTA 2012. CCIS, vol. 322, pp. 413–422. Springer, Heidelberg (2012)

Queries Based Workload Management System for the Cloud Environment

Eman A. Maghawry, Rasha M. Ismail, Nagwa L. Badr, and Mohamed Fahmy Tolba

Faculty of Computer and Information Sciences, Ain Shams University, Cairo, Egypt
{e_maghawry,rashaismail}@yahoo.com, nagwabadr@cis.asu.edu.eg,
fahmytolba@gmail.com

Abstract. Workload management for concurrent queries is one of the challenging aspects of executing queries over the cloud computing environment. The core problem is to manage any unpredictable load imbalance with respect to varying resource capabilities and performances. Key challenges raised by this problem are how to increase control over the running resources to improve the overall performance and response time of the query execution. This paper proposes an efficient workload management system for controlling the queries execution over cloud environment. The paper presents an architecture to improve the query response time by detecting any load imbalance over the resources. Also, responding to the queries dynamically by rebalancing the query executions across the resources. The results show that applying this Workload Management System improves the query response time by 68%.

Keywords: Cloud Computing, Query workload, Query execution.

1   Introduction

Cloud Computing is becoming an emerging computing paradigm with its dynamic usage of scalable virtualized resources. It provides services to various remote clients with different requirements. As data continues to grow, it enables the remote clients to store their data on its storage environment with different clients' expectations over the internet. As clouds are built over wide area networks, the use of large scale computer clusters often built from low cost hardware and network equipment, where resources are allocated dynamically amongst users of the cluster [1]. Therefore the cloud storage environment has resulted in an increasing demand to co-ordinate access to the shared resources to improve the overall performance.

Managing the query workload in cloud computing environment is a challenge to satisfy the cloud users. The workloads produced by queries can change very quickly; consequently this can lead to decreasing the overall performance (e.g. query processing time) depending on the number and the type of requests made by remote users. In this case, a cloud service provider must manage the unpredictable workloads, through making decisions about which requests from which users are to be executed on which computational resources. Furthermore, the importance of managing the

A.E. Hassanien et al. (Eds.): AMLTA 2014, CCIS 488, pp. 77–86, 2014.

workload arises by the demand to revise resource allocation decisions dynamically. These decisions are based on the progress feedback of the workload or the behavior of the resources to recover any load imbalance that may occur. This can lead to improving the overall performance during queries execution over the distributed resources. Where the workload contains or consists of database queries, adaptive query processing changes the way in which a query is being evaluated while the query is executing over the computational resources [2].

The challenge in this paper is how to provide a fast and efficient monitoring process to the queries executed over the distributed running resource. Furthermore, responding to any failure or load imbalance occurs during the queries execution. This is done by generating an assessment plan that redistributes the queries execution over the replicated resources. This paper focuses on presenting an enhancement of the workload management sub-system of our previous architecture that was presented in [3] to overcome the challenge of slow query response time. It is beneficial to manage the queries execution after implementing the query optimization sub-system. The main objective of this paper is to minimize the overall queries response time in our query processing architecture that was presented in [3].

The paper is organized as follows. Section 2 reviews related works. Section 3 presents the proposed query workload architecture. Section 4 describes the mechanism used to implement the second module in the proposed architecture. Section 5 presents the experimental environment. Section 6 presents the evaluation of the results. Conclusion and future work are discussed in Section 7.

2   Related Work

This section reviews related work on the topic of query processing and database workloads regarding how to characterize workloads and monitor query progress. In [4] they present a modeling approach to estimate the impact of concurrency on query performance for analytical workloads. Their solution relies on the analysis of query behavior in isolation, pairwise query interactions and sampling techniques to predict resource contention. In [5] they proposed a query optimization technique for query processing to improve the query response time. Furthermore, proposing a selection module for query execution by selecting a subset of resources and applying ranking function to improve execution performance for individual queries, however their technique didn't consider other queries running at the same time and load imbalance that may occur during the queries execution. In [6] their approach based on the fact that multiple requests that are executed concurrently, they may have a positive impact on the execution time of the workload. They applied a monitoring approach to derive those impacts. In [7] they propose an architecture for Adaptive Query Processing (AQP), its components communicate with each other asynchronously according to the publish/subscribe model in order to dynamically rebalance intra-operator parallelism across Grid nodes for both stateful and stateless operations. In [8] authors presented a joint query support in CloudTPS, a middleware layer which stands between a Web application and its data store. The system enforces strong data consistency and scales linearly under a demanding workload composed of join queries and read-write transactions. A workload management is proposed in [9] for controlling the execution of individual queries, they implemented an experimental system that includes a dynamic execution controller that leverages fuzzy logic. Several techniques have been proposed in [10], for dynamically re-distributing processor load assignments throughout a computation to take into account of varying resource capabilities. Also they proposed a novel approach to adaptive load balancing, based on incremental replication of an operator state. In [2] they describe the use of utility functions to co-ordinate adaptations that assign resources to query fragments from multiple queries. As well as, demonstrating how a common framework can be used to support different objectives, specifically to minimize overall query response times. In [11] authors presented a Merge-Partition (MP) query reconstruction algorithm. Their algorithm is able to exploit data sharing opportunities among the concurrent sub-queries. This can reduce the average communication overhead.

Although previous researches address several issues in query processing and opti-mization, our proposed architecture combines the query optimization and query re-source allocation with monitoring the concurrent queries execution over the running resources. Furthermore, responding to any load imbalance by applying our workload management system on the cloud environment.

3   The Proposed System's Architecture

Our proposed architecture overcomes the challenge of a low query response time by optimizing the sub-queries. Furthermore, it assigns the sub-queries to the appropriate resources. It manages queries execution to respond to any load imbalance before re-turning the queries results to the users. Our proposed architecture which is shown in Fig. 1, involves three main sub-systems [3]:

1. **Query Optimization Sub-system:** Accepts queries from users and then detects the data sharing among the sub-queries of submitted queries from users, therefore specifying the ordering of the queries execution. Finally, it allocates the query to the appropriate resources. The output of this sub-system is the list of the resources that are responsible for the execution of the queries.
2. **Workload Manager Sub-system:** Manages the queries execution on resources and responds to any load imbalance that may occur throughout the execution. This paper will be focused on presenting this sub-system. The main advantages of this sub-system are the improvement of the query execution performance and overall query response time. It holds the following main processes:

   — *Observer:* Collects information about throughput and utilization values of each running resource during the queries execution. If these values exceed specific thresholds this means there is a fault with the current execution, therefore the observer notifies the planner about occurring load imbalance on a specific resource.

- *Planner:* Performs the assessment phase; it creates an assessment plan to recover the load on the resources during the queries execution by assigning the failure queries on another suitable replica.
- *Responder:* Receives the assessment plan from the planner to respond to the failure that occurs during the queries execution.

3. **Integrator Sub-system:** Is responsible for collecting the queries results from the resources and partioning the merged queries results that are presented in [11] and finally, retrieving the results to the users.

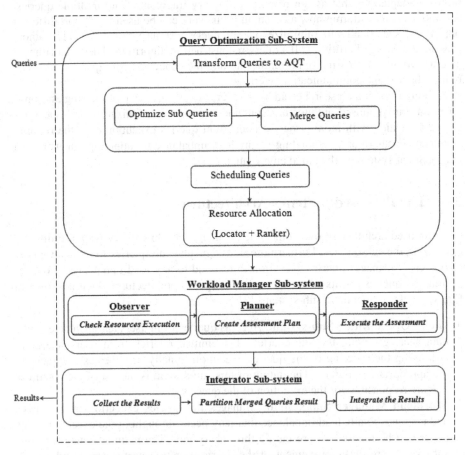

Fig. 1. The Proposed Enhanced Query Workload Architecture

4   The Implementation of the Proposed System

The proposed system accepts queries from the remote users and then applies the query optimization sub-system which is presented and implemented in [3]. During the queries execution, the importance of managing the workload arises by dynamically revising the resources allocation. So the workload management sub-system is implemented through three main processes:

1. Observer: Dynamically revises and monitors the resources during queries execution by the following steps:

- Firstly, by collecting the performance information about the running resources every 15 seconds which is ideal [12] for benchmarking scenarios. The performance information is the percentage of processing time which each resource spends on processing queries. The other measurement is the query throughput which is the number of requests received per second. Such information gives a good indicator of just how much load is being placed on the system, which has a direct correlation to how much load is being placed on the processor [13].
- Secondly, the observer checks the utilization of the resources processor. In other means whether the query throughput exceeds the specific baseline value and if the processing time on specific resource exceeds 80% [13].
- Finally, it notifies the planner with the updated information of the loaded resource to generate an assessment plan to handle the failure that may occur during the execution.

2. Planner: Is implemented to get the notifications from the observer when failure or load may occur on a specific resource. Therefore it creates an assessment plan for handling this load during the queries execution. The following steps are used to implement the planner:

- Firstly, it collects the performance information about the replicas of the loaded resource, and then it determines the most available unloaded replica to execute the queries.
- Secondly, it generates the assessment plan based on determining the suitable replica that can execute these queries.
- Finally, it notifies the responder by the assessment plan with the new queries allocation distribution.

3. Responder: Is implemented to receive and implement the assessment plan from the planner to respond to the load imbalance during the queries execution. The following steps are used:

- Firstly, it uses the assessment plan to specify the queries with a failure execution, therefore killing the queries execution on the loaded resource.
- Secondly, it loads the information about the corresponding replica.
- Finally, it executes the queries on the corresponding replica.

5   Experimental Environment

The TPC-H database [14] is used as our dataset (scale factor 1) to test our work. The TPC-H database has eight relations: REGION, NATION, CUSTOMER, SUPPLIER, PART, PARTSUPP, ORDERS, and LINEITEM.

The cloud environment is simulated with the help of a VMWare workstation. A VMWare workstation is the global leader in virtualization and cloud infrastructure [15].

Eight virtual machines are deployed as in Fig. 2, which shows their capabilities and the relations distribution amongst them with an assumption about partitioning the relations horizontally to two parts (ex: Lineitem_P1, Lineitem_P2).

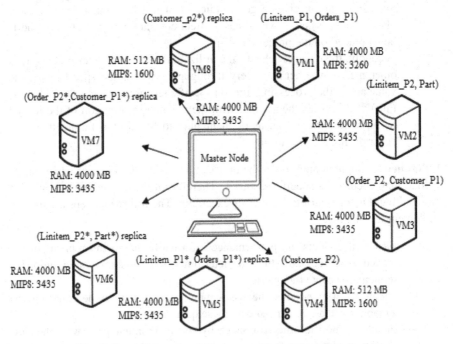

Fig. 2. Capabilities of each Resource with Relations Distribution

Twenty queries that are using the following relations LINEITEM, ORDERS, PART, CUSTOMER are used to test our experiment. Table 1, shows an example of the queries that have been used. The threshold of query throughput is specified by estimating the average transactions per second during peak activity, then using this value as a baseline to compare the query throughput during any stage of the execution.

Microsoft Windows Server 2008 and Microsoft Structured Query Language Server (MS SQL server) is used to deploy the TPC-H database.

Table 1. Example of five queries used in our experiment

Queries
select p_type,l_extendedprice,l_discount from lineitem, part where l_partkey = p_partkey and l_shipdate >= '01-09-1995'
select p_type,l_extendedprice,l_discount from lineitem, part where l_partkey = p_partkey and l_shipdate < '01-09-1995'
select l_returnflag, l_linestatus from lineitem where l_shipdate < '1998/12/08'
select   l_shipmode from orders, lineitem where o_orderkey = l_orderkey and l_commitdate < l_receiptdate and l_shipdate < l_commitdate and l_receiptdate >= '1997-01-01' and l_receiptdate < '1998-01-01'
select o_orderpriority from orders where o_orderdate >= '1993-07-01' and o_orderdate < '1993-07-01'

6   Evaluations

Table 2 shows examples of the resources that are assigned to execute the merged and non-merged queries after applying our query optimization sub-system which is presented in [3]. After executing twenty queries, VM1 transactions per second exceeds 54, which is the baseline value and its processor utilization exceeds 80% that means there is a load on VM1 during the queries execution. By applying our Workload Management Sub-system, the latest queries with failure execution on VM1 are assigned to VM5 which is the replica of VM1.

In order to evaluate our system the average execution time is computed and compared with applying our proposed system and the query processing technique in [5]. Fig.3 shows that the execution time of the queries with applying our proposed system can reduce the queries execution time, as our proposed system combines the query optimization and query resource allocation of submitted concurrent queries with monitoring these queries execution by the Workload Management sub-system which is not considered in [5].

This experiment was executed for three times and the average response time is calculated. The results show that using our proposed Workload Management Sub-system reduces the queries execution time over the technique presented in [5] by 68%. The results are shown in Table 3.

Table 2. The resources selected to execute the merged and non merged queries

Queries	Assigning Virtual Machines
select p_type,l_extendedprice,l_discount from lineitem,part where l_partkey = p_partkey and l_shipdate >= '01- 09-1995 or l_shipdate < '01-09-1995	VM1 – VM2
select l_returnflag, l_linestatus from lineitem where l_shipdate < '1998/12/08'	VM1 – VM2
select l_shipmode from orders, lineitem where o_orderkey = l_orderkey and l_commitdate < l_receiptdate and l_shipdate < l_commitdate and l_receiptdate >= '1997-01-01' and l_receiptdate < '1998-01-01'	VM1 – VM3
select o_orderpriority from orders where o_orderdate >= '1993-07-01' and o_orderdate < '1993-07-01'	VM1– VM3

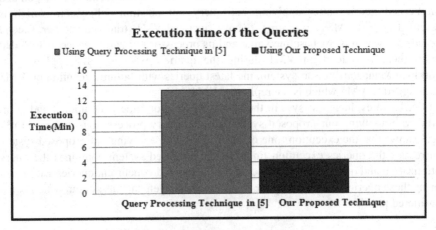

Fig. 3. The average queries execution time

Table 3. The average of measuring the queries execution time

Approaches	Average Execution Time(Min)
Using Query Processing Technique in [5]	13.41261
Using our Proposed Technique	4.290434

7 Conclusion and Future Work

In this paper, a workload management technique is proposed to minimize the overall query execution time over cloud computing environment. This is done by monitoring the running resources performance such as processing time and throughput over the cloud computing environment. Furthermore, responding to any load imbalance may occur across these running resources. The results show that applying our proposed system which combines the query optimization and query resource allocation with monitoring the concurrent queries execution by the Workload Management technique over the running resources in cloud computing environment, improves the response time of the concurrent queries. In our future work, the integration sub-system of the proposed architecture will be implemented to collect and partition the merged queries result from the resources and improve the query scheduling mechanism.

References

1. Yang, D., Li, J., Han, X., Wang, J.: Ad Hoc Aggregation Query Processing Algorithms based on Bit-store in a Data Intensive Cloud. J. Future Generat. Comput. Syst. 29, 725–1735 (2013)
2. Paton, N.W., de Aragão, M.A.T., Fernandes, A.A.A.: Utility-driven Adap-tive Query Workload Execution. Future Generation Computer Systems 28, 1070–1079 (2012)
3. Maghawry, E.A., Ismail, R.M., Badr, N.L., Tolba, M.F.: An Enhanced Resource Allocation Approach for Optimizing a Sub-query on Cloud. In: Hassanien, A.E., Salem, A.-B.M., Ramadan, R., Kim, T.-h. (eds.) AMLTA 2012. CCIS, vol. 322, pp. 413–422. Springer, Heidelberg (2012)
4. Duggan, J., Cetintemel, U., Papaemmanouil, O., Upfal, E.: Performance Pre-diction for Concurrent Database Workloads. In: SIGMOD, Athens, pp. 337–348 (2011)
5. Liu, S., Karimi, A.H.: Grid Query Optimizer to Improve Query Processing in Grids. Future Generation Computer Systems 24, 342–353 (2008)
6. Albuitiu, M.C., Kemper, A.: Synergy based Workload Management. In: Proceedings of the VLDB PhD Workshop, Lyon (2009)
7. Gounaris, A., Smith, J., Paton, N.W., Sakellariou, R., Fernandes, A.A.A., Watson, P.: Adapting to Changing Resource Performance in Grid Query Processing. In: Pierson, J.-M. (ed.) VLDB DMG 2005. LNCS, vol. 3836, pp. 30–44. Springer, Heidelberg (2006)

8. Wei, Z., Pierre, G., Chi, C.: Scalable Join Queries in Cloud Data Stores. In: IEEE/ACM International Symposium on Cluster, Cloud and Grid Computing, Ottawa, pp. 547–555 (2012)

9. Krompass, S., Kuno, H., Dayal, U., Kemper, A.: Dynamic Workload Man-agement for Very Large Data Warehouses: Juggling Feathers and Bowling Balls. In: 33rd International Conference on VLDB, Vienna, Austria, pp. 1105–1115 (2007)

10. Paton, N.W., Buenabad, J.C., Chen, M., Raman, V., Swart, G., Narang, I., Yellin, D.M., Fernandes, A.A.A.: Autonomic Query Parallelization using Non-dedicated Computers: An Evaluation of Adaptivity Options. In: VLDB, vol. 18, pp. 119–140 (2009)

11. Chen, G., Wu, Y., Liu, J., Yang, G., Zheng, W.: Optimization of Sub-query Proc-essing in Distributed Data Integration Systems. Journal of Network and Computer Applications 34, 1035–1042 (2011)

12. Performance Monitoring, https://software.intel.com/en-us/articles/use-windows-performance-monitor-for-infrastructure-health

13. Fritchey, G., Dam, S.: SQL Server 2008 Query Performance Tuning Distilled, 2nd edn., USA (2009)

14. Transaction Processing and Database Benchmark, http://www.tpc.org/tpch/

15. VMware, http://www.vmware.com/

A Large Dataset Enhanced Watermarking Service for Cloud Environments

Nour Zawawi, Mohamed Hamdy, Rania El-Gohary, and Mohamed Fahmy Tolba

Ain Shams University, Egypt
{nourzawawi,fahmytolba}@gmail.com,
dr.raniaelgohary@fcis.asu.edu.eg, m.hamdy@cis.asu.edu.eg

Abstract. Preventing data abuses in cloud remains an essential point of the research. Proving the integrity and non-repudiation for large datasets over the cloud has an increasing attention of database community. Having security services based on watermarking techniques that enable permanent preservation for data tuples in terms of integrity and recovery for cloud environments presents the milestone of establishing trust between the data owners and the database cloud services. In this paper, an enhanced secure database service for Cloud environments (EWRDN) is proposed. It based over enhancements on WRDN as a data watermarking approach. The proposed service guarantees data integrity, privacy, and non-repudiation recovering data to its origin. Moreover, it gives data owner more controlling capabilities for their data by enabling tracing users' activities. Two compression categories to recover data to its origin introduced for the proposed service. Two compression technique (the arithmetic encoding and the transform encoding) chosen to represent each type. For large data sets, it has been proven that, the arithmetic encoding has a fixed recovery ratio equal to one. At the same time, the transform encoding saves space and consumed less time to recover data. Moreover, testing the performance is done of the proposed service versus a large number of tuples, large data set. The performance quantified in terms of processing time and the required memory resources. The enhanced EWRDN service has shown a good performance in our experiments.

Keywords: Copyright protection, Digital Watermarking, Security Service, Data Compression, Large dataset.

1 Introduction

Introducing database secure services in the cloud represents the key function of establishing the trust for clients to save their critical and confidential data on the Cloud. While, demand for the use of large data sets is growing, pirated copying has become a severe threat. To fight against pirated copying, database watermarking promises a solution for protecting data by embedding secret codes (watermarks) into the tables inside the databases. The digital watermarking for integrity verification is called fragile watermarking as compared to robust watermarking for copyright protection.

A.E. Hassanien et al. (Eds.): AMLTA 2014, CCIS 488, pp. 87–96, 2014.

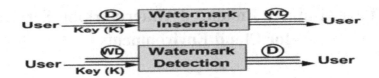

Fig. 1. Basic Watermarking Technique

Figure 1, shows the basic database watermarking techniques. It consists of two phases: I- Watermark Insertion and, II – Watermark Detection. During watermark insertion phase, user adds his data set (D) with his private key (K). Then, the watermark will be calculated and the watermarked data set (WD) is then available. In the detection phase, the user needs to add the suspicious dataset (WD) with his private key (K). The embedded watermark (if present) is extracted and compared with the original watermark information. So, the original data (D) will be extracted with proof of ownership.

A Novel Watermarking Approach for Data Integrity and Non Repudiation in Rational Databases (WRDN) introduced [1]. The main idea is to apply WRDN as a trusted security service on cloud computing. But some problems arise. These problems can be summarized as hiding and locking technique. Moreover, one needs to have the ability to recover data if unauthorized changes or errors appeared. To overcome these problems, some enhancement of WRDN model was made. An Enhancement model for WRDN (EWRDN) is presented [4]. It does not prevent copying, but it deters illegal copying by providing a means of establishing the ownership of a redistributed copy.

Today's availability, performance and security are the three main problems when it comes to cloud adoption. At the same time, performance needed to be measured in terms of time and space. In cloud environments, managing millions or even more simultaneous users and their data is normal. The scalability of cloud services remains an important issue. The better an application's scalability, the more users it can handle simultaneously [5]. In this paper, an important question is answered about the time needed to recover data, since; the challenge of the large data set (huge amount of tuples) is always motivating data based service design for cloud environment.

With this flexibility, the challenge in deciding which data to be compressed appears. The factors that influence this decision are the data types and confidentiality. EWRDN uses lossless method to compress data. It assumes that all the data in the data set are critical and needed to be recovered as its origin. At the same time, a loose method can produce a much smaller compressed file than any lossless method, while still meeting the user requirements.

The rest of this paper is organized as follows: Section 2 gives related work overview. The Enhanced Watermark Approach for Secure Database Service (EWRDN) insertion and detection algorithm is presented in Section 3. Section 4 introduces performance analysis of EWRDN. Finally, the conclusion of this paper with summaries and suggestions for future work are introduced in Section 5.

2   Related Work

The approaches handle database attacks can be summarized as: the first algorithm is the distortion based algorithm, which introduces small changes in data values during embedding phase where changes are tolerable and should not make the data useless. The watermarking scheme proposed by [6], also known as AHK, is one of the pioneering research in database watermarking. The fundamental assumption is that the watermarked database can tolerate a small amount of errors in numeric data. Although the basic assumption of AHK scheme is that the relation has a primary key whose value does not change, Li et al. [7] suggest three different schemes to obtain virtual primary keys for a relation without primary key. Sion et. Al [9-6] use the most significant bits of the normalized dataset instead of primary key. Database watermarking based on cloud model is proposed by Zhang et al. [8].

The second algorithm is the distortion free, where the proposed model is considered to be one of them. There is no modification made to any data item and the digital watermarking is used for integrity verification. The watermarking scheme proposed by Y. Li et al. [11] was the first distortion free algorithm made. The Basic idea was that all tuples are securely partitioned into (g) groups. A different watermark is embedded in each group such that any modifications can be detected and localized into the group level with high probabilities. The watermarking scheme proposed by Li and Deng [12] is applicable for marking any type of data. The interesting feature of this scheme is that it does not use any secret key. Moreover, the unique watermarking key is used in both the creation and the verification phases. While, Kamel [13] suggested a way to improve the detection rate of malicious alteration by watermarking not only the relational tables (data records) but also all relevant indexes by proposing a fragile watermarking technique for protecting data integrity in databases and more specifically in R-tree data structures. The approach proposed in [17] aims to generate fake tuples and insert them erroneously into the database. The fake tuple creation algorithm takes care of candidate key attributes and sensitivity level of non-candidate attributes, while in [19]they add only one hidden column, using a secret formula to relational database that contains only numeric values. Moreover, it locks this calculated column from any attacks or manipulations. However, the work done by [20] uses the same schema made in [19] but by applying it over a non-numeric data over the watermarking on a new row.

3 EWRDN as a Watermarking Approach

A Novel Watermarking Approach for Data Integrity and Non Repudiation in Rational Databases (WRDN) is introduced in [1]. WRDN proves data ownership and integrity of database. It survives against two types of attacks that face the database (Insertion, Deletion). It is based on adding a watermark over a hidden column then locks this column. It is designed to be a part of Database Management System (DBMS). So, there are no fears over watermark data detection. The main idea of this paper is creating a data security service over the cloud. Unfortunately, applying WRDN directly to be a cloud service is not feasible due to the following reasons:

- There are no guaranties that cloud provider will apply the same mechanism and cover the watermark column.
- In cloud the data and the users could not been in the same country. So, the cloud service provider will also have the authority and the ability to unlock and view the watermark column.

In this paper, new abilities have been added to EWRDN to provide solutions that prove data integrity and ownership for large data set over the cloud. This is the ability to prove data tampering by tracing authorized users activities and recording them. If unauthorized changes or errors appeared it recover data to its origin. Moreover, it gives database owner more control over his data. By, tracing authorized users' activities and record it in order to differentiate them from unauthorized ones. Figure 2 illustrate EWRDN architecture.

EWRDN needs to calculate the corresponding watermark for each tuple then saves it. Then, it adds a user signature on each attribute which has been changed or added. Finally, compresses new signature data and saves them over the database. It lock watermark column and compressed records using a secret key (K) known only by the data owner. EWRDN proves important features of database security like Non Repudiation, Integrity, Copyright protection and Recovery.

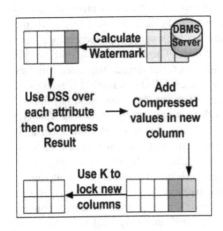

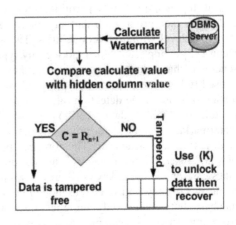

(a) Watermarking Insertion Technique   (b) Watermarking Detection Technique

Fig. 2. Insertion and Detection Mechanism of EWRDN

3.1   Insertion Technique

Figure 2(a), shows EWRDN insertion technique. In the beginning one needs to have two hidden columns. No one of the database users knows about them and has no control over them. First, calculate the watermark value for each tuple which depends on a mathematical function known only by data owner. Then, use user's private key (PrK) to add signature over each attribute [21]. The following experiments used Digital Signature Standards (DSS) [21] second; it uses compression technique [22,23] on new signature attribute to compress value. Add compressed values into a new column. Finally, both watermarking and compressed column need to be locked using (K) which is a private key that is only known to the database owner.

3.2   Detection Technique

Figure 2(b), shows EWRDN detection technique. First, calculate the watermark value. Then, unlock the watermark and compressed column using private key (K). Finally, it needs to compare the calculated watermark results with the original watermark values. If they match, then data is tampered free, and the data owner has the ability to trace users work and updated data. Otherwise, data is tampered to recover data, decompress values in the corresponding row. Then, apply signature algorithm (DSS) used to separate data from user signature then restore data to the corresponding attributes.

3.3 Role of Compression Techniques

Data compression is the reduction in data size in order to save space or transmission time. It can be performed on just the data content or on the entire transmission unit depending on a number of factors. EWRDN method is proposed [4]. It is based over the idea of using compression technique to be able to recover data. Data compression techniques consist of two types:

- Lossless Compression: The values after decompression are the accurate copy of the original data.
- Loose Compression: The values after decompression gives an approximation or close values of the original data.

At previous work, EWRDN introduced lossless method for data compression. It assumes that all data in the data set are critical and needed to be recovered as its origin. In practice, not all of the data are critical. This means some attributes can be partially restored. The enhancements in this papers show that data owners can specify their preferences of the required restoration features about the data. Some loose method can produce high data restoration capabilities compared to lossless methods. This may be accepted by the data owners and meeting their preferences. In results' section of this paper, it is going to be shown that loose compression methods produce better compression performance in terms of time and space than lossless methods with an unnoticeable loss in quality. Users or clients can specify their data restoration guarantees based on a predefined fees' schema. Moreover, the experiments test is based over a large data set. So, one need to improve time performance and saves more space. Besides, one needs to calculate time performance for each compression technique. A method of each compression technique has been chosen. Arithmetic Coding represent lossless compression [2, 23] and Transform Coding represent loose compression [3, 24].

The idea behind arithmetic coding is to have a range of probability line from zeros to one, and assign to every symbol a range in this line based on its probability. Each time the probability increased the higher range which assigns to it. Finally, one has to start encoding symbols. It has a good compression ratio (better than Huffman coding), with an entropy around the Shannon Ideal value [2]. Transform Coding is based on utilizing redundancy in the data in order to be able to transform it to values, X_i. That is why one needs to compress the data by using fewer bits to represent the differences (Quantization) [3].

4   Performance Analysis

The performance experiments have been done using 100 thousands tuples, where there are 31 attributes for each tuple. Data Cleaning and Reduction is applied over the data. Then, data preprocessed is applied, no noisy and consistent [25]. As a result, it became 50 thousands tuples and 30 attributes. The experiment is conducted using an AMDFX-8350 processor running at 4 GHz, Cash 16 With 8 Core Motherboard Giga-byte GA-F2A85X-UP4, 4 GHz Ram Bus 1600 Hard 500 GHZ with Sapphire HD 7870 2GP DDR5. The system has been tested over two types of experiments: adding new records or updating attributes. To measure the performance of EWRDN system, two factors need to be considered: 1- The time needed for EWRDN to add watermark and recover data; 2- The amount of space needed to apply EWRDN system.

4.1   Result Analysis

Analyzing the results of EWRDN scheme is made by Bernoulli trials and binomial probability. Discussing the results is based on Robustness condition which is based on two parameters false hit and false miss.

- False hit is the probability of a valid watermark being detected from non-watermarked data. On EWRDN, it never happened because each data has its own watermark. That is due to hiding the watermark data. Therefore, all detected strings will match their watermark, and the false hit is zero.
- False miss is the probability of not detecting a valid watermark from watermarked data that has been modified in attacks. Two cases are considered when trying to calculate the false miss.
  - Deletion: The watermark value associated with the deleted tuples will not be deleted. However, the other tuples will not be affected. So, false miss is zero.
  - Updating: Suppose an attacker update tuples or attributes. Watermark detection will never return a false answer, because new added values will fail to have corresponding watermarks.

For more details about EWRDN service analysis and a comparison between WRDN and EWRDN refer to [4].

4.2   Time

There are two types of time that affect the model performance: Processing and Recovery Time. In case of processing time, EWRDN will have a static time performance of O (n) for each tuple inserted or updated, where n is the number of tuples available in the dataset. While, in case of recovery time, one needs to calculate the number of altered tuples.

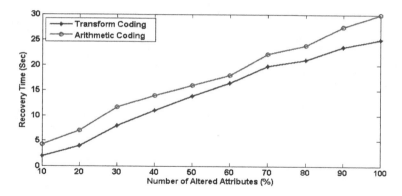

Fig. 3. The Arithmetic and Transform Recovery Time

Figures 3 illustrates the difference between arithmetic and transform coding recovery time which depends on the number of tuples changed. It has been clear that arithmetic coding takes more time to recover data due to its large symbol tables. At the same time, transform coding does not need to use large symbol tables. There are nearly five seconds different between the two techniques. At the beginning, there are nearly two seconds different between the two techniques. But, as the number of altered tuples increased, the time difference between the two techniques increased. Arithmetic coding consumes more time for large data set to recover data. It takes 30 seconds to recover 50 thousands tuples. At the same time, transform coding takes 25 seconds to recover data for the same data set.

4.3 Space

Space complexity depends on compressed values and compression ratio. At the same time, the compression ratio differs according to type of compression. The arithmetic encoding as a representative of the first compression type has a complexity of $O(\alpha.n)$ where α is the compression ratio and n is the number of attributes. The value of α has a range from]0,1]. On the other hand, the transform encoding as a representation of the second compression type has a complexity of $O(\bar{\alpha}.n)$ where $\bar{\alpha}$ is the compression ratio and n is the number of attributes. The value of has a range from]0,1[. As mentioned, it has been found that the compression ratio is better in case of using the transform encoding but with a cost of loss of information. The different representation of the compression ratio α and $\bar{\alpha}$ represented in figure 4 and figure 5. Compression ratios α and $\bar{\alpha}$ are computed as follows:

$$CompressionRatio = \frac{CompressedSize}{UncompressedSize} \qquad (1)$$

After applying Equation 1, it will be found that arithmetic coding has a compression ratio which is equal to 0.56. While, transform coding has a compression ratio is equal to 0.65. It has been illustrated in Figure 4. As a result, arithmetic coding consumed more space than transform coding.

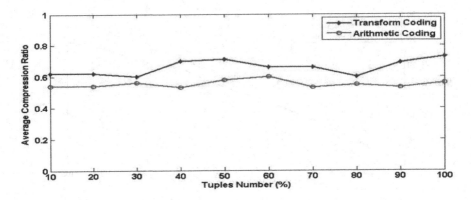

Fig. 4. Compression ratios α and ᾱ

4.4   Recovery Ratio

As the main idea behind the proposed system is to recover data in case an error or unauthorized changes appear. So, a comparison between the recovery ratio of the two of arithmetic and transform encoding need to be made. The arithmetic encoding uses the same number of bytes to compress data; so, it decompresses data to its origin. On the other hand, transform encoding is based on decreasing the number of bytes used to save data in less space. In the proposed experiments, one has nearly the numbers into two fractions of number and the integer's number has been left the same with no changes.

Figure 5 illustrate the relation between recovery ratio in both arithmetic and transform coding. Arithmetic coding has a fixed Recovery Ratio number equal to 1 which means that, in any case scenarios the data will be recovered to its origin. At the same time, transform coding recovery ratio depends on the types of numbers being compressed. It means that data will not been recovered to its origin, it will only recover 85% of original data.

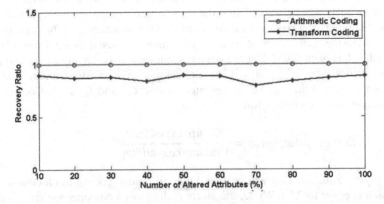

Fig. 5. Arithmetic Coding and Transform Recovery Ratio

5   Conclusion

This paper presented an Enhanced Watermarking Approach for Secure database Service (EWRDN Service) in action versus a large data set. The proposed service of EWRDN gives data owner more control of his data. Based on a traditional database watermarking technique (WRDN), EWRDN adds a watermark(s) to save copyright of user data. At the same time, it adds a user signature to trace user activity. Then, it compresses the signed data to recovery data if errors or unauthorized changes. It has been shown that, as large as, dataset is, the performance of proposed service (EWRDN) stayed in order. Different data compression techniques are applied for the purpose of evaluation of the proposed enhancements for EWRDN. The arithmetic encoding is chosen to represent lossless data compression methods. On the other hand, the transform encoding is chosen to represent loose data compression methods. The comparison between the two encoding methods has been elaborated two reflect how they can be fitted to EWRDN. It has been proved that, the transform encoding has a better recovery time and saves more space than the arithmetic encoding. Moreover, applying a loose compression technique (like the transform encoding) showed that time required for data recovery is good as compared to lossless compression technique (like the arithmetic encoding). It saves 5 seconds to recover a total 50 thousands tuples. At the same time, the arithmetic encoding has a recovery ratio equal to 1. It means that, the arithmetic encoding recovers data as its origin. Meanwhile, the transform encoding recovery ratio equal to 0.85. Our proposed enhancement showed that EWRDN can be considered as a hybrid approach, where users compress critical data using lossless techniques and others using loose technique based on their preferences. Moreover, importance of using loose compression techniques to enhance the proposed cloud service has been introduced and investigated.

References

1. Zawawi, N., El-Gohary, R., Hamdy, M., Tolba, M.F.: A novel watermarking approach for data integrity and non-repudiation in rational databases. In: Hassanien, A.E., Salem, A.-B.M., Ramadan, R., Kim, T.-h. (eds.) AMLTA 2012. CCIS, vol. 322, pp. 532–542. Springer, Heidelberg (2012)
2. Bodden, E., Clasen, M., Kneis, J.: Arithmetic coding revealed a guided tour from theory to praxis. Technical report. McGill University (2007)
3. Brandt, J.: Transform coding for fast approximate nearest neighbor search in high dimensions. In: IEEE Conference on Computer Vision and Pattern Recognition (CVPR), pp. 1815–1822 (2010)
4. El-Zawawi, N., Hamdy, M., El-Gohary, R., Tolba, M.F.: A database watermarking service with a trusted authority architecture for cloud environment. International Journal of Computer Applications 69(13), 1–9 (2013)
5. Armbrust, M., Fox, A., Griffith, R., Joseph, A.D., Katz, R., Konwinski, A., Lee, G., Patterson, D., Rabkin, A., Stoica, I., Zaharia, M.: A view of cloud computing. Commun. ACM 53, 50–58 (2010)
6. Agrawal, R., Kiernan, J.: Watermarking relational databases. In: Very Large DataBase, VLDB, pp. 155–166. Morgan Kaufmann (2002)

7. Li, Y., Swarup, V., Jajodia, S.: Constructing a virtual primary key for fingerprinting relational data. In: Digital Rights Management Workshop 2003 (2003)
8. Zhang, Y., Niu, X., Zhao, D., Li, J., Liu, S.: Relational databases watermark technique based on content characteristic. In: International Conference on Innovative Computing, Information and Control (ICICIC 2006) (2006)
9. Sion, R., Atallah, M., Prabhakar, S.: Rights protection for relational data. IEEE Transactions on Knowledge and Data Engineering 16(12), 1509–1525 (2004)
10. Sion, R., Atallah, M., Prabhakar, S.: Rights protection for categorical data. IEEE Transactions on Knowledge and Data Engineering 17(7), 912–926 (2005)
11. Li, Y., Guo, H., Jajodia, S.: Tamper detection and localization for categorical data using fragile watermarks. ACM (2004)
12. Li, Y., Deng, R.H.: Publicly verifiable ownership protection for relational databases. In: ACM Symposium on Information, Computer and Communications Security, ASIACCS (2006)
13. Kamel, I.: A schema for protecting the integrity of databases. Computers & Security 28, 698–709 (2009)
14. Bhattacharya, S., Cortesi, A.: A distortion free watermark framework for relational databases. In: International Conference on Software and Data Technologies, ICSOFT (2009)
15. Bhattacharya, S., Cortesi, A.: A generic distortion free watermarking technique for relational databases. In: Prakash, A., Sen Gupta, I. (eds.) ICISS 2009. LNCS, vol. 5905, pp. 252–264. Springer, Heidelberg (2009)
16. Bhattacharya, S., Cortesi, A.: Database authentication by distortion freewatermarking. In: International Conference on Software and Data Technologies. ICSOFT, vol. 5, pp. 219–226 (2010)
17. Pournaghshband, V.: A new watermarking approach for relational data. In: ACM Southeast Regional Conference, pp. 127–131 (2008)
18. Prasannakumari, V.: A roubust tamper proof watermarking for data integrity in relational database. Research Journal of Information Technology 1, 115–121 (2009)
19. Gamal, G.H., Rashad, M.Z., Mohamed, M.A.: A simple watermark technique for relational database. Mansoura Journal for Computer Science and Information System (2008)
20. El-Bakry, H., Hamada, M.: A novel watermark technique for relational databases. In: Wang, F.L., Deng, H., Gao, Y., Lei, J. (eds.) AICI 2010, Part II. LNCS, vol. 6320, pp. 226–232. Springer, Heidelberg (2010)
21. Sako, K.: Digital signature schemes. In: Encyclopedia of Cryptography and Security, 2nd edn., pp. 343–344. Springer (2011)
22. Sayood, K.: Introduction to Data Compression. Elsevier (2012)
23. Sayood, K.: Lossless Compression Handbook, ch. 5, pp. 101–152. Academic Press (2004)
24. Mohammed, A.A., Hussein, J.A.: Hybrid transform coding scheme for medical image application. In: IEEE International Symposium on Signal Processing and Information Technology (ISSPIT), pp. 237–240 (2010)
25. Han, J., Kamber, M., Pei, J.: Data Mining Concepts & Techniques, ch. 3, 3rd edn. Elsevier, Morgan Kaufmann (2011)

Enhancing Mobile Devices Capabilities
in Low Bandwidth Networks with Cloud Computing

Mostafa A. Elgendy[1], Ahmed Shawish[2], and Mahmoud I. Moussa[1]

[1] Faculty of Computers and Informatics, Benha University
[2] Faculty of Computers and Information Science, Ain Shams University, Egypt
{Mostafa.elgendy,Mahmoud.mossa}@fci.bu.edu.eg,
Ahmed.gawish@fcis.asu.edu.eg

Abstract. Recently as smart phones have merged into heavy applications like video editing and face recognition. These kinds of applications need intensive computational power, memory, and battery. A lot of researches solve this problem by offloading applications to run on the Cloud due to its intensive storage and computation resources. However, none of the available solutions consider the low bandwidth case of the networks as well as the communication and network overhead. In such case, it would be more efficient to execute the application locally on the Smartphone rather than offloading it on the Cloud. In this paper, we propose a new framework to support offloading heavy applications in low bandwidth network case, where a compression step is proposed for the favor of minimizing the offloading size and time. In this framework, the mobile application is divided into a group of services, where execution-time is calculated for each service apart and under three different scenarios. An offloading decision is then smartly taken based on real-time comparisons between being executed locally, or compressed and then offloaded, or offloaded directly without compression. The extensive simulation studies show that both heavy and light applications can benefit from the proposed framework in case of low bandwidth as well as saving energy and improving performance compared to the previous techniques.

Keywords: Smartphones, Android, Offloading, Mobile Cloud computing, Compression, Network bandwidth.

1   Introduction

Recently smartphones have merged into heavy applications such as natural language translators, speech recognizers, optical character recognizers, image processors and search, online games, video processing and editing, navigation, face recognition and augmented reality, however these applications consumes most of the mobile battery, memory, and computational resources.

Cloud Computing has been introduced as an approach to save resources and extend the battery life time of such vital device. Mobile applications can augment their capabilities with unlimited computing power and storage space by offloading some services to run on the Cloud; as result saving time and computation power [1, 2].

A.E. Hassanien et al. (Eds.): AMLTA 2014, CCIS 488, pp. 97–108, 2014.

All of the available researches proposed frameworks to offload the mobile services to be executed on the Cloud as they focus on saving the mobile battery, power and computational resources [3- 11]. However, in low bandwidth networks scenario, all of them execute the application locally on the Smartphone rather than offloading it. The reason behind this decision was simply based on the fact that low bandwidth scenario result in the increases of both transmission time and packet loss probability and hence waste of the limited Smartphone's resources in repetitive re-transmission failure. Here, we argue that offloading decision is not a blind decision problem. In fact it should be smartly taken based on real-time parameter like the expect execution time of the task. In addition, we argue that offloading is still acceptable even in low bandwidth if we minimize the amount of data that should be offloaded and hence minimize the transmission time itself as well as the failure probability.

This paper proposed a framework to support offloading heavy applications in low bandwidth network case. Any mobile application can be easily divided into a group of services some of them run on mobile and others offloaded to run on the Cloud without modifying application source code. In this framework, we introduce an offloading model that decide at runtime whether to execute service locally on the mobile device or it's better to offload those services to run on the Cloud. Moreover, the framework works very well in low bandwidth networks and highly utilizes the Cloud computational resources by compressing the data before sending it onto the Cloud.

The extensive simulation studies show that both heavy and light applications can benefit from the proposed framework in case of low bandwidth as it compress data before sending to the Cloud and this save power and improving performance compare to previous techniques. Also android developers can use the proposed framework very easily by adding framework library into their projects and by adding builders to the project building process.

The rest of this paper organized as follows. Section 2 introduces the background and shows related work. Section 3 describes the proposed framework and covers all the implementation details of the framework. Section 4 discusses the results of the extensive simulation studies. The paper is finally concluded and future work is presented in Section 5.

2   Background

This section gives some details about mobile environment and application development process. In addition, it provides a complete review on the related work done in the offloading context.

2.1   Mobile Environment and Application Development

Android IPC. The main Android applications components are *Activities, Services, Content Providers,* and *Broadcast Receivers*, which have their own specific lifecycle within the system. Android IPC handle the communication between activity and service. First an activity can be bounded to any service, then Android IPC handle the communication using a predefined interface called AIDL and a stub/proxy pair generated by the Android pre-compiler based on this interface [12].

2.2 Related Work

There are a lot of researches in Mobile Cloud Computing which can be categorized in two categories. One category is to use Cloud as a storage service to save mobile storage. Mobile user can send all their data to any Cloud storage service provider and at any time if these data are needed, it can be accessed through network. A number of online file storage services are available on Cloud server like Amazon S3 [16], and DropBox [17]. Another category is to use Cloud extensive computing power to execute mobile applications services to increase performance and save mobile power and memory resources [3, 4], [11]. In the second category a common approach for remote service execution is to partition mobile application into some services that executes locally on mobile and some other intensive services which are offloaded to be executed on the Cloud, this is called application partitioning. One of the drawbacks of this approach is how to handle service offloading in low bandwidth network which will be discussed in the following related work.

Kumar [1] tries to measure the energy consumed by mobile and added network energy consumed to it, and measure the energy consumed by Cloud and compared for deciding whether a task offloading reduced energy or not. *Kumar* also conclude that offloading data intensive tasks to the Cloud depends on the network bandwidth as if the network is low, it will better to execute service locally on the mobile and if the network is high, it will better to execute service remotely on the Cloud. However in low bandwidth networks application may get rid of Cloud by compressing data before offloading, as result execution time and power consumption can be save. *Phone2Cloud* [10] use a naive history-based method to predict average execution time of an application on smartphone. It monitor network bandwidth and leverages average CPU workload got from the resource monitor and input size of the application to predict execution time using the history log. However in data intensive application and low bandwidth network, *Phone2Cloud* always prefer to run service locally on the mobile. *Phone2Cloud* can improve his framework by compressing data before offloading to the Cloud in low bandwidth networks. *Cuckoo* [7] implements a framework that automatically offloads heavy back-end tasks to execute on the Cloud. *Cuckoo* use the very simple model which always prefers remote execution. This work can be improved if framework tries to use some metrics in taking offloading decision like service processing time instead of offloading all the time. Moreover, in some case such as low bandwidth networks the time for communicating and transmit data on network and execute service on the Cloud is larger than the time to executing services on mobile, so *Cuckoo* decide to run services locally on the mobile. *Cuckoo* can compress data before sending to the Cloud, as result saving time in low bandwidth networks. *[6, 8, 9]* using a models which depends on network bandwidth and other metrics for deciding whether to offloads heavy tasks to execute on the Cloud. The models prefer to execute service locally on the mobile in law bandwidth networks. However in some applications which need to send a lot of data over network while offloading, data will be batter compressed before sending to the Cloud. **Xinwen Zhang** [5] designs architecture to enable elastic applications to be launched on a mobile device or in the Cloud. *Xinwen Zhang* enables flexible and optimized elasticity using multiple factors including device status, Cloud status, application performance measures, and user preferences. *Xinwen Zhang* can switch between different network interfaces, however this can't solve the problem low bandwidth networks as it will always run the service locally on the mobile.

As we can easily notice, all of the available frameworks favor to execute the application locally on the Smartphone rather than offloading it on the Cloud in low bandwidth networks scenario. Although, this behavior may look reasonable, it is not carefully taken based real-time parameters and careful consideration of pros and cons that may change from one case to another. In addition, transmitting data over a low bandwidth network has not been addressed as it should be where no solution has been proposed to overcome this problem. In this paper, we prove that a careful consideration to both of the above mentioned drawbacks can lead to a significant enhancement in the Mobile-Cloud Computing in low bandwidth scenario.

3   Proposed Framework Design

3.1   Framework Architecture

The proposed framework made some contributions in developing mobile applications: i) allowing mobile application to use Cloud storage and computational resources in running some services of mobile application so saving mobile device memory, power and battery resources - ii) using a dynamic offloading model in deciding whether to offload services or not, so Mobile applications became smarter as it take decision based on the available resources - iii) allowing intelligent usage of mobile resources specially in low bandwidth networks which lead to a tradeoff between energy consumed and procession time of services - and finally - v) automating a large part of the development process and integrating easily in development tools; so developing Android application become easily. As shown in Fig. 1. The proposed framework consists of four main components *i) Decision Manager - ii) Offloading Manager - iii) Execution Profile - and iv) Cloud Manager*. The first three components are deployed on the mobile and the *Cloud Manager* component is deployed on the Cloud. To use the framework, it's initially assumed that the application should be structured using Android AIDL services pattern.

Offloading Manager. *Offloading Manager* is responsible for executing the application services based on the decision taken by *Decision Manager*. If the decision is to execute the service locally on the mobile, then *Offloading Manager* calls the local service implementation from the mobile side. However if the decision is offloading the service for execution on the Cloud, then the *Offloading Manager* connect to the *Cloud Manager* and send any data needed to execute the service, Then it waits until the *Cloud Manager* execute the service on the Cloud and send the result back to the mobile side. At the end *Offloading Manager* is responsible for receiving the returned results and delivering it to the application.

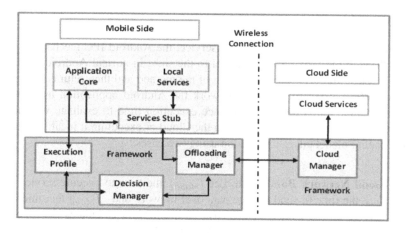

Fig. 1. Framework Architecture

Execution Profile. *Execution Profile* is a profile created for each service by *Decision Manager* at the first of its run to store service execution time. It store service execution time of executing sample example of the service in the following three scenarios. The first scenario is when service executed locally on the mobile. The second scenario is when service offloaded and executed on the Cloud. Note in this case it store only Cloud execution time removing the time to send data over network, as the time of sending data over network depends on network bandwidth, so it can be calculated when trying to call the service. The third scenario is when compressing service data, then offloading it for execution on the Cloud. In subsequent runs of the *Decision Manager,* data stored in *Execution Profile* and network bandwidth will be used in taking the offloading decision.

Decision Manager. *Decision Manager* uses a dynamic offloading model to decide at runtime weather the service will be offloaded to the Cloud or executed locally on the mobile. First it get the network bandwidth, then it read stored data about execution time from the *execution profile* for the three running scenarios. Finally it uses the offloading model algorithm described in Section 3.2 to make an offloading decision. When it decides to run the application locally or remotely, it calls *Offloading Manager* which is reasonable for service execution.

Cloud Manager. *Cloud Manager* is reasonable for service execution on the Cloud. In the first run it receives the Jar file which contains the remote implementation of all application services from *Offloading Manager* and install it the Cloud. At any time when the *Offloading Manager* try to call service from the Cloud, *Cloud Manager* receive all required data to execute the service, execute it and return the result to *Offloading Manager.*

3.2   Offloading Model

When an activity invokes a method of a service, the Android IPC mechanism directs this call through the proxy and the kernel to the stub. In normal Android application the stub invokes the local implementation of the method and then returns the result to the proxy. When using proposed framework the Android application uses dynamic offloading model which depend on service execution time to evaluate whether it is beneficial to offload the method to run on the Cloud or executing it locally on mobile. This offloading model takes his decision using two cases.

Case 1: Good Network Bandwidth. Let T_{Mobile} the time to execute service on the mobile, T_{Net} the time to send service data over network and T_{Cloud} the time to execute the service on the Cloud. In this case the framework calculates the time to offload and execute the service remotely on the Cloud T_{Remote}.

$$T_{Remote} = (T_{Net} + T_{Cloud}) \tag{1}$$

Offloading of service task is beneficial if the following is true.

$$T_{Mobile} > T_{Remote} \tag{2}$$

Note that when network bandwidth is high, the T_{Net} will be small which lead to minimize T_{Remote}. Minimizing this time mean that the time to execute service remotely becomes small. So there is a lot of chance that the service will to be executed on the Cloud.

Case 2: Low Network Bandwidth. In low bandwidth network the time to send data over network will increased compared to this time in high bandwidth. Increasing this time mean that the chance to offload the service become small compared to the time to execute service locally. Proposed framework can minimize this time by compressing data before sending over network to the Cloud, so minimize the overall time to execute service remotely. Finally this leads to increase the chance to execute service on the Cloud. Let $T_{Compress}$ the time to compress service data before sending to the Cloud. As in this case data needed to be compressed before sending to the Cloud, so the framework calculate the time to offload and execute the service remotely on the Cloud T_{Remote}.

$$T_{Remote} = (T_{Compress} + T_{Net} + T_{Cloud}) \tag{3}$$

Offloading of service task is beneficial the following equation is true.

$$T_{Mobile} > T_{Remote} - T_{Th} \tag{4}$$

Note that T_{Th} is threshold the user set based on his acceptable difference between two times. Fig.2. shows the flowchart of the framework offloading algorithm.

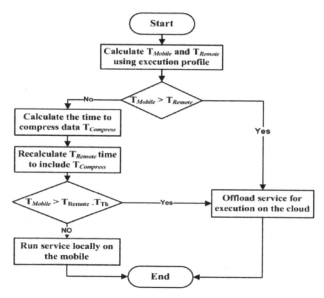

Fig. 2. Workflow of offloading model

3.3   Cloud Side

Cloud Manager is written with pure Java, so any application can offload its computation to any resource running a Java Virtual machine, either being machines in a commercial Cloud such as Amazon EC2[14] or private Cloud such as laptops and desktops. *Cloud Manager* handles all offloading requests from the clients, installation of offloaded services and their initialization, libraries needed and. Finally *Cloud Manager* invokes services when *Offloading Manager* needs to call them. Note that at first run of user application it sends the jar file created by *Jar Creator* to the Cloud.so all mobile services become available for execution on the Cloud.

3.4   Integration into Build Process

In any Android application the connection between the activities and AIDL services processed as follow: When an activity needs to invoke a method in a service, it makes call to the matching method in the proxy. The proxy is responsible for connecting to service to call the need method. The proxy doesn't connect to service directly but, it connects to stub which call the local service and return the result to the proxy. The proxy takes this result and passes it to the caller activity. The framework is deployed in the application layer without modifying the underlying Android platform. The framework provided three Eclipse builders that can be inserted into an Android project's build configuration in Eclipse.

Stub Modifier. The first builder is called the *Stub Modifier* and has to be invoked after the Android Pre Compiler, but before the Java Builder. The *Stub Modifier* will rewrite the generated Stub for each AIDL interface, so that at runtime it connected to the *Decision Manager* to take offloading decision whether a method will be invoked locally on mobile or remotely on the Cloud.

Remote Creator. The second builder called *Remote Creator* used to derive a dummy remote implementation from the available AIDL interface for each service. Now the application with two copies of a service during the build process: i) the first copy of the service added by Android called the local service that executes on the mobile. -ii) the second copy of the service added by framework using Remote Creator and contains the same implementation as the local services and called remote service. This second copy will be executed on the Cloud, so developer can change its implementation to use all Cloud resources like parallel processing

Jar Creator. The third builder called *Jar Creator* used to build a Java Archive File (jar) which contains the remote implementation. This jar file will be installed on the Cloud. The *Remote Creator* and the *Jar Creator* have to be invoked after the Java Builder, but before the Package Builder, so that the jar will be part of the Android Package file that results from the build process.

3.5   Communication: IBIS

In order to execute methods on a remote resource, the phone has to communicate with the Cloud resource using Ibis communication middleware. Ibis offered a simple and effective interface that abstracts from the actual used network, being Wi-Fi, Cellular or Bluetooth [13]. The Ibis communication middleware is an open source scientific software package developed for high performance distributed computing in Java. Since it is written in Java, it can also run on Android devices. The Ibis middleware consists of two subsystems, the Ibis Distributed Deployment System, which deploys applications on remote resources and the Ibis High-Performance Programming System, which handles the communication between the individual parts of a distributed application. The framework has been implemented on top of the Ibis High Performance Programming System, which offered an interface for distributed applications [7].

4   Simulation Studies

To evaluate the proposed framework, a face detection application was used. It's an application that allow user to select image from gallery or take any person photo, then the application execute face detection service locally on mobile or remotely on the Cloud using framework offloading model. After that detection service return an array of all detected faces. Finally the application use this array to draws a rectangle around each detected face as shown in Fig.3. This application uses JavaCV library to detect image faces. JavaCV is a wrapper that allows accessing the OpenCV library directly from within Java Virtual Machine (JVM) and Android platform.

Fig. 3. Screenshot of face detection application

4.1   Simulation Setup

Hardware. On the mobile side a Samsung Galaxy S Advance GT-I9070 mobile was used. The mobile uses Android operating system in version 4.1.2, integrates with Wi-Fi interface, and a battery capacity of 1500mAh. It has CPU with 1 GHz, 1.97 GB system storage and 3.92 GB USB storage at 3.7 volts. On the Cloud side a laptop with a core I3 2.13 processor, 4 Giga ram acted as a Cloud provider. We evaluate the execution time, power consumption and CPU consumption for our application. To measure the power consumption and CPU consumption, a software called *little eye V2.4.0.0* is used [15].

4.2   Result and Discussion

Five images were used in the evaluation of the face detection application. In each experiment, the application was evaluated in three scenarios; the first one represents the execution of the face detection service on the mobile device, the second represents the offloading of the service for execution on the Cloud and the third represents compressing data before offloading the service for execution on the Cloud.

Fig.4. shows the execution time in the three scenarios in low bandwidth networks. The x -axis show the size of the images in kilo bytes and the y-axis show the processing time in seconds. It can be easily noted that the execution time on the Cloud without compression is greater than execution time on mobile and on the Cloud with compression as it takes more time to transfer data through the low bandwidth network. For example, the image with size 9830.4 kb takes about 12 seconds when executed on Cloud while it takes about 7 second when executed on the mobile. Similarly, as the image size increases, the execution time on the Cloud without compression increases compared to the other two scenarios. It also noted that the execution time on the mobile is nearly equal to the compression scenario. For example, the image with size 9830.4 kb takes about 7 second when executed on the mobile and almost the same when offloaded on the Cloud with compression. From this result, we conclude that compressing data and offloading it will give the same performance as processing the requested service on the mobile; nevertheless it will save the mobile resources.

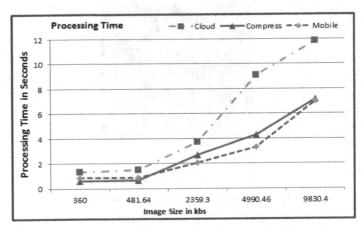

Fig. 4. Processing Time of the application under the three scenarios

Fig.5. shows the CPU consumption percentage in the three scenarios. The results demonstrate the aggressive consumption of the mobile resources in case of executing such heavy service locally on the mobile, while the efficiency of offloading service to save such resources. For example an image with size 9830.4 kb consumes about 48% of the mobile CPU in the first scenario while consuming 10% and 16% in the second and third scenarios, respectively. It also noted that the execution of face detection service on mobile consumed about 34% of CPU on average, while this percentage is minimized to 12.45% in the compression offloading scenario and 7.4 % in the offloading scenario without compression. From this result, it can be concluded that in low bandwidth networks if the user priority is to save the mobile CPU consumption, then it's better to offload service to the Cloud with or without compression.

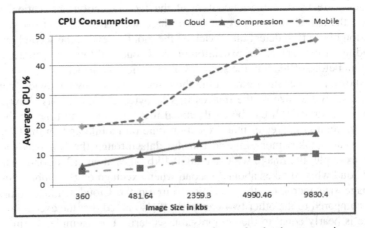

Fig. 5. CPU Consumption of the application under the three scenarios

Fig.6. describes the power consumption in the three scenarios, respectively. The results match well with the conclusion of the previous one; offloading data to the Cloud or compressing data and then offloading to the Cloud is a better choice in case of heavy services if the network bandwidth is low.

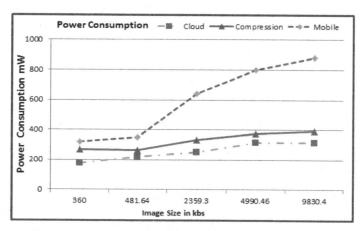

Fig. 6. Power Consumption of the application under the three scenarios

In general, the extensive simulation studies report that in low bandwidth network it is better to compress the data before offloading to the Cloud. Moreover, it is worth to note that the proposed framework supports automatic offloading of multiple Android services thanks to the integration with the popular open source Android framework and the Eclipse development tool. The proposed framework help the developers to easily adds offloading components automatically and efficiently solve a group of drawbacks in the current available techniques. For example, by adopting a runtime offloading model, our framework is smarter than Cuckoo [7] that uses a very simple heuristic approach to always send services to be executed on Cloud without any decision. With respect to the solutions provided by [1], [5, 6], [8, 9, 10] our framework is still better as it overcome offloading in low bandwidth.

5   Conclusion

This paper proposed a framework that helps smartphone to handle heavy applications especially in low bandwidth networks. In this framework, any mobile application is divided into a group of services, and then each of these services are either executed locally on the mobile or remotely on the Cloud with or without compression using a novel dynamic offloading decision model. The extensive simulation studies report the ability of the proposed framework to efficiently utilize the available smartphone's resources based on a smart offloading model that save mobile resources especially in low bandwidth networks.

Our future work will focus on enabling parallelization of the offloaded services also we can take some others metrics in our consideration when taking offloading decision like memory consumption.

References

1. Kumar, K., Lu, Y.H.: Cloud Computing for Mobile Users: Can Offloading Computation Save Energy? Computer 43(4), 51–56 (2010), doi:10.1109/MC(2010)
2. Mell, P., Grance, T.: The NIST definition of Cloud computing. NIST Special Publication 800(145), 7 (2011)
3. Kovachev, D., Cao, Y., Klamma, R.: Mobile Cloud Computing: A Comparison of Application Models. In: Computing Research Repository, abs-1107-4940 (2011)
4. Khan, A.R., Othman, M., Madani, S.A., Khan, S.U.: A Survey of Mobile Cloud Computing Application Models. IEEE Communications Surveys & Tutorials 16(1), 393–413 (2014)
5. Zhang, X., Kunjithapatham, A., Jeong, S., Gibbs, S.: Towards an elastic application model for augmenting the computing capabilities of mobile devices with Cloud computing. Mobile Networks and Applications 16(3), 270–284 (2011)
6. Kovachev, D., Yu, T., Klamma, R.: Adaptive Computation Offloading from Mobile Devices into the Cloud. In: 2012 IEEE 10th International Symposium on Parallel and Distributed Processing with Applications (ISPA), July 10-13, pp. 784–791 (2012)
7. Kemp, R., Palmer, N., Kielmann, T., Bal, H.: Cuckoo: A Computation Offloading Framework for Smartphones. Mobile Computing, Applications, and Services 76, 59–79 (2012)
8. Chen, E., Ogata, S., Horikawa, K.: Offloading Android applications to the Cloud without customizing Android. In: IEEE International Conference on Pervasive Computing and Communications Workshops, March 19-23, pp. 788–793 (2012)
9. Namboodiri, V., Ghose, T.: To Cloud or not to Cloud: A mobile device perspective on energy consumption of applications. In: 2012 IEEE International Symposium on World of Wireless, Mobile and Multimedia Networks (WoWMoM), June 25-28, pp. 1–9 (2012)
10. Xia, F., Ding, F., Li, J., Kong, X., Yang, L.T., Ma, J.: Phone2Cloud Exploiting computation offloading for energy saving on smartphones in mobile Cloud computing. Information Systems Frontiers 16(1), 95–111 (2014)
11. Shiraz, M., Gani, A., Khokhar, R.H., Buyya, R.: A Review on Distributed Application Processing Frameworks in Smart Mobile Devices for Mobile Cloud Computing. IEEE Communications Surveys & Tutorials 15(3), 1294–1313 (2013)
12. Android Developer AIDL,
    http://developer.Android.com/guide/components/aidl.html
13. Nieuwpoort, R.V.V., Maassen, J., Hofman, R., Kielmann, T., Bal, H.E.: Ibis: an Efficient Java-based Grid Programming Environment. In: Joint ACM Java Grande - ISCOPE 2002 Conference, pp. 18–27 (2002)
14. Amazon Elastic Computing, http://aws.amazon.com/ec2/
15. Little eye, http://www.littleeye.co/
16. Amazon s3, http://status.aws.amazon.com/s3-20080720.html
17. Dropbox, http://www.dropbox.com

Digital Pathological Services Capability Framework

Ammar Adl[1,2], Iman B. Shaheed[3], M.I. Shaalan[3], A.K. Al-Mokaddem[3], and Aboul Ella Hassanien[1,2]

[1] Cairo University, Faculty of Computers and Information, Cairo, Egypt
[2] Scientific Research Group in Egypt (SRGE)
www.egyptscience.net
[3] Department of Pathology, Faculty of veterinary Medicine, Cairo University, Giza, Egypt

Abstract. Pathology digital lab is the modern, flexible and time effective research assistant. The process of creating pathology slides contains five creation steps from the tissue samples collection till clearing and staining stage. The reservation and sharing of such slides using classical models limit the ability of pathologists to benefit from important and rare slides. The virtual lab with its digital slides conquers those limitations and adds more intelligence to research and diagnosis fields. Having the digital slides, it is easy to save, share, search, apply automatic diagnosis through pattern recognition techniques, getting alerts for new slides and much more. The target of this work is to present the virtual lab design with its functionalities by explaining the glass slides creation process and then digitalize through scanners and the digital lab platform.

1   Introduction

Tissue-based diagnosis is characterized by its high specificity and sensitivity when compared to other diagnostic medical disciplines such as radiology, clinical pathology, or endoscopic imaging techniques. It can be considered as the focus of all modern developments in medicine, which can be subsumed under the headlines of molecular biology including molecular genetics, electronic or digital communication and computerized medicine [1,2]. Telecommunication techniques have also become of great interest to diagnostic pathologists.

Digital microscopy creates the digital representation of the whole microscopic slides at decent quality, which can be dynamically viewed, navigated and magnified through the computer screen, and shared with others to eliminate any spatial or temporal limitations. Digital microscopy offers unique features which are not available for conventional optical microscopy. Digital slides of tissue sections or cells can be shared, through local networks or internet, with many pathologists worldwide. That unlimited access to slides makes digital microscopy an efficient tool for tele-pathology yielding primary diagnosis, tele-consultation for second opinion, graduate teaching, continuous education, proficiency testing, external quality assurance and inter-laboratory process validation. The aim of this project

A.E. Hassanien et al. (Eds.): AMLTA 2014, CCIS 488, pp. 109–118, 2014.

110   A. Adl et al.

is to establish and prepare the first pathology digital lab in the Middle East. This
project will enhance the education process and improve the diagnosis through
tele-consultant [3].

1.1 Problem Statement and Objectives

Nowadays, pathology courses are taught using traditional methods in all facul-
ties of veterinary medicine in Egypt. It is still using blackboards, self-illuminated
light microscopes monitor presentations. Those methods make it hard to illustrate pathology efficiently due to many reasons: (1) The large number of students in one class, (2) The lack of adequate number of slides and microscopes,
(3) Glass slides get lost, break, and fade over time, (4) Rare glass slides cannot
be duplicated and made available to all students, (5) Student glass slides are
often incomplete and not identical, which creates discrepancies in testing and
scores, and (6) Independent, non-laboratory study time is limited by access to
glass slides and microscopes.

The objectives of this study is to establish and prepare the first pathology
digital lab in the Middle East, which enhances the education process and improves how microscopic material is presented to students. It will increase the
efficiency and accessibility of virtual slides compared to traditional microscopy.
Students will be skilled in presenting morphologic findings on "slides" in small
group case analysis sessions. Also, it will be a new tool for delivery of pathology
slides for teaching and diagnostic purposes [4].

2 Slides Creation Process

Slides creation is achieved through five main stages; First is the tissue samples collection and/or cell isolation. Then comes the tissue processing phase including fixation, dehydration, Infiltration and embedding in paraffin. The third
stage is the sectioning with microtome followed by mounting on microscope
slides. Finally comes the clearing and staining stage. The process is illustrated in
Figure 1.

2.1 Tissue Samples Collection and/or Cells Isolation

In this stage, tissue specimens were collected from freshly died animals or from
abattoir to be fixed later [5]. Cell lines or stem cells are isolated, incubated in
suitable media and examined for viability.

2.2 Tissue Processing

Fixation. The fixation process must be started as quickly as possible after the
removal of the sample [6]. It starts by filling a labeled vial about 2/3 full with
the fixative. Then placing the tissue specimen in cassettes and then into the vial
containing the fixative. Finally storing the vial on the bench top at the work
area.

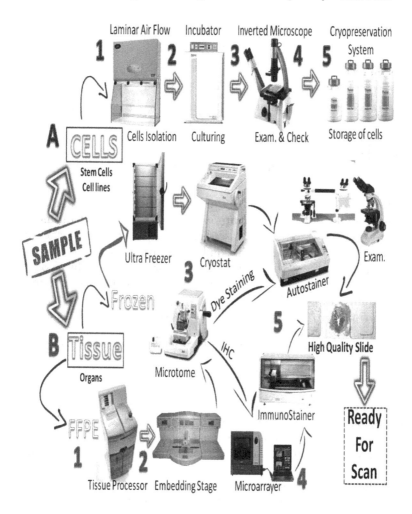

Fig. 1. Pathology Lab Process

Dehydration. Dehydration is the process of removing water from the tissue block. It occurs after the fixation process. Ethyl alcohol is a favored reagent because it is miscible in paraffin. The tissue must be dehydrated slowly to avoid tissue distortion. Rather, dehydration is carried out in a slow, step-wise manner by passing the tissue block through a series of solutions of increasing alcohol concentration. In this way the water is fully leached out and replaced with alcohol. The needed supplies (per group of two) are 70%, 85%, 95% and 100% ethyl alcohol solutions. The process goes as following:

- Pass tissue capsules through the series of solutions.
- For each step, place approximately 50 ml a fresh ethyl alcohol solution into plastic vial.

– Decant the used solutions into the organic solvent disposal container in the hood.

And the step incubation period is:

1. 70% ethyl alcohol for 1 hour
2. 70% ethyl alcohol for 1 hour
3. 85% ethyl alcohol for 1 hour
4. 95% ethyl alcohol for 1 hour
5. 100% ethyl alcohol for 1 hour
6. 100% ethyl alcohol for 1 hour

Infiltration and Embedding in Paraffin. Prior to sectioning, the tissue block must be infiltrated with a material that acts as support during the sectioning process. For the method described here, paraffin serves this purpose. During infiltration, the paraffin will equilibrate within the tissue block, eventually occupying all of the space in the tissue that originally held by alcohol. After infiltration, the tissue is allowed to solidify in a mold, embedded within a small cube of paraffin. The needed supplies for this process are melted paraffin in metal pitchers Petri plates, 4 base molds. The infiltration process goes as following:

– Discard the 100% alcohol from the last dehydration step, and fill the vial about full with melted paraffin.
– Allow tissue to equilibrate for 1 hour in an incubator set at 85C.

The equilibration process is allowing a solution to reach a stable concentration within a tissue. Thus, for example, after 1 hour the alcohol will have reached 70% within the tissue block. The process goes as follows: (1) Pour the paraffin into the container labeled for paraffin disposal, and (2) Repeat step 1 using fresh melted paraffin.

Finally the embedding process goes by placing base-pieces for two embedding molds in a plastic Petri plate label the plate along the edge with some name. Then, decant the paraffin from the second infiltration step into the waste container. The process should be performed quickly but carefully. Use forceps to transfer the tissue blocks to the well of separate base mold, snap the base of tissue cassette into the base mold and then fill the mold with paraffin. Allow the paraffin to solidify at room temperature. If the paraffin begins to solidify homogeneously around the tissue block, allow the paraffin in the base mold to melt in the incubator, and then allow it to solidify.

2.3   Sectioning

Sectioning is accomplished by using a cutting apparatus called a microtome. The microtome will drive a knife across the surface of the paraffin cube and produce a series of thin sections of very precise thickness. The objective is to produce a continuous "ribbon" of sections adhering to one another by their leading and trailing edges. The thickness of the sections can be preset, and a thickness between 5 - 10 m is optimal for viewing with a light microscope. The sections can then be mounted on individual microscope slides.

Preparation and mounting of the embedded tissue block on the microtome is very important to successful sectioning. The paraffin surrounding the tissue block must be trimmed firstly, and then secured to a holder which is then mounted on the microtome. Frozen samples can also be cut but using cryostat instead of microtome.

2.4   Mounting of Sections on Microscope Slides

In this procedure, the sections are permanently attached to microscope slides. If "serial" sections are desired, (i.e., sections that reveal sequential layers of the tissue structure) then sectioning must be performed carefully and systematically. The needed supplies (per group of two) are microscope slides, Coplin jar, Slide storage box, Hematoxylin and Eosin stain.

Preparing the Microscope Slides. To prepare the microscope slides, label the 5 microscope slides at one end with some name and tissue type, and number them sequentially from 1 to 5 using a diamond pen. Then wash the microscope slides with soap and water, and rinse free of soap with tap water. Place the slides in a coplin jar and rinse several times with H2O. Finally, handle the slides only by their edges, place the slides in your slide storage box, and let them dry.

Mounting Sections on Microscope Slides. It will be apparent during sectioning that the sections are not perfectly flat, but rather slightly crinkled. This is normal, and the sections will become flattened by floating them on water held at 45C. The solution also contains an adhesive, which causes the tissue section to bind to the slide. The process goes as follows:

1. Carefully transfer the sections to a solution held in a 45C water bath. Within a few seconds you should see the sections flatten and the wrinkles disappear.
2. Dip a clean microscope slide into the adhesive solution, and slowly pull it upward, out of the solution, allowing sections to adhere to the surface. Make sure that the slide is oriented with the label facing upward.
3. Dry the bottom of the slide and carefully blot excess adhesive from around the sections.
4. Let the slides dry in the storage box.

Tissue microarray has the advantage of getting multiple tissue films on the same slide, thus is cost and time saving [7].

2.5   Clearing and Staining

The clearing process is the removal of paraffin, before a section can be stained. After clearing, only the tissue remains adhering to the slide. Clearing is accomplished by passing the mounted sections through the solvent clearance that dissolves the paraffin. Staining of histological sections allows observation of features otherwise it is not distinguishable.

For routine histological work, it is customary to use two dyes, one that stains certain components a bright color and the other, called the counterstain, that stains other cellular structures a contrasting color. While literally hundreds of staining techniques have been developed, the two most widely used stains for routine work are hematoxylin and eosin. Hematoxylin stains negatively charged structures, such as DNA, a blue color. Eosin imparts a red color to most of the other cell components. To produce permanent staining with hematoxylin, the dye must be oxidized to "hematin", which is achieved by treating the tissue sections with Scott's solution [8].

3 Pathology Virtual Lab: Vision and Functionalities

3.1 Overview

Virtual lab is a computer driven technology depends on image processing and management used to gather and analyze information gathered from a digitally mapped slides. The main part of the digital lab is "Virtual microscopy". It is the process of converting the glass slides into digital slides. Digital slides are high resolution images for glass slides. It can be viewed, managed and shared through a computer software. The software is not concerned only with viewing the slides, but also with applying automatic diagnosis filters and business intelligence algorithms to extend and predict expected results based on a historical database of older experiments. Virtual lab environment contains scanning, viewing, managing, analyzing, integration and sharing. The scanning process results in a scratches free high quality images of the glass slides using advanced scanners. The viewing process is the process of presenting, searching and using digital slides over networks and internet. The managing process includes intelligent retrieval, archiving and data maintaining functions. The analyzing process is concerned with pattern recognition and visual searching tools combined with machine learning techniques used to present smart diagnosis results and other medically important analysis. The integration and sharing processes are concerned with integrating the lab with active medical systems, and sharing slides among different labs or even pathologists. Figure 2 shows virtual lab overview.

3.2 Vision

Pathology 2.0 is the term denoting the integration of digital pathology lab with regular work flow systems to replace the conventional diagnosis process [9]. It enables rare cases consultations in a matter of hours [10], and if it is published online, it will avail the opportunity to get experts support around the world. The improvements in speed of scanning, compression algorithms and high resolutions displays, even mobile high resolution screens, give a promising edge for the development of strong fully fledged pathology virtual lab.

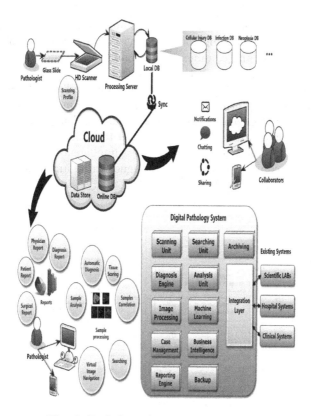

Fig. 2. Pathological Digital Lab Overview

3.3   Functionalities

Pathology virtual lab has many functionalities that are categorized in grouping categories. Scanning category contains all the functions related to slide scanning, scanner profiles, batch scanning and bar-code information. Management category contains slide management, tracking features, image exportation, annotation saving. Integration category contains interfacing with external databases and systems. Sharing category contains sharing with collaborators, chatting, real-time case viewing. Analysis category contains image analysis, diagnosis, scoring, auto focus and control detection. Workflow category contains archiving images, remote reading, remote consultation, case history viewing and searching. Finally the reports category contains structured and unstructured reports, final diagnosis reports, surgical pathology reports and molecular pathology report.

3.4   Design

Pathology virtual lab component diagram is introduced in Figure 3. The diagram presents 8 main packages for the components. The core package contains the components presenting the main business components for a virtual lab.

It contains scanner component with its profiler and scanning units. Also, the image viewer component responsible for image management functionalities such as zooming, tissue markers and comments. File manager is responsible for retrieval system functions. Case manager is the component responsible for managing and profiling a slide or group of slides using scoring unit and sample analysis component. The last component of the code package is the mobile client for better engagement with updates and on the fly notifications. Database package holds the differen types of database components used to build the pathology virtual labs. Other important packages are the business intelligence and reporting packages which include components for smart diagnosis and analysis system with a decision support reporting tools.

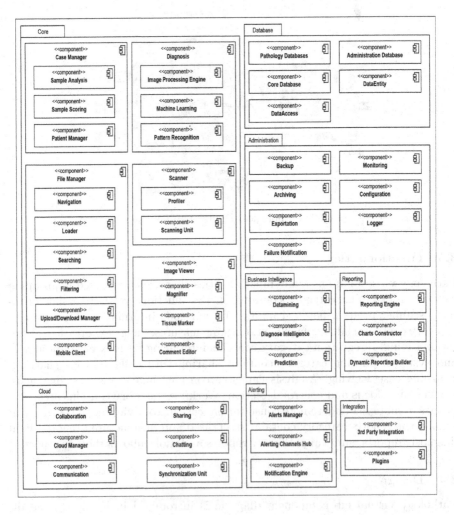

Fig. 3. Pathology Lab Component Diagram

Collaboration is served in the component view through the integration, alerting and cloud packages containing components for real-time alerting and cloud chatting and synchronization. With the integration plugins, pathologists can collaborate with already existing medical systems and share recent results actively with such systems. Finally the administration package containing the housekeeping components concerned with system maintenance and configurations.

4 Conclusion and Future Works

Digital Pathology describes the creation, viewing, management, sharing, analysis, and interpretation of digital images of glass slides and includes workflow considerations unique to a digital imaging environment. Nowadays digital pathology is growing rapidly as a method of viewing microscopic glass slides virtually. Virtual slides solved a number of problems in their learning, while providing good to excellent image quality. We believe that the use of high-quality learning resources such as virtual slides can ensure that microscopic examination of tissues remains both meaningful and interesting. Digital images can be used for a wide variety of purposes, from clinical practice, to continuous education, proficiency testing, primary diagnosis, tele-consultation, quality assurance and research. It enables diagnostic collaboration and the promise of workflow enhancements/efficiencies. As far we known this technology did not exist in Middle East labs either in medicine or veterinary. Establishment of digital lab will introduce these advanced technology and vital new teaching tool to Egypt. It will be the first pioneer lab in Middle East and Africa for digital pathology.

In this work, the importance of pathology virtual lab and digital pathology was highlighted. After creating the glass slides, it is scanned with high quality scanners and then the high resolution images, sometimes its size reaches Gigabytes, are sent to media servers where the data is saved in network repositories. A pathologist at this point just needs an account to login and view, search, apply pattern recognition diagnosis filters and much more. Such system have many building blocks categorized into these main three categories; Hardware, Databases and Software. The hardware used to construct the lab mainly contains scanners and high resolution screens, devices used to create the original glass slides are not mentioned here. The databases are the online repositories of the digital slides and the software is the platform that manage those slides. The components inside each of those categories don't just include imagery and file management systems, but also contain decision support business intelligence platform with detailed reports about slides analysis. Availing data in this flexible way helps pathologists to make time efficient advances and collaborate their efforts.

References

1. Kayser, K., Kayser, G., Radziszowski, D., Oehmann, A.: From telepathology to virtual pathology institution: the new world of digital pathology. Rom. J. Morphol. Embryol. 45, 3–9 (1999)

2. Kayser, K., Borkenfeld, S., Kayser, G.: Recent development and perspectives of virtual slides (vs) and telepathology in europe. Diagnostic Pathology 8(suppl. 1), S2 (2013)
3. Krenacs, T., Zsakovics, I., Micsik, T., Fonyad, L., Varga, S.V., Ficsor, L., Kiszler, G., Molnar, B.: Digital microscopy: the upcoming revolution in histopathology teaching, diagnostics, research and quality assurance. Microscopy: Science, Technology, Applications and Education 2, 965–977 (2010)
4. Hamilton, P.W., Wang, Y., McCullough, S.J.: Virtual microscopy and digital pathology in training and education. Apmis 120(4), 305–315 (2012)
5. Wong, B., Edward, O., Warren, E.: Tissue sampling methods and standards for vertebrate genomics. Giga Science 1(8), 1–12 (2012)
6. Wilson, L.B.: A method for the rapid preparation of fresh tissues for the microscope. Journal of the American Medical Association 45(23), 1737–1737 (1905)
7. Jawhar, N.M.: Tissue microarray: a rapidly evolving diagnostic and research tool. Annals of Saudi Medicine 29(2), 123 (2009)
8. Bancroft, J.D., Gamble, M.: Theory and practice of histological techniques. Elsevier Health Sciences (2008)
9. Stathonikos, N., Veta, M., Huisman, A., van Diest, P.J.: Going fully digital: Perspective of a dutch academic pathology lab. Journal of Pathology Informatics 4 (2013)
10. Al-Janabi, S., Huisman, A., Van Diest, P.J.: Digital pathology: current status and future perspectives. Histopathology 61(1), 1–9 (2012)

Part III
Machine Learning
in Watermarking / Authentication
and Virtual Machine

Highly Secured Multilayered Motion Vector Watermarking

Suvojit Acharjee[1], Sayan Chakraborty[2], Sourav Samanta[3], Ahmad Taher Azar[4],
Aboul Ella Hassanien[5], and Nilanjan Dey[2]

[1] Dept. of ETCE, Jadavpur University, Kolkata, West Bengal, India
[2] Dept. of CSE, Bengal College of Engineering & Technology, Durgapur, West Bengal, India
[3] Dept. of CSE, University Institute of Technology, Burdwan, West Bengal, India
[4] Faculty of Computers and Information, Benha University, Egypt
[5] Faculty of Computers and Information, Cairo University, Egypt

Abstract. With the recent development in multimedia, video has become a powerful medium of information. To exploit the temporal redundancy during video compression, motion vector estimation is required. Now-a-days internet and digital media has become very popular, which made data authentication and data security a challenging task. Digital watermarking was introduced to provide data authentication. Though, it was not enough to prevent the unauthorized access of data by third parties. To prevent unauthorized access, data is encrypted using a secret key known only to the user. This process is known as cryptography. In this paper, an algorithm has been proposed to embed the watermark inside calculated motion vector. The position of watermark bit inside the motion vector will depend on a key provided by user. The correlation values between the four original and recovered experimental video frames are 0.97, 0.98, 0.98 and 0.91 respectively whereas structural similarity index metric (SSIM) between them are 0.97, 0.87, 0.90 and 0.68, respectively. The high correlation values and SSIM shows the effectiveness of the proposed method.

Keywords: Watermarking, Cryptography, encryption, motion vector, medical videos, Watermarked motion vector.

1 Introduction

Security refers to the degree of resistance. As described by Institute for Security and Open Methodologies (ISECOM), in the open source security testing methodology manual (OSSTMM), security means "a form of protection where a separation is created between the assets and the threat." Recently, the growth of digital media and connectivity made data security the most challenging task. Now-a-days from medical reports to flight tickets everything is available electronically. The data sharing has also become very easy with the advancement of internet and digital media. Such advancement leads to the increase of risk in unauthorized access, malicious attack on data and third parties uses. To provide authentication of the digital data, digital watermarking [1] was introduced. Digital watermarking [2, 3] refers to the process of embedding [4] a user signal inside the data to

A.E. Hassanien et al. (Eds.): AMLTA 2014, CCIS 488, pp. 121–134, 2014.

prove the ownership of the user. The secret signal may be a digital signature, picture or voice of the user etc. The watermarked signal can be visible or invisible over the data. Watermarking [5] can be applied to any type of signal (i.e. image, video etc.), keeping in mind that the technique must not change the quality of the data. While many different watermarking [6] algorithms were used with different performances, watermarking provided content protection and digital rights management, making it the most popular solution towards digital data and author identity protection. Application of watermarking is very wide which includes copyright protection and control, ownership identification etc. Some of the previous works in this domain includes a new digital watermarking algorithm proposed by A. Tirkel *et al.* [7]. This work included watermarking on bit plane manipulation of the LSB, which offered easy and rapid decoding. In 1999, Neil F. Johnson [8] *et al.* introduced a method of recovering watermark from images after attacks on those images. This work included attacks on some techniques against watermark, in such a way that the secret information hidden cannot be recovered after the attack. Later, Matheson [9] *et al.* discussed digital watermarking and its use on secured digital content. In 2001, Lu *et al.* proposed a watermarking [10] scheme for image authentication and protection purpose. In this work, two watermarks were hidden using the host image's quantization of wavelet coefficients and the watermarks were extracted without the help of original image. In 2009, Kumari [11] *et al.* proposed a watermarking technique which was applied on gray level images. The watermark was embedding using LSB and it did not affect the gray value of the image. This approach offered high robustness and security. Maity [12] *et al.* proposed a technique on spatial domain watermarking. The required block was selected with the help of the variance of the block. As a result, average brightness of the watermarked varied a little.

But watermarking [13] cannot prevent the unauthorized user from viewing the data which is a serious threat to privacy of the user. The recent news reports show the breach in privacy of users by constant monitoring of different national agency. The best way to address this issue is to encrypt the data using secret key which will be known only to the user. This concept is known as cryptography. Previously, in 1976 Diffie *et al.* gave new directions [14] in cryptography. In this work, they suggested different paths to fix these currently open issues on network threats. It also discussed the theories of communication and computation and their role in providing various tools to solve cryptographic problems. In 1990, Omura proposed novel applications of cryptography [15] in digital communications. In this work, they discussed various cryptography tools and mechanism to improve the data security. In 2003, Gabidulin *et al.* introduced reducible rank codes [16] and also discussed how they can be useful to cryptography. In 2007, Kaps *et al.* proposed cryptography [17] on a speck of dust where they discussed the privacy and security needed for small and tiny wireless and RFID devices, and solved the cryptographic problems on such devices.

In the domain of multimedia, video is one of the most powerful medium of information. It has a wide range of application from television signal to medical diagnostics. Apart from providing entertainment, video also provides us the important information of different continents of the world. Also, video contents can help us for proper diagnostic of disease. Another use of video can be keeping an eye on the suspicious activity in a crowded area through surveillance. The application of the video is increasing day by day. Video refers to a series of image taken in a very short time which demands large memory or bandwidth to store or transfer the video. This calls for video compression, which is a very important step for proper organization of video data. Video compression generally exploits the temporal redundancy between two frames. Moving pictures expert group (MPEG) provides the standard for video compression as described in fig.1. Motion vector [18] estimation is the most important and computationally expensive part of video compression. Motion vector examines the movement of the object from previous frame to present fame. Motion vector has a vertical and a horizontal component which describes the movement of the object.

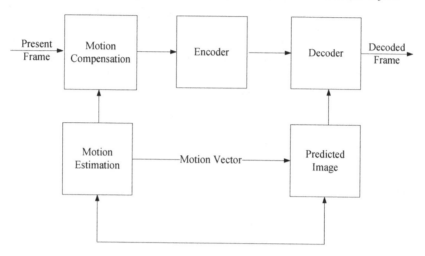

Fig. 1. Block Diagram of MPEG standard

Some of the previous work on motion vector includes an algorithm proposed by Netravali *et al.* [19] which calculated motion vector for every pixel present in the frame. Later, Bargman *et al.* [20] proposed a fix on the Pel recursive algorithm which is also used on motion vector. Block matching algorithm can follow either square search pattern or diamond search pattern. In another work, Zeng *et al.* [21] discussed the watermarking technique which was done on motion compensated image. The current work proposes an algorithm to embed the watermark inside calculated motion vector. In section 2, methodology is discussed. Section 3 illustrates the result and discusses them. Paper concludes in section 4.

2   Methodology

Motion vector is the most computationally expensive step in video compression. Inspired by different algorithms for motion vector estimation, the following algorithms are used in this proposed work which calculated the motion vector for every pixel:

2.1   Motion Vector Estimation

The algorithm to perform motion vector estimation is described below:

begin
The pixel whose best match needs to be found is set as centre pixel in current frame and reference frame.

A rectangular search window with a distance d pixel on all sides of the centre pixel in the reference frame is created. d varies from 3 to 15 pixels.

Three numbers between 1 to $((2\times d)+1)^2$ is randomly generated.

The appropriated pixel from the search window in reference frame is picked according to the randomly generated number.

The difference in intensity between these pixels and the centre pixel of current frame is calculated using equation one.

The pixel value with the smallest intensity difference from the centre pixel is selected, which is the nearest pixel from the randomly selected pixels.

if the pixel difference = negligible and lies inside threshold range.

The motion vector is calculated,

else,

The same search location is kept.

The process is repeated.

end if

Post process.

end

2.2   Selection of Key for Encryption

For selection of the encryption key, the following steps are considered.

The fifth, sixth and seventh bits are unused in binary representation of motion vector (ranges between -15 to 15).

begin
All the six free bits which includes both vertical and horizontal components are numbered.

A key of n number of digits (valued between 1 to 6) is selected.

end

2.3   Watermark Embedding

begin
Position of the watermark is selected based upon the key.

for i=1:n, n = number of bits

n number of bit is shifted from the watermark image pixel to embed the watermark.

Bitwise OR operation between motion vector and bitwise shifted watermark pixel is performed.

end for.

The new motion vector is saved.

end

2.4   Watermark Extraction

begin
Position (n) of the watermark bit is selected.
 Bitwise and operation between n bits shifted 1 and the motion vector is performed.
  the recovered watermark image is compared with the original watermark image.
  if recovered watermark = original watermark
   then
    key is correct.
  else
   Repeat
  end if
The frame is recovered.
end

3   Explanation of the Proposed Method

Motion vector estimation is performed by the proposed method. The pixel whose best match needs to be found has been set as the centre pixel in current frame and reference frame. A rectangular search window with a distance d pixel has been created on all sides of the centre pixel in the reference frame. The value of d can vary from 3 to 15 pixels. For this work, d has been set to 15. It should be noted that for better result d should have high value. Although, for faster computation d must be lower. Further, the system randomly generated three numbers between 1 to $((2 \times d)+1)^2$. In this work, the range is from 1 to 961. In fig. 2 the search area is shown for d=3. The appropriated pixel from the search window in reference frame has been picked according to the randomly generated number. The number of each pixel is shown in fig. 2. Then the absolute difference in intensity between these pixels and the centre pixel of current frame has been calculated using equation 1. Later, the pixel value with the smallest intensity difference with centre pixel has been selected. The current system considered this pixel as the nearest pixel from the randomly selected pixels. If the pixel difference is within the threshold value such that even a human eye cannot differentiate the difference, then the pixel is considered as the best match and calculation of the motion vector is done. Else, the system kept the same search location and repeated the same process again. If the proposed automated system cannot find the best match, then the process has to be repeated until it reaches maximum iteration. The maximum iteration depends on the size of the search window. Number of iteration is dependent on search window. Finally, the proposed system also demonstrates that if the system cannot find the motion vector, then the motion vector needs to be set as 0 for both horizontal and vertical component.

$$diff = |\, cur(i, j) - Ref(i, j)\,|$$

$$(1)$$

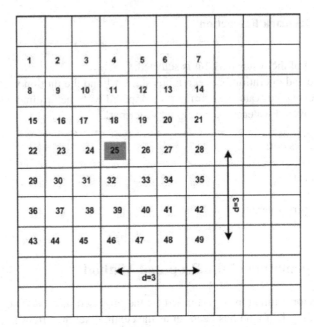

Fig. 2. Search window for d=3

In this work, a binary image has been used for watermarking. The size of the image has been set such that it is equal to the size of the frame. As the watermark image is binary, only a single bit is enough to represent the pixel of watermark image. The range of motion vector is between -15 to 15 which signified that motion vector can be represented using four LSB. The MSB has been used to represent the sign. Hence, fifth, sixth and seventh bit of vertical and horizontal component remains unused. Thus, the six bits were utilized to embed watermark. In fig. 3, the bit structure of the vertical and horizontal component of motion vector is illustrated. Here S represents sign bit, U represents unused bit and M is the motion vector.

Vertical	S	U	U	U	M	M	M	M
Horizontal	S	U	U	U	M	M	M	M

Fig. 3. Bit structure of vertical and horizontal components of motion vector

The three unused bits in vertical component and horizontal component has been numbered or ranked accordingly (Fig. 4). Unused bits in vertical component have been numbered with 1, 2 and 3. Unused bits of horizontal component have been numbered as 4, 5, 6.

Vertical	S	3	2	1	M	M	M	M
Horizontal	S	6	5	4	M	M	M	M

Fig. 4. Ranking of unused bits, vertical and Horizontal components of motion vector

The system maintains the key for cryptography with minimum six digits. The strength of the key increases with the length of the key. Every number has been valued inside the range of 1 to 6 as the system only has six bits to represent the watermark image.

Position of the watermark has been selected based upon the key. As an example, if the value of key is 1, then watermark will be embedded in fifth bit of the vertical component. To embed the watermark a bitwise shift operation has been performed by shifting n number of bits on the watermark image pixel. Then a bitwise OR operation has been performed between motion vector and bitwise shifted watermark pixel. Value of the key plays a key role here. For e.g. if key is from 1 to 3 then n=n+4 else n=n+1. If the key ranges within 1 to 3, OR operation needs to be performed between the vertical components of motion vector followed by watermarking using bit shift. If key is beyond the predefined range, OR operation needs to be performed between the horizontal components of motion vector followed by watermarking using bit shift. New motion vector has been saved for transmissions. Position of the watermark bit has been selected using the given key from the user. Later, bitwise and operation has been performed between n bit shifted 1 and the motion vector. Value of n and the motion vector component was decided as described in step 2 of watermark embedding. After the completion of these processes, the watermark image and motion vector has been recovered. It has been observed, if the watermark image is same as the original watermark image then the provided key needs to be declared as correct. This also signifies that the recovered motion vector is correct. In the final step, the frame has been recovered.

4   Result

The proposed algorithm is evaluated using four video sequences. One of the video is an echocardiograph video with 59 frames and length of the video is one second. Rest of the three videos (Table Tennis [22], Foreman [23], and Flower [24]) has been mentioned by MPEG standards as standard test videos. Matlab R2012a has been extensively used to implement the process and to read the videos. Quality of recovered watermark image and recovered frames has been measured using PSNR [25], MSSIM [26] and Pearson's correlation coefficient [25]. Peak signal to noise ratio (PSNR) has been presented in logarithmic decibel scale. PSNR has been very easily defined using mean square error (MSE). MSE [27] can be defined using equation (1) and PSNR can be calculated using equation (2).

$$MSE = \frac{1}{(row \times col)} \sum_{i=1}^{row} \sum_{j=1}^{col} [Orginal(i,j) - recovered(i,j)]^2 \tag{2}$$

$$PSNR = 10 \log_{10} \left(\frac{Maximum\ Intensity^2}{MSE} \right) \tag{3}$$

Pearson's Correlation coefficient can be calculated using equation (3).

$$\rho = \frac{\sum_{i=1}^{row} \sum_{j=1}^{col} \left(Ori(i,j) - \mu_{Ori} \right) \left(Rec(i,j) - \mu_{Rec} \right)}{\sqrt{\left(\sum_{i=1}^{row} \sum_{j=1}^{col} \left(Ori(i,j) - \mu_{Ori} \right)^2 \sum_{i=1}^{row} \sum_{j=1}^{col} \left(Rec(i,j) - \mu_{Rec} \right)^2 \right)}} \tag{4}$$

Structure similarity index metric [26] (SSIM) is a new similarity index metric between two images which can compare the structure of the images. It can be calculated using equation (4).

$$SSIM(O,R) = \frac{\left(2\mu_O\mu_R + K_1\right)\left(2\sigma_{OR} + K_2\right)}{\left(\mu_O^2 + \mu_R^2 + K_1\right)\left(\sigma_O^2 + \sigma_R^2 + K_2\right)} \tag{5}$$

Where μ_O and μ_R are the mean intensity of Original Image O and Recovered Image R. σ_O and σ_R are the variance of the images. K_1 and K_2 are the constant to stabilize the weak denominator. For two identical images SSIM [26] and correlation should be 1 and PSNR should be infinity.

Table 1 shows the average, median and standard deviation of different quality metrics measured between the recovered watermark image and original watermark image in four different videos. From the table, it can be observed that mean value of MSSIM and correlation is 1and the PSNR is infinite which suggests that recovered watermark and original watermark [27, 28, 29] image are identical.

Table 1. Quality evaluation of original watermark and recovered watermark image

Video Name	Echocardiograph	Foreman	Flower	Table Tennis
MSSIM				
Mean	1	1	1	1
Median	1	1	1	1
Standard Deviation	0	0	0	0
Correlation				
Mean	1	1	1	1
Median	1	1	1	1
Standard Deviation	0	0	0	0
PSNR				
Mean	Infinite	Infinite	Infinite	Infinite
Median	Infinite	Infinite	Infinite	Infinite
Standard Deviation	0	0	0	0

Table 2 shows the quality metrics measured between the original frames and recovered frames. It can be observed that the mean correlation and MSSIM are always high with infinite PSNR which suggests that the recovered frames are capable of retaining almost same quality as original frames. From this table it can be also observed that the standard deviation is very low which refers to every recovered frame having almost similar quality with a very low deviation. It proves the consistency of the system.

Table 2. Quality evaluation between original frame and recovered frame

Video Name	Echocardiograph	Foreman	Flower	Table Tennis
MSSIM				
Mean	0.9733	0.8699	0.9020	0.6798
Median	0.9766	0.8954	0.9708	0.6992
Standard Deviation	0.0099	0.0781	0.0855	0.1189
Correlation				
Mean	0.9695	0.9816	0.9761	0.9120
Median	0.9701	0.9928	0.9824	0.9449
Standard Deviation	0.0031	0.0285	0.0158	0.1011
PSNR				
Mean	Infinite	Infinite	Infinite	Infinite
Median	Infinite	Infinite	Infinite	Infinite
Standard Deviation	0	0	0	0

The quality assessment using MSSIM and correlation of the proposed system is reported in fig. 5 and fig. 6. Along x axis it represents different videos in the below sequence:

(i) Echo Cardiograph (ii) Flower (iii) Foreman (iv) Table Tennis

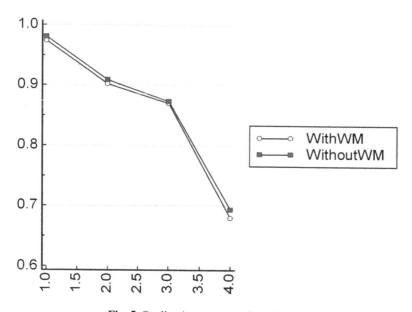

Fig. 5. Quality Assessment using MSSIM

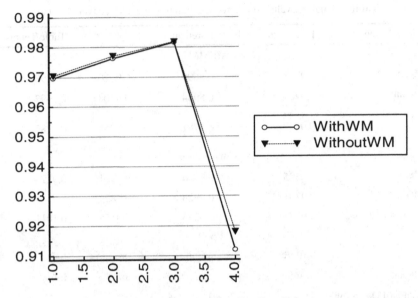

Fig. 6. Quality Assessment using Correlation

Fig. 7 and fig. 8 shows the normal distribution plot of MSSIM and correlation measured between original frames and recovered frames using proposed algorithm for 59 frames of echocardiography video.

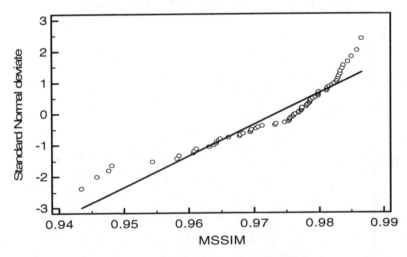

Fig. 7. Normal Distribution plot of MSSIM

Fig. 7 and Fig. 8 illustrates the performance metric MSSIM and correlation remained almost same for every frame which suggests the output of the proposed system is consistent. Also fig. 5 and fig. 6 shows the embedding of the watermark signal in the motion vector [31, 32, 33] did not reduce the quality of the images.

This algorithm also employed an encryption, based on the security key. Key can be provided by user or it may be assigned automatically. The value of every digit of the key should lie in the range of 1 to 6, which means there are 6 options available for a single position of the key. Table 3 shows the number of probable combination of the key with an increase in the digits in the key. When key has 6 digits, the number of probable combination is higher which makes the system far more secure. Hence, the minimum digit requirement of the key is 6. If the length of the key is increased, then it will also make the decoding more challenging.

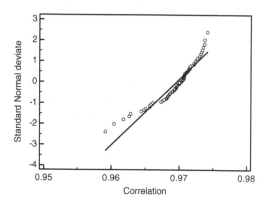

Fig. 8. Normal Distribution plot of Correlation

Table 3. Number of digits in a key and total number of possible combination of the key

Number of Digits in a key	Total number of possible combination of the key
1	6
2	36
3	216
4	1296
5	7776
6	46656
7	279936
8	2239488

During this complete process, the security and authenticity concern of the data is addressed. Thus, an additional layer of security is added in the system.

5   Conclusion

Previously, lot of work has proposed various techniques to embed the watermark signal in compressed video which addressed the authenticity concern of the data. Though, those techniques could not hold the unauthorized access of the video which was a serious threat to privacy. The proposed system aimed to increase the security of the video signal by introducing a key based encryption to decide the position of the watermark signal inside the motion vector. Firstly, every video has been defragmented into multiple image frames on which, motion vector calculation has been performed. Using user provided key, binary watermark has been embedded on the obtained video frames. Thus, the system successfully generated the watermark embedded motion vector. Further, this embedded motion vector has been decoded by user provided key, and motion vector was calculated. This process led to recovery of the video frames which have been post-processed and analyzed. From the visual representation and statistical analysis of the result, it can be said that embedding watermark into the motion vector did not reduce the quality of the output, which establishes the robustness and imperceptibility of this present work. High SSIM and correlation values of recovered frames establish the robustness and imperceptibility of the proposed system. Future work may include the proposed system being implemented using other motion vector estimation algorithms. A performance study can be done using different motion vector estimation algorithm in the proposed system.

References

1. Dey, N., Acharjee, S., Biswas, D., Das, A., Chaudhuri, S.S.: Medical Information Embedding in Compressed Watermarked Intravascular Ultrasound Video. Scientific Bulletin of the Politehnica University of Timisoara - Transactions on Electronics and Communications 57(71), 1–7 (2012)
2. Dey, N., Mukhopadhyay, S., Das, A., Chaudhuri, S.S.: Using DWT analysis of P, QRS and T Components and Cardiac Output Modified by Blind Watermarking Technique within the Electrocardiogram Signal for Authentication in the Wireless Telecardiology. I. J. Image, Graphics and Signal Processing 4(7), 33–46 (2012)
3. Dey, N., Roy, A.B., Das, A., Chaudhuri, S.S.: Stationary Wavelet Transformation Based Self-recovery of Blind-Watermark from Electrocardiogram Signal in Wireless Telecardiology. In: Thampi, S.M., Zomaya, A.Y., Strufe, T., Alcaraz Calero, J.M., Thomas, T. (eds.) SNDS 2012. CCIS, vol. 335, pp. 347–357. Springer, Heidelberg (2012)
4. ChakrabortyS., M.P., Pal, A.K., Biswas, D., Dey, N.: Reversible color image watermarking using trigonometric functions. In: International Conference on Electronic Systems, Signal Processing and Computing Technologies, pp. 105–110 (2014)
5. Dey, N., Das, P., Das, A., Chaudhuri, S.S.: DWT-DCT-SVD based intravascular ultrasound video watermarking. In: Second World Congress on Information and Communication Technologies, pp. 224–229 (2012)
6. Dey, N., Das, P., Das, A., Chaudhuri, S.S.: DWT-DCT-SVD Based Blind Watermarking Technique of Gray Scale Image in Electrooculogram Signal. In: International Conference on Intelligent Systems Design and Applications, pp. 680–685 (2012)
7. Tirkel, A.Z., Rankin, G.A., Schyndel, R.M.V., Ho, W.J., Mee, N.R.A., Osborne, C.F.: Electronic Water Mark, DICTA 1993, pp. 666–673. Macquarie University (1993)
8. Johnson, N.F.: An Introduction to Watermark Recovery from Images. In: Proceedings of the SANS Intrusion Detection and Response, pp. 9–13 (1999)

9. Matheson, L.R., Mitchell, S.G., Tarjan, R.E., Zane, F., Shamoon, T.G.: Robustness and Security of Digital Watermarks. IEEE Transactions on Image Processing 10(10), 227–240 (2001)
10. Lu, C.S., Liao, H.Y.M.: Multipurpose Watermarking for Image Authentication and Protection. IEEE Transactions on Image Processing 10(10), 1579–1592 (2001)
11. Kumari, G.S.N., Kumar, B.V., Sumalatha, L., Krishna, V.V.: Secure and Robust Digital Watermarking on Grey Level Images. International Journal of Advanced Science and Technology 11, 1–8 (2009)
12. Maity, S.P., Kundu, M.K.: Robust and Blind Spatial Watermarking in Digital Image. In: Proc. 3rd Indian Conf. on Computer Vision, Graphics and Image Processing, pp. 388–393 (2002)
13. Dey, N., Maji, P., Das, P., Das, A., Chaudhuri, S.S.: An Edge Based Watermarking Technique of Medical Images without Devalorizing Diagnostic Parameters. In: International Conference on Advances in technology and Engineering, pp. 1–5 (2013)
14. Diffie, W., Hellman, M.E.: New directions in cryptography. IEEE Transactions on Information Theory 22(6), 644–654 (1976)
15. Omura, J.: Novel applications of cryptography in digital communications. IEEE Communications Magazine 28(5), 21–29 (1990)
16. Gabidulin, E.M., Ourivski, A.V., Honary, B., Ammar, B.: Reducible rank codes and their applications to cryptography. IEEE Transactions on Information Theory 49(12), 3289–3293 (2003)
17. Kaps, J.P., Arlington, V.A., Gaubatz, G., Sunar, B.: Cryptography on a Speck of Dust. Computer 40(2), 38–44 (2007)
18. Acharjee, S.: Motion Vector Estimation for Video Compression, MTechdiss (2013), https://dspace.jdvu.ac.in/handle/123456789/27771
19. Netravali, A.N., Robbins, J.D.: Motion compensated television coding-part1. Bell Syst. Tech. J. 58, 631–670 (1979)
20. Bergmann, H.C.: Displacement estimation based on the correlation of image segments. In: International Conference on Image Processing, pp. 215–219 (1982)
21. Zeng, X., Zhenyong, C., Ming, C., Zhang, X.: Reversible Video Watermarking Using Motion Estimation and Prediction Error Expansion. J. Inf. Sci. Eng. 27(2), 465–479 (2009)
22. http://www.cipr.rpi.edu/resource/sequences/sequences/sif/yuv/sif_yuv_tennis.tgz
23. http://www.cipr.rpi.edu/sequences/sif/yuv/sif_yuv_foreman.tgz
24. http://www.cipr.rpi.edu/resource/sequences/
25. Chu, T.C., Ranson, W.F., Sutton, M.A.: Applications of digital-image-correlation techniques. Experimental Mechanics 25(3), 232–244 (1985)
26. Choi, S.S., Cha, S.H., Tappert, C.: A survey of binary similarity and distance measures. Journal of Systemics, Cybernetics and Informatics 8(1), 43–48 (2010)
27. Eldar, Y.C., Nehorai, A.: Mean-Squared Error Beamforming for Signal Estimation: A Competitive Approach. In: Robust Adaptive Beamforming, pp. 5143–5154 (2005)
28. Tian, J.: Wavelet based reversible Watermarking for authentication. In: Security and Watermarking of Multimedia Contents, vol. 4675, pp. 667–671 (2002)
29. Weng, S., Zhao, Y., Pan, J.S., Ni, R.: Reversible Watermarking Based on Invariability and Adjustment on Pixel Pairs. IEEE Signal Processing Letters 15, 721–724 (2008)

134   S. Acharjee et al.

30. Wu, H.T., Cheung, Y.M.: Reversible Watermarking by Modulation and Security Enhancement. IEEE Transactions on Instrumentation and Measurement 59(1), 221–228 (2010)
31. Acharjee, S., Chaudhuri, S.S.: A new fast motion vector estimation algorithm for video compression. In: International Conference on Informatics, Electronics & Vision (ICIEV), pp. 1216–1219 (2012)
32. Acharjee, S., Dey, N., Biswas, D., Das, P., Chaudhuri, S.S.: A novel Block Matching Algorithmic Approach with smaller block size for motion vector estimation in video compression. In: 12th International Conference on Intelligent Systems Design and Applications, pp. 668–672 (2012)
33. Acharjee, S., Dey, N., Biswas, D., Das, P., Chaudhuri, S.S.: A novel Block Matching Algorithmic Approach with smaller block size for motion vector estimation in video compression. In: 12th International Conference on Intelligent Systems Design and Applications, pp. 668–672 (2012)

Investigation of Touch-Based User Authentication Features Using Android Smartphone

Ala Abdulhakim Alariki and Azizah Abdul Manaf

University Technology Malaysia,
54100, Kuala Lumpur, Malaysia
aaaala3@live.utm.my, azizah07@ic.utm.my

Abstract. Currently most mobile phones use touch screen, In addition the behavior of user's touch gesture is significantly important in interaction with the phone. Due to increasing demand for safer access in touch screen mobile phones, old strategies like pins, tokens, or passwords have failed to stay abreast of the challenges. However, we study user authentication scheme based on this touch dynamics features for accurate user authentication. We developed the software needed to collect readings from touch screen of mobile phone running the Android operation system. Several touch dynamics features are examined to explore the efficiency of feature and each category (similar processing and representation form). Also, the impact of normalization and seven feature selection algorithms are examined. After applying Exhaustive Search reduction technique to the observation vectors composed of all 12 extracted features, nine features are retained.

Keywords: touch gesture authentication, user authentication, Android.

1   Introduction

Mobile phones using touchscreens such as smartphones based on the Android OS have been pervasively integrated into our daily work and everyday lives. In recent market analysis, it was found that 640 million tablets and 1.5 billion smartphones will be in use worldwide by 2015 [1]. In addition, 65% of mobile devices use touch screen, and this percentage seems to be increasing [2]. As the usage of touch screen on mobile devices increase, users will tend to store private information in them which will make them susceptible to external threat. New mobile malwares were found in 2010 based on McAfee's threat report. Such malware causes serious harm to users such as; discharge of non-public identification data. A requirement for improved authentication strategies exists in an exceedingly wide selection of sensible phones on demand [3]. Currently, there are three user techniques in mobile phones which are: passwords, physiological biometrics and behavioral biometrics. Password authentication usually utilize a Personal Identification Number (PIN) to authenticate the genuine user. Passwords and patterns are the most commonly used methods for user authentication [4]. However, passwords fail to keep up with the challenges presented because they can be lost or stolen by way of "shoulder surfing", which compromises the system security [5].

A.E. Hassanien et al. (Eds.): AMLTA 2014, CCIS 488, pp. 135–144, 2014.

On the other hand, most of the previous study [6],[7],[8], [9] and [10] has reported that biometric-based person recognition is a good alternative to classical systems and overcomes the difficulties of password and token approaches. Biometric authentication is further classified into Physiological and Behavioral types. Physiological Biometric refers to what the person is, and behavioral Biometrics is related to what a person does. Physical biometrics use human traits such as iris, retina, face and fingerprint. These physical biometrics traits will achieve very accurate performance, but have large drawbacks in that they are expensive, difficult and intrusive for collectability, Low degree of user acceptability, requires additional tools that are non-standardized, and not easily revoked if compromised [11].

In contrast, behavioral biometric methods, use measurements from human actions such as keystroke dynamics or mouse dynamics. Both of these dynamics have been actively studied in the context of desktop computers, but only keystroke dynamics has been explored on mobile phones [12]. Authentication based on keystroke dynamics on mobile phones learns legitimate users' behavior and verifies a user periodically or continuously, which overcomes the drawback of physiological biometrics which only authenticates the user at the beginning of a session. In addition, it is simple to implement because it supports mainly a software implementation [13]. With the increased popularity of touchscreen mobile phones, touch behavior is becoming more and more important, as many smartphones now feature touchscreens as the main input method [2]. Our motivation is therefore to develop a user authentication scheme based on touch gestures on mobile phones.

In this paper, we employ behavioral biometric methods and mainly focus on a novel user behavioral biometric, namely touch dynamics, which refers to collecting detailed information about individual touches, such as touch direction, finger pressure, finger size and acceleration. Similar to keystroke dynamics and mouse dynamics, our scheme also does not require any special hardware device for data collection. In particular, our scheme extracts and constructs 6 features related to the touch dynamics of a user as an authentication touch gesture. Section 2 presents the state of the art on touch gesture behavioral authentication. Section 3 presents data collection and database. In section 4, the methodology regarding the selected classifier and features used in this experiment is explained. The results concerning the impact of normalization and features selection are discussed in section 5.

2   Related Work

In this section, we report some of recent researches on touch gesture-based behavioral biometrics authentication. In [14] authors looked at the different sensors provided by mobile phones, and show that data collected from these sensors can distinguish mobile users by analyzing the user's touch interaction with the device. Based on fifteen minutes of real-world device interaction from six test subjects, the researcher was able to correctly identify the test subject that generated a 90 second sample 83% of the time using a subset of features extracted from the data. However, they evaluated the experiment with six participants only. Including more users and larger sample sizes is needed in order to make a more robust determination on the ability to identify users based on their behavior metric data.

In [2] authors proposed a unique user authentication theme supporting touch dynamics that uses a group of behavioral options associated with touch dynamics for correct user authentication. The experimental results show that a neural network classifier is well-suited to completely identify different users with a mean error rate of 7.8% regarding our designated options. However, by victimization different classification methodology techniques, involving many participants and assembling a lot of touch gesture knowledge would facilitate higher accuracy and performance.

In [15] authors showed that multi-touch gestures contain sufficient biometric info, ensuing from variation in pure hand mathematics and muscle behavior, to permit discrimination between users. A multi-touch gesture is essentially a time-series of a set of x-y coordinates of finger touch points. With score-based classifiers using the time feature they achieved only 4.46 % EER. Further, with the combination of three commonly used gestures: pinch, zoom, and rotate, 58% EER was achieved using a score-based classifier. However, to improve the authentication accuracy and performance the incorporation of more features like finger pressure and finger size is needed.

In [16] authors expanded on their work with a larger study of youngsters and adults playacting similar touch and surface gesture interaction tasks on mobile devices. A total of thirty participants (16 youngsters, 14 adults) participated in the study. The extracted options included x-coordinate, y-coordinate, time, touch pressure, and touch size for every up, down and move event gesture created by the user. They have found frequent intentional and unintentional touches outside of screen targets for children, and age-related challenges in recognizing children's gestures, both of which will impact the success of children's interactions. For example, youngsters miss a bigger proportion of targets than do adults, a bigger quantity of holdover touches when an onscreen target has been chosen. In future studies, they plan to explore different classification methodology to facilitate and boost accuracy.

In [17] authors developed an application for the Android mobile platform to collect data on the way individuals draw lock patterns on touch screen. The data collected on the participants' finger movement times were used to calculate common standard metrics used to assess biometric systems The EER makes it easier to compare the performance of various biometric systems or classifiers, and the lower its value the better the classifier. Their result showed a relatively low EER of 10.39% was achieved by analyzing the data from 32 individuals using a Random Forest classifier when combining the three different lock patterns and without any analytical enhancements to the data.

In [18] authors proposed to authenticate people within a biometric technique consisting of recognizing a person performing a 3-D gesture with one of his/her hands while holding a mobile device that integrates an accelerometer. The requirement for the mobile device to be valid for this technique is that it must include a 3-axis accelerometer embedded in it, so that the movement involved in the gesture can be registered. The robustness of this biometric technique has been studied by analyzing a database of 25 users with real falsifications. Equal Error Rates of 2.01 and 4.82%

have been obtained in a zero-effort and an active impostor attack, respectively. Involving more participants and collecting more touch gesture data would help in obtaining an even better understanding of the performance of the scheme.

In summary, most of the schemes faced accuracy problem based on score classifiers of EER, FAR and FRR. Furthermore, gathering more touch dynamics features such as touch direction, finger pressure, finger size and acceleration would improve authentication accuracy and verification performance. Implementing an artificial intelligence classification method other than neural network will help to get more accuracy result. If prototype with more touch gesture features can be developed and this prototype can be tested in a group of users by using one of the artificial intelligence classification techniques, accuracy may be increased leading to a more robust authentication with good performance.

3   Description of Data Sets

Since there is no common and standardized touch-based user authentication features database available, we made and collected our home-made dataset. Data collection is the first component of our system, which will collect and capture user touch input. We developed the software needed to collect readings from touch screen of mobile phone running the Android operation system. The application is developed by using eclipse Java language and implemented in Android-compatible phone. This application, through a simple graphical user interface, permitted us to record the user's name and six sketches in the touch screen. The mobile phone type used for this experiment is Galaxy S3 mini with the following specification in figure 1.

Before performing the experiment, all participants are required to practice the basic operations on the mobile device. After that, we presented the goal of our experiment, which includes the types of collecting features and the operations of program. In our experiment 25 users with an age range of 20 to 40 years had drawn a gesture six times. We recorded raw input data from touch screen such as finger pressure, finger size, time and acceleration. All these data are exported to a database where the six features are stored and used for later evaluation. This description of the data collection also shows that no special hardware is required four this application.

Mobile	OS	CPU	Display size	Display Type
Samsung Galaxy S3 Mini	v4.1 (Jelly Bean)	1 GHz dual-core Cortex-A9	480 x 800 pixels (~233 ppi pixel density)	Super AMOLED capacitive touchscreen

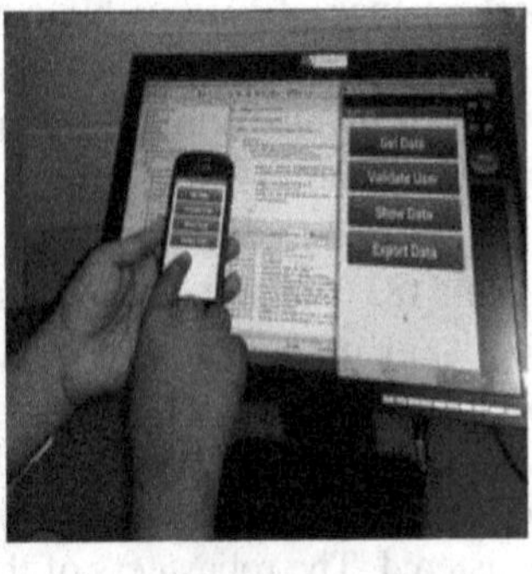

Fig. 1. Experiment setting (a) Hardware Specification (b) Main Application Interface

Table 1. Sample raw data collected from touchscreen input

Features\ Name	USER 1	USER 2	USER 3	USER 4	USER 5
Pressure up	0.058823533	0.06666667	0.086274512	0.050980397	0.082352944
Pressure down	0.156862751	0.10980393	0.098039225	0.047058828	0.421568635
Size UP	0.003921569	0.007843138	0.007843138	0.007843138	0.417843138
Size down	0.011764707	0.015686275	0.007843138	0	0.411764707
Time (ms)	1935	831	1141	961	636
Acceleration	0.0007546888	0.000280611	0.00064399	0.000176541	0.001095353
Distance	282.5466653	193.7787312	838.3987142	163.039698	443.0660592
Speed	0.146018948	0.233187402	0.734792913	0.169656293	0.696644747
Touch Major Down	2.28515625	3.046875	1.5234375	0	1.28515625
Touch Major Up	0.76171875	1.5234375	1.5234375	1.5234375	1.4234375
Touch Minor Down	0.76171875	1.015625	0.5078125	0	0.46171875
Touch Minor Up	0.25390625	0.5078125	0.5078125	0.5078125	0.15078125

Table 1 shows a sample of raw data collected from touch screen and recorded by the phone. Each record consists of at least the following six fields: finger pressure up, finger pressure down, finger size up, finger size down, duration of each touch input and acceleration.

4   Methodology

4.1   Features Extraction

The features are the project parameters which allow the system to accept or reject an input gesture. Feature extraction is an automated process of locating and encoding distinctive characteristics from biometric data in order to generate a template called feature extraction. The main task of touch gesture application is to extract touch dynamic features from the collected raw data. The features in this experiment were obtained using the Android application programming interface MotionEvent function library. Opposed to other Smartphone operating systems, the Android platform allows the retrieval of detailed information about a touch even [14]. The extracted features in this experiment are shown in table 2.

4.2   Classification

Classification is to find the best class that is closest to the classified pattern. The goal of classifying the selected features is, to map every pattern of readings into an output class, and the percent of correct mapping forms the accuracy of the classifier. We adopted the implementation of classification algorithms in Weka suite of machine learning software [19]. To get an impression of how well which classifier works, we used the Weka machine learning toolkit to compare a large amount of classifiers on a subset of our data from the large scale user study. Based on these preliminary experiments we concentrated on the Random Forest classifier and Naïve Bayes to differentiate and classify multiple users. Both Naïve Bayes and Random Forest have been widely used in many real-life classification problems, such as image classification [20], object class segmentation [21] and many other applications.

Table 2. Features extraction

Features	Extraction Method
Finger Pressure	To extract finger pressure value we used android library **getPressure( )** method measured by kilopascals.
Finger Size	To extract finger size value we used android library **getSize( )** method measured by pixels.
Time (ms)	The press time and the release time were obtained android library using the **getDownTime( )** and **getEvent- Time( )** methods, respectively.
Acceleration	To calculate the acceleration this formula (**Speed /Time**)
Distance	$$\sqrt{(x_2 - x_1)^2 + (y_2 - y_1)^2}$$
Speed	To calculate the speed we used this formula (**Distance/Time**)
Touch Major	Returns the length of the major axis of an ellipse that describes the touch area at the point of contact for the given pointer index. To extract Touch Major we used android library **getTouchMajor (int pointerIndex)** method.
Touch Minor	Returns the length of the minor axis of an ellipse that describes the touch area at the point of contact for the given pointer index. To extract Touch Minor we used android library **getTouchMinor (int pointerIndex)** method.

5   Results

In this section we will present the experimental result on the database. For the evaluation of the application the experiment performed using WEKA tool under 10-folds cross-validation testing option. The extracted features are examined from perspectives such as: normalization and features reduction.

5.1   Normalization

Typically, feature normalizing is a scaling down transformation of the features which are useful for classification algorithms. However, within a feature there is often a large difference between the maximum and minimum values, e.g. 0.01 and 1000. Features are normalized by scaling its values so that they fall within a small-specified range, such as 0.0 to 1.0. In our experiment the finger pressure and finger size values separately returned by getPressure( ) and getSize( ) are normalized within a range [0, 1] (please see

Android function library [22] for more details). The rest of the features that need to be normalized in the experiments were performed with min-max normalization method.

$$V' = ((V\text{-}minA) / (maxA - minA)) * (new\text{-}maxA - new\text{-}minA) + new\text{-}minA \qquad (1)$$

Min-max normalization performs a linear transformation on the original data. Suppose that mina and max are the minimum and the maximum values for attribute A. Min-max normalization maps a value V of A to V' in the range [new-minA, new-maxA] .

5.2 Features Selection

Feature selection is a process that selects a subset of original features. The main idea of feature subset selection is to remove redundant or irrelevant features from the data set as they can lead to a reduction of the classification accuracy or clustering quality and to an unnecessary increase of computational cost [23].

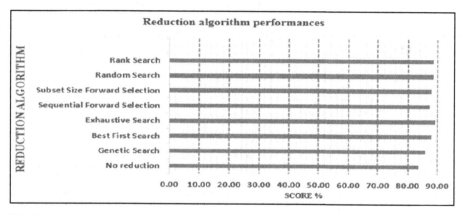

Fig. 2. Performances of reduction algorithm using Min-max normalization and Random Forest classifier

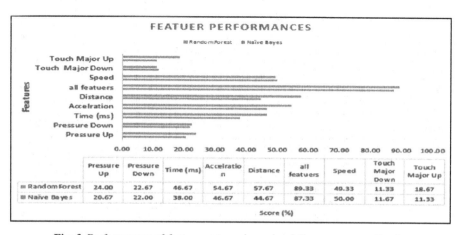

	Pressure Up	Pressure Down	Time (ms)	Acceleration	Distance	all features	Speed	Touch Major Down	Touch Major Up
Random Forest	24.00	22.67	46.67	54.67	57.67	89.33	49.33	11.33	18.67
Naïve Bayes	20.67	22.00	38.00	46.67	44.67	87.33	50.00	11.67	11.33

Fig. 3. Performances of feature categories using Min-max normalization

Here, eight algorithms are investigated to study impact of dimension reduction on the recognition performance. The twelve normalized proposed features (12) are used with Genetic Search, Best First Search, Exhaustive Search, Sequential Forward Selection, Subset Size Forward Selection, Random Search and Rank Search. Results illustrated in Fig. 2 shows that Exhaustive Search has (89.33%) recognition rate. Score gain is related to the fact that the redundant and insignificant feature parameters are removed from the feature set.

5.3   Comparison of Features

Descriptor performance in aggregated and individual forms for Random Forest classifier and Naïve Bayes normalization are shown in Fig 3. After applying Exhaustive Search Selection reduction technique to observation vectors composed of all the proposed features (12), nine features are retained. In total 150 instances belong to 25 users, and each user has 6 instances. The evaluation result shows that the best performance is attributed to the combination of all features which correctly classified 134 instances with 89.33% accuracy. Finally, the selected parameters from all proposed features in the direct

Table 3. Confusion matrix using Exhaustive Search SFS, Random Forest and Min-max normalization

88.67%	ahmad soleymani	paria	shahin.moosavi	mohammad sadeq	jeff	nadia	rahman	syahmi	myrna	zamri	mohamed munser	alias mansor	suba	Atabak	azizah abdul manaf	ira	arvin	abdulla sharaf addin	Moamar Elyazgi	bjia	azin	reza	mahdi	abdalrahman salem	ringa
ahmad soleymani	5	0	0	0	0	0	0	0	0	0	0	0	0	0	0	0	0	0	0	0	0	0	0	0	1
paria	0	5	0	1	0	0	0	0	0	0	0	0	0	0	0	0	0	0	0	0	0	0	0	0	0
shahin.moosavi	0	0	5	1	0	0	0	0	0	0	0	0	0	0	0	0	0	0	0	0	0	0	0	0	0
mohammad sadeq	0	1	0	4	0	0	0	0	0	0	0	0	0	0	0	0	0	0	0	0	0	0	1	0	0
jeff	0	0	0	0	6	0	0	0	0	0	0	0	0	0	0	0	0	0	0	0	0	0	0	0	0
nadia	0	0	0	0	0	5	0	0	0	1	0	0	0	0	0	0	0	0	0	0	0	0	0	0	0
rahman	0	0	0	0	0	0	6	0	0	0	0	0	0	0	0	0	0	0	0	0	0	0	0	0	0
syahmi	0	0	0	0	0	0	0	5	0	0	0	0	0	0	0	1	0	0	0	0	0	0	0	0	0
myrna	0	0	0	0	0	0	0	0	5	0	0	0	0	0	0	0	0	0	0	0	0	0	1	0	0
zamri	0	0	0	0	0	0	0	0	0	6	0	0	0	0	0	0	0	0	0	0	0	0	0	0	0
mohamed munser	0	0	0	0	0	0	1	0	0	0	5	0	0	0	0	0	0	0	0	0	0	0	0	0	0
alias mansor	0	0	0	0	0	0	0	0	0	0	1	5	0	0	0	0	0	0	0	0	0	0	0	0	0
suba	0	0	0	0	0	0	0	0	0	0	0	0	5	0	0	0	0	1	0	0	0	0	0	0	0
Atabak	0	0	0	0	0	0	0	0	0	0	0	0	0	6	0	0	0	0	0	0	0	0	0	0	0
azizah abdul manaf	0	0	0	0	0	0	0	0	0	0	0	0	0	0	6	0	0	0	0	0	0	0	0	0	0
ira	0	0	0	0	0	0	0	0	0	0	0	0	0	0	0	6	0	0	0	0	0	0	0	0	0
arvin	0	0	0	0	0	0	0	0	0	0	0	0	0	0	0	0	6	0	0	0	0	0	0	0	0
abdulla sharaf addin	0	0	0	0	0	0	0	0	0	0	0	0	0	0	0	1	0	5	0	0	0	0	0	0	0
Moamar Elyazgi	0	0	0	0	0	0	0	0	0	0	0	0	1	0	0	0	0	0	5	0	0	0	0	0	0
bjia	0	0	0	0	0	0	0	0	0	0	0	0	0	0	0	0	0	0	0	6	0	0	0	0	0
azin	0	0	1	0	0	0	0	0	0	0	0	0	0	0	0	0	0	0	0	0	5	0	0	0	0
reza	0	0	1	0	0	0	0	0	0	0	0	0	0	0	0	0	0	0	0	0	1	4	0	0	0
mahdi	0	0	0	0	0	0	0	0	0	0	0	0	0	0	0	0	0	0	0	0	0	0	6	0	0
abdalrahman salem	0	0	0	0	0	0	0	0	0	0	0	0	0	0	0	0	0	0	0	0	0	0	0	6	0
ringa	1	0	0	0	0	0	0	0	0	0	0	0	0	0	0	0	0	0	0	0	0	0	0	0	5

classification scheme using Exhaustive Search Selection are: finger pressure up, finger pressure down, time, acceleration, distance, speed, touch major up and touch major down. Results are illustrated in Table 3 with more details for each class.

6   Conclusion

Popularity of touch screen in mobile phone make touch behavior more important than ever compared to the biometric authentication techniques. A gesture based authentication system has made it almost impossible for a shoulder surfer to replay the password, although the entire gesture was observed. Subtleties like force, speed, flexibility, pressure, and individual anatomical differences would prevent the casual observer from accessing the password. This paper examined the interesting topic of accurate authentication features, among the commonly available touch motion features supported platforms today. We focused principally on comparing the contribution of sub-class features belonging to the same category (based on the same estimation method) to global vector composed of several component computed with different processing techniques.

Several touch dynamics features are examined to explore the efficiency of each feature and each category (similar processing and representation form). In addition, the impact of the normalization and seven feature selection algorithms are examined. After applying Exhaustive Search reduction technique to the observation vectors composed of all 12 extracted features, nine features were retained. The presented system correctly classified 134 instances with 89.33% by combination of all features and both touch major up and touch major down features have the lowest recognition rate with 18.67 and 11.33% respectively.

Acknowledgment. The authors would like to express greatest appreciation to MOE, UTM and ERGS GRANT vot no: R.K130000.7838.4L043.

References

1. Feng, T., Liu, Z., Kwon, K.-A., Shi, W., Carbunar, B., Jiang, Y., et al.: Continuous mobile authentication using touchscreen gestures. In: 2012 IEEE Conference on Technologies for Homeland Security (HST), pp. 451–456. IEEE (2012)
2. Meng, Y., Wong, D.S., Schlegel, R., Kwok, L.-f.: Touch Gestures Based Biometric Authentication Scheme for Touchscreen Mobile Phones. In: Kutyłowski, M., Yung, M. (eds.) Inscrypt 2012. LNCS, vol. 7763, pp. 331–350. Springer, Heidelberg (2013)
3. Crawford, H.: Keystroke dynamics: Characteristics and opportunities. In: 2010 Eighth Annual International Conference on Privacy Security and Trust (PST), pp. 205–212. IEEE (2010)
4. Weiss, R., De Luca, A.: PassShapes: utilizing stroke based authentication to increase password memorability. In: Proceedings of the 5th Nordic Conference on Human-computer Interaction: Building Bridges, pp. 383–392. ACM (2008)

5. Zheng, N., Bai, K., Huang, H., Wang, H.: You Are How You Touch: User Verification on Smartphones via Tapping Behaviors. In: William, Mary (eds.) Department of Computer Science, pp. 1–13. WM-CS (2012)
6. Sesa-Nogueras, E., Faundez-Zanuy, M.: Biometric recognition using online uppercase handwritten text. Pattern Recognition 45(1), 128–144 (2012)
7. El-Abed, M., Giot, R., Hemery, B., Rosenberger, C.: A study of users' acceptance and satisfaction of biometric systems. In: 2010 IEEE International Carnahan Conference on Security Technology (ICCST), pp. 170–178. IEEE (2010)
8. Jain, A.K., Kumar, A.: Biometrics of next generation: An overview, pp. 1–36. Springer, Berlin (2010)
9. Shanmugapriya, D., Padmavathi, G.: An Efficient Feature Selection Technique for User Authentication using Keystroke Dynamics. IJCSNS International Journal of Computer Science and Network Security 11(10), 191–195 (2011)
10. Karnan, M., Akila, M., Krishnaraj, N.: Biometric personal authentication using keystroke dynamics: A review. Applied Soft Computing 11(2), 1565–1573 (2011)
11. Ngugi, B., Tremaine, M., Tarasewich, P.: Biometric keypads: Improving accuracy through optimal PIN selection. Decision Support Systems 50(4), 769–776 (2011)
12. Zheng, N., Paloski, A., Wang, H.: An efficient user verification system via mouse movements. In: Proceedings of the 18th ACM Conference on Computer and Communications Security, pp. 139–150. ACM (2011)
13. Yampolskiy, R.V., Govindaraju, V.: Taxonomy of behavioural biometrics. Behavioral Biometrics for Human Identification, 1–43 (2010)
14. Wolff, M.: Behavioral Biometric Identification on Mobile Devices. In: Schmorrow, D.D., Fidopiastis, C.M. (eds.) AC/HCII 2013. LNCS (LNAI), vol. 8027, pp. 783–791. Springer, Heidelberg (2013)
15. Sae-Bae, N., Ahmed, K., Isbister, K., Memon, N.: Biometric-rich gestures: a novel approach to authentication on multi-touch devices. In: Proceedings of the 2012 ACM Annual Conference on Human Factors in Computing Systems, pp. 977–986. ACM (2012)
16. Anthony, L., Brown, Q., Nias, J., Tate, B., Mohan, S.: Interaction and recognition challenges in interpreting children's touch and gesture input on mobile devices. In: Proceedings of the 2012 ACM International Conference on Interactive Tabletops and Surfaces, pp. 225–234. ACM (2012)
17. Angulo, J., Wästlund, E.: Exploring touch-screen biometrics for user identification on smart phones. In: Camenisch, J., Crispo, B., Fischer-Hübner, S., Leenes, R., Russello, G. (eds.) Privacy and Identity 2011. IFIP AICT, vol. 375, pp. 130–143. Springer, Heidelberg (2012)
18. Guerra-Casanova, J., Sánchez-Ávila, C., Bailador, G., de Santos Sierra, A.: Authentication in mobile devices through hand gesture recognition. International Journal of Information Security 11(2), 65–83 (2012)
19. Hall, M., Frank, E., Holmes, G., Pfahringer, B., Reutemann, P., Witten, I.H.: The WEKA data mining software: an update. ACM SIGKDD Explorations Newsletter 11(1), 10–18 (2009)
20. Bosch, A., Zisserman, A., Munoz, X.: Image classification using random forests and ferns. In: International Conference on Computer Vision, ICCV 2007, pp. 14–20 (2007)
21. Schroff, F., Criminisi, A., Zisserman, A.: Object class segmentation using random forests. In: British Machine Vision Conference, BMVC, pp. 1–4 (2008)
22. Meier, R.: Professional Android 4 application development. John Wiley & Sons Publishing, Inc. (2012)
23. Janecek, A., Gansterer, W.N., Demel, M., Ecker, G.: On the Relationship Between Feature Selection and Classification Accuracy. In: FSDM, pp. 90–105. Citeseer (2008)

Protective Frameworks and Schemes to Detect and Prevent High Rate DoS/DDoS and Flash Crowd Attacks: A Comprehensive Review

Mohammed A. Saleh[1] and Azizah Abdul Manaf[2]

[1] Faculty of Computing, Universiti Teknologi Malaysia (UTM), Johor Baru, Malaysia
mabbsaleh@gmail.com
[2] Advanced Informatics School (AIS), Universiti Teknologi Malaysia (UTM),
Kuala Lumpur, Malaysia
azizaham.kl@utm.my

Abstract. As the dependency on web technology increases every day, there is on the other side an increase in destructive attempts to disrupt an essential web technology, which yields an improper service. Denial of Service (DoS) attack and its large counterpart Distributed Denial of Service (DDoS) and Flash Crowd attacks are among the most dangerous internet attacks, which overwhelm the web server, thereby slow it down, and eventually take it down completely. This review paper evaluates and describes the effectiveness of different existing Frameworks and Schemes for Detecting and Preventing High Rate DoS/DDoS and Flash Crowd Attacks. Firstly, the review paper describes them according to the similar category, and then it compares them based on the predefined metrics. Finally, advantages and disadvantages for each category are described.

Keywords: DoS, DDoS, High Rate DDoS, HTTP-based DDoS, Flash Crowd, Protective Framework, Protective Scheme.

1 Introduction

Denial of Service (DoS) attack and its large counterparts Distributed Denial of Service (DDoS) and Distributed Reflected Denial of Service (DRDoS) attacks are among the most dangerous internet attacks that overwhelm the web server, thus slow it down, and in the end take it down completely. Researchers outlined that DoS, DDoS and DRDoS attacks were responsible for a massive damage that was up to billions of dollars per incident [1]. Denial of Service (DoS) attack is an effort by a single machine, namely an attacker to make a target, whether is server or network resource, unavailable for its customers, which yields to prevent those users from accessing the service. DoS attack consists of highly damageable attacks to collapse or degrade the quality of service [2]. While, Distributed Denial of Service (DDoS) attack is an attempt to flood a victim, whether it is a machine or network, through a big volume of traffic that is generated by large number of machines that are combined together to generate that

A.E. Hassanien et al. (Eds.): AMLTA 2014, CCIS 488, pp. 145–152, 2014.
© Springer International Publishing Switzerland 2014

volume of traffic. Furthermore, to diffuse source of attack, these machines are part of different networks, so it is hard to trace back IP sources of the attack, and then to block them accordingly [3, 4]. DDoS attack uses a large number of compromised hosts called zombies or Bots that are collected by planting malicious software on the unprotected computers. Then, these hosts (zombies or bots) are grouped in together to shape one huge network called a Botnet that awaits a command from the attacker to launch the DDoS attack [5-8].

Flash Crowd is a sudden high request in a service that is caused by legitimate users who simultaneously request the server at the same period, which eventually forces the server to decease its performance. It occurs once a big amount of service's customers access a server at the same time. Flash Crowd may overwhelms the server, and therefore causes a Denial of Service (DoS) attack, which results in either a delay of response or a complete take down. Flash crowd could happen due to some of an exciting event that has just occurred. Likewise, it could be due to the broadcasting of a new created service or a free hot software download [9-11]. From perspectives of service requesters (clients), regardless whether they are legitimates or illegitimates, flash crowd may not be counted as an attack, but on the contrary, it counted as an attack from the perspective of victim (service provider), since it affects the web server negatively. Whereas a High Rate Distributed Denial of Service (HR-DDoS) attack is synonym to the traditional DDoS and DRDoS attacks. They all happen when attackers exceed and violate the adopted threshold value [12, 13].

2   Metrics for Evaluating Frameworks and Schemes for Detecting and Preventing High Rate DoS/DDoS and Flash Crowd Attacks

Indeed, various metrics for evaluating schemes and frameworks are already indentified by previous work [14] in terms of detecting and preventing denial of service attack (DDoS), as follows.

1. Edge Router Involvement: An ideal protective framework or scheme should consider an Edge Router involvement.
2. Number of Required Attacking Packets: A protective framework or scheme should be able to detect and prevent the attacks based on few packets.
3. Processing Overhead: An additional processing overhead is occurred through measuring flow of network packets, and calculating various statistical parameters. An ideal protective framework or scheme should be able to incur minimal processing overhead.
4. Storage Requirement: An additional amount of memory is required to store certain information in order to detect and prevent the attacks. An ideal protective framework or scheme should be able to acquire minimum amount of memory.
5. Ease of Implementation: An ideal protective framework or scheme should be designed in a way that it could be easily implemented.

6. Scalability: It refers to the extra configurations that are required on the devices when there is an expansion to the environment. An ideal protective framework or scheme should be able to scalable, and the devices' configuration should be independent to each other.
7. Bandwidth Overhead: Additional network traffic is considered bandwidth overhead. Large bandwidth overhead is undesirable, since it may exhaust the network capacity.

3   Available Existing Frameworks and Schemes for Detecting and Preventing High Rate DoS/DDoS and Flash Crowd Attacks

3.1   Using User Web-Behavior Modeling

In this subsection, methodologies of available existing frameworks and schemes have adopted User Web-Behavior Modeling to detect and prevent High Rate DoS/DDoS and Flash Crowd attacks. For instance, Yi and Shensheng [15] proposed a new model called dynamic hidden semi-Markov, which aims to model the user behavior in various times, and based on this model an anomaly detection scheme is developed and implemented for the sake of detecting web based distributed denial of service attacks. The dynamic hidden semi-Markov model introduces an effective algorithm to provide the online automatic updates for the model's parameters, and it uses Markov states to shape the users' click behavior.

Likewise, scheme is proposed by Chengxu and Kesong [16] called HTTP request transition matrix scheme to detect HTTP application based DDoS attacks. The proposed scheme assumes the normal human user will focus on interesting pages on the web application in order to describe users' bowering behavior, and therefore it forms the patterns of transition probability from one page to another page on same web application. In contrast, since the Bot does not know the most interesting pages, it sends random requests to web application, which leads to very small transition probability to its requests sequence. Detecting of HTTP application based DDoS attacks is based on threshold value, which is determined by three parameters; frequency vector, transition probability matrix and bot request sequence probability.

Similarly, research done by Jin, Xiaolong [17] proposed two different DDoS detection schemes; Large Deviation Measuring Click Ratio based Web Access Behavior (LD-IID) scheme and Large Deviation Measuring Web Access Behavior based on Markov Process (LD-MP) scheme. The research characterizes user's access behavior for each detection scheme through two different models; Click-ratio based model and Markov process based model respectively. For both DDoS attack detection schemes, a large deviation theory is adopted in order to estimate the probability of each ongoing user's accesses.

3.2   Using Entropy Variations (Entropy Based Threshold Values)

In this subsection, methodologies of available existing frameworks and schemes have adopted using of Entropy Variations in order to detect and prevent High Rate DoS/DDoS and Flash Crowd attacks. For example, Jie, Zheng [18] proposed an advanced entropy based DDoS detection scheme to determine the most suitable threshold value for detecting DDoS attacks accurately. The proposed detection scheme divides various DDoS attacks into different groups, and it treats each group with different method. An advanced entropy based DDoS detection scheme calculates the entropy to get the normal network behavior in case of no attack.

Likewise, Oshima, Nakashima [19] proposed an early DoS/DDoS attacks detection method based on the concept of short-term entropy to focus on an early DoS/DDoS attacks detection. The proposed method calculates the entropy values average of two different network's conditions; when the network under normal condition, and when the network under attacking condition. Then, based on these two entropy values, it computes the z score based on formula (1) to test the significance (where $\alpha = 5\%$, $r = 25$), and validity of hypothesis between two averages. Calculating of entropy average when a network is under normal condition is computed with an effective window width; 50 for DDoS and 500 for slow DoS attacks. It firstly handles the incoming packets and arranges them based on arriving time, and splits them according to the size of window width. Secondly, it extracts packet's characteristics, such as the source IP address and the destination port number to calculate the entropy value by using formula (2) and formula (3). As is deduced by the proposed method, the entropy value is low under DoS attacks, while is high under DDoS attacks.

$$H = -\sum_{i=1}^{n} Pi log_2 Pi \tag{1}$$

$$Pi = \frac{xi}{w}, where\ xi\ packets\ frequency \tag{2}$$

$$z = \frac{|\bar{H}_N - \bar{H}_A|}{\sqrt{\sigma_n^2/n} + \sqrt{\sigma_r^2/r}} \tag{3}$$

3.3   Using Network Behavior Analysis

In this subsection, methodologies of available existing frameworks and schemes have adopted using of Network Behavior Analysis to detect and prevent High Rate DoS/DDoS and Flash Crowd attacks. For example, Lei, Xiaolong [20] developed a mechanism based on network's traffic behavior analysis to inspect abnormal traffic that is caused by DDoS attacks. The developed mechanism has employed Hurst parameter together with the variance and the autocorrelation as the key performance measures to detect network traffic anomalies according to the statistical results. The detection device of the mechanism is expected to be set up in front of network switch or router. The main objective of this device is to response to these abnormal activities (anomalies) in network traffic in the real time. To achieve this aim, the mechanism has employed three important measures, namely traffic intensity, packet number and suspicious traffic duration. The mechanism has adopted threshold's values for them in order to decide whether the traffic that is passing through network need for additional evaluation, or not.

Similarly, work is done by Zhongmin and Xinsheng [21] proposed a detection method based on the analysis of network's traffic distribution by using of the correlation of IP addresses. The proposed method records the network traffic at edge router close to the target (victim), and then it uses correlation coefficient periodically in several adjacent time Intervals. In normal cases, accesses of the IP address to the target network will be relatively stable, therefore the change of the correlation coefficient of IP packets is also be relatively stable. In contrast, during network attacks, the attacker launches a large number of requests to attack the target network.   In this case, by calculating and analyzing correlation coefficients of network traffic in sliding window, the method detects network DDoS attacks.

4   Comparison of Frameworks and Schemes for Detecting and Preventing High Rate DoS/DDoS and Flash Crowd Attacks

Metric Name	Web-Behavior Modeling	Entropy Variations	Network Behavior Analysis
Edge Router Involvement	Moderate	High	High
Number of Required Attacking Packets	Moderate	Moderate	High
Processing Overhead	Moderate	Low	High
Storage Requirement	Low	Low	High
Ease of Implementation	Low	High	Moderate
Scalability	Moderate	High	High
Bandwidth Overhead	Moderate	Low	High
Advantages	• It is compatible with the existing protocols • It consumes low memory storage • It is resistant to IP sources spoofing attack.	• It is compatible with the existing protocols • It is highly scalable • It does not causing processing overhead • It does not causing bandwidth overhead	• It is compatible with the existing protocols • No additional requirements and configurations to involve Edge Router

		• It consumes low memory storage • It is easily to implement	• It is highly scalable
Disadvantages	• Additional requirements and configurations to involve Edge Router • It does causing moderate processing overhead • It does causing moderate bandwidth overhead • It does not highly scalable • It does require fair network traffic to detect and prevent High Rate and Flash Crowd attacks • It is extremely difficult to be implemented.	• Additional requirements and configurations to involve Edge Router • Requires fair network traffic to detect and prevent High Rate and Flash Crowd attacks • It can be easily decoyed if attacking IP sources are spoofed.	• It does causing processing overhead • It does causing bandwidth overhead • It does require all network traffic to detect and prevent High Rate and Flash Crowd attacks • It is not very easy to be implemented • It can be easily decoyed if attacking IP sources are spoofed. • It consumes high memory storage.

5   Conclusion

This review paper discussed a comprehensive survey on different Frameworks and Schemes for Detecting and Preventing High Rate DoS/DDoS and Flash Crowd attacks. As from the review, it found that all of exiting frameworks and schemes are compatible with existing protocols and they could involve an Edge Router, as well. In contrast, they are vary on the other metrics like scalability, easy of implement, processing overhead, bandwidth overhead, and storage (memory) requirements. In addition, it found that Web-Behavior Modeling frameworks and schemes are only resistant to IP sources spoofing attack among the other protective frameworks and schemes. Lastly, merits and demerits for each category are described.

Acknowledgement. This work of research has been done in Universiti Teknologi Malaysia (UTM) under support of Ministry of Education of Malaysia, vote number R.K130000.7838.4F287.

References

1. Wei, Y., et al.: Localization Attacks to Internet Threat Monitors: Modeling and Counter-measures. IEEE Transactions on Computers 59(12), 1655–1668 (2010)
2. Rahmani, H., Sahli, N., Kammoun, F.: Joint Entropy Analysis Model for DDoS Attack Detection. In: IAS 2009 Fifth International Conference on in Information Assurance and Security (2009)
3. Vijayasarathy, R., Raghavan, S.V., Ravindran, B.: A System Approach to Network Modeling for DDoS Detection using a Naive Bayesian Classifier. In: Third International Conference on in Communication Systems and Networks (COMSNETS) (2011)
4. Subbulakshmi, T., Guru, I.A.A., Shalinie, S.M.: Attack Source Identification at Router Level in Real Time using Marking Algorithm Deployed in Programmable Routers. In: International Conference on in Recent Trends in Information Technology (ICRTIT) (2011)
5. Oshima, S., Nakashima, T., Sueyoshi, T.: The Evaluation of an Anomaly Detection System Based on Chi-square Method. In: 26th International Conference on in Advanced Information Networking and Applications Workshops (WAINA) (2012)
6. Kambhampati, V., Papadopoulos, C., Massey, D.: A Taxonomy of Capabilities Based DDoS Defense Architectures. In: 9th IEEE/ACS International Conference on Computer Systems and Applications (AICCSA) (2011)
7. Wang, Y., Tefera, S.H., Beshah, Y.K.: Understanding Botnet: From Mathematical Modelling to Integrated Detection and Mitigation Framework. In: 13th ACIS International Conference on in Software Engineering, Artificial Intelligence, Networking and Parallel & Distributed Computing (SNPD) (2012)
8. Kline, E., Afanasyev, A., Reiher, P.: Shield: DoS Filtering using Traffic Deflecting. In: IEEE 19th International Conference on Network Protocols (ICNP) (2011)
9. Thapngam, T., Shui, Y., Wanlei, Z.: DDoS Discrimination by Linear Discriminant Analysis (LDA). In: International Conference on Computing, Networking and Communications (ICNC) (2012)
10. Qi, C., et al.: CBF: A Packet Filtering Method for DDoS Attack Defense in Cloud Environment. In: IEEE Ninth International Conference on Dependable, Autonomic and Secure Computing (DASC) (2011)
11. Haiqin, L., Yan, S., Min Sik, K.: Fine-Grained DDoS Detection Scheme Based on Bidirectional Count Sketch. In: Proceedings of 20th International Conference on Computer Communications and Networks (ICCCN) (2011)
12. Ying, X., et al.: Detecting Application Denial-of-Service Attacks: A Group-Testing-Based Approach. IEEE Transactions on Parallel and Distributed Systems 21(8), 1203–1216 (2010)
13. Yang, X., Ke, L., Wanlei, Z.: Low-Rate DDoS Attacks Detection and Traceback by Using New Information Metrics. IEEE Transactions on Information Forensics and Security 6(2), 426–437 (2011)
14. Kumar, K., Sangal, A.L., Bhandari, A.: Traceback Techniques against DDOS Attacks: A Comprehensive Review. In: 2nd International Conference on Computer and Communication Technology (ICCCT) (2011)

15. Yi, X., Shensheng, T.: Online Anomaly Detection Based on Web Usage Mining. In: IEEE 26th Internationalin Parallel and Distributed Processing Symposium Workshops & PhD Forum (IPDPSW) (2012)
16. Chengxu, Y., Kesong, Z.: Detection of Application Layer Distributed Denial of Service. In: International Conference on Computer Science and Network Technology (ICCSNT) (2011)
17. Jin, W., Xiaolong, Y., Keping, L.: Web DDoS Detection Schemes Based on Measuring User's Access Behavior with Large Deviation. In: IEEE Global Telecommunications Conference (GLOBECOM 2011) (2011)
18. Jie, Z., et al.: An Advanced Entropy-based DDOS Detection Scheme. International Conference on Information Networking and Automation (ICINA) (2010)
19. Oshima, S., Nakashima, T., Sueyoshi, T.: Early DoS/DDoS Detection Method using Short-term Statistics. In: International Conference on Complex, Intelligent and Software Intensive Systems (CISIS) (2010)
20. Lei, L., et al.: Real-Time Diagnosis of Network Anomaly Based on Statistical Traffic Analysis. In: IEEE 11th International Conference on Trust, Security and Privacy in Computing and Communications (TrustCom) (2012)
21. Zhongmin, W., Xinsheng, W.: DDoS Attack Detection Algorithm based on the Correlation of IP Address Analysis. International Conference on Electrical and Control Engineering (ICECE) (2011)

Virtual Machine Placement Based on Ant Colony Optimization for Minimizing Resource Wastage

Medhat A. Tawfeek, Ashraf B. El-Sisi, Arabi E. Keshk, and F. A. Torkey

Faculty of Computers and Information, Menoufia University, Egypt
{medhattaw,ashrafelsisim,arabikeshk,torkey1951}@yahoo.com

Abstract. Cloud computing is concept of computing technology in which user uses remote server for maintain their data and application. Resources in cloud computing are demand driven utilized in forms of virtual machines to facilitate the execution of complicated tasks. Virtual machine placement is the process of mapping virtual machines to physical machines. This is an active research topic and different strategies have been adopted in literature for this problem. In this paper, the problem of virtual machine placement is formulated as a multi-objective optimization problem aiming to simultaneously optimize total processing resource wastage and total memory resource wastage. After that ant colony optimization algorithm is proposed for solving the formulated problem. The main goal of the proposed algorithm is to search the solution space more efficiently and obtain a set of non-dominated solutions called the Pareto set. The proposed algorithm has been compared with the well-known algorithms for virtual machine placement problem existing in the literature. The comparison results elucidate that the proposed algorithm is more efficient and significantly outperforms the compared methods on the basis of CPU resource wastage and memory resource wastage.

Keywords: Cloud computing; Ant colony optimization; Virtual machine placement.

1   Introduction

Current cloud computing providers have several data centers at different geographical locations over the internet in order to optimally serve costumers needs around the world [1]. One of the most important key technologies in cloud computing is virtualization. Virtualization enables dynamic sharing of physical resources and hides the technical complexity from users [2]. The problem of virtual machine (VM) placement has become a challenging problem for improving resource utilization in cloud infrastructures [3]. Several researches have been addressed the importance of needing efficient approaches for VM placement [4, 5]. Many meta-heuristic algorithms such as Ant Colony Optimization (ACO) and genetic algorithms will be suitable for VM placement problem [6]. ACO is a metaheuristic inspired by the observation of real ant colonies and has been successfully applied to numerous optimization problems [7]. The basic idea of ACO is to simulate the foraging behavior of ant colonies. When an

A.E. Hassanien et al. (Eds.): AMLTA 2014, CCIS 488, pp. 153–164, 2014.

ants group tries to search for the food, they use a special kind of chemical pheromone to communicate with each other. Initially ants start searching their foods randomly. Once the ants find a path to food source, they leave pheromone on the path. An ant can follow the trails of the other ants to the food source by sensing pheromone on the ground. As this process continues, most of the ants attract to choose the shortest path as there have been a huge amount of pheromones accumulated on this path [2]. In this paper the problem of VM placement is formulated as a multi-objective optimization problem aiming to simultaneously optimize total CPU resource wastage and memory resource wastage. ACO algorithm is proposed and designed to deal effectively with the potential large solution space for large scale data centers. The performance of the proposed algorithm is compared to First-fit-decreasing-CPU (FFD-CPU), First-fit-decreasing- memory (FFD-MEM), Best-fit-decreasing-CPU (BFD-CPU), Best-fit-decreasing-memory (BFD-MEM) algorithms in [8, 9] and Virtual machine placement based on ant colony system (VMPACS) algorithm in [3]. The results show that the proposed algorithm can compete efficiently with other approaches to the VM placement problem on basis of different metrics such as residual processing resource and residual memory resource. The rest of this paper is organized as follows. Section 2 scans the related work. Section 3 provides the processing wastage and memory wastage models that will be used by the proposed algorithm. Section 4 formulates the virtual machine placement problem. In section 5, the details about proposed algorithm are covered. The implementation and simulation results are seen in section 6. Finally, Section 7 concludes this paper.

2   Related Work

Because of the fundamental significance of placement optimization for virtual machines (VMs) on physical machines, extensive study has been made in this field and bunch of algorithms exist in the literature. The importance of needing efficient methods for placing VMs appropriately is seen in [4, 5]. A simple process for VM placement starts by choosing a target server with compatible virtualization software, comparable CPU types, similar network connectivity, and the usage of shared storage. Then the first virtual machine will be placed on the first server and the second virtual machine will be on the same server only if it can satisfy the resource requirements. If not, a new server is added and the VM will be placed on it. These steps will be continued until each of the VMs has been placed [10]. Traditional analytical approaches based on linear and quadratic programming are used in [11] to minimize the number of used nodes. The linear programming formulations of server consolidation problems were covered in [12]. This approach restricts the number of virtual machines in a single physical server ensuring that some virtual machines are assigned to different physical servers and limits the total number of migrations. Genetic algorithm was proposed to adaptively self-reconfigure the VMs in cloud data centers consisting of heterogeneous nodes [13]. The first-fit decreasing (FFD) algorithm and the Best-fit decreasing (BFD) algorithm were used to pack list of VMs into a minimal number of physical machines and deal with the VM placement problem as a bin packing problem

[8, 9]. A single-objective algorithm based on Max-Min Ant System (MMAS) metaheuristic to minimize the required number of physical machines was proposed in [14]. The majority of the studies on VM placement focus on a single criteria optimization however many real world problems require taking into account multi criteria optimization. For this reason, recent research tends to look at the multiple-objective situation like multi-objective grouping genetic algorithm (MGGA) [6] and VMPACS algorithm [3]. MGGA is a modified genetic algorithm with fuzzy multi-objective evaluation that was proposed for efficiently searching the solution space and combining possibly conflicting objectives [6]. This system tackled power-cost under a fixed performance constraint by minimizing migration costs while packing VMs in a small number of machines. VMPACS uses multi-objective for the virtual machine placement problem. The main goal of VMPACS algorithm is to obtain a set of non-dominated solutions that simultaneously minimize total resource wastage and power consumption. The performance of the VMPACS algorithm outperformed a multi-objective grouping genetic algorithm (MGGA), and a single-objective ACO algorithm [3]. In this paper, Virtual machine placement based on ACO algorithm is proposed and designed to optimize total processing resource wastage and total memory resource wastage.

3   The Proposed Memory and CPU Wastage Model

In a cloud environment, there are pools of server nodes. The processing and memory resource wastage from each server may vary greatly with different VM placement solutions. The VMPACS algorithm deals with CPU and memory resource wastage as a single objective function. But in this paper, resource wastage will be handled at the multiple-objective situation: one objective function for processing resources and another for memory resources. To fully utilize the available resources, the Eq. (1) is used to calculate the potential wasted processing (CPU) resources. The Eq. (2) is used to calculate the potential wasted memory resources. The terms "host" and "server" will be used interchangeably. The same thing will be for the terms "CPU" and "processing". Therefore, the problem of VM placement is formulated to optimize CPU resource wastage and memory resource wastage.

$$W_{pj} = \frac{T_{pj} - U_{pj}}{U_{pj}} \tag{1}$$

$$W_{mj} = \frac{T_{mj} - U_{mj}}{U_{mj}} \tag{2}$$

Where, W_{pj} and W_{mj} denote the CPU processing resource wastage and memory resource wastage of the j-th server respectively. T_{pj} is the threshold of CPU processing utilization associated with the j-th server, T_{mj} is the threshold of memory utilization associated with the j-th server, U_{pj} is total used CPU processing and U_{mj} is total used memory within the j-th server. The main idea from the above thresholds is that 100% utilization can cause severe performance degradation and the VM live migration technology consumes some amount of processing capability on the migrating node. The goal of processing resource wastage is to measure the total CPU resources wastage with respect to total CPU resources usage. When the amount of used CPU resources increased and the amount of CPU resources wastage decreased the W_{pj} will be decreased and the similar thing to the memory resource wastage. The goal from the proposed algorithm is that W_{pj} and W_{mj} must be decreased as much as it can.

4   The Problem Formulation

Suppose that we are given n VMs to be placed on m servers. The variable $i \in I$ is used to index the VM and the variable $j \in J$ is used to index the server assuming that none of the VMs requires more resource than can be provided by a single server. D_{pi} is the CPU demand of each VM and D_{mi} is the memory demand of each VM. Two binary variables x_{ij} and y_j are used such that the binary variable x_{ij} indicates if VM_i is assigned to $server_j$ or not and the binary variable y_j indicates whether $server_j$ is in use or not. The proposed algorithm objective is to simultaneously minimize the memory resource wastage and the CPU resource wastage. The VM placement optimization problem can be formalized as follows.

$$Min \sum_{j=1}^{m} W_{pj} = \sum_{j=1}^{m} \left(y_j \times \frac{(T_{pj} - \sum_{i=1}^{n}(x_{ij} \times D_{pi}))}{\sum_{i=1}^{n}(x_{ij} \times D_{pi})} \right) \tag{3}$$

$$Min \sum_{j=1}^{m} W_{mj} = \sum_{j=1}^{m} \left(y_j \times \frac{(T_{mj} - \sum_{i=1}^{n}(x_{ij} \times D_{mi}))}{\sum_{i=1}^{n}(x_{ij} \times D_{mi})} \right) \tag{4}$$

Subject to:

$$\sum_{j=1}^{m} x_{ij} = 1 \ \forall i \in I \tag{5}$$

$$\sum_{i=1}^{n}(x_{ij} \times D_{pi}) \leq T_{pj} \times y_j \tag{6}$$

$$\sum_{i=1}^{n}(x_{ij} \times D_{mi}) \leq T_{mj} \times y_j \tag{7}$$

$$x_{ij}, y_j \in \{0,1\} \ \forall i \in I \ and \ \forall j \in J \tag{8}$$

The constraint (5) assigns a VM to only one of the servers (not more than one server). Constraint (6) model the CPU constraint of the server. Constraint (7) model the memory constraint of the server. Constraint (8) defines the domain of the variables of the problem.

5   The Proposed ACO Algorithm for Multi-Objective VM Placement

Given a set of n virtual machines and a set of m physical machines, there are a total of m^n possible VM placement solutions [3]. . For example, the number of possible solutions to place 15 VMs on 10 physical machines just computes 10^{15} =1,000,000,000,000,000 possible solution. Even if one million solutions could be evaluated per second, examining all 10^{15} possible solutions would require more than 31 years. It is therefore typically impractical to make a complete enumeration of all possible solutions to find the best solutions. This section shows how to apply an ACO algorithm to efficiently search for good solutions in large solution spaces and solve the formulated problem in Section 4. The pseudo code of the proposed ACO algorithm is as follows.

Input: Set of VMs and set of hosts with thresholds of resources utilization
Output: The Pareto set of best solutions for VMs placement
Steps:
 1. **Initialize**:
        Set t=0.
        Set Pareto set empty.
        Set Initial value $\tau_{ij}(t) = \tau_0$ for each path between VM and host.
 2.   **For** k=1 to NOA (number of ants) do
                Sort the host list in random order.
                Do ants_trip **while** any VM isn't placed.
                        Get the new host from the host list.
                        For each remaining VM that can be placed into the current host.
                                Compute the probabilistic transition rule to select VM to be placed according to probabilistic transition rule.
                        End For
                End Do
                Compute the value of the two objectives for the current solution.
                If the current solution is not dominated by any other solutions in Pareto set **Then**
                        Add this solution to Pareto set
                        Eliminate the solutions dominated by the added one from Pareto set.
                End If
        End For
 3. **Apply** local pheromone update according to local pheromone updating rule.
 4.   **Apply** global pheromone update according to global pheromone updating rule
 5.   **Increment** t by one.
 6.   **If** $(t < t_{max})$
                Goto step 2
        Else
                Return Pareto set (the set of non dominated solutions).
        End If

In an initialization phase of the proposed algorithm, the parameters are initialized, Pareto set is empty and all the pheromone trails are initialized. Iterations are indexed by t, $0 \le t < t_{max}$, where t_{max} is the maximum number of iterations allowed. In the iterative phase each ant receives all VM requests, introduces a physical server one by one and starts assigning VMs to hosts. This is achieved by the use of a probabilistic transition rule, which describes the desirability for an ant to choose a suitable VM as the next one to place into current host. This rule is based on the information about the current pheromone concentration on the movement and a heuristic which guides the ants toward choosing the most appropriate VM depending on the objective functions. After all ants have constructed their solutions, local pheromone updating and a global pheromone updating are performed. Local pheromone updating depends on the founded solution in this iteration. Global pheromone updating depends on current Pareto set solutions. The main operations of the ACO procedure are initializing pheromone, probabilistic transition rule that is used for choosing next VM to be placed on host, local pheromone updating, global pheromone updating and Pareto set as following:

5.1   Initializing Pheromone

The proposed algorithm starts with a pheromone trails matrix. The amount of virtual pheromone trail $\tau_{ij}(t)$ on the edge connects VM i to host j (will be defined as the favorability of packing VM i into host j). The initial amount of pheromone on edges is assumed to be a small positive constant τ_0 which means a homogeneous distribution of pheromone at time (iteration) t = 0.

5.2   Probabilistic Transition Rule

At each step of the construction of a solution, a VM is chosen relative to a transition probability which depends on two factors: a pheromone factor and a heuristic factor as in Eq. (9).

$$p_{ij}^k(t) = \begin{cases} \arg\max_{s \in allowed_{VM}} \{(\alpha \times \tau_{sj}(t)) + ((1-\alpha) \times \eta_{sj})\} & \text{if } q \le q_0 \\ J & \text{if } q > q_0 \end{cases} \quad (9)$$

Where, α is a parameter that controls the relative importance of pheromone trail and the visibility information (heuristic factor). q is a random number uniformly distributed in [0, 1] and If q is greater than q_0, this process is called exploration; otherwise it is called exploitation [3]. q_0 is a fixed parameter $(0 \le q_0 \le 1)$ determined by the relative importance of exploitation of accumulated knowledge problem versus exploration. $\tau_{sj}(t)$ shows the pheromone concentration at the time t on the path between VM s and host j, $allowed_{VM}$ is the set of VMs that are qualified for inclusion in the current host j computed as follows.

$$allowed_{VM} = \{i \in \{1,...,n\} \mid (\sum_{r=1}^{m} x_{ir} = 0) \wedge ((\sum_{u=1}^{n}(x_{uj} \times D_{pu}) + D_{pi}) \le T_{pj}) \wedge$$

$$((\sum_{u=1}^{n}(x_{uj} \times D_{mu}) + D_{mi}) \le T_{mj})\}$$

η_{sj} is the heuristic information. This information indicates the desirability of assigning VM i to host j. The heuristic information considers the partial contribution of each move to the multi-objective function value (minimize the CPU resource wastage and the memory resource wastage). The partial contribution of assigning VM i to host j for the first objective function can therefore be calculated as in Eq. (10). It is shown from Eq. (10) the desirability of assigning VM i to host j will be increased when the amount of used CPU resources increased and the amount of CPU resources wastage decreased.

$$\eta_{ij,1} = \frac{1}{\sum_{v=1}^{j} W_{pj}} \tag{10}$$

Similarly to the first objective function, the partial contribution of the second objective function that is related to the memory resources wastage can be calculated by Eq. (11).

$$\eta_{ij,2} = \frac{1}{\sum_{v=1}^{j} W_{mj}} \tag{11}$$

There are several ways to combine desirability in multi-objective problem to find the total desirability of each ant movement [3]. In this paper we propose the following formula to calculate the total desirability of assigning VM i to host j for the two objective functions:

$$\eta_{ij} = \eta_{ij,1} * \eta_{ij,2} \tag{12}$$

Finally, J is a random variable selected according to the following random-proportional rule probability distribution Eq. (13) which is the probability that ant k chooses the next VM to be placed.

$$J = \begin{cases} \dfrac{(\alpha \times \tau_{ij}(t)) + ((1-\alpha) \times \eta_{ij})}{\sum_{s \in allowed_{VM}} ((\alpha \times \tau_{sj}(t)) + ((1-\alpha) \times \eta_{sj}))} & \text{if } j \in allowed_{VM} \\ 0, & \text{otherwise} \end{cases} \tag{13}$$

5.3 Local Pheromone Updating Rule

After iteration, local pheromone updating which is applied to all edges is refreshed by Eq. (14).

$$\tau_{ij}(t+1) = (1-\rho)\tau_{ij}(t) + \Delta\tau_{ij}(t) \tag{14}$$

Where, ρ is the trail decay, $0 < \rho < 1$ and $\Delta\tau_{ij}(t)$ is computed by Eq. (15).

$$\Delta\tau_{ij}(t) = \sum_{k=1}^{NOA} \Delta\tau_{ij}^{k}(t) \tag{15}$$

Where, $\Delta\tau_{ij}^{k}(t)$ is a quantity of pheromone of ant k computed by Eq. (16).

$$\Delta\tau_{ij}^k(t) = \begin{cases} \dfrac{Q}{L^k(t)} & if\,(i,j) \in T^k(t) \\ 0 & if\,(i,j) \notin T^k(t) \end{cases} \qquad (16)$$

Where, $T^k(t)$ is the tour or solution done by ant k at iteration t, $L^k(t)$ is length of tour (the total memory resource wastage and the total CPU resource wastage of this tour) that is computed by Eq. (17), and Q is an adaptive parameter. It is very helpful in the cloud environment to depend on the result in the past task scheduling.

$$L^k(t) = \sum_{j=0}^{um} W_{pj} \times \sum_{j=0}^{um} W_{mj} \qquad (17)$$

Where, um is the total number of used severs in this tour.

5.4   Global Pheromone Updating

When all ants complete a traverse, global pheromone updating reinforces pheromone on the edges belonging to the all non-dominated or Pareto solutions computed by Eq. (18).

$$\tau_{ij}(t+1) = \tau_{ij}(t) + \Delta\tau_{ij}^+(t) \qquad (18)$$

Where $\Delta\tau_{ij}^+(t)$ is computed by Eq. (19).

$$\Delta\tau_{ij}^+(t) = \sum_{sol=1}^{nos} \Delta\tau_{ij}^{sol}(t) \qquad (19)$$

Where, nos is the number of solutions in Pareto set and $\Delta\tau_{ij}^{sol}(t)$ is a quantity of pheromone of the solution sol in Pareto set computed by Eq. (20) on each edge (i,j) that it has used.

$$\Delta\tau_{ij}^{sol}(t) = \begin{cases} \dfrac{Q}{L^{sol}(t)} & if\,(i,j) \in T^{sol}(t) \\ 0 & if\,(i,j) \notin T^{sol}(t) \end{cases} \qquad (20)$$

Where, $L^{sol}(t)$ is the length of the non-dominated solution tour (T^{sol}). This global updating rule tries to increase the learning of ants

5.5   Non-Dominated (Pareto) Solution

The majority of existing multi-objective optimization problems uses the concept of dominance during selection. The goal is to efficiently obtain a set of non-dominated solutions (the Pareto set) and can be evaluated using the procedure in [15].

6   Implementation and Experimental Results

The researcher has used CloudSim for experimenting in simulated cloud environment because CloudSim can be used to model data centers, host, service brokers, scheduling and allocation policies of a large scaled cloud platform [16, 17]. The experiments are implemented using CloudSim platform. We randomly generated problem instances. The instances were a demand set of CPU and memory utilizations for different numbers of VMs. The number of servers was set to the number of VMs in order to support the worst VM placement scenario, in which only one VM is assigned per server. After the VM placement algorithm was finished, if there were several non-dominated solutions, one solution belonging to the set of non-dominated solutions was randomly chosen. The VM placement algorithms that are compared in the experiments include: FFD-CPU, FFD-MEM, BFD-CPU, BFD-MEM, VMPACS algorithm and the proposed algorithm.   FFD places items in a decreasing order of size after that and at each step; the next item is placed to the first available host.   FFD-CPU is the FFD solution sorted by virtual-machine CPU processing requirements and FFD-MEM represents the FFD solution sorted by memory requirements.   BFD likes FFD but it places a virtual machine in the fullest server that still has enough capacity. BFD-CPU and BFD-MEM are the BFD solutions sorted by CPU processing requirements and memory requirements, respectively.

The programs for the compared algorithms were coded in the Java language under CloudSim platform that ran on an Intel Pentium processor with 1.7 GHz CPU and 256 MB RAM. . The parameter settings of VMPACS algorithm are as follows.   NOA (number of ants) = 10, t_{max} (number of iterations) = 100, α = 0.45, ρ_l = ρ_g = 0.35, and q_0 = 0.8   as in [3]. The parameters (α, ρ, t_{max}, m the number of ants and Q) considered here are those that affect directly the computation of the algorithm. Several values for each parameter were tested while all the others were held constant on 100 VMs.   The parameter settings of the proposed algorithm were determined to be α = 0.4, ρ= 0.3, t_{max}=100, NOA = 8, Q=1.0, τ_0 = .01 and q_0 = 0.9. VM placement under the proposed and VMPACS algorithms combines the partial solution information under construction and the feed information of the reserved time of a non-dominated solution in the external set and simultaneously incorporates continuous updating of pheromone, therefore they can find more appropriate VM placement and achieve better performance.   The CPU resources wastage and the memory resources wastage of the same experiment are shown in Fig. 1 and Fig. 2 respectively. It is shown from Fig. 1 and Fig. 2 that the proposed algorithm can find the solutions with a smaller number of used servers and high resource utilization compared to FFD-CPU, FFD-MEM, BFD-CPU, BFD-MEM and VMPACS algorithms. The proposed algorithm also produces the lowest resource wastage in many cases of different number of virtual machines. The proposed algorithm outperforms FFD-CPU, FFD-MEM, BFD-CPU and BFD-MEM algorithms because it is able to search the solution space more efficiently based on models for minimizing total CPU processing resources wastage, memory resources wastage and the total power consumption.   The proposed algorithm outperforms VMPACS algorithm because the resource wastage model in VMPACS algorithm deals with CPU and memory as commensurable objective function measured in the same units but the proposed algorithm deals with the CPU and memory as non-commensurable and use two objective functions.

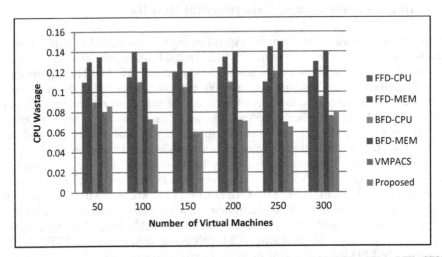

Fig. 1. The total CPU processing resources wastage for FFD-CPU, FFD-MEM, BFD-CPU, BFD-MEM, VMPACS and proposed algorithm

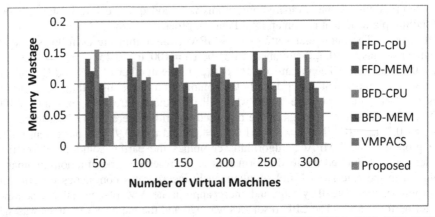

Fig. 2. The total memory resources wastage for FFD-CPU, FFD-MEM, BFD-CPU, BFD-MEM, VMPACS and proposed algorithm

Fig. 3 plots the time required to generate set of non-dominated solutions for all algorithms with the same procedure used in [15]. The proposed algorithm takes less than one minute to solve the difficult 300-VM and 300-host placement problem. Therefore, we can say that the proposed algorithm and VMPACS algorithm are suitable for large data centers moreover the proposed algorithm take less time to placement process in each size of VMs than VMPACS. The reason is that the proposed algorithm use less computational approaches in updating local and global pheromone than VMPACS.

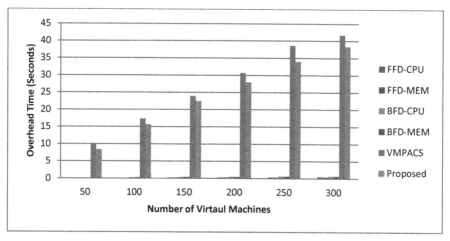

Fig. 3. The overhead time required for FFD-CPU, FFD-MEM, BFD-CPU, BFD-MEM, VMPACS and proposed algorithm

7   Conclusions and Future Work

In this paper, the problem of virtual machine placement is formulated as a multi-objective optimization problem aiming to simultaneously optimize total memory resource wastage and total processing resource wastage. After that, ant colony optimization algorithm was proposed to solve the formulated problem. The proposed algorithm developed to effectively deal with the potential large solution space for large scale data centers and to obtain a set of non-dominated solutions. The proposed algorithm compared with VMPACS algorithm and well-known bin-packing algorithms such as FFD-CPU, FFD-MEM, BFD-CPU and BFD-MEM have been evaluated with the number of virtual machines varying from 50 to 300. The results demonstrate that proposed algorithm is the superior and outperformed the compared approaches. In the future work, the proposed algorithm will be compared with other meta heuristics algorithms to ensure its high performance. Moreover, the proposed algorithm will be optimized to handle peak temperature among the servers and the carbon emission from the proposed algorithm will be considered. In addition, using   rough sets and ant colony optimization in feature extraction may provide more challenging and may allow us to refine our learning algorithms and/or approaches to the minimizing resource wastage [18].

References

1. Akande, A.O., April, N.A., Belle, J.V.: Management Issues with Cloud Computing. In: Proceedings of the Second International Conference on Innovative Computing and Cloud Computing, New York, NY, USA, pp. 119–124 (2013)

2. Tawfeek, M.A., El-Sisi, A., Keshk, A.E., Torkey, F.A.: Cloud task scheduling based on ant colony optimization. In: International Conference on Computer Engineering & Systems ICCES, Egypt (2013)
3. Gao, Y., et al.: A multi-objective ant colony system algorithm for virtual machine placement in cloud computing. J. Comput. System Sci. 79(8), 1230–1242 (2013)
4. Cardosa, M., Korupolu, M., Singh, A.: Shares and utilities based power consolidation in virtualized server environments. In: Proceedings of IFIP/IEEE Integrated Network Management (IM 2009), pp. 327–334 (2009)
5. Grit, L., Irwin, D., Yumerefendi, A., Chase, J.: Virtual machine hosting for networked clusters: Building the foundations for autonomic orchestration. In: Proceedings of the 2nd International Workshop on Virtualization Technology in Distributed Computing (2006)
6. Xu, J., Fortes, J.A.B.: Multi-Objective Virtual Machine Placement in Virtualized Data Center Environments. In: IEEE/ACM International Conference on Cyber, Physical and Social Computing (CPSCom), Green Computing and Communications (GreenCom), pp. 179–188 (2010)
7. Pacini, E., Mateos, C., Garino, C.G.: Distributed job scheduling based on Swarm Intelligence: A survey. Computers and Electrical Engineering 40(1), 252–269 (2014)
8. Dosa, G., Li, R., Han, X., Tuza, Z.: Tight absolute bound for First Fit Decreasing bin-packing. Theoretical Computer Science 510(0), 13–61 (2013)
9. Wang, J., et al.: Best fit decreasing based defragmentation algorithm in semi-dynamic elastic optical path networks. In: Communications and Photonics Conference (ACP), Asia, pp. 1–3, (2012)
10. Khanna, G., Beaty, K., Kar, G., Kochut, A.: Application performance management in virtualized server environments. In: Proceedings of the 10th IEEE/IFIP Network Operations and Management Symposium (NOMS), pp. 373–381 (2006)
11. Chaisiri, S., Lee, B., Niyato, D.: Optimal virtual machine placement across multiple cloud providers. In: IEEE Asia-Pacific Services Computing Conference (APSCC), pp. 103–110 (2009)
12. Speitkamp, B., Bichler, M.: A Mathematical Programming Approach for Server Consolidation Problems in Virtualized Data Centers. IEEE Transactions on Services Computing 3(4), 266–278 (2010)
13. Mi, H., et al.: Online Self-Reconfiguration with Performance Guarantee for Energy-Efficient Large-Scale Cloud Computing Data Centers. In: IEEE International Conference on Services Computing (SCC), pp. 514–521 (2010)
14. Feller, E., Rilling, L., Morin, C.: Energy-Aware Ant Colony Based Workload Placement in Clouds. In: 12th IEEE/ACM International Conference on Grid Computing (GRID), pp. 26–33 (2011)
15. Deb, K.: Multi-Objective Genetic Algorithms: Problem Difficulties and Construction of Test Problems. Evolutionary Computation 7, 205–230 (1999)
16. Calheiros, R.N., et al.: CloudSim: a toolkit for modeling and simulation of cloud computing environments and evaluation of resource provisioning algorithms. Software: Practice and Experience 41(1), 23–50 (2011)
17. Sakellari, G., Loukas, G.: A survey of mathematical models, simulation approaches and testbeds used for research in cloud computing. Simulation Modelling Practice and Theory 39, 92–103 (2013)
18. Hassanien, A.E., Suraj, Z., Slezak, D., Lingras, P.: Rough computing: Theories, technologies and applications. IGI Publishing Hershey, PA (2008)

A Hybrid Chaos and Fuzzy C-Means Clustering Technique for Watermarking Authentication

Kamal Hamouda[1], Mohammed Elmogy[2], and B.S. El-Desouky[3]

[1] Department of Mathematics, Statistics Division and Computer Science,
Mansoura University, Egypt
abo_malek98@yahoo.com
[2] Information Technology Dept., Faculty of Computers and Information,
Mansoura University, Egypt
melmogy@mans.edu.eg
[3] Department of Mathematics, Faculty of Science, Mansoura University,
35516 Mansoura, Egypt
b_desouky@yahoo.com

Abstract. In the last two decades, different watermarking schemes are proposed for image authentication and copyright protection. The accuracy of both tamper detection and localization are considered as two important aspects of authentication of watermarking schemes. In this paper, a novel hybrid Chaos and Fuzzy C-Means (FCM) clustering technique is proposed for watermarking authentication. The proposed technique is used to improve the security of the watermarking system and it can be applied to any image with different sizes. The Chaotic maps are sensitive to initial values, so watermark can't be established without the same initial values. Therefore, it gives high security to scheme. The Fuzzy C-Means clustering technique is used to make the watermark dependent on the plain image. The experimental results show that the proposed scheme achieves superior tamper detection and localization accuracies under different types of image attacks.

Keywords: Digital watermarking, Image authentication, Chaotic map, Fuzzy C-Means (FCM) clustering, Tamper detection.

1   Introduction

By virtue of the widespread and rapid growth of Internet technologies, people can exchange information easily and rapidly. Digital media such as text, audio, images, and video are commonly transmitted via the Internet. Trustworthy, digital multimedia plays an important role in applications such as news reporting, intelligence information gathering, criminal investigation, security surveillance, and health care. Many image processing software are developed, therefore the digital data can be easily manipulated, tampered, and distributed with the help of the powerful image processing tools. The digital multimedia authentication and copyright have become an important issue, so digital watermarking has been proposed. For instance, the ease and

A.E. Hassanien et al. (Eds.): AMLTA 2014, CCIS 488, pp. 165–176, 2014.

extent of such manipulations emphasize the need for image authentication techniques in applications where verification of integrity and authenticity of the image content is essential. The digital multimedia authentication and copyright can be applied by the knowledge called digital watermarking. It is the process that embeds data called a watermark into a multimedia object. This process is imperceptible to human observer in such a way that the watermark can be later on detected or extracted for object assertion purposes by computer algorithm [1].

Digital watermarking can be classified according to the invisible watermark into three different categories [2]: robust, fragile, and semi-fragile watermarking. A robust digital watermarking is performed to assure that the encapsulated information cannot be destroyed during any computer attack. A fragile digital watermarking is performed to ensure that detecting the presence of alterations in the image and the encapsulated information can be easily destroyed during any malicious attacks and non-malicious attacks. In other words, a fragile watermarking can detect any modification to the image. A semi-fragile digital watermarking is like the fragile one. It can be destroyed by certain types of attacks, e.g., addition or removal while resisting minor changes called non-malicious attacks e.g., JPEG compression. In other words, it can localize and detect modifications to some important applications for robust watermarking, such as fingerprinting, data mining, copyright protection, and ownership of the digital media. The main application field of fragile and semi-fragile watermarking is the image and video content authentication. So, image authentication watermarking techniques can be classified into three different categories [2]: fragile, semi-fragile, and content-based fragile watermarking. Content-based watermarking can detect only the significant changes in the image when we permit content saving processing, for example coding and scanning. A fragile watermarking constitutes three components [3]: embedding of the watermark, tamper detection, and tamper localization. The embedding of the watermark is a procedure which applies a watermark on the source image before its distribution. The marked image is substantially identical to the original image for an external user. The tamper detection is principally based on statistical processes. The tamper localization locates the tampered areas on the image. A result of the tamper localization process can be a two level image showing the ground of the tampered regions. The watermarking extraction techniques are classified into three different categories [4]: non-blind, semi-blind, and blind. Non-blind methods need the original image which limit their usage since the original image is complicated to obtain on occasion. Semi-blind methods do not require the original image as they need the watermark or little side information. Blind methods require neither the original image nor the watermark.

The remainder of this paper is organized as follows. In section 2, some current related work on Chaos and FCM clustering techniques are discussed. In section 3, the proposed watermarking scheme is discussed which includes authentication data embedding procedure and tamper detection procedure. The experimental results are presented in section 4. Finally, the conclusion is briefly described in section 5.

2   Related Work

In last decades, there are several watermarking schemes have been proposed for image authentication. In this section, we will take a look at some of published work in the field of watermarking. For example, Chen and Wang [5] proposed a block-based fragile watermarking scheme for image authentication. The FCM clustering technique is proposed to construct the relationship between image blocks and provide more accurate tamper detection and localization. The effectiveness of their proposed scheme can effectively detect and locate unauthorized manipulations caused by a series of attacks. They compared their work with others with respect to the image quality, tamper detection without post processing, and tamper detection with post processing scheme. On the other hand, their proposed scheme deal only with square images and they did not mention how to detect the tamper in the processed image.

Di Martino and Sessa [6] proposed a fuzzy transform for a fragile watermarking tamper detection with compressed images. A fuzzy transform is applied to compress each blocks and get on the outcome reduce the memory necessary to the storage of the image data set. Also, the FCM clustering technique is proposed to construct the relationship between image blocks and provide more accurate tamper detection and localization. The proposed scheme can resist various types of computer attacks and it is good in terms of accuracy for tamper detection.

Suthaharan [7] proposed a fragile image watermarking technique for pixel level tamper detection and resistance. A Logistic map is proposed to take advantage of its sensitivity to a small change in the initial condition. It uses five most significant bits of pixels to generate watermark bits and embeds them in the three least significant bits. It improves tamper detection and resistance capabilities. Also, the confusion process is used to induce complexity in the relation map between the distribution of the watermark and the value of the user-defined key. He presented two new approaches called nonaggressive and aggressive tamper detection algorithms.

Nesakumari and Maruthuperumal [4] proposed a Chaotic system which depends on the normalized image watermarking scheme. Their normalization transform which consists of the Arnold cat map transforming and Logistic map has better robustness. It overcomes geometrical attacks. Their proposed scheme deals only with the square images.

Tong et al. [8] proposed a watermarking approach to provide an enhanced tampering localization and self-recovery. They divided the image into non-overlapping 2×2 blocks. A Chaotic map is applied to improve the rate of tampering localization and security. Furthermore, a sister block embedding scheme is proposed to improve the recovery effect after tampering. Their proposed scheme can be more secure and has better effect on tampering detection and recovery even though the tampered area is relatively large.

Jiang and Chen [9] proposed a Chaos in interpolatory orthogonal multi-wavelets domain for a digital watermarking algorithm. The Logistic Chaotic mapping and Arnold transforming are employed to scramble the original watermarking image in order to improve the security of the watermark embedding algorithm. The watermark information is embedded into the middle frequency interpolatory orthogonal multi-wavelets transforming coefficients of the image to find the multi-resolution decomposition coefficients from the samples of the signal rather than the inner products.

Teng et al. [10] proposed a cryptanalysis and an improvement of a Chaotic system based on the fragile watermarking scheme to detect the tampered areas. The improvement measure is presented to enhance the security of the fragile watermarking scheme. Their experimental results showed that the improvement measure is presented to enhance the security of the fragile watermarking scheme.

Our proposed watermarking scheme is based on the Arnold cat map (ACM) [11, 12, 13], the Logistic map [14], and the FCM clustering technique [15]. We divide the image of size N × M into non-overlapping blocks according to the greatest common divisor of N and M. The places of the pixels of the image are change by using ACM and a Chaotic image pattern. They are generated by using Logistic map to improve the security of the proposed scheme. The Fuzzy C-Means clustering technique is used to make the watermark dependent on the plain image. A watermark is obtained by using exclusive-or (XOR) operation between Chaotic image pattern obtained by using logistic map (the binary watermark) and a binary logo obtain by FCM clustering. The image is divided into 8-bit planes and the LSB plane is used for watermark embedding.

3   The Proposed Scheme

The proposed scheme consists of two procedures: authentication data embedding procedure and tamper detection procedure. Details of the proposed scheme are described in the following subsections.

3.1   Watermark Embedding

The inputs of the watermark embedding phase are a grayscale original image G of size $N \times M$ pixels, the number of clusters C, the weighting exponent m, the termination threshold ε For FCM, a and b parameters, the number of iterations k of the Arnold's cat map, μ , and x_0 of the Logistic map and a binary watermark WM of size $N \times M$. Whereas, the output of this phase is a watermarked image G_W and U is a fuzzy partition matrix or called membership matrix. As shown in Fig. 1, the watermark embedding phase consists of 6 steps as follows:

In step1, let a grayscale original image G of size $N \times M$ pixels is presented. We calculate the greatest common divisor of N and M called $dpixel$. After that, we divide the original image G into non-overlapping $dpixel \times dpixel$ blocks $B_j (1 \leq j \leq (\frac{N}{dpixel} \times \frac{M}{dpixel})$. Moreover, each block image B_j is scrambled alone using ACM k times. Then, each block image in new image denoted by BG_{scr} of size $N \times M$ is collected.

In step2, BG_{scr} of size $N \times M$ is rearranging to get a new matrix denoted by $B'G_{scr}$ of size $(N \times M) \times 1$ by applying the FCM clustering. Any pixel is considered as a block in $B'G_{scr}$ to acquire a membership matrix $U \in [0,1]$ of size $C \times (N \times M)$ and a set of Cluster centers V. Then, we obtain a new membership matrix $\ddot{U} \in [0,1]$ of size $C \times (N \times M)$ by rearranging the C membership degrees of each column in matrix U in a descending order. After that, generate a feature sequence $F \in \{0,1\}$ of size $1 \times (N \times M)$ by the following equation:

$$F = \left[\ddot{U}_{1j} - \ddot{U}_{cj} \right], 1 \leq j \leq N \times M, \tag{1}$$

where $\ddot{U}_{1j}$ and $\ddot{U}_{cj}$ represent to maximum and minimum values of jth column in $\ddot{U}$, respectively, and,$[.]$ represent to $round$ operation. Rearrange F to get the pattern corresponding with the image BG_{scr} of size $N \times M$.

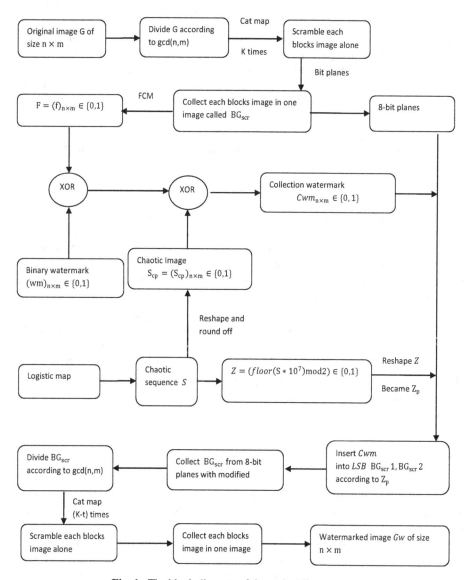

Fig. 1. The block diagram of the embedding process

In step3, we apply the logistic map with initial condition μ and x_0 to generate a Chaotic sequence S of size $1 \times (N \times M)$. The Chaotic sequence is rounded off and rearranged to get the Chaotic image pattern $S_{cp} \in \{0, 1\}$ of size $N \times M$. Also, an integer sequence $Z \in \{0, 1\}$ of length $1 \times (N \times M)$ through S is generated using the following formula:

$$Z = \lfloor S * 10^7 \rfloor \bmod(2), \tag{2}$$

where mod and $\lfloor . \rfloor$ represent to the rest of the divisible and floor operation. We rearrange integer sequence Z to get the pattern $Z_p \in \{0, 1\}$ corresponding with the image of size $N \times M$.

In step4, the scrambled image BG_{scr} is divided into 8-bit planes, from the most significant bit plane $BG_{scr}8$ to the least significant bit plane $BG_{scr}1$. A bit can contain different amounts of information depending on its position in the pixel. The higher 4 bits (8th, 7th, 6th and 5th) carry 94.125% of the total information of the image. On the other hand, the lower 4 bits (4th, 3rd, 2nd and 1st) carry less than 6% of the image information.

In step5, a binary Chaotic watermark $C_{WM} \in \{0, 1\}$ is obtained of size $N \times M$ by using exclusive-or operation between the watermark WM, F and S_{cp} as the following formula:

$$Cwm = F \oplus WM \oplus S_{cp}. \tag{3}$$

The watermark Cwm is dependent on the plain original image due to the FCM clustering (F). So, different original image generate different watermark and any modification on the watermarked image will affect the watermark even if the watermark is obtained before. The lower two bit planes of $BG_{scr}1$ and $BG_{scr}2$ are replaced by Cwm according to Z_p. If $Z_p = 0$, we replace the correspondence position bit in $BG_{scr}1$; if $Z_p = 1$, we replace the correspondence position bit in $BG_{scr}2$.

In step6, all 8-bit planes of image BG_{scr} of size $N \times M$ are collected with new modified in $BG_{scr}1$ and $BG_{scr}2$. We take the greatest common divisor of N and M called $dpixel$. After that, we divide BG_{scr} into non-overlapping $dpixel \times dpixel$ blocks $B_j (1 \leq j \leq (\frac{N}{dpixel} \times \frac{M}{dpixel}))$. Moreover, each block image B_j is scrambled alone, using ACM $(T - k)$ times. Then; we collect each blocks image in new image called watermarked image G_W of size $N \times M$, where T is the period of cat map depend on a, b and k.

3.2   Tamper Detection

The inputs of the tamper detection are a watermarked image G_W of size $N \times M$ pixels, the fuzzy partition matrix U or called membership matrix, the number of iterations k of the Arnold's cat map, μ and x_0 of the Logistic map, and a binary watermark WM of size $N \times M$. Also, the outputs of the tamper locate the tamper region. As shown in Fig. 2, the tamper detection phase consists of 3 steps as follows.

In step1, we have a grayscale original image G_W of size $N \times M$ pixels. We take the greatest common divisor of N and M called $dpixel$. After that, the image G_W is divided into non-overlapping $dpixel \times dpixel$ blocks $B_j (1 \leq j \leq (\frac{N}{dpixel} \times \frac{M}{dpixel})$. Moreover, each block image B_j is scrambled alone, using ACM k times. Then, we collect each blocks image in new image denoted by BGw_{scr} of size $N \times M$. Finally, we divide the scrambled image BGw_{scr} into 8-bit planes.

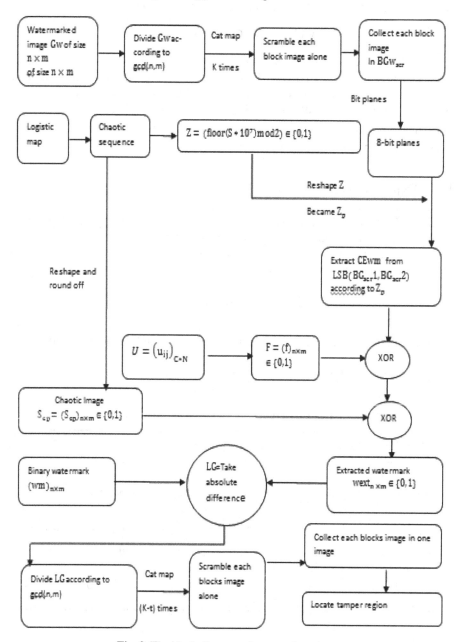

Fig. 2. The block diagram of tamper detection

In step2, we obtain the same Chaotic image pattern Z_p and S_{cp} as in step3 and F as in step2 of embedding algorithm. Then, we extract direct watermark CEwm $\in \{0, 1\}$ of size N × M   from LSB($BGw_{scr}1, BGw_{scr}2$) according to Z_p. We generate extracted watermark Wext $\in \{0, 1\}$ of size N × M by using exclusive-or operation between the watermark WM, F and S_{cp} as the following formula:

$$Wext = F \oplus CEwm \oplus S_{cp}. \qquad (4)$$

In step3, we take the absolute difference of extracted watermark Wext and a binary watermark WM denoted by $LG \in \{0, 1\}$ of size N × M. We take the greatest common divisor of N and M called $dpixel$. After that, we divide LG into non-overlapping $dpixel \times dpixel$ blocks $B_j (1 \leq j \leq (\frac{N}{dpixel} \times \frac{M}{dpixel})$. Moreover, each block image B_j is scrambled alone, using ACM (T-k) times. Then, we collect each blocks image in new image called locate the tamper region.

4   Experimental Results

A series of experiments were conducted to demonstrate the validity of the proposed fragile watermarking scheme. They are conducted to quantitatively evaluate the performance of the proposed scheme. The peak signal-to-noise ratio (PSNR) was adapted to measure the image quality of a watermarked image[10]:

$$PSNR = 10 \times \log_{10} \frac{255^2}{MSR} (dB), \qquad (5)$$

where MSR is the mean squared error between the original image G and the watermarked image Gw which can be calculated from the following equation:

$$MSR = \frac{1}{M \times N} \sum_{i=1}^{M} \sum_{j=1}^{M} (G_{ij} - Gw_{ij})^2. \qquad (6)$$

Table 1. The PSNR for various standard images

Image name	size	PSNR	Image name	size	PSNR
cameraman	256 × 256	50.3050	sailboat	512 × 512	50.2383
pool	256 × 256	49.6409	doll	320 × 240	50.0682
airfield	1042 × 1024	50.1700	field	320 × 240	49.9811
airplane	512 × 512	50.1988	sofa	320 × 240	50.1785
baboon	512 × 512	50.2674	arctichare	400 × 594	48.9529
boat	512 × 512	50.2769	cablecar	480 × 512	50.1948
fruits	512 × 512	50.2872	mountain	480 × 640	49.6236
goldhill	512 × 512	50.1944	watch	383 × 510	50.4494
peppers	512 × 512	50.2386			

Table 2. The comparison results of PSNR of watermarked images

Image name	Chen and Wang	Bakrawy et al	Proposed scheme
pool	44.48	46.53	49.6409
sofa	44.14	45.99	50.1785
doll	44.20	46.03	50.0682
field	44.07	46.59	49.9811

In this scheme, a grayscale original image G of size $N \times M$ pixels may be $N = M$ called square image. To precede all of the experiments, the values for the weighting exponent m, and the termination threshold ε and the number of clusters C were set to 2, $1e - 5$ and 5, respectively. The parameters of ACM are a = 1, b = 1 and $k = 75$ for square image only. Also, it must be greater than T, so we select value according to divide the original image of different dimensions. The parameters of logistic map are chosen as $\mu = 3.854$ and $x_0 = 0.654$. The size a binary watermark WM adapted with size of original image G.

The peak signal-to-noise ratio (PSNR) for the watermarked images is shown in Table 1. Note that, if N or M is prime, then the greatest common divisor of N and M equal to one, so take k is any value and taking into account that the T must be greater than k.

The comparison results of image quality are summarized for the result or PSNR proposed scheme, Chen and Wang's approach [5] and Bakrawy et al.'s approach [17] show in Table 2.

(a) Original cameraman   (b) Watermarked cameraman image.   (c)Tampered image.   (d) Detected tampered region.

Fig. 3. Text addition attack

 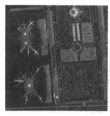  

(a) Original airfield image.   (b) Watermarked airfield image.   (c)Tampered image.   (d) Detected tampered region.

Fig. 4. Copy and paste attack

| (a) Original goldhill image. | (b) Watermarked goldhill image. | (c)Tampered image. | (d) Detected tampered region. |

Fig. 5. Content removal attack

| (a) Watermarked airfield image. | (b) Watermarked airfield image. | (c)Tampered image. | (d) Detected tampered region. |

Fig. 6. Collage attack

| (a) Original fruit image. | (b) Watermarked fruit image. | (c)Tampered image. | (d) Detected tampered region. |

Fig. 7. Gaussian noise attack

| (a) Watermarked baboon image. | (b)Tampered image (cropping). | (c) Cropping with original image size. | (d) Detected tampered region. |

Fig. 8. Copy and paste attack

4.1   Performance Under Attacks

In the experiments, the Original cameraman image is shown in Fig. 3 (a), the watermarked cameraman image is shown in Fig. 3 (b) and it is modified by adding the text "cameraman" at the center of the image is shown in Fig. 3(c). Also, the Original airfield image is shown in Fig.4 (a), the watermarked airfield image is shown in Fig.4 (b) and it is modified by inserting two more planes in the image, where the planes are copied from the same watermarked image is shown in Fig. 4 (c). After that, the Original goldhill image is shown in Fig. 5 (a), the watermarked goldhill image is shown in Fig. 5(b) and it is removed without degrading the image quality. We have removed window of the house from the watermarked image is shown in Fig. 5 (c). Further, the counterfeit image is shown in Fig. 6 (c) was constructed by copying the plane from the watermarked airplane image is shown in Fig. 6(a) and inserting it in sky of watermarked cablecar image is shown in Fig. 6 (b). Moreover, the Original fruit image is shown in Fig. 7 (a), the watermarked fruit image is shown in Fig. 7 (b). We apply the Gaussian noise in the watermarked image is shown in Fig. 7 (c). Finally, the watermarked baboon image is shown in Fig. 8 (b). We apply the cropping attack in the watermarked image is shown in Fig. 8 (b) and Fig. 8 (c).

5   Conclusion

The digital multimedia authentication and copyright have become an important issue, so digital watermarking has been proposed. The digital multimedia authentication and copyright can be applied by the knowledge called digital watermarking. It is the process that embeds data called a watermark into a multimedia object. This process is imperceptible to human observer in such a way that the watermark can be later on detected or extracted for object assertion purposes by computer algorithm.

In this paper, a novel watermarking authentication based on hybrid Chaos and FCM clustering is proposed. Instead of using any secret keys, we used Chaotic maps as keys to make the scheme highly secure and sensitive to initial values. Since the watermark is dependent on the plain image, it is very difficult to find initial values for Chaotic maps and the technique of a FCM clustering to assume the watermark. Experimental results show that our scheme has high fidelity and it is capable of localizing modified regions in watermarked image with different sizes. In the future work, we will work on how to increases the security of our proposed schema.

References

1. Rawat, S., Raman, B.: A chaotic system based fragile watermarking scheme for image tamper detection. AEU - Int. J. Electron. Commun. 65, 840–847 (2011)
2. Chen, Y.-L., Yau, H.-T., Yang, G.-J.: A Maximum Entropy-Based Chaotic Time-Variant Fragile Watermarking Scheme for Image Tampering Detection. Entropy 15, 3170–3185 (2013)

3. Fridrich, J., Goljan, M., Baldoza, A.C.: New fragile authentication watermark for images. In: Proceedings of the International Conference on Image Processing 2000, vol. 1, pp. 446–449 (2000)
4. NesaKumari, G.R., Maruthuperumal, S.: Normalized Image Watermarking Scheme Using Chaotic System. Int. J. Inf. Netw. Secur. IJINS 1, 255–264 (2012)
5. Chen, W.-C., Wang, M.-S.: A fuzzy c-means clustering-based fragile watermarking scheme for image authentication. Expert Syst. Appl. 36, 1300–1307 (2009)
6. Di Martino, F., Sessa, S.: Fragile watermarking tamper detection with images compressed by fuzzy transform. Inf. Sci. 195, 62–90 (2012)
7. Suthaharan, S.: Logistic Map-Based Fragile Watermarking for Pixel Level Tamper Detection and Resistance. EURASIP J. Inf. Secur. 2010, 1–7 (2010)
8. Tong, X., Liu, Y., Zhang, M., Chen, Y.: A novel chaos-based fragile watermarking for image tampering detection and self-recovery. Signal Process. Image Commun. 28, 301–308 (2013)
9. Li, J., Chen, L.: A Digital Watermarking Algorithm Based on Chaos in Interpolatory Orthogonal Multiwavelets Domain. Comput. Inf. Sci. 6 (2013)
10. Teng, L., Wang, X., Wang, X.: Cryptanalysis and improvement of a chaotic system based fragile watermarking scheme. AEU - Int. J. Electron. Commun. 67, 540–547 (2013)
11. Dawei, Z., Guanrong, C., Wenbo, L.: A chaos-based robust wavelet-domain watermarking algorithm. Chaos Solitons Fractals 22, 47–54 (2004)
12. Zhang, C., Wang, J., Wang, X.: Digital Image Watermarking Algorithm with Double Encryption by Arnold Transform and Logistic. In: Fourth International Conference on Networked Computing and Advanced Information Management, NCM 2008, pp. 329–334 (2008)
13. Ahmad, M., Gupta, C., Varshney, A.: Digital image encryption based on chaotic map for secure transmission. In: International Multimedia, Signal Processing and Communication Technologies, IMPACT 2009, pp. 292–295 (2009)
14. Persohn, K.J., Povinelli, R.J.: Analyzing logistic map pseudorandom number generators for periodicity induced by finite precision floating-point representation. Chaos Solitons Fractals 45, 238–245 (2012)
15. Bezdek, J.C.: Pattern recognition with fuzzy objective function algorithms. Kluwer Academic Publishers (1981)
16. Dunn, J.C.: A Fuzzy Relative of the ISODATA Process and Its Use in Detecting Compact Well-Separated Clusters. J. Cybern. 3, 32–57 (1973)
17. El Bakrawy, L.M., Ghali, N., Ella Hassanien, A., Kim, T.: A rough k-means fragile watermarking approach for image authentication. In: 2011 Federated Conference on Computer Science and Information Systems (FedCSIS), pp. 19–23 (2011)

Fan Search for Image Copy-Move Forgery Detection

Sondos M. Fadl, Noura A. Semary, and Mohiy M. Hadhoud

Faculty of Computers and Information, Menofia University, Egypt
{sondos.magdy,noura.samri,mmhadhoud}@ci.menofia.edu.eg

Abstract. Image forgery detection is currently one of the interested research fields of image processing. Copy-Move (*CM*) forgery is one of the most commonly techniques. In this paper, we propose an efficient methodology for fast *CM* forgery detection. The proposed method accelerates blocking matching strategy. Firstly, the image is divided into fixed-size overlapping blocks then Discrete Cosine Transform (*DCT*) is applied to each block to represent its features, which are used to indirectly compare the blocks. After sorting the blocks based on *DCT* coefficients, a distance is measured between nearby blocks to denote their similarity. The proposed Fan Search (FS) algorithm starts once a duplicated block is detected. Instead of exhaustive search for all blocks, the nearby blocks of the detected block are examined first in a spiral order. The experimental results demonstrate that the proposed method can detect the duplicated regions efficiently, and reduce processing time up to 75% less than other previous works.

Keywords: Image tampering, Copy-move forgery, Image fakery detection, Image falsification detection, Blind image forensics.

1   Introduction

As image is better than thousands of words, World Wide Web (*WWW*) nowadays contains a large amount of digital images used for effective communication process. It becomes very ease for anyone to edit photographs by using available image editing tools. Detection of digital image forgery is an important task in many fields such as journalism that form public opinion to the community, crime etc. In addition, fake photos are used to the defaming business and political opinions. This makes it necessary to verify the authenticity of the images.

Image forgery can be classified into three types: (1) image enhancement; such as blurring, contrast or brightness alteration etc., (2) image compositing; that mixes between two or more different images and (3) copy-move forgery; where some parts of an image are copied and pasted to another place of the same image. Also the techniques of image forgery detection can be categorized into two methods: active and passive. Active methods such as watermarking [1] and illegal image copy

A.E. Hassanien et al. (Eds.): AMLTA 2014, CCIS 488, pp. 177–186, 2014.

detection [2-4] depend on prior information about the original image that in many cases is not available. Passive or blind methods should be used to authenticate the image originality such as Pixel-based, Format-based, Physically-based, Geometric-based, Statistical-based and Camera-based techniques [11].

In this paper, *CM* forgery type is considered. Generally, the most performed operations in (*CM*) forgery are either hiding a region in the image, or adding a new object into the image. To detect the duplicated *CM* regions, Block Matching is one of used methods, where the image is dividing into fixed-size overlapping blocks then each block is matched with all other blocks in the same image. Some of features are extracted as representative of blocks that are used for matching with the other blocks. The result is matched blocks with considering distance threshold. However, this method is time consuming strategy. There are various other approaches to enhance the block matching complexity. In general, *CM* forgery detection consists of basic steps; feature extraction, matching, and copy decision based on the similarity information between blocks. In the feature extraction step, features are selected from each block, which are used to compare the blocks.

Huang et al. [5] used *DCT* (discrete Cosine transform) as the discriminative features. The main benefit of using it is the strong energy compaction property of *DCT* that any type of manipulations such as *JPEG* compression and noise addition will not affect the coefficients energy. However the above method fails for any type of geometric transformations of the block such as rotation, scaling etc. Popescu et al. [6] used a principal component analysis (*PCA*) on image blocks as features. In Tripathi's et al. work [8], after dividing the image to blocks, each block has been divided into four equal-sized sub-blocks to give features of each block. This method is good for some geometrical transformations like scaling, rotation.

In this work, we accelerate the *CM* forgery detection time by examining the neighbours of first detected blocks in a fan like order. The proposed system can also give accurate results even in wake of Gaussian blur and *JPEG* compression.

The rest of the paper is organized as follows; Section 2 explains the details of the proposed system. Section 3 presents the experimental results while Section 4 concludes the paper

2   Proposed Method

CM in digital image could be performed for at least one region in the tampered image. The task of detection is to determine whether an image contains duplicated regions. For instance, Figure 1 depicts a famous example of *CM* which not only altered a photograph but also the history as well. The image is since 1930, where a commissar was removed from the original image after falling out of favour with Stalin.

Fig. 1. An example of *CM* forgery; (left) the altered image, (right) the original image

Since the shape and the size of the regions are unknown, it is definitely computationally impossible to examine every possible pairs of regions with different shapes and sizes. It is more effective to divide an image into fixed-sized overlapping blocks.

The proposed system (Figure 2) is applied on grayscale images only. Thus, if the input image is RGB, it converts the image into the corresponding gray scale version by (1).

$$I = 0.228R + 0.587G + 0.114B \qquad (1)$$

In the proposed system the image is divided into 1-pixel overlapped blocks. Features are extracted from each block by applying *DCT*. Zigzag order is used to convert the block into vector, and then some coefficients have been truncate to reduce the dimension of the features and reduce processing time. The blocks are lexicographically sorted by radix sort [9] based on its features and the difference between each nearby pairs of blocks is computed. If the difference is less than some threshold, then two blocks are considered to be similar. Spatial distance between these blocks is calculated to reduce false detection. Once a match is detected, the sequential blocks comparison is stopped and fan search (*FS*) method is applied in a spiral (Fan) order for the 8-directions block neighbors. Each step of the system will be discussed in more details in this section.

2.1   Blocking Stage

For an image of size $M \times N$, the image could be divided into small fixed-size overlapping blocks of $b \times b$ pixels, resulting in B blocks where

$$B = (M-b+1) \times (N-b+1) \qquad (2)$$

2.2   Features Extraction

Each block is represented by the quantized *DCT* coefficients. Assuming the size of the block is b, there are b^2 elements in the transformed coefficients matrix. A zigzag scan is performed to convert the blocks into vector of sorted frequencies (low frequency, middle frequency and high frequency) as shown in Figure 3.

Similar blocks have equal *DCT* coefficients. Low frequencies coefficients are sufficient for determining this equality. So, the *DCT* coefficients vector is truncated up to *L* coefficients to reduce the feature vector length and the processing time. All blocks features are lexicographically sorted by radix sort [9] based on their *L* features and saved in the sorted matrix *Fsorted*.

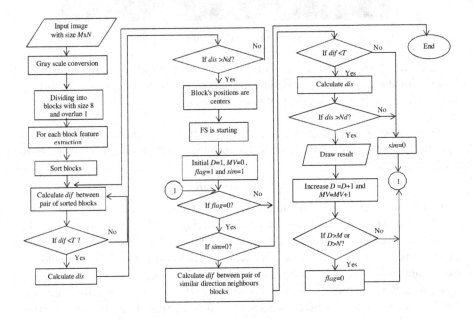

Fig. 2. Flowchart of proposed CM detection method

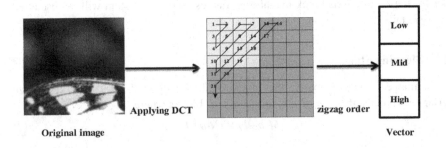

Fig. 3. Formation of column vector

2.3   Duplicated Blocks Detection

Assume two adjacent blocks feature vectors Fs_i and Fs_{i+1}, the difference between both feature vectors is calculated by (3):

$$dif = \sum_{j=1}^{L} |Fs_i^j - Fs_{i+1}^j|/L, \qquad (3)$$

where L is the length of the feature vector. If dif is less than a threshold T, then two blocks are supposed to be similar. In addition, spatial distance is tested by (4) to eliminate the false positives.

$$dis = \sqrt{(B_i^x - B_{i+1}^x)^2 + (B_i^y - B_{i+1}^y)^2} \tag{4}$$

where (B_i^x, B_i^y) is the position of block i and (B_{i+1}^x, B_{i+1}^y) is the position of block $i+1$, Consider only $dis > Nd$.

2.4   Fan Search Method

Once a match is detected, the matched blocks are retrieved as C_1 and C_2, and the position of both block are set to (x_1,y_1) for C_1 and (x_2,y_2) for C_2. FS searching algorithm is applied as the following steps:

1- *Flag* and *Sim* are initialized by 1, which refers to the possibility of comparison between the blocks.
2- Distance (D) and Move variable (MV) are initialize by $D=1$, $MV=0$.
3- Eight neighbours blocks $(P_1, P_2, \dots$ and $P_8)$ blocks are retrieved for C_1 and C_2.
4- Compare between blocks pairs with similar direction as in section 2.3.
5- If blocks are similar, mark the regions in the map image and increase MV by 1. New eight neighbours are identified. If the new eight neighbours are the same eight neighbours examined before, D is increasing by 1, new eight neighbours are identified and go to step 4.
6- If D is greater than image size (M or N) in any direction, set *Flag* by 0 and stop the comparison.
7- If blocks are not similar, set $Sim=0$ and stop the comparison.

Here is an example, if first center is $C_1 = (x_1, y_1)$ and second center is $C_2 = (x_2, y_2)$, Then distance between center and neighbours is $(D = 1,2, \dots 9)$. In order to search in all neighbours, assume move variable which is $(MV = 0,1, \dots)$. Eight parameters, $P_1, P_2, \dots$ and P_8 defines as:

$$P_1 = (x_1 - D, y_1 - D + MV), P_2 = (x_1 - D, y_1 + MV),$$
$$P_3 = (x_1 - D + MV, y_1 + D), P_4 = (x_1 + MV, y_1 + D),$$
$$P_5 = (x_1 + D, y_1 + D - MV), P_6 = (x_1 + D, y_1 - MV),$$
$$P_7 = (x_1 + D - MV, y_1 - D) \text{ and } P_8 = (x_1 - MV, y_1 - D).$$

The blocks of similar direction will be compared. If D is greater than M or N then *flag*=0, and if this blocks are not similar, then set *sim*=0. If *flag*=0 or *sim*=0 in the direction, then this direction is stopped. If blocks are similar, mark the regions in the map image. Also, isolated regions of area that is less than a threshold Ath is eliminated by opening operation. Finally, similar detected regions are marked in the altered image. Figure 4 illustrates the FS process.

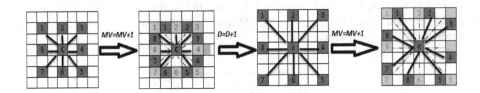

Fig. 4. Example on FS algorithm

3   Experimental Results and Discussion

In this section, we first introduce the experiment procedure and parameters and then, discuss the results to prove the robustness and sensitivity of our algorithm.

3.1   Experiment Method and Procedure

The experiments were carried out on the Matlab R2012a running on 4 GB RAM and 2.30 GHZ processor machine. All images used in the experiment were taken from the data set in [12]. All the images were 128×128 pixels gray image saved in BMP format. All the parameter in the experiment were set as: $b=8$, $T = 0.3$, $Nd = 16$, $Ath = 10$ and $L=9$.

3.2   Visual Result

We have performed four types of experiments to examine the robustness of our system.

Experiment 1: In this experiment, a random rectangular region is copied and pasted in a non-overlapping place. The images shows in Figure 5 present the results of *CM* detection marked on the tampered images. Each row is composed of three images: original image, tampered image and result image from left to right. The four images indicate four possible positions of duplicated regions: horizontal, vertical, diagonal and ant diagonal. Table 1 shows the performance time of *FS* compared to other methods [5], [8], and [10] that applied to 100 images. Note that, the proposed method (*FS*) decreased the processing time up to 75% faster. Precision and Recall has been calculated as the metrics for quantifying the accuracy of forgery detection. We have used 50 images for calculating average precision and recall (see Table 2). From the results, it seems that the proposed system performs other systems in time and detection accuracy.

The precision and recall are defined as follows:

$$Precision = \frac{(Forged\ Region \cap Detected\ Region)}{Detected\ Region} \tag{5}$$

$$Recall = \frac{(Forged\ Region \cap Detected\ Region)}{Forged\ Region} \tag{6}$$

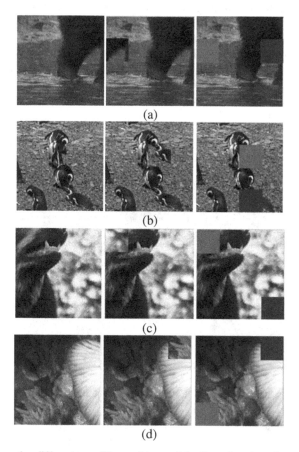

Fig. 5. Shown are the different possible positions of duplicated regions in (a) horizontal, (b) vertical, (c) diagonal, and (d) anti diagonal directions

Table 1. The performance time of different methods

Time(s)	G. Lynch [10]	Y. Huang [5]	R. Tripathi [8]	Proposed FS
	7.68	4.7005	6.4018	1.2981

Table 2. The precision and recall for different methods

	G. Lynch [10]	Y. Huang [5]	R. Tripathi [8]	Proposed FS
Precision	97%	99%	80%	99%
Recall	95%	99%	75%	98%

Experiment 2: An irregular region is copied and pasted in a non-overlapping area. Figure 6 shows the detection result of the system.

Experiment 3: The tempered image is distorted by blurring and *JPEG* encoder before *CV* detection. We have used 100 original images to make 100 tampered images by copying an irregular region at a random location and pasting onto a non-overlapping region using Photoshop editing tool. The tampered images together with their original were then distorted by different processing operations: *JPEG* compression with different qualities ($Q=90$ and $Q=70$) and Gaussian blurring (window size was 3×3 and $\sigma = 1$). Table 3 shows the average precision for different operations.

Experiment 4: The system has been examined with different values of *T*. Table 4 presents the number of true positives and false positives for various *T*, calculated on the images shown in Figure 7 which tempered area contains 564 blocks.

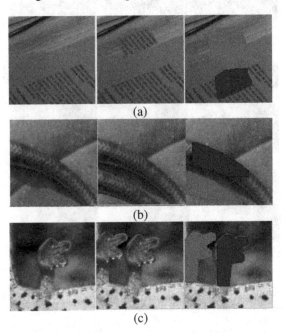

(a)

(b)

(c)

Fig. 6. Result of irregularly regions *CM* forgery detection

Table 3. The precision for different operations

Precision	BMP	JPEG (Q=90)	JPEG (Q=70)	Blurring
	99%	70%	60%	95%

Table 4. The true positive and false positive for different threshold

	Detection rates	
Threshold (*T*)	True Positive	False Positive
0.1	100 %	0 %
0.3	99 %	1 %
0.5	96.5 %	3.5 %
0.7	91.7 %	8.3 %

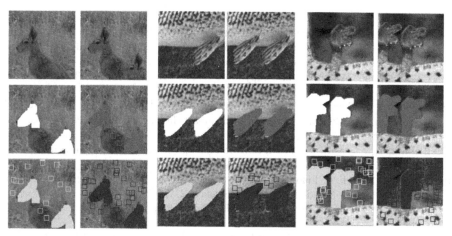

Fig. 7. Results of some images with variant values of T; first row presents the original images (left) and tempered images (right). Second row; White highlights (left) refers to $T=0.1$, and Red highlights (right) refers to $T=0.3$. Third row; Green highlights (left) refers to $T=0.5$, and Blue highlights (right) refers to $T=0.7$.

4   Conclusion and Future Work

In this paper, we have proposed a fast and efficient method for accelerating CM forgery detection. Instead of exhaustive block matching, fan search method is proposed. FS fasten CM forgery detection by comparing only the neighbors of suspected similar blocks. Our method detects the duplicated regions of tempered images even under the influence of blurring and JPEG compression. Results are not only much accurate but also the processing time is accelerated. The experimental results show that the proposed method has the ability to detect CM forgery in an image faster than other systems by about 75%. The method is to be improved for detecting duplicated region under the influence of geometric transformations such as rotation, scale etc.

References

1. Khana, A., Malika, S.A., Alib, A., Chamlawia, R., Hussaina, M., Mahmoodc, M.T., Usmand, I.: Intelligent reversible watermarking and authentication: hiding depth map information for 3D cameras. Information Sciences 216, 155–175 (2012)
2. Hsiao, J., Chen, C., Chien, L., Chen, M.: A new approach to image copy detection based on extended feature sets. IEEE Transactions on Image Processing 16(8), 2069–2079 (2007)
3. Ling, H., Cheng, H., Ma, Q., Zou, F., Yan, W.: Efficient image copy detection using multiscale fingerprints. IEEE Magazine of Multimedia 19(1), 60–69 (2012)
4. Nikolopoulos, S., Zafeiriou, S., Nikolaidis, N., Pitas, I.: Image replica detection system utilizing R-trees and linear discriminant analysis. Pattern Recognition 43(3), 636–649 (2010)

5. Huang, Y., Lu, W., Sun, W., Long, D.: Improved DCT-based detection of copy-move forgery in images. Forensic Science International 206(1), 178–184 (2011)
6. Popescu, A.C., Farid, H.: Exposing digital forgeries by detecting duplicated image regions. Dept. Comput. Sci., Dartmouth College, Tech. Rep. TR2004-515 (2004)
7. Lin, H., Wang, C., Kao, Y.: Fast copy-move forgery detection. WSEAS Transactions on Signal Processing 5(5), 188–197 (2009)
8. Tripathi, R.C., Singh, V.K.: Fast and efficient region duplication detection in digital images using sub-blocking method. International Journal of Advanced Science and Technology 35, 93–102 (2011)
9. Blelloch, G., Zagha, M.: Radix sort for vector multiprocessors. In: Proceedings of the 1991 ACM/IEEE Conference on Supercomputing, pp. 666–675. ACM (1991)
10. Lynch, G., Shih, F.Y., Liao, H.Y.M.: An efficient expanding block algorithm for image copy-move forgery detection. Information Sciences 239, 253–265 (2013)
11. Fridrich, J.: Digital image forensics. IEEE Signal Processing Magazine 26(2), 26–37 (2009)
12. Ng, T., Hsu, J., Chang, S.: Columbia Image Splicing Detection Evaluation Dataset, http://www.ee.columbia.edu/ln/dvmm/downloads/AuthSplicedDataSet/AuthSplicedDataSet

Part IV
Features Extraction and Classification

Features Extraction and Classification of EEG Signals Using Empirical Mode Decomposition and Support Vector Machine

Noran M. El-Kafrawy, Doaa Hegazy, and Mohamed Fahmy Tolba

Faculty of Computer and Information Sciences, Ain Shams University, Cairo, Egypt
nel-kafrawy@acm.org, doaa_hegazy@fcis.asu.edu.eg, fahmytolba@gmail.com

Abstract. Interpreting brain waves can be so important and useful in many ways. Having more control on your devices, helping disabled people, or just getting personalized systems that depend on your mood are only some examples of what it can be used for. An important issue in designing a brain-computer interface (BCI) is interpreting the signals. There are many different mental tasks to be considered. In this paper we focus on interpreting left, right, foot and tongue imagery tasks. We use Empirical Mode Decomposition (EMD) for feature extraction and Support Vector Machine (SVM) with Radial Basis Function (RBF) kernel for classification. We evaluate our system on the dataset 2a from BCI competition IV, and very promising classification accuracy that reached 100% is obtained.

1 Introduction

Consider the potential to manipulate computers or machinery with nothing more than thoughts. Our brains are filled with neurons. Every time we think, move, feel or remember something, our neurons are at work. That work is carried out by small electric signals that zip from neuron to neuron. Although the paths the signals take are insulated, some of the electric signal escapes. Scientists can detect those signals, interpret what they mean and use them to direct a device of some kind [1]. There are several applications that could make use of these data. Such as: assistive technology, virtual reality [2], game controlling [3] and robotics [4, 5]. We are interested in understanding these waves to be able to use them as an input to other systems. Several studies have been made in this field with varying results. We will focus on features extraction and classification of the brain waves.

In this paper we propose a method for classification of the EEG signals. This method consists of three steps: (1) Pre-Processing for eliminating unwanted noise. (2) Features Extraction using Empirical Mode Decomposition (EMD) [6]. (3) Classification using Support Vector Machine (SVM) [7]. The method is applied to real human data, and better results relative to the state of art algorithms are obtained.

In section two, we give related work and the available techniques used to solve the problem of feature extraction and recognition of EEG signals. Details of our

A.E. Hassanien et al. (Eds.): AMLTA 2014, CCIS 488, pp. 189–198, 2014.

proposed model are presented in section three. The data set used, numerical experiments and results are shown in section four, where we show the results obtained by the selected method. Finally, conclusions are driven and presented in section five.

2   Related Work

Many researchers use different techniques in feature extraction [8, 9, 10, 11]. Toka Fatehi et. al [12] gave an overview of different methods extracting features from an EEG signal. These methods include: Time Analysis, Frequency Analysis, Time-Frequency Analysis and Time-Frequency-Space Analysis.

In [12], it is shown that the best classification results were obtained from the space-time-frequency analysis.

However, feature extraction methods aren't restricted on Fourier Transform presented in [12]. In [13], Hualou Liang et. al. showed that the Fourier-based methods are designed for the frequency analysis of stationary time series, while in neurobiology the researchers deal with time series data that are non-stationary. Thus, Fourier-based methods can have limited use in revealing the underlying neurophysiological variations in such data [13]. Moreover, it is mentioned in [13] that the major drawback of Fourier-based approaches is that the basis functions are fixed, and therefore cannot capture any time-varying characteristic of neural signals.

In [14], Wang et. al. addressed the problem of analyzing EEG signals and classifying them into different classes of different motor imaginary tasks. They used Hilbert-Huang transformation method in feature extraction: It consists of two main steps (1) EMD. (2) The Hilbert spectral analysis. They used BP neural network in classification. They classified three different tasks (left hand, right hand and foot) and their best classification rate was 93.8%.

Demir, B. et. al. addressed in [15] a similar problem. They analyzed images using EMD (the same method for feature extraction used in [14]). Their best classification rate was 100%. However, our proposed method mostly differ from this mentioned in [14] in the classification part. We use the SVM with radial basis function (RBF) kernel to classify the feature vector obtained from the EMD process.

In the model proposed in this paper, EMD is used for features extraction. EMD was first introduced by Huang et. al. [6] as a signal decomposition method. EMD provides an alternative to traditional time-frequency methods and its main function is to decompose a signal into a collection of oscillatory modes, called intrinsic mode functions (IMFs). As mentioned in [16], EMD has a useful feature that it relies on a fully data-driven mechanism that does not require any prior known basis. By this feature it differs from the traditional signal analysis tools, such as Fourier or wavelet-based methods, which require some predefined basis functions to represent a signal.

Classification of EGG signal is an emerging and important field in signal processing. Different methods for classifying are proposed and used in the literature. In [17], Lotte et. al. presented a review on classification algorithms for

EEG-based brain-computer interfaces. Moreover, in [17], a survey of the classification algorithms used to design BCI systems is presented. They are divided into five different categories: Linear Classifiers, Neural Networks, Nonlinear Bayesian Classifiers, Nearest Neighbor Classifiers and Combination of Classifiers.

They also provid some guidelines to choose a classifier. Several measures of performance have been proposed in BCI, such as accuracy of classification, Kappa coefficient, Mutual Information, sensitivity and specificity. The most common one is the accuracy of classification and the percentage of correctly classified feature vectors, and this is the only considered measure.

Synchronous BCI is the most widely spread. In [17], three kinds of algorithms proved to be efficient in this context, namely, SVM, dynamic classifiers and combination of classifiers. SVM is very efficient regardless of the number of classes; this is because it has good properties like: regularization, simplicity and immunity to the curse-of-dimensionality. It is said to be the most appropriate classifier to deal with feature vectors of high dimensionality (e.g.: large number of time segments).

3   Proposed Work

Our system starts by reading a previously saved database of EEG signals. These signals have four different motor imaginary tasks (left hand, right hand, both feet and tongue). Our job is to classify these signals into four different classes. We apply EMD to extract features from these signals (IMFs), then we use the SVM classifier with the RBF kernel. Figure 1 shows the main steps of the proposed EEG signal classification model.

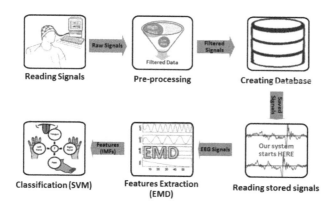

Fig. 1. Block Diagram of the Work

3.1   Pre-Processing

The signals are sampled with 250 Hz and bandpass-filtered between 0.5 Hz and 100 Hz. The sensitivity of the amplifier was set to 100 μV. An additional 50 Hz notch filter was enabled to suppress line noise.

3.2   Feature Extraction: Empirical Mode Decomposition

A Single Channel Case. EMD is a method of breaking down a signal without leaving the time domain. It is useful for analyzing natural signals (like brain waves – EEG). It decomposes the signal into intrinsic mode functions (IMFs), which is easier to analyze.
An IMF is a function with the following properties:

1. Has only one extreme between zero crossings, and
2. Has a mean value of zero.

The EMD uses a process called sifting to decompose the signal into IMFs. The following steps [18] explain its procedure. For a signal $x(t)$, let m_1 be the mean of its upper and lower envelopes as determined from a cubic-spline interpolation of local maxima and minima as shown in Figure 2.

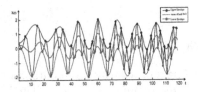

Fig. 2. Iteration 0: Upper & lower envelopes with their mean (m_1)

The difference between the data and m_1 is the first component h_1.

$$h_1 = x(t) - m_1 \tag{1}$$

The sifting process is repeated with h_1 treated as the main data. m_{11} is the mean of the upper and lower envelopes of h_1.

$$h_{11} = h_1 - m_{11} \tag{2}$$

This sifting process is repeated k times, until h_{1k} is itself an IMF.

$$h_{1(k-1)} - m_{1k} = h_{1k} \tag{3}$$

$$c_1 = h_{1k} \tag{4}$$

c_1 contains the shortest period component of the signal. It is separated from the rest of the data:

$$x(t) - c_1 = r_1 \tag{5}$$

And this process is repeated on $r_j : r_1 - c_2 = r_2, \cdots, r_{n-1} - c_n = r_n$ until we get all the possible IMFs of the signal as shown in Figure 5.

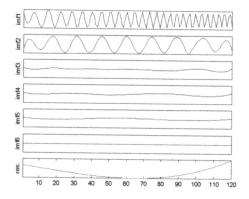

Fig. 3. IMFs obtained using EMD

Characteristic features are obtained from the first two IMFs. As they contain the most features (high frequencies).These are used as input feature vectors for the classifier.

Multichannel Case. To analyze multichannel EEG signals recorded in our experiment synchronously, we decompose each channel separately to prevent any possible oscillatory information leaking among the channels[19]. We then calculate the 1st two IMFs of each channel, sum them and use the sum as a feature vector for the SVM classifier.

Figure 4 represents an event recorded from five channels. Each of them will be separately analyzed.

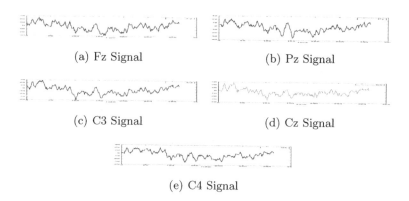

(a) Fz Signal   (b) Pz Signal

(c) C3 Signal   (d) Cz Signal

(e) C4 Signal

Fig. 4. Signals obtained from Fz, Pz, C3, Cz, C4 nodes correspondingly

Figure 5 shows the IMFs obtained after applying EMD algorithm on the signals shown in Figure 4.

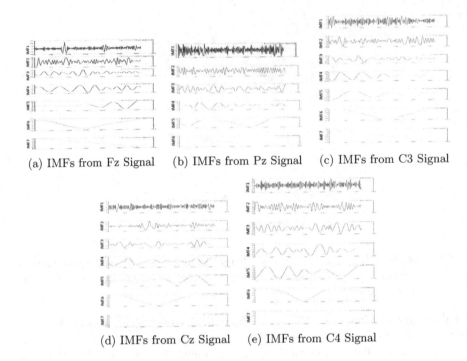

(a) IMFs from Fz Signal   (b) IMFs from Pz Signal   (c) IMFs from C3 Signal

(d) IMFs from Cz Signal   (e) IMFs from C4 Signal

Fig. 5. IMFs obtained from each signal after applying the EMD algorithm

Figure 6 shows the summation of the 1st two IMFs of each node, which contains the highest frequencies and used as a feature vector for the SVM classifier.

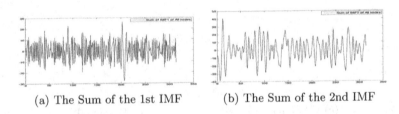

(a) The Sum of the 1st IMF   (b) The Sum of the 2nd IMF

Fig. 6. The summation of the 1st two IMFs of each node, which are used as a feature vector for the classifier

3.3   Classification: Support Vector Machine

Due to the use of a large number of time segments, our feature vectors are of very high dimensionality. SVM is the most appropriate classifier to deal with such feature vectors [17]. There are several kernel functions that could be used with SVM: Linear, Quadratic, Radial Basis Function (RBF), Polynomial and Multi-layer Preceptron (MLP). As mentioned in [18], the Guassian or Radial

Basis Function (RBF) kernel are the most used in BCI classification tasks. The corresponding SVM is known as Gaussian SVM or RBF SVM and is given as:

$$k(x,y) = \exp\left(\frac{-\|x-y\|^2}{2\sigma^2}\right) \tag{6}$$

In(6) x & y can be recognized as the squared Euclidean distance between the two feature vectors, while σ is a free parameter.

4   Results and Discussion

4.1   Data Set

The publically online available BCI competition 2008 dataset [20] is used in this work. The data set consists of 9 subjects. These subjects has four different motor imaginary tasks: the imagination of left hand movement (class 1), right hand (class 2), both feet (class 3), and tongue (class 4). Each subject recorded two sessions on two different days. Each session consists of 6 runs separated by short breaks. One run consists of 48 trials (12 for each of the four possible classes), This yields to a total of 288 trials per session. Twenty-two electrodes were used to record the EEG signals; the montage of these electrodes is corresponding to the international 10-20 system. As stated in [20], the subjects were sitting in a comfortable armchair in front of the screen of the computer. At the beginning of a trial ($t = 0s$), a fixation cross was shown on the black screen and a short acoustic warning tone was presented. After two seconds ($t = 2s$), a cue in the form of an arrow pointing either to the left, right, down or up (corresponding to one of the four classes left hand, right hand, foot or tongue) appeared and stayed on the screen for $1.255s$. This prompted the subjects to perform the desired motor imagery task. There was no feedback provided. The subjects were asked to carry out the motor imagery task until the fixation cross disappeared from the screen at $t = 6s$. A short break followed where the screen was black again.

4.2   Experimental Evaluation

In our experiment we used 80% of the total number of feature vectors as a training sample, while the remaining 20% as a test sample. Given a total number of features equal to fifty-four [1], the training sample is of size forty-three feature vectors, while the test sample is eleven. The feature vector in our problem is of a high dimensionality (three-hundred-fourteen) due to the use of a large number of time segments.

We applied EMD in order to get the required features (IMFs). We used the first and the second IMFs as feature vectors: as they contain the highest frequencies and thus the most features. Ten-fold Cross validation was first applied

---

[1] The number of training and test samples as well as feature size are experimentaly evaluated. However, not shown here due to the limited space.

to select the best parameters to train the SVM classifier. These parameters were chosen after several searches, this is to insure a global minimum (some of the results for the parameters are shown in Table 1). The default parameters [1,1] are close to the optimal for this data and partition. We used these optimal values in training the new SVM classifier and obtaining our final results. SVM classifier was then applied on the feature vectors using these optimal parameters. The best results were obtained in all subjects at input size fifty-four.

Table 1. Selecting optimal values for the SVM parameters

RBF_Sigma	Boxconstraint
0.9802	0.9658
1.5673	1.1059

Table 2 shows that the classification accuracy of the four imagery tasks reached 100% which is considered as a major progress considering the classification of EEG signals. The rest of the subjects also have 100% classification rate.

Table 2. The classification rate of imagery movement tasks for different subjects at input size = 54

Subject	Imagery Tasks			
	Left Hand	Right Hand	Foot	Tongue
Subject 1	100%	100%	100%	100%
Subject 2	100%	100%	100%	100%

To evaluate our system, we compare our results with those of [14]. We wanted to test our system using the same dataset. However, it was not applicable to get the data set used in [14] to use in our experiments as it is not available online. However we are testing for the same activities, the method mentioned in [14] uses also three different activities (Left hand, right hand and foot), so the dataset is expected to be similar. The comparison results are shown in Table 3, which reveals the outstanding performance of our system.

Table 3. Classification rate comparison

Motor Imagery Task / System	Left Hand	Right Hand	Foot	Tongue
Our System	100%	100%	100%	100%
Jiang Wang System	83.3%	82.1%	93.8%	

5   Conclusions

In this paper, EMD is applied to analyze EEG signals in four different motor imagery tasks. Seven subjects were used in our experiment. They imagined the movements of left hand, right hand, foot and tongue. An SVM classifier with RBF kernel was used to classify the recorded signals. The classification accuracy obtained, which reached 100% shows that the methodology used is very efficient for EEG signals classification. However, the classification performance on a huge number of samples is to be investigated in the future.

References

[1] Grabianowski (ed.): How brain-computer interfaces work (November 2007)
[2] Leeb, R., Friedman, D., Müller-Putz, G.R., Scherer, R., Slater, M., Pfurtscheller, G.: Self-paced (asynchronous) bci control of a wheelchair in virtual environments: a case study with a tetraplegics. In: Computational Intelligence and Neuroscience, p. 79642 (2007)
[3] Scherer, R., Lee, F., Schlogl, A., Leeb, R., Bischof, H., Pfurtscheller, G.: Eeg-based interaction with virtual worlds: A self-paced three class brain-computer interface. In: Computational Intelligence and Neuroscience (2007)
[4] Plant, K.A., Ponnapalli, P.V.S., Southall, D.M.: Mobile robots and eeg - a review. In: SGAI Conf., pp. 363–368 (2007)
[5] Satti, A.R., Coyle, D., Prasad, G.: Self-paced brain-controlled wheelchair methodology with shared and automated assistive control. In: IEEE Symposium on Computational Intelligence, Cognitive Algorithms, Mind, and Brain (2011)
[6] Huang, N.E., Shen, Z., Long, S.R., Wu, M.C., Shih, H.H., Zheng, Q., Yen, N.C., Tung, C.C., Liu, H.H.: The empirical mode decomposition and the Hilbert spectrum for nonlinear and non-stationary time series analysis. Proceedings of the Royal Society of London. Series A: Mathematical, Physical and Engineering Sciences 454, 903–995 (1971)
[7] Osuna, E., Freund, R., Girosi, F.: Support vector machines: Training and applications. Technical report, Cambridge, MA, USA (1997)
[8] Wang, J., Xu, G., Wang, L., Zhang, H.: Feature extraction of brain-computer interface based on improved multivariate adaptive autoregressive models. In: 2010 3rd International Conference on Biomedical Engineering and Informatics (BMEI), vol. 2, pp. 895–898 (October 2010)
[9] Herman, P., Prasad, G., McGinnity, T.M., Coyle, D.: Comparative analysis of spectral approaches to feature extraction for eeg-based motor imagery classification. IEEE Transactions on Neural Systems and Rehabilitation Engineering 16(4), 317–326 (2008)
[10] Jangraw, D.C., Sajda, P.: Feature selection for gaze, pupillary, and eeg signals evoked in a 3d environment. In: Proceedings of the 6th Workshop on Eye Gaze in Intelligent Human Machine Interaction: Gaze in Multimodal Interaction, GazeIn 2013, pp. 45–50. ACM, New York (2013)
[11] Hosni, S.M., Gadallah, M.E., Bahgat, S.F., AbdelWahab, M.S.: Classification of eeg signals using different feature extraction techniques for mental-task bci. In: International Conference on Computer Engineering Systems, ICCES 2007, pp. 220–226 (2007)

[12] Suleiman, A.-B.R., Fatehi, T.A.-H.: Features extraction techniques of eeg signal for bci applications, p. 5 (2011)

[13] Liang, H., Bressler, S.L., Desimone, R., Fries, P.: Empirical mode decomposition: a method for analyzing neural data. Neurocomputing 65-66, 801–807 (2005)

[14] Wang, J., Xu, G., Wang, J., Yang, S., Yan, W.: Application of hilbert-huang transform for the study of motor imagery tasks. In: 30th Annual International Conference of the IEEE Engineering in Medicine and Biology Society, EMBS 2008, pp. 3848–3851 (2008)

[15] Demir, B., Erturk, S., Gullu, M.K.: Hyperspectral image classification using denoising of intrinsic mode functions. IEEE Geoscience and Remote Sensing Letters 8(2), 220–224 (2011)

[16] Weng, B., Barner, K.: Optimal signal reconstruction using the empirical mode decomposition. EURASIP Journal on Advances in Signal Processing 2008(1), 845294 (2008)

[17] Lotte, F., Congedo, M., Lécuyer, A., Lamarche, F., Arnaldi, B.: A review of classification algorithms for eeg-based brain-computer interfaces. Journal of Neural Engineering 4(2), R1–R13 (2007)

[18] Lambert, M., Engroff, A., Dyer, M., Byer, B.: Empirical mode decomposition (December 2002)

[19] Rutkowski, T.M., Mandic, D.P., Cichocki, A., Przybyszewski, A.W.: Emd approach to multichannel eeg data - the amplitude and phase synchrony analysis technique. In: Huang, D.-S., Wunsch II, D.C., Levine, D.S., Jo, K.-H. (eds.) ICIC 2008. LNCS, vol. 5226, pp. 122–129. Springer, Heidelberg (2008)

[20] Brunner, C., Leeb, R., Müller-Putz, G., Schlögl, A., Pfurtscheller, G.: BCI Competition - Graz data set A (2008)

Fiducial Based Approach to ECG Biometrics Using Limited Fiducial Points

M. Tantawi, A. Salem, and Mohamed Fahmy Tolba

Faculty of computer and information sciences, Ain Shams University, Cairo, Egypt
manalmt2012@hotmail.com

Abstract. The majority of electrocardiogram (ECG) based biometric systems utilize fiducial based features, derived from 11 landmarks (three peaks, two valleys and six onsets and offsets) detected from each ECG heartbeat. The onsets & offsets landmarks may be obscured by a variety of noise sources. Hence, sophisticated algorithms are usually needed for the detection of these points, which in turn increase computational load and also the results may be suboptimal. This work proposes the utilization of a reduced set of 23 features named 'PV set', which only requires the detection of the five major peaks/valleys instead of all the 11 landmarks. The performance of the 'PV set' is evaluated in comparison with a super set of 36 fiducial features (including PV set) that based on all the 11 landmarks, in addition to IG and RS sets which are subsets of the superset selected based on Rough sets (RS) and information gain (IG) criterion respectively. The evaluation was drawn based on measuring quantities, such as subject identification (SI) accuracy, heartbeat recognition (HR) accuracy and receiver operating characteristic (ROC) curves. The proposed PV set achieved comparable results to the other sets and better results at high noise levels, yielding a reliable and computationally cheaper solution.

Keywords: Biometrics, Electrocardiogram, Fiducial feature selection, Information gain, Radial Basis Function, Rough set.

1 Introduction

Biometrics is a process of human identification and/or verification based on physiological or behavioral characteristics of individuals [1]. Physiological biometrics encompasses anatomical traits that are unique to an individual such as: retina, iris and fingerprint, while behavioral biometrics considers with functional traits such as: signature, keystroke and gait. A new alternative branch of biometrics which has been growing up over the past two decades is the utilization of biological signals such as the electroencephalogram (EEG) and electrocardiogram (ECG) as biometric traits. The potential benefit of such signals is that they are difficult to be spoofed or falsified and can be a liveliness indicator. Moreover, they are more acceptable by users and lower cost relative to physiological traits. Finally they are potentially more reliable than *behavioral* traits (i.e. no one can exactly reproduce the same signature or keystroke typing pattern all the time) [1].

A.E. Hassanien et al. (Eds.): AMLTA 2014, CCIS 488, pp. 199–210, 2014.

The ECG is a recording of the electrical activity of the heart over time. Each recorded heartbeat displays characteristic features which are common to all normal subjects, reflecting the underlying cardio physiology of the subject. Most obvious amongst the characteristic features are a series of deflections that represent the contraction and relaxation of the heart chambers. These deflections manifest as waves that are readily visible in the ECG. In particular, a normal ECG heartbeat will contain three sequential prominent waves labeled P, QRS and T waves [1-3].These waves (amongst other features) have been utilized in a clinical environment to assess the cardio-physiological state of health of an individual for over a century. In 1977, Forsen and colleagues [4] utilized the ECG as a means of identifying/authenticating users in the context of modern day biometrics. The results from Forsen's work were promising, and spawned a new approach to biometrics, as is evidenced by the number of investigators considering this issue in the literature. Generally speaking, ECG based biometric systems can be categorized to fiducial [5-12] and non- fiducial [2, 13-20] systems based on the nature of the considered features.

The fiducial features represent the temporal distances, angles and amplitude differences between 11 fiducial points, as shown in figure 1. These 11 points include three peaks (P, R & T), two valleys (Q & S) and six onsets & offsets of the three waves of each heartbeat. Although the fiducial features are simply computed, they presuppose that the 11 fiducial points are accurately identifiable [2]. Hence, consistently accurate and efficient means for identifying ECG landmarks is critical to the efficacy of the features, subsequently the success of the identification process. Nevertheless, the detection process is not an easy task, especially for the onsets and the offsets. These points are prone to error because they are the least prominent points. Thus, they are the most susceptible to noise, that can be resulted from the sensitivity of the ECG recording device to its surrounding environment, digitization, power line interference, body movements, respiration and the attachment of the electrodes to the skin surface. Hence, more sophisticated algorithms are needed to discern these points. However, these algorithms increase the computational load and they are not perfect, since there is no universally acknowledged rule for defining exactly where the wave boundaries lie. Consequently, for the same ECG records, even cardiologists could not give us markers exactly in the same locations for the wave boundaries. Therefore, there are no standards for the fiducial detection algorithms to be improved on their basis [1-3]. One solution is the utilization of non-fiducial features, such as the coefficients of a wavelet transform or a similar method. The non-fiducial approach obviates the needed for fiducial detection, except for the R peak. However, the features derivation requires more expensive computation and there is no potential for mapping the features back to the underlying physiology of the phenomenon.

Alternatively, this study proposes considering the temporal, amplitude and angle features that only require peaks (P, R and T) and valleys (Q and S) in their derivation. The peaks and valleys points distinguish themselves by their prominence from the baseline (sharpness) which makes them easier to detect and less sensitive to noise than the onsets and offsets. The proposed fiducial set named 'PV set' obviates the need to detect the onsets and the offsets. In other words, the PV set includes all the fiducial features that do not need the detection of any of the onsets or the offsets points for their derivation. Thus, it requires only the detection of five landmarks instead of 11.The efficacy and sufficiency of the PV was examined against a superset of 36 fiducial features (including PV set) that based on all the 11 points. Moreover, the efficacy of PV set was compared to two subsets of the super set which were selected based on rough sets (RS) [21,30] and an information gain (IG) [22] based approaches. The four sets of features (super set, RS, IG and PV) were examined with a Radial Basis Function (RBF) classifier for comparison over the same ECG datasets. The comparison was drawn on the basis of subject identification (SI) accuracy, heartbeat recognition (HR) accuracy and receiver operating characteristic (ROC) curves.

2   Methodology

The proposed methodology in this paper includes 4 stages: 1) preprocessing and detection of fiducial points; 2) feature extraction and normalization; 3) feature selection; and 4) classification (subject identification).

2.1   Preprocessing and Detection of Fiducial Points

A Butterworth filter of second order with cutoff frequencies of 0.5 and 40 Hz is utilized for reduction of noise and elimination of baseline wandering. After data preprocessing, the detection of fiducial points is a preliminary and crucial stage that includes detecting 11 points that represent the peak and the end points of each of the three complexes QRS, P and T (Figure 1); Details about the algorithms employed in this paper for detection of fiducial points can be found in [7].

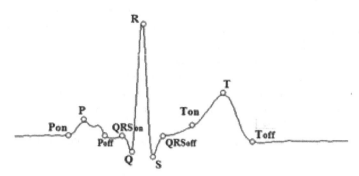

Fig. 1. A single heart beat with the 11 major fiducial landmarks labeled according to industry standard Nomenclature

2.2   Feature Extraction and Normalization

A super set of 36 features that represents the majority of features utilized in the literature is extracted from each heartbeat. As shown in table 1, these features encompass

21 temporal features (distances between fiducial points), 12 amplitude features and 3 angle features. The temporal features are normalized using the full heartbeat duration to control for variations across subjects in terms of heart rate variability. In addition, the amplitude features are normalized by the R amplitude to avoid signal attenuation, and the angle features are used as raw features. In order to have the same impact on the classifier, the range of all features is scaled to a range [0-1].

Table 1. The labels for the 36 fiducial features of the super set

Temporal features	1) QR	6) QS	11) RT	16) P_bQ	21) PT
	2) RS	7) S T	12) PS	17) ST_e	
	3) $P_b P_e$	8) $P_e R$	13) QT	18) $P_e QRS_b$	
	4) $T_b T_e$	9) RT_b	14) $P_b R$	19) $QRS_e T_b$	
	5) PQ	10) PR	15) RT_e	20) $P_e T_e$	
Amplitude features	22) RQ	25) PQ	28) RT	31) TQ	
	23) PT	26) P↑	29) QS	32) TS	
	24) PS	27) RP	30) T↑	33) RS	
Angle Features	34) < PQR	35) <QRS	36) < RST		

2.3   Feature Selection

The goal of feature selection is to reduce the time complexity of the algorithm and to improve the performance of the classifier by removing inaccurate or irrelevant information that may confuse the classifier. In this work, three sets of features were selected from the superset: 1) 'RS' set, where features were selected on the basis of Rough Sets (RS) [21]; 2) 'IG' set, where features were selected on the basis of Information Gain (IG) criterion [22]; 3) 'PV' set, which encompasses features that are derived from five fiducial points (peaks and valleys) for their derivation. Since the detection of peaks and valleys is more accurate and easier than the onsets and the offsets, the proposed 'PV' set will include only features that need Peaks (R, P and T) and Valleys (Q and S) to be computed. Thus, the new set needs only five fiducial points instead of 11. Table 2 shows the PV set. By comparing this table to table 1 (superset), we can realize that PV set is a subset of the superset that encompasses 23 out of the 36 features. The 23 features are as follows: 10 temporal, features that represent all the distances that can be computed between peaks and valleys; 10 amplitude features that represent all the amplitude differences that can be computed between peaks and valleys along with all the 3 angle features.

Table 2. The 'PV' feature set (a subset of table 1) proposed in this study

	Index of selected features
PV set (23 features)	1, 2, 5, 6, 7, 10, 11, 12, 13, 21, 22, 23, 24, 25, 27, 28, 29, 31, 32, 33, 34, 35, 36

2.4   Classification

After the reduced feature sets have been acquired, the considered set of features of a heartbeat is fed into the classifier in order to perform the classification task (i.e. associate a heartbeat to a particular subject ECG record). Due to its superiority with the full set of features in our previous work [7, 8], Radial Basis Functions (RBF) neural network was employed here also as a classifier. The RBF network is based on the simple idea that an arbitrary function y(x) can be approximated as the linear superposition of a set of localized basis functions φ(x) [23]. The RBF is composed of three different layers: the input layer in which the number of nodes is equal to the dimension of input vector. In the hidden layer, the input vector is transformed by a radial basis activation function (Gaussian function):

$$\varphi(x, c_j) = \exp\left(\frac{-1}{2\sigma^2}\|x - c_j\|^2\right) \tag{1}$$

where $\|x-c_j\|$ denotes the Euclidean distance between the input data sample vector x and the center c_j of Gaussian function of the jth hidden node; finally the outer layer with a linear activation function, the kth output is computed by equation

$$F_k(x) = \sum_{j=1}^{m} w_{kj}\ \varphi(x, c_j) \tag{2}$$

w_{kj} represents a weight synapse associates with the jth hidden unit and the kth output unit with m hidden units [23]. The orthogonal least square algorithm [24] is applied to choose the centers, which is a very crucial issue in RBF training due to its significant impact on the network performance.

3   Experiments and Results

Physionet databases, such as PTB [25], MIT_BIH [26, 27] and Fantasia [28] were employed for evaluating the different sets of features considered in this work. Only one ECG record is provided for each subject by MIT_BIH (24 subjects) and Fantasia databases (40 subjects). While, PTB is divided into two sets:1) PTB_1 includes 13 subjects with more than one record and some of them are few years apart; 2) PTB_2 includes 38 subjects with only one record. Duration of these records varies from 1.5-2 minutes. Thus, for more reliability and robustness PTB_1 dataset was employed for training and testing, while the others were utilized for impostor rejection test.

   In the following experiments, the PTB_1 dataset was partitioned into three sets, one for training and two for testing. The training set contains 13 records, one for each subject. 'Test set 1' contains 13 records (one for each subject) recorded in the same day of the training records but in different sessions, while 'Test set 2' contains nine records (belonging to six subjects) and they were recorded over a period of months to years.

   After filtering, 10 beats (≈ 8secs) were randomly chosen from each record in the training set. The super set of 36 features was extracted and normalized from each beat (as shown in section 2.2). Thereafter, the RS, IG and PV feature sets were selected and fed into the RBF classifier. The evaluation of the different feature sets includes measuring the following quantities:

SI & HR accuracies: Subject Identification (SI) accuracy is defined as the percentage of subjects correctly identified by the system. The Heartbeat Recognition (HR) accuracy is defined as the percentage of heartbeats correctly recognized for each subject. A subject was considered correctly identified if more than half of his\her beats were correctly classified to him\her and a heartbeat is recognized by a majority voting of the classifier outputs. These quantities were measured for both test set1 and test set2.

Receiver Operating Characteristic (ROC) curves: the ROC curve is a visual representation of the trade-off between false acceptance rate (FAR) and false rejection rate (FRR) as the system threshold on match score is varied [29]. FAR is the probability of access attempts by unauthorized individuals which are nevertheless successful. Meanwhile, FRR is the probability of access attempts by enrolled individuals who are nevertheless rejected [29].However, in this work, two thresholds was employed. One is typically used (call it $\Theta 1$) which is the value above which a heartbeat is considered classified (instead of just majority voting), while another threshold (call it $\Theta 2$) represents the minimum HR accuracy needed for a subject to be considered identified (instead of just 50% as before). The ROC is acquired by varying $\Theta 2$, while the optimum value for $\Theta 1$ is found empirically as the value that provides the best ROC curve while varying $\Theta 2$. At each considered value for $\Theta 2$, the FRR is computed based on Test set 1 only because the records of Test set 2 are for only six subjects. For more reliability and robustness, the FAR is computed using a set of imposters (approximately 100 subjects) that were gathered from three different databases: including 38 subjects of PTB_2, 24 subjects of MIT_BIH and 40 subjects of Fantasia. The conducted experiments and their results are provided in details in the following subsections.

3.1   SI & HR Accuracies

The RBF was trained using the PTB_1 training set and was tested using Test set 1and Test set 2 for all the considered feature sets. There are two parameters to be adjusted for the classifier, which are the spread (width Gaussian of functions) and mean squared error (MSE), both of which were set based on empirical investigations. Table 3 provides a ranked list of the labels of the selected features by IG and RS (PASH algorithm). While, table 4 summarizes the SI and HR accuracies for both test sets along with the optimum parameters for the RBF classifier with each feature set.

Table 3. The ranked list of features selected by IG (first row) and RS (second row), peaks & valleys based features considered by IG and RS are shown in bold

Feature selection method	Index of selected features
IG (21 features)	36, 27, 10, 22, 32, 28, 23, 30, 24, 14, 31, 5, 35, 16, 12, 8, 29, 6, 7, 21, 20
RS (21 features)	36, 12, 33, 3, 22, 23, 5, 10, 24, 29, 18, 27, 32, 8, 14, 16, 35, 21, 20, 31, 7

Table 4. The SI and HR results of the four feature sets

	RBF parameters	Test Set 1		Test Set 2	
		SI Accuracy	HR Accuracy	SI Accuracy	HR Accuracy
Superset	Spread =0.8 MSE=0.001	100%	96.7%	100%	81.3%
RS set	Spread =0.54 MSE=0.003	100%	97%	100%	76.3%
IG set	Spread =0.53 MSE=0.003	100%	96%	100%	81.8%
PV set	Spread =0.54 MSE=0.003	100%	96%	100%	81.8%

3.2   ROC Curves

Four ROC curves were obtained while varying threshold Θ_2 with the four fiducial feature sets: super set, IG, RS and the proposed PV sets, see fig. 2. The set of impostors was gathered from the other databases in the same way mentioned earlier in this section. The conducted experiments revealed that the optimum value of Θ_1 (where the best ROC curves have been achieved) is 0.57.

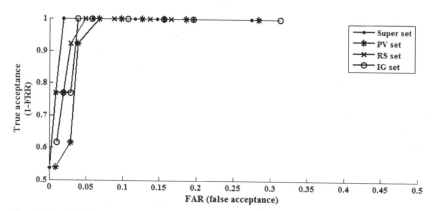

Fig. 2. The ROC curves for the four considered feature sets: super set, IG, RS and PV set

3.3   Sensitivity to Noise

In order to measure the sensitivity of the feature sets to noise, a subset of 8 records (belonging to 8 subjects) of test set 1 is chosen. Four versions named test11, test12, test13 and test14 are generated from this subset by adding different amounts of Gaussian white noise 20, 15, 10 and 5 respectively to the original test records. The amount of added noise is represented in this paper in decibels (db). In order to measure the error caused by the added noise, the 11 fiducial points were detected from the original test set using the detection algorithms considered in this work. Thereafter, the positions of these detected points were taken as a ground truth. The mean and the standard deviation of the difference between the detected points of each version and the original test set 1 were computed in milliseconds. Tables 5& 6 show the results of each of the 11 points with each of the four versions of the noisy testing sets. Lately, The RBF classifier was trained using PTB_1 training set and tested using the noisy versions of test set 1. The SI and HR were computed for the four feature sets as shown in table 7.

Table 5. A Summary of the mean 'm' and standard deviation 's' of the error in the location for each of the peaks and valleys points as a result of the level of noise added to each of the four generated noisy versions of test set 1. Results are given in milliseconds in the form of m ± s.

	R	Q	S	P	T
Test11 (20 db)	0.04±0.4	-0.01±0.2	0.17±0.7	-0.07±1.9	-0.27±2.4
Test12 (15 db)	0.004±0.5	-0.1±1.7	0.11±1	-0.4±2	0.01±3.27
Test13 (10 db)	-0.02±0.6	-0.2±2	0.24±1.48	-3.42±3	-0.03±5
Test14 (5 db)	0.06±0.9	-0.5±3	0.4±2.1	-14.7±14.2	-0.9±7.6

Table 6. A Summary of the mean 'm' and standard deviation 's' of the error for each of the onsets and offsets caused by noise added to each of the four generated noisy versions of test set 1. Results are given in milliseconds in the form of m ± s.

	P_{on}	P_{off}	QRS_{on}	QRS_{off}	T_{on}	T_{off}
Test11 (20 db)	-3.7±9.6	0.1±4.8	-0.3±2.48	0.08±1	-0.35±6.8	0.88±4.1
Test12 (15 db)	-7±12	-0.23±6.1	-0.78±3.1	0.23±1.4	-3±9	1.13±5.4
Test13 (10 db)	-12.2±16	-0.7±8	-2.6±4.87	0.23±2	-4.4±13.5	5.6±8.8
Test14 (5 db)	-19.9±19	-8.7±12.8	-3.2±6.4	0.48±2.6	-9.5±7.6	7.5±12

Table 7. A Summary of the SI and HR accuracies achieved for the four noisy versions of test set1 by each of the four feature sets. (m/n) means m of n subjects were successfully identified.

	Super set		RS set		IG set		PV set	
	SI	HR	SI	HR	SI	HR	SI	HR
Test11 (20 db)	100 %	95%	100%	95%	100%	96%	100%	90.1%
Test12 (15 db)	100 %	93%	100%	93%	100%	92%	100%	88.8%
Test13 (10 db)	100 %	88.8%	100%	86%	100%	84.6%	100%	81.1%
Test14 (5 db)	50%	78%	75%	72%	75%	69%	87.5%	70%

4   Discussion

The efficacy of a classifier is clearly dependent upon the accuracy and relevance of the features contained within the model used for the classification task. Any inaccuracies in extracting features will yield suboptimal classification, regardless of the relevance of the features. In fiducial based ECG biometrics, the critical step is in the determination of fiducial points. Broadly speaking, most fiducial based approaches utilize a set of 11 points, which include determining the six onset and offset locations of the three major waves (P, QRS and T) contained in each heartbeat. The detection of these points is prone to error because these are the least prominent points in any of the ECG data and are close to the same resolution as the noise in the data. To overcome this problem, significant processing strategies are deployed, which are, in addition to being computationally expensive, suboptimal in many cases. Notwithstanding the non-fiducial approach, which has its own computational issues, is there another way to utilize the feature space of an ECG trace?

This work proposed the deployment of a potentially more precisely acquired subset of fiducial features (the 'PV set'), which can be acquired using standard, computationally efficient approaches. The PV features are 23 temporal, angle and amplitude features that can be derived by only detecting peaks (P, R & T) and valleys (Q &S) from the three major waves of each heartbeat in the ECG trace. These are the most prominent aspects of the ECG. Thus, they are distinguishable from the variety of noises that regularly occur in ECGs. The efficacy of the PV set is evaluated against the super set of fiducial features and two reduced sets (IG and RS sets) which were selected on the basis of information gain and rough sets criteria.

From table 3, the following observations can be deduced on the three subsets (PV, RS and IG): 1) the subsets represent roughly 60% only of the super set; 2) roughly 75% of the RS and IG features (16 out of 21) are peaks and valleys based features (belong to PV set); 3) there is a major overlap between the RS and IG sets (18 out of 21 features) provides 'support' for the significance of these 18 features, where 14 of them belong to PV set; 4) the ranked lists of features selected by IG and RS provided by table 1 also indicate that most of the top places are occupied by peaks and valleys based features. It is worth mentioning that these observations were also true when the IG & RS selection algorithms were applied to Physionet datasets rather than PTB_1 for verification. Hence, these observations support the significance and rich information content of the PV set. Moreover, the conduced experiments revealed that the five peaks and valleys points required by the PV set only need less than quarter of the time needed for detecting the whole 11 fiducial points required by IG and RS sets. The time needed to detect the whole 11 points from 10 heartbeat equals to 0.08 seconds, while the time needed to detect only the five peaks and valleys is equal to 0.014 seconds (machine core i7, windows 7).It should be noted that the fiducial detection algorithms utilized in this work are commonly used and successfully evaluated in comparison with the other detection methods in our previous work [7].

The four considered feature sets provided very close SI and average HR accuracies (table 4). However, the results of the sensitivity to noise experiment (section 3.3) demonstrated that the detection error caused by adding noise increases as the noise increases for the onsets and the offsets. While, for peaks and valleys, it is still the minimum, for the 'P' fiducial point in the last case (5db). This can be explained by the fact that the P peak has usually small amplitude, thus in case of high levels of noise, the p wave may become completely embedded within the, rendering it difficult to detect. Consequently this experiment provides evidence that the peaks and valleys that are less sensitive to noise subsequently can provide a more accurate feature set. Moreover, the results of testing the RBF classifier along with each of the considered feature sets with the noisy versions of test set 1 (Table 7) demonstrated that although the proposed PV set provided the same SI accuracy along with slightly less average HR for an identified subject for the first three cases (20, 15 and 10db) compared to the IG and RS sets, it provided the best SI accuracy in case of very low signal to noise ratio (fourth case: 5db shown in bold). Regarding the impostor rejection test, the proposed PV, RS and IG sets provided very close ROC curves to each other and close to the best ROC curve obtained by the superset (fig. 2).

The Physionet datasets include many subjects. However, except for the subjects of PTB_1, each subject has only one record which limits the stability test. Hence, we are collecting subject data of our own, which includes test/re-test acquisition over five time points (t0, 1 month, 2 months, 6 months, and 12 months It is hoped that having such a dataset will provide the means to more thoroughly evaluate the temporal stability of the classification scheme introduced in this work.

5   Conclusions and Future Work

To conclude, a new criterion for fiducial feature selection is introduced and successfully implemented. The approach obviates the difficulties of acquiring onsets/offsets, which may be obscured by noise typically contained within an ECG recording. Instead, the approach relies solely on the detection of the three peaks (P, R & T) and two valleys (Q & S). These five points are easier to detect (due to their sharpness) and less sensitive to noise than the six onsets and offsets. This work further substantiates these claims by providing direct empirical evidence in the context of an ECG biometrics paradigm. Although the PV set represents only peaks and valleys based features, it provided comparable results to the super set and the two considered subsets, and the best results at high noise levels. Moreover, the detection of the five landmarks required for the derivation of the PV set need roughly quarter of the time needed for the detection of the whole 11 landmarks required for the derivation of the other sets. Thus, a significant decrease in the time consumption occurs, which is a potential benefit for a real time system. This work will be explored further by applying this approach to additional datasets (we are compiling our own longitudinal dataset), in order to provide a more thorough evaluation of the stability and scalability of the proposed approach.

References

1. Sufi, F., Khalil, I., Hu, J.: ECG based Authentication, ECG-Based Authentication. In: Stavroulakis, P., Stamp, M. (eds.) Handbook of Information and Communication Security, pp. 309–331. Springer, Berlin (2010)
2. Wang, Y., Agrafioti, F., Hatzinakos, D., Plataniotis, K.: Analysis of human electrocardiogram for biometric recognition. J. Advances in Signal Processing 1, 1–6 (2008)
3. Agrafioti, F., Gao, J., Hatzinakos, D.: Heart Biometrics: Theory, Methods and Applications. In: Yang, J. (ed.) Biometrics: Book 3, pp. 199–216. Intech (2011)
4. Forsen, G., Nelson, M., Staron, R.: Personal attributes authentication techniques. In: Griffin, A.F.B. (ed.) RADC report RADC-TR-77-1033 (1977)
5. Biel, L., Petersson, O., Philipson, L.: ECG Analysis: a new approach in human identification. IEEE Trans. Instrum. Meas. 50(3), 808–812 (2001)
6. Fatemian, S., Hatzinakos, D.: A new ECG feature extractor for biometric recognition. In: Proc. 16th Ann. Internat. Conf. on Digital Signal Processing, pp. 323–328. IEEE Press, Piscataway (2009)
7. Tantawi, M., Revett, K., Tolba, M.F., Salem, A.: On the Applicability of the Physionet Electro-cardiogram (ECG) Repository as a Source of Test Cases for ECG Based Biometrics. Int. J. Cognitive Biometrics 1(1), 66–97 (2012)
8. Tantawi, M., Revett, K., Tolba, M.F., Salem, A.: Fiducial Feature Reduction Analysis for Electrocardiogram (ECG) Based Biometric Recognition. Int. J. Intelligent Information Systems 40(1), 17–39 (2013)
9. Singla, S., Sharma, A.: ECG based biometrics verification system using LabVIEW. Songklanakarin J. Sci. Technol. 32(3), 241–246 (2010)
10. Gahi, Y., Lamrani, A., Zoglat, A., Guennoun, M., Kapralos, B., El-Khatib, K.: Biometric Identification System Based on Electrocardiogram Data. In: New Technologies, Mobility and Security, NTMS 2008, pp. 1–5 (2008)
11. Singh, Y.N., Gupta, P.: ECG to Individual Identification. In: Proc. of the 2nd IEEE BTAS Conf., pp. 1–8 (2008)
12. Israel, S.A., Irvine, J.M., Cheng, A., Wiederhold, M.D., Wiederhold, K.: ECG to identify individuals. Pattern Recognition 38(1), 133–142 (2005)
13. Wan, Y., Yao, J.: A Neural Network to Identify Human Subjects with Electrocardiogram Signals. In: Proc. of the World Congress on Engineering and Computer Science 2008, San Francisco, USA (2008)
14. Wao, J., Wan, Y.: Improving Computing Efficiency of a Wavelet Method Using ECG as a Biometric Modality. Int. J. Computer Netw. Security 2(1), 15–20 (2010)
15. Coutinho, D., Fred, A., Figueiredo, M.: One-lead ECG-based personal identification using Ziv-Merhav cross parsing, in proc. In: 20th Int. Conf. on Pattern Recognition, pp. 3858–3861 (2010)
16. Ghofrani, N., Bostani, R.: Reliable features for an ECG-based biometric system. In: Proc. 17th Iranian Conf. of Biomedical Engineering, pp. 1–5 (2010)
17. Venkatesh, N., Jayaraman, S.: Human electrocardiogram for biometrics using DTW and FLDA. In: Proc. 20th Internat. Conf. on Pattern Recognition (ICPR), pp. 3838–3841 (2010)
18. Ye, C., Coimbra, M., Kumar, B.: Investigation of human identification using two-lead electrocardiogram (ECG) signals. In: Proc. 4th Int. Conf. on Biometrics: Theory Applications and Systems, pp. 1–8 (2010)

19. Safie, S., Soraghan, J., Petropoulakis, L.: Electrocardiogram (ECG) Biometric Authentication Using Pulse Active Ratio (PAR). Peer-review for IEEE Trans. Inf. Forensics and Security 6(4), 1315–1322 (2011)
20. Safie, S., Soraghan, J., Petropoulakis, L.: ECG based biometric for doubly secure authentication. In: Proc. 19th European Signal Processing Conf. (EUSIPCO), Barcelona, Spain, pp. 2274–2278 (2011)
21. Zhang, M., Yao, J.: A Rough Sets Based Approach to Feature Selection. In: Proc. 23nd Ann. Int. Conf. of NAFIPS, Banff, Canada, pp. 434–439 (2004)
22. Mitchel, T.: Machine learning, 2nd edn. McGraw-Hill, New York (1997)
23. Haykin, S.: Neural networks: A comprehansive Foundation, 2nd edn. Prentice Hall (1999)
24. Chen, S., Chng, E.: Regularized Orthogonal Least Squares Algorithm for Constructing Radial Basis Function Networks. Internat. J. Control 64(5), 829–837 (1996)
25. Oeff, M., Koch, H., Bousseljot, R., Kreiseler, D.: the PTB Diagnostic ECG Database, National Metrology Institute of Germany (October 2013),
    http://www.physionet.org/physiobank/database/ptbdb/
26. The MIT-BIH Normal Sinus Rhythm Database (October 2013),
    http://www.physionet.org/physiobank/database/nsrdb/
27. The MIT_BIH Long Term Database (October 2013),
    http://www.physionet.org/physiobank/database/ltdb/
28. The Fantasia Database (October 2013),
    http://www.physionet.org/physiobank/database/fantasia/
29. Revett, K.: Behavioral Biometrics: A Remote Access Approach. John Wiley & Sons (2008) ISBN: 978-0-470-518830
30. Hassanien, A.E., Suraj, Z., Slezak, D., Lingras, P.: Rough computing: Theories, technologies and applications. IGI Publishing Hershey, PA (2008)

Selecting Relevant Features for Classifier Optimization

Mvurya Mgala and Audrey Mbogho

Department of Computer Science, University of Cape Town,
ICT4D Research Centre, 7701 Cape Town, South Africa
{mmvurya,ambogho}@cs.uct.ac.za

Abstract. Feature selection is an important data pre-processing step that comes before applying a machine learning algorithm. It removes irrelevant and redundant attributes from the dataset with an aim of improving the algorithm performance. There exist feature selection methods which focus on discovering features that are most suitable. These methods include wrappers, a subroutine of the learning algorithm itself, and filters, which discover features according to heuristics, based on the data characteristics and not tied to a specific algorithm. This paper improves the filter approach by enabling it to select strongly relevant and weakly relevant features and gives room to the researcher to decide which of the weakly relevant features to include. This new approach brings clarity and understandability to the feature selection preprocessing step.

Keywords: feature selection, information gain, wrapper, filter, descriptive statistics.

1 Introduction

The trend in education is to achieve universal primary education where children are able to complete a full course of primary schooling. In most developing countries, thousands of children complete primary schools with low grades and are forced to drop out of the school system at an age with no skills for meaningful employment. Education stakeholders; education officers, parents and teachers would like to intervene to assist such children, the challenge is to identify this children early enough because of the large numbers of pupil. The teachers in many cases are overwhelmed and cannot offer individual attention to such children whose low performance may need more than just extra lessons. It is necessary to explore methods that can discover knowledge from pupil data that allow classification of the children into categories such as those that need high intervention and low intervention. This study seeks to determine the most relevant factors that contribute to academic performance for the purpose of developing an academic prediction model.

Many factors contribute to the challenge of applying machine learning to educational data in rural Africa where education is still based on the traditional

A.E. Hassanien et al. (Eds.): AMLTA 2014, CCIS 488, pp. 211–222, 2014.

classroom teaching because of lack of infrastructure. The quality of data is one such challenge, given that data has to be gathered through surveys and hard copy secondary data. Such data will most likely have irrelevant features, noisy and unreliable entries, making knowledge discovery during training difficult. Feature selection can be seen as the process of eliminating as much of the redundant data as possible so as to remain with an optimum subset of features [1]. Algorithms that select features as preprocessing before learning are categorized as wrappers [5]; they employ a statistical subroutine such as cross validation and are embedded in the learning algorithm. The approach is useful except for the fact that the process is very slow because the learning algorithm has to loop many times.

The other approach is called filters [5]; features are filtered out independent of any learning algorithm, usually before learning commences. Filters have proved to be quicker than wrappers and can therefore be applied to large data sets with many features. One other advantage they have is that they can be used with any algorithm unlike the wrappers which have to be re-run when one is changing algorithms.

This paper presents an enhanced filter approach to feature selection by combining the information gain approach with descriptive statistics. In descriptive statistics, box plots are used to select the features.

The next section discusses related work. In section 3 we describe a filter approach adopted in this work and the descriptive statistics. Section 4 presents experimental results for both the filter approach and the descriptive statistics. The last section concludes and discusses future work.

2   Related Work

A study conducted by Hall [3] on feature selection for discrete and numeric class machine learning, revealed filters to be more practical than wrappers because they are much faster. Experiments conducted using a correlation-based filter algorithm as a pre-processing step for Nave Bayes, Instance-based learning decision trees, locally weighted regression and model trees show the approach to be an effective feature selector. It reduces data dimensionality by more than sixty percent in most cases without negatively affecting accuracy. Also decision trees and model trees built from the preprocessed data are often significantly smaller.

Ye and Liu [8] conducted a study on efficient feature selection via analysis of relevance and redundancy. They demonstrated that feature relevance alone is insufficient for feature selection of high dimensional data. Based on the previous definition of feature relevance by Kohavi et al. [5], features can be classified into strongly relevant, weakly relevant and irrelevant. Strong relevance indicates that the feature is always necessary for the optimal subset; it cannot be removed without affecting the original conditional class distribution. Weak relevance suggests that the feature is not necessary but may become necessary for optimal subset at certain conditions. Irrelevance indicates that the feature is not necessary at all.

An optimal subset therefore should include all strongly relevant features none of irrelevant and a subset of weakly relevant features. Ye and Liu proposed a new framework of efficient feature selection via relevance and redundancy analysis. They devised a feature selection algorithm that demonstrated efficiency and effectiveness in supervised learning.

Another study that used the filter approach is by Kotsiantis et al. [4]. Their results show an improvement in the accuracy of the algorithms after running the experiments without some of the attributes rated as having no influence.

As a way of comparing the two approaches to features selection, we consider a study conducted using the wrapper approach by Bratu et al. [1]. Their work analyzed the wrapper approach for feature selection with the purpose of boosting classification accuracy. Results show that they were able to reduce the number of attributes considerably by over (50%) which speeded up training and improved classification.

These studies show that there is no universally best feature selection method which produces the highest and most accurate improvement on any dataset. This study proposes a framework of selecting strongly relevant features and some of the marginal (weakly relevant) features, and as a way of saying we agree to the "no-free lunch" theorem of feature selection, we allow the researcher to decide on which weakly relevant features to include.

3   Feature Selection

This section discusses the two techniques we considered for feature selection, namely, correlation-based feature selection and descriptive statistics.

3.1   Correlation-Based Feature Selection

Correlation [8] is applied widely in machine learning to determine relevance. In this section we describe the correlation based filter approach to feature selection.

There are two types of measure for correlation between random variables: linear and non-linear. In linear correlation, the well-known measure is linear correlation coefficient. However, it is not safe to always assume linear correlation between features in real world. Linear correlation measures may fail to capture correlation measures that are non-linear in nature. Many measures among the non-linear correlation measures are based on the information theoretical concept of entropy. Defined as a measure of the uncertainty of random variables, the entropy of a variable X is defined as:

$$H(X) = -\sum_i P(x_i) log_2(P(x_i)) . \tag{1}$$

The entropy of variable X after observing another variable Y is defined as:

$$H(X|Y) = -\sum_j P(y_i) \sum_i P(x_i|y_i) log_2(P(x_i|y_i)) . \qquad (2)$$

Where $P(x_i)$ are the prior probabilities for all values of X and $P(x_i|y_i)$ is the posterior probabilities of X given the values of Y. The amount by which the entropy of X decreases reflects additional information about X provided by Y and is called information gain [7]. Mitchell [6] defines information gain as a statistical measure that determines how well an attribute separates the training data according to the target classes.

It is expressed as:

$$IG(X|Y) = H(X) - B(X|Y) . \qquad (3)$$

According to this measure a feature Y is regarded more correlated to X than to another feature Z, if:

$$IG(X|Y) > IG(Z|Y) . \qquad (4)$$

This study adopts the information gain measure to determine the features that correlate more and rank them according to equation 4. Fig. 1 shows the ranked features.

3.2   Descriptive Statistics

The boxplots [10] give a summary of the descriptive statistics. The box represents the interquartile range bounded by the data values that correspond to the 25th and 75th quartiles. Fifty percent of the data values fall within this box and its length represents the interquartile range. The line within the box is the median. The whiskers are the largest and the smallest data values that are not outliers. Data values that are between 1.5 and 3 interquartile ranges below or above the 25th or the 75th quartiles are considered outliers and are represented with an open circle. Data values that are more than 3 interquartile ranges below and above the 25th and 75th quartiles are called extreme values and are represented with asterisk. Using the boxplots one can see the median clearly. If the median is positioned towards the lower end of the data, it suggests that the data is positively skewed.

4   Methods

The idea of combining two feature selection methods is tested on data collected for the purpose of predicting the academic performance of primary school pupils in a rural county in Kenya. A total of 2546 records are gathered from 55 primary schools. The database contains pupils previous test marks, personal, family and school related information and the national examination marks. A total of 23 features are gathered through semi-structured interviews with education officers and head teachers and from literature as possible causes of low academic performance. These features are: total test marks, sex, religion, age, distance to school, pupil absenteeism, study time, pupil discipline, command in speaking English, pupil education attitude, pupil motivation, parent encouragement, parents stability, family finance ability, parents education qualification, family size, parents involvement, community involvement, teacher attitude, teacher commitment, teacher absenteeism, school facilities and teacher shortage. The information gain algorithm in the Weka machine learning environment [2] is adopted for part one of the experiments.

The results of the ranked features are illustrated in Fig. 1. Features are ranked according to equation 4, and those that have a high information gain are selected as the optimum subset.

Part two of the experiment involves selecting the features using boxplots. They are created from the same dataset using a statistical application as discussed in section 3.2 above. The results section presents a detailed explanation of how to detect correlations.

Data is digitized using a statistical package. As part of pre-processing, records with missing test marks or final examination marks are deleted. Records only missing some pupil response values are filled, noisy data is removed, spelling mistakes and wrong entries are corrected. Final examination marks columns that had both numbers and letters in the same cell are separated. Table 1 shows all the features gathered, it is the initial stage in preprocessing where pupil responses are coded into digits.

5   Results

This section describes the results obtained after applying the two feature selection approaches.

5.1   Information Gain Feature Selection

Fig. 1 shows a chart of all the features of the dataset that are fed into the feature selection algorithm with their corresponding information gains. The larger the value of information gain, the more strongly relevant the feature is to the training. The features as given by the information gain algorithm are; test scores (0.21653), pupil sex (0.02252), shortage of teachers (0.01807), pupil motivation (0.01613), family income (0.0133), pupil age (0.01185), study time (0.0108), teacher attitude (0.00972), pupil absenteeism (0.00725), teacher

commitment (0.00612), parents encouragement (0.00611), pupil education attitude (0.0045), School facilities (0.00895), Command to speak English (0.008),
Distance to school (0.00584), Pupil discipline (0.00584). As seen in Figure 1 test
score gives the longest bar because it is overly co-related with the final exam
mark.

6   Descriptive Statistics Feature Selection

Figures 2–7 illustrate the various boxplots for each of the selected feature. Different categories in each feature are plotted against a standard scale, the final
examination marks (KCPE_TOT). A description of each boxplot is given.

Table 1. Features and their numeric codes

Variable	Description	Domain
AGE	Pupil's age	normal:1, overage: 2
SEX	Pupil's sex	female:1, male: 2
DIST	Distance from home	1, 2, 3, 4, 5 km
ABS	Days absent from school per week	0, 1, 2, 3 times
STUD_T	Time to study at home	0, 1, 2, 3 hours
DISPL	Pupil disciplined how often	1, 2, 3 or more, 4: very often
COM_ENG	Pupil's command of English	speak local language:1, uncertain:2, speak English always:3
PUP_M	Pupil motivated?	motivated:1, neutral:2, not motivated:3
P_ENC	Parent encouragement	encouraging:1, neutral:2, not encouraging:3
P_ATT	Pupil education attitude	positive:1, neutral:2, negative:3
F_HARM	Parents' state of harmony	yes:1, neutral:2, no:3
F_FIN	Parent can pay secondary school fees	yes:1, neutral:2, no:3
PQ	Parent qualification	degree:4, diploma:3, secondary:2, primary:1, none:0
F_SIZ	Family size	3-5:1, 6-10:2, 11 or more:3
P_PART	Parent participates in educ.	yes:1, neutral:2, no:3
C_PART	Community participation	yes:1, neutral:2, no:3
T_ATT	Teacher attitute toward pupils	positive:1, neutral:2, negative:3
T_COMM	Teacher committed to teaching	yes:1, neutral:2, no:3
T_ABS	Teacher absent	never:1, neutral:2, always:3
S_FAC	Lack of school facilities	inadequate:1, neutral:2, sufficient:3
L_TEAC	Lack of teachers	inadequate:1, neutral:2, sufficient:3
T_MARKS	Test scores	400-500:1, 350-399: 2, 300-349: 3, 250-299: 4, 200-249: 5, 0-199:6

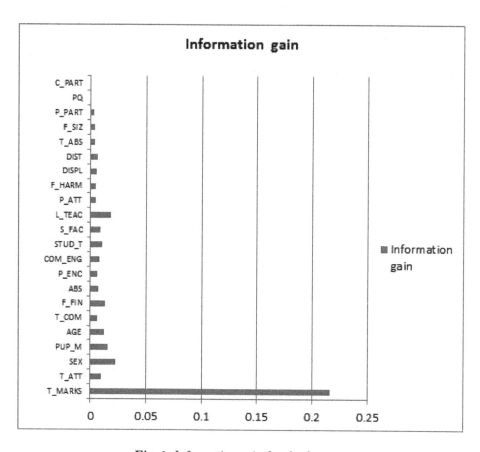

Fig. 1. Information gain for the features

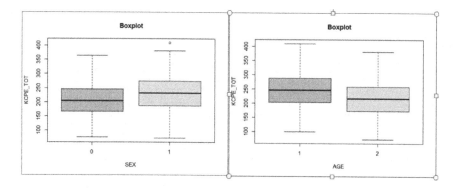

Fig. 2. Final marks against sex and age

The left half of Fig. 2 is a boxplot showing the distribution of the boys (1) and the girls (0). The plot shows that the boys have a higher median of the total score, suggesting sex co-relates with total score.The right half of Fig. 2 is a plot of the pupils ages, normal age (1) and overage (2); those with normal age obtain a higher medium of the total score. This suggests age co-relates with total score.

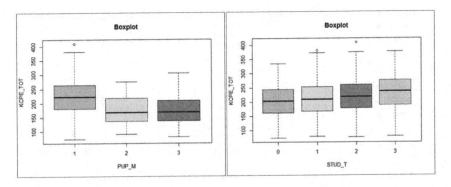

Fig. 3. Final marks against pupil motivation and study time

The left half of Fig. 3 is a boxplot of the pupil motivation on the total score scale. The median score is higher for pupils with high motivation (1). Suggesting pupils motivation co-relates with total score.The right half of Fig. 3 is a boxplot of study time on the total score scale. It is noticed that the median score increases as the amount of study time. This suggests there is a co-relation between study time and total score.

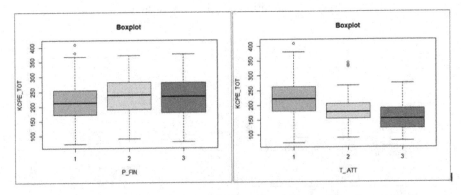

Fig. 4. Final marks against family financial ability and teacher attitude

The left half of Fig. 4 is a boxplot of parents financial ability and the total score. Financial ability (1) does not suggest any co-relation with total score. The poor families (3) however could act as a motivation to work harder. The neutral group (2) could fall either side. On the right is a boxplot of teacher attitude and the total score. Good attitude (1) corresponds to a higher median score, suggesting, there exists a co-relation between teacher attitude and total score.

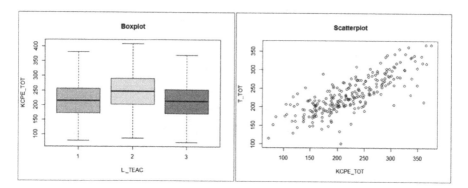

Fig. 5. Final marks against teacher shortage and test marks

Fig. 5 shows a boxplot of teacher shortage and total score. Shortage of teachers (1) and no shortage (3) seem to have the same score median, suggesting there is a no co-relation. The right side is a scatter plot of total score and the test score for three previous years. The plots show a co-relation between the two.

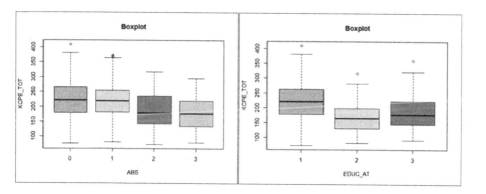

Fig. 6. Final marks against pubil absenteeism and pupil attitude

Fig. 6 is a boxplot of pupil absence from school and total score. Zero days absent (0) and one day absent (1) indicate a higher score median, suggesting a co-relation between absence from school with total score. As seen, the score median decreases as days absent increase. The right half is a boxplot of pupils attitude towards education and total score. A good attitude (1) corresponds to a higher total score median, showing a correlation between these two variables.

Fig. 7 shows a boxplot of parents encouragement and total score pupil who are encouraged (1) have a higher total score median, suggesting parents encouragement co-relates with total score.Fig. 7also shows a boxplot of teacher commitment and total score. Commitment of teachers (1) indicates a higher median total score, implying a co-relation exists.

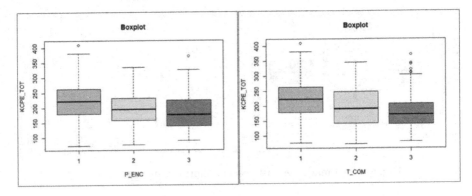

Fig. 7. Final marks against parental encouragement and teacher commitment

6.1   Optimum Subset Verification

Table 2 shows the results of experiments carried out using the two different subsets, 8 features obtained using the information gain approach; total test marks, sex, age, pupil motivation, study time, family finances, teacher attitude and shortage of teachers. Ten features from descriptive statistics; total test marks, sex, age, pupil motivation, study time, teacher attitude, pupil absenteeism, pupil education attitude, parents encouragement and teacher commitment. The subsets are input into the algorithms in turn to obtain the percentage accuracy by each algorithm to determine which of the two an optimum dataset is. The conclusion discusses the finding.

Table 2. Comparison of subsets

Algorithm	Information Gain Subset Accuracy (%)	Descriptive Statistics Subset Accuracy
LWL	72.8772	73.7016
RepTree	76.2984	75.6389
Logistic	76.7931	76.5045
J48	76.2984	75.7214
Random Forest	74.7321	74.0725
Bayes Net	75.4740	76.5458
SMO	74.9794	74.9794
Average	75.3504	75.3092

7   Conclusion and Future Works

Data preprocessing is known to improve the efficiency and effectiveness of learning algorithms. Combining techniques of feature selection has proved to be a better approach to confirm the selected optimum subset. Usually, an optimum subset is a combination of strongly relevant features and some weakly relevant features. The challenge is to identify which weakly relevant features to include given that the irrelevant features are easily eliminated. This study used both information gain and descriptive statistics approaches to select the optimum subset features. Information gain approach selected 8 features out of a total of 22 features, namely; total test marks, sex, age, shortage of teachers, pupil motivation, family finances, study time and teacher attitude. The descriptive statistics approach selected 10 out of 22, namely; total test marks, sex, age, pupil motivation, study time, teacher attitude, pupil absenteeism, pupil education attitude, parents encouragement, and teacher commitment.

Experiments carried out using the two subsets on 7 different algorithms reveal a marginal difference on the average percentage prediction accuracy. Information gain approach gave 75.3504% while descriptive statistics gave 75.3092%. We conclude that the features that are shared by both subsets are the strongly relevant features, these are; total test marks, sex, age, pupil motivation, study time and teacher attitude. The other features appearing in either subset are weakly relevant, these are; shortage of teachers, family finances, pupil absenteeism, pupil education attitude, parents encouragement and teacher commitment. Any of these can be added to the list since they only weakly influence learning. These findings provide a foundation for further work on enhancing the effectiveness of an algorithm for predicting academic performance of primary school pupils in rural Africa.

In our future works, rough sets-based feature extraction, rule generation and classification will provide more challenging and may allow us to refine our learning algorithms and/or approaches to the selecting relevant features for classifier optimization [9].

Acknowledgements. Our thanks go to the Hasso Plattner Institute for funding this work, to the university of Cape Town for providing a conducive environment for doing this research, and to the Kenyan NACOSTI for allowing us to collect data.

References

1. Bratu, C.V., Muresan, T., Potolea, R.: Improving Classification Accuracy through Feature Selection. In: Proceedings of the 4th IEEE International Conference on Intelligent Computer Communication and Processing, pp. 25–32. IEEE Press, New York (2008)
2. Garner, S.R.: Weka, The Waikato Environment for Knowledge Analysis. In: Proceedings of the New Zealand Computer Science Research Students Conference, New Zealand, pp. 57–64 (1995)

3. Hall, M.A.: Feature Selection for Discrete and Numeric Class Machine Learning. Technical report, University of Waikato, Department of Computer Science, New Zealand (1999)
4. Kotsiantis, S., Pierrakeas, C., Pintelas, P.: Efficiency of Machine Learning Techniques in Predicting Students Performance in Distance Learning Systems. Applied Artificial Intelligence 18(5), 411–426 (2004)
5. Kohavi, R., John, G., Long, R., Manley, D., Pfleger, K.: MLC++: A Machine Learning Library in C++. In: Proceedings of the Sixth International Conference on Tools with Artificial Intelligence, pp. 740–743. IEEE Press, New York (1994)
6. Mitchell, T.M.: Machine Learning. McGraw-Hill, Inc., New York (1997)
7. Quinlan, J.R.: C4.5: Programs for Machine Learning. Morgan Kaufmann Publishers Inc., San Francisco (1993)
8. Yu, L., Liu, H.: Efficient Feature Selection via Analysis of Relevance and Redundancy. Journal of Machine Learning Research 5, 1205–1224 (2004)
9. Hassanien, A.E., Suraj, Z., Slezak, D., Lingras, P.: Rough computing: Theories, technologies and applications. IGI Publishing Hershey, PA (2008)
10. Kessler, F.C.: Understanding Descriptive Statistics,
    http://nationalatlas.gov/articles/mapping/a_statistics.html

Satellite Super Resolution Image Reconstruction Based on Parallel Support Vector Regression

Marwa Moustafa[2], Hala M. Ebied[1], Ashraf Helmy[2],
Taymoor M. Nazamy[1], and Mohamed Fahmy Tolba[1]

[1] Faculty of Computer and Information
Sciences, Ain Shams University, Cairo, Egypt
halam@fcis.asu.edu.eg, fahmytolba@cis.asu.edu.eg
[2] Data Reception, Analysis and Receiving Station Affairs, National Authority for Remote,
Sensing and Space Science, Cairo, Egypt
{marwa,akhelmy}@narss.sci.eg

Abstract. Super Resolution (SR) refers to the reconstruction of a high resolution image from one or more low resolution images for the same scene. The reconstruction process is considered an inverse problem to the observation model. In this paper the SR problem is formulated by using Support Vector Regression (SVR). SVR is a very expensive computationally algorithm, thus it could be accelerated by using the computational power of a Graphics Processing Unit (GPU). The proposed parallel SVR has been implemented using NVidia's compute device unified architecture (CUDA). An experiment has been done for a real satellite image. The experimental result demonstrates the speedup of the presented GPU implementation and compared with the serial CPU implementation and state-of-the-art techniques. The speedup of the presented SVR GPU-based implementation is up to approximately 50 times faster than the corresponding optimized CPU.

Keywords: Super-Resolution, Support Vector Regression, CUDA, GPU.

1 Introduction

Different satellite sensors acquire information on the earth's surface with different spatial resolutions. The remote sensing images received from sensors are usually varied in their characteristics according to blurred or degraded images by some factors. The satellite images represent a huge source of information in modern applications. High resolution images are essential in up-to date geospatial applications. Super Resolution (SR) provides an economic technique to provide these HR images. SR reconstructed techniques are considered a promising approach to these HR images economically. The SR techniques can be classified into two classes [1]; Non-iterative and Statistical based methods

Non-iterative methods also called classical methods [2], such as Bilinear, bi-cubic and B-spline kernels can be used in both multi or single low resolution based cases. The multiple low resolution images are first aligned using some registration

A.E. Hassanien et al. (Eds.): AMLTA 2014, CCIS 488, pp. 223–235, 2014.

algorithms to produce an HR image grid [1]. The reconstruction techniques suffer from ill conditioned registration, blurred, aliased LR images and an insufficient number of low-resolution images of the same scene. In large SR scale or low resolution, images are inadequate; SR using only low resolution images will easily fail. In a single image approach, SR reconstruction is also called magnification, image scaling, interpolation, zooming and enlargement [3,4]. The interpolation methods are used to fill the missing pixels in the HR image grid [1]. Edge discrimination and artifacts are considered the main disadvantages of using the classical method, although it's robust and easly implemented.

Recently, statistical based methods are used to break the limitation of classical SR methods. The learning based techniques can be classified into supervised learning and reinforcement learning according to the model used. The statistical learning is also known as a machine learning based method, bring prior knowledge in the reconstruction process. The goal of these methods is to learn the mapping between low and high images then the learned mapping is adopted to obtain a high resolution image when given a low resolution image [5]. The learning based scheme outperforms the classical methods even if the number of low resolution images is reduced to one [2].

The Support Vector machine is one of the most widely used in many remote sensing applications [6]. In this paper, a parallel implementation of Support Vector Regression (SVR) is proposed using GUP. Recently, Support Vector Regression (SVR), the use of SVM for regression, has been used to generate super-resolution images [7, 8]. The performance of GPU based implementation was evaluated and compared with the CPU native implementation which is presented in [7] and [8].

The remainder of the paper is organized as follows: Section 2 reviews related researches on statistical learning methods. Section 3 discusses Support Vector Machine and reconstruction algorithm based SVR. Section 4 presents the GPU implementation of the SVR. Experiments and results are discussed in section 5. Finally, Section 6 concludes some remarks and provides recommendations for future work.

2   Related Works

The statistical learning methods are a machine learning field based on statistical and functional analysis. The goal of this method is to learn correlation between the input and output data. Statistical methods are used effectively in real world applications such as; medical imaging, speech recognition and computer vision.

The Support Vector Machine (SVM) [9] is a popular approach to solving classification and regression problems. SVM is based on the statistical learning theory. Support Vector Regression allows the ability to predict functional outputs without any prior knowledge or assumption of the training data. Linear or nonlinear kernel is used to fit the data. SVR has been used in various applications of data mining, bioinformatics and financial forecasting. Li and Simske [10] proposed a learning based method based on local similarity across seemingly different images. The results obtained from this study have shown that the proposed method outperforms the competing Support

Vector Regression (SVR) method and the Kernel Regression method in terms of Peak Signal-to-Noise Ratio (PSNR). The Results showed that the performance of SVR depends on the size of the data set. Also, the Accuracy of the testing phase depended on the agreement between the training images and the testing images. An intensive computing is a challenge when used a large data set which considered as another limitation.

Ni and Nguyen, [11] defines an optimal kernel that is achieved by formulating the kernel learning problem in a SVR form as a convex optimization problem, specifically a semi-definite programming (SDP) problem. They have added an additional constraint to reduce the SDP to a Quadratically Constrained Quadratic Programming (QCQP) problem. The study idea was improved by observing structural properties in the Discrete Cosine Transform (DCT) domain to aid in the learning the regression. Further improvement involves a combination of classifications and SVR-based techniques extending works in resolution synthesis, while [12] using SVM regression in both the spatial and DCT domains. The study results have a better mean-squared error. With the addition of structure in the DCT coefficients, DCT domain image super resolution is further improved.

Le and Bhanu [7] proposed single image super resolution based on the Support Vector Regression. By combining the pixel intensity values with local gradient information, the learned model by SVR from low resolution image to high resolution image is useful and robust for image super resolution. The study conducted the experiments on different types of images and the results are promising. They have compared the proposed method to the previous works; the SR image that was produced by the proposed image has a better PNSR value than the state-of-the-art approaches. Ho and Zeng [8] have proposed a simple but effective method to generate high-resolution images from a low resolution one by utilizing the image's own features or information. In this study, the edge region was separated from the smooth region so as to apply different processing methods in these two regions. The smooth pixel is reconstructed by some classical interpolation schemes, whereas the edge pixel is handled by training an SVR model. The resulted images are sharper and have a less embossed effect around the edges. SVR has high prediction accuracy. The training phase of a nonlinear kernel is considered a computationally expensive task notably for larger datasets.

Hu etal. [21] have presented an approach that is mainly based on simple linear algebra computations. The proposed technique involves learning multiple adaptive interpolation kernels. Assume that each high resolution image patch can be sparsely represented by several simple image structures and that each structure can be assigned a suitable interpolation kernel. The experimental results of the proposed technique validate its performance compared with the state-of-the-art super-resolution reconstruction algorithms.

Purkait et al. [22] have developed a novel fuzzy rule based prediction framework for high quality image zooming. In this study a patch based image zooming technique has been produced, where each low resolution (LR) image patch is replaced by an estimated high resolution (HR) patch. This technique is generated a large number of LR-HR patch pairs from a collection of natural images, then group them into different clusters and generate a fuzzy rule for each of these clusters. Experimental results show that the proposed technique is capable of reconstructing thin lines, edges, fine details and textures within the image efficiently. Chen et al.[23] have presented a Bayesian based super resolution algorithm that uses approximations of symmetric alpha stable (SS) Markov Random Fields (MRF) as prior. The approximated SS prior is employed to perform maximum posteriori (MAP) estimation for the high resolution (HR) image reconstruction process. Compared with other state-of the-art prior models, experiments confirm a better fit is achieved by the proposed model to the actual data distribution and the consequent improvement in terms of visual quality.

One of the limitations of statistical methods is its intensive computation. The computation of these statistical methods requires expensive matrix manipulations especially in satellite images. The performance of the prediction model depended on the training set size and the agreement between the training images and the testing images.

3   Reconstruction Technique

Drucker et al. developed both support vector classifier and support vector regression techniques in 1997 [13]. Statistical machine learning algorithm has been recently used in solving SR problem. SR uses support vector regression to map the relationship between the low resolution image and high resolution image by using different non-linear function model with estimated parameter.

3.1   Support Vector Regression

Support vector regression (SVR) is defined as a mapping function $\phi : x \rightarrow F$ in the SVM, to map the data x into a higher dimensional space F in which the data can be linearly separated [10]. SVR uses nonlinear functions to operate in feature space to linearly estimate an unknown function in output space. Consider the training set with N input-output pairs as [10]:

$$\Omega = \{(x_1, y_1), (x_2, y_2), \ldots, (x_N, y_N)\} \tag{1}$$

Where $x_i \in x$ is the input vector from the input image and $y_i \in y$ is the output values in the ground-truth image. Then estimate the function $f : x \rightarrow y$ by utilizing the feature space through the following optimization problem:

$$\min_{w,b,\xi,\xi^*} \frac{1}{2} w^T w + C \sum_{i=1}^{n} \xi_i + c \sum_{i=1}^{n} \xi_i^* \tag{2}$$

$$\text{Subject to} \begin{cases} \mathbf{W}^T \phi(x_i) + b - y_{i \leq} \varepsilon + \xi_i \\ y_i - (w, \varphi(x_i) > +b) \leq \varepsilon + \xi_i^*, \\ \xi_i, \xi_i^*, \geq 0, i = 1, \ldots n \end{cases}$$

Where, ξ and ξ^* are the slack variables, C is a constant and determines the tradeoff between the flatness of the mapping function. The w is a high dimensional vector because φ maps data to a higher dimensional space, thus, the dual problem is solved.

$$\min_{\alpha,\alpha^*} \frac{1}{2}(\alpha - \alpha^*)^T Q(\alpha - \alpha^*) + \varepsilon \sum_{i=1}^{n}\left(\alpha_i - \alpha_i^*\right) - \sum_{i=1}^{n} z_i\left(\alpha_i - \alpha_i^*\right) \tag{3}$$

subject, to $\sum_{i=1}^{n}\left(\alpha_i - \alpha_i^*\right) = 0, 0 \le \alpha_i, \alpha_i^* \le C, i = 1,....n$ Where C is the cost of error and ε is the width of the tube. The model generated by the SVR depends only on a subset of the training data since the cost function ignores the training data within the threshold of ε. The function $Q(x_p, x_q) = k(x_p, x_q) = <\phi(x_p), \phi(x_q)>$ is called the kernel function [8].Commonly used kernel functions are linear, polynomial, Gaussian, sigmoid etc. [9]. The new data point's output is predicted with the function:

$$\sum_{i=1}^{n}\left(\alpha_i^* - \alpha_i\right)K(x_i, x) + b \tag{4}$$

The standard implementation of SVM requires solving several quadratic programming (QPs) problems with a linear constraint [16]. The Computational and memory requirements are very expensive. Different methods have been used to optimize the QPs. One method is the Interior Point Method [16], which was suggested by Von Neumann, it finds a solution to the Karush–Kuhn–Tucker conditions of the primal and dual problems by using Newton like iterations. Another method is the Sequential Minimal Optimization (SMO) algorithm which is used to break the problem into 2-dimensional sub-problems [14, 15].

3.2   Reconstruction Algorithm Based SVR

The SR algorithm proposed in [7, 8] consists of two processes: the training phase and the testing phase as shown in Figure 1. The training phase of the SVR learned model is used to construct a HR image to be used in the testing process. In the training phase, the HR image is blurred and then downscaled by factor 2 to produce the LR image. The initial estimation of an HR image is constructed using bi-cubic interpolation by factor 2, more details can be found [7, 8].

4   GPU Implementation for SVR

Originally, A Graphics Processing Unit (GPU) was designed to accelerate the output image in a frame buffer that's intended for output to display. Recently, it has been used for general purpose computation especially with NVIDIA's Computer Unified Device Architecture (CUDA). GPU Architecture consists of firs: A device that contains an array of independent streaming multiprocessors (SMs). Second, memory sets. Each SM can execute a large number of threads simultaneously. A group of threads is organized into a thread block, while thread blocks are grouped into a block structure. The CUDA Function, referred as Kernel, executed the same instruction by using a large number of threads in parallel fashion. The main function running on the CPU is called the host.

Figure 2 outlines the sequential modified SMO algorithm. SMO is a simple algorithm that can quickly solve the Quadratic Programming of SVM without extra matrix storage. The implementation of the modified SMO on GPU is proposed to be based on distributing the process of matrix multiplication among the GPU threads. Figure 3 shows the parallel implementation of modified SMO's on GPU. An RBF kernel is chosen to be implemented due to its effectiveness in remote sensing applications

$$K(x, x_i) = \exp(-\lambda |x - x_i|^2) \tag{5}$$

where x and x_i are two samples of the feature vectors in input space.

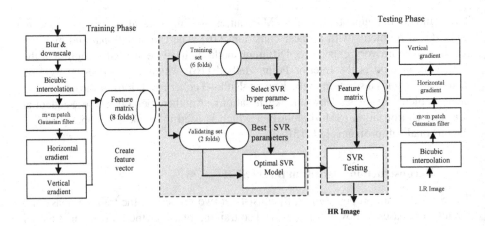

Fig. 1. Full Image SVR Model for Training phase and Testing phase

In the training phase, the CPU Host computes the $x_i \bullet x_i$; then transfers the result to GPU memory. The CUBLAS and CULA libraries [24] are used to compute the $x_i \bullet y_i$ on GPU kernel, and another GPU kernel performs the subtract $(x \bullet x - y \bullet y)$, multiplies by λ and then exponentiation the result. The Shared memory has been used to minimize the data transfer between GPU device memory and CPU host memory.

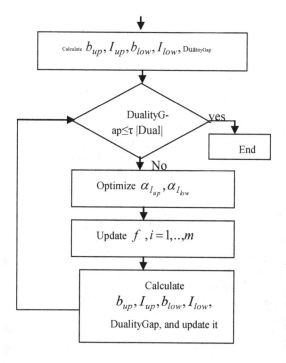

Fig. 2. SMO serial Implementation

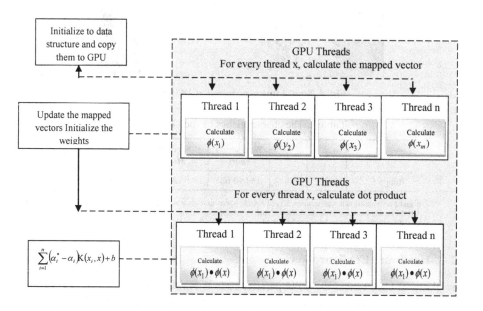

Fig. 3. Parallel GPU Support Vector Regression

5   Experimental Results

5.1   Dataset

The dataset used to perform these experiments is the SPOT-4 image. The French SPOT-4 earth observation satellite was launched on March 24, 1998. It offers a resolution of 10 meters for the panchromatic band [19] and covers wavelengths between 0.61 and 0.68 μm. Fig. 4 shows a panchromatic image acquired at18/08/2011. The scene characteristics are shown in table 1. Fourteen 400 x 400 pixels were cropped from the original SPOT-4 image to be used for both the training and testing phases. The cropped images were chosen to represent the diversity of features such as water, deserts, urban, roads and agriculture areas.

Fig. 4. Spot-4 image acquired at 18/08/2011

Table 1. The characteristics of spot-4 image

Images	K	J	Wavelength (μm)	Spectral Mode	Processing level
Image I	113	288	0.61 -0.68	A	1B

K and J are numbers indicating columns (K) and rows(J), where the scene is located in the global SPOT-5 image reference grid.
A is the camera that acquired the panchromatic band
1B represents the image after being geo-referenced.

5.2   Hardware Configuration

The platform used to test the GPU parallel code is NVIDIA NVS 5200M. This architecture, GF117, is based on the optimized Fermi architecture of the GF108 chip and offers 96 cores, 16 TMUs and 4 ROPs. Each core is clocked twice as fast as the rest of the graphics chip. In our GUP implementation of Support Vector Regression, we used the same parameter values obtained during the serial implementation. In calculating the execution time, data transfer time from the host memory to device memory has been ignored. Table 2 defines the hardware specification that has been used in implementing our experiments.

Table 2. Hardware Specification used in the experiment

CPU	Intel ® Core (TM) i7-3740QM
GPU	NVIDIA NVS 5200M @ 625 MHz, 96 cores; 1024 MB memory size
Operating system	Windows 7 x64
Compiler	Visual Studio 2012
CUDA runtime version	CUDA v5.5

5.3   Performance Criteria

Peak signal-to-noise ratio [18] is used to measure the quality of the output images based on pixel distance. The high value of PSNR means good image resolution.

$$PNSR = 10\log_{10} \frac{\sum_{i=1}^{N}\sum_{j=1}^{M} 255^2}{\sum_{i=1}^{N}\sum_{j=1}^{M} (x(i,j) - y(i,j))^2} \qquad (6)$$

where x denote the original image and y denotes the SR image.

The speedup Sp is calculated to show that a GPU implementation is faster than the sequential implementation on CPU.

$$S_p = \frac{Executio \ Time \ of \ the \ sequentioal \ a\lg orithm}{Execution \ time \ of \ parallel \ a\lg orithm} \qquad (7)$$

5.4   Results

Both CPU serial codes and GPU parallel codes for SVR Super resolution algorithms have been implemented, and the execution time is compared. LIBSVM [15] with Native C/C++ have been used in the sequential implementation of the SR reconstruction algorithm. Many parameters affect the performance of the SVR; C, ε, and δ. In the first experiment, we used cross-validation techniques to divide the training set into three set; training, validation and test tests. An eight from fourteen sample images were used with the cross-validation process. Two images were retained as the internal validation data for testing the model, and the remaining 6 images were used as training data. The cross-validation process was repeated 4 times. Next, the best accuracy of the 4 validations was used as an estimation to save the optimal model with the optimal parameters. The optimal parameters obtained from cross-validation were (C = 260, ε = 1, δ = 2).

In the training phase, eight 200x200 HR images were blurred and downscaled by factor 2 to create a low resolution image. After that, the bi-cubic interpolation is applied to the low resolution image to generate estimation for the desired size of the HR image. Next, the feature matrix is created by applying the SVR super resolution technique. In the testing phase, six 400x400 images were downscaled by factor 2 to create a low resolution image. The same procedure was followed to obtain the 400x400 HR images. Figure 5 shows the high resolution image results from the re-construction algorithm.

In the second experiment, we used our GUP-implementation for the Support Vector Regression with the same parameter values obtained during the sequential implementation. In calculating the execution time, data transfer time from the host memory to device memory has been ignored. The resulted images from the algorithm are sharper than the state of art bi-cubic interpolation. Artifact and effect around the edges has been reduced in the SR images obtained from the proposed algorithm.

Fig. 5. High Resolution Images produced from using SVR SR Algorithm

Table 3 illustrates the execution times during the training phase of SVR using CPU sequential implementation and using CUDA on GPU parallel implementation, for different image types. The speedup of the SR reconstruction algorithm is shown in fig. 6. In the training phase, approximately 50x speedup has been gained while in the test phase, the average acceleration was about 27x. The GPU parallel implementation decreases the execution time of the SVR method.

Table 3. Execution times (ms) of SVR using sequential implementation on CPU and using parallel implementation with CUDA on GPU for different image types.

	Method	CPU	GPU
Training Phase	Using eight image	1173910	29347.75
Testing Phase	Image 1	23035	418.818
	Image 2	10858	904.833
	Image 3	5105	500.490
	Image 4	15941	520.947
	Image 5	12680	845.333
	Image 6	17690	3.23705

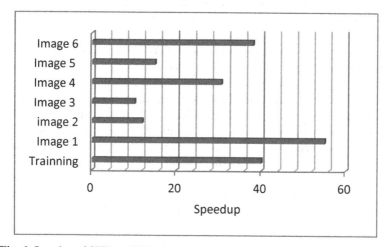

Fig. 6. Speedup of SVR on GPU with respect to sequential implementation on CPU

Table 4. Acuuracy of the SVR SR Algorithm and bicubic method using PSNR.

Image	Bicubic Algorithm	CPU SVR Algorithm	GPU SVR Algorithm
Image 1	33.2	34.54	33.57
Image 2	32.24	33.59	33.24
Image 3	31.57	32.15	32
Image 4	29.83	31.57	31.3
Image 5	32.49	33.95	33.6
Image 6	31.62	29.89	29.7

The last exterminate is performed to compare the accuracy of the sequential implementation and parallel implantation of the SR method with the classical method Bi-cubic using the PSNR. Table 4 shows the PSNR for the super resolution images that are produced from the SVR SR algorithm and the Bicubicclassic method. The difference between PNSR values obtained for parallel implementation and serial

Implementation is hardly noticeable. One can observe from table 4 that, SVR parallel implementation outperforms the SVR sequential implementation and bi-cubic algorithm. The SVR based super resolution reconstruction algorithm is considered a good choice in the case of using satellite images.

6   Conclusion and Future Works

This paper introduced a parallel implementation to a super resolution technique based on Support Vector Regression. CUDA, NVIDIA's parallel computing architecture, is used to implement kernels in both training and testing phases. The paper also draws a comparison between the serial and parallel implementations in the super resolution technique based on the Support Vector Regression. Sharper images were produced from the SVR super resolution algorithm. Visually the around edge effect has been reduced in the obtained HR images.

The experiments are conducted using two SPOT-4 images. Fourteen sub-images were clipped by a window the size of 400×400 pixels to represents different features. The proposed SVR parallel implementation is performed using several training tasks simultaneously. All tasks can share the same cache memory for storing kernel values. Experimental result shows that our Support Vector Regression algorithm speedup by 50x over CPU serial implementation in the training phase and on average 27x for the testing phase. The accuracy of the HR images using SVR SR parallel technique is better than the images obtained from the bi-cubic interpolation method. In our future works, an improved Support Vector Machine based on Rough Set for construction cost will provide more challenging and may allow us to refine our learning algorithms and/or approaches to the resolution image reconstruction [25]

References

1. Yang, J., Huang, T.: Image super-resolution: Historical overview and future challenges. Super-Resolution Imaging, pp. 1–24. CRC Press (2010)
2. Kang, L.-W., et al.: Self-learning-based single image super-resolution of a highly compressed image. In: 5th International Workshop on Multimedia Signal Processing (MMSP) (2013)
3. Candocia, F.M., Principe, J.C.: Super resolution of images with learned multiple reconstruction kernels. In: Guan, L., Kung, S.Y., Larsen, J. (eds.) Multimedia Image and Video Processing, pp. 219–243. CRC Press, New York (2000)
4. van Ouwerkerk, J.D.: Image super-resolution survey. Image and Vision Computing 24, 1039–1052 (2000)
5. Miravet, C., Rodriguez, F.B.: A two-step neural-network based algorithm for fast image super-resolution. Image and Vision Computing 25, 1449–1473 (2007)
6. Mountrakis, G., Im, J., Ogole, C.: Support vector machines in remote sensing: A review. ISPRS Journal of Photogrammetry and Remote Sensing 66, 247–259 (2011)
7. An, L., Bhanu, B.: Improved Image Super-Resolution by Support Vector Regression. In: International Joint Conference on Neural Networks, San Jose, California, USA, July 31-August 5 (2011)

8. Ho, T., Zeng, B.: Super-resolution images by support vector regression on edge pixels. In: International Symposium on Intelligent Signal Processing and Communication Systems, November 28-December 1, pp. 674–677 (2007)
9. Smolaandand, A.J., Schölkopf, B.: A tutorial on support vector regression. Statistics and Computing 14, 199–222 (2004)
10. Li, D., Simske, S.: Example Based Single-frame Image Super-resolution by Support Vector Regression. Journal of Pattern Recognition Research 1, 104–118 (2010)
11. Ni, K.S., Nguyen, T.Q.: Image Super resolution Using Support Vector Regression. IEEE Transactions on Image Processing 16(6) (June 2007)
12. Karl, S.N., Kumar, S., Vasconcelos, N., Nguyen, T.Q.: Single image super resolution based on support vector regression. In: ICASSP (2006)
13. Drucker, H., Burges, J.C., Kaufman, L., Smola, A., Vapnik, V.: Support Vector Regression Machines. In: Advances in Neural Information Processing Systems, NIPS 1996, vol. 9, pp. 155–161. MIT Press (1997)
14. Chang, C.-C., Lin, C.-J.: LIBSVM: a library for support vector machines (2001), Software available at: http://www.csie.ntu.edu.tw/~cjlin/
15. Fan, R.E., Chen, P.H., Lin, C.J.: Working set selection using second order information for training SVM. Journal of Machine Learning Research 6, 1889–1918 (2005)
16. Hua, T.L., Liu, L.X., Kai, S.Z., Yang, W.Y.: GPU Acceleration of Interior Point Methods in Large Scale SVM Training.12th IEEE International Conference on Security and Privacy in Computing and Communications (TrustCom), pp. 863–870 (2013)
17. Do, T.N., Nguyen, V.H., Poulet, F.: A Fast Parallel SVM Algorithm for Massive Classification Tasks. Sciences Communications 14, 419–428 (2008)
18. Wang, Z., Bovik, A.C., Sheikh, H.R., Simoncelli, E.P.: Image quality assessment: From error visibility to structural similarity. IEEE Transactions on Image Processing 13(4), 600–612 (2004)
19. Achard, F., Malingreau, J.P., Phulpin, T., Saint, G., Saugier, B., Segun, B., Madjar, D.V.: The Vegetation Instrument on Board SPOT-4 - A Mission for Global Monitoring of the Continental Biosphere. LERTS brochure, Toulouse (1990)
20. Zhao, H., Yu, J.: A Modified SMO Algorithm for SVM Regression and its Application in Quality Prediction of HP-LDPE. In: Wang, L., Chen, K., S. Ong, Y. (eds.) ICNC 2005. LNCS, vol. 3610, pp. 630–639. Springer, Heidelberg (2005)
21. Hu, X., Peng, S., Hwang, W.L.: Learning adaptive interpolation kernels for fast single-image super resolution. Signal Image and Video Processing, 1077–1086 (2014)
22. Purkait, P., Pal, N.R., Chanda, B.: A Fuzzy-Rule-Based Approach for Single Frame Super Resolution. IEEE Transactions on Image Processing 23(5), 2277–2290 (2014)
23. Chen, J., Yanez, J.N., Achim, A.: Bayesian Video Super-resolution with Heavy Tailed Prior Models. IEEE Transactions on Circuits and Systems for Video Technology 99, 1–1 (2014)
24. CULA (GPU-Accelerated LAPACK), http://www.culatools.com/
25. Hassanien, A.E., Suraj, Z., Slezak, D., Lingras, P.: Rough computing: Theories, technologies and applications. IGI Publishing Hershey, PA (2008)

Cattle Identification Based on Muzzle Images Using Gabor Features and SVM Classifier

Alaa Tharwat[1,4], Tarek Gaber[2,4], and Aboul Ella Hassanien[3,4]

[1] Faculty of Eng, Suez Canal University, Ismailia, Egypt
[2] Faculty of Computers and Informatics, Suez Canal University, Ismailia, Egypt
[3] Faculty of Computers and Information, Cairo University, Cairo, Egypt
[4] Scientific Research Group in Egypt (SRGE)
http://www.egyptscience.net

Abstract. The accuracy of animal identification plays an important role for producers to make management decisions about their individual animal or about their complete herd. The animal identification is also important to animal traceability systems as ensure the integrity of the food chain. Usually, recording and reading of tags-based systems are used to identify animal, but only effective in eradication programs of national disease. Recently, animal biometric-based solutions, e.g. muzzle imaging system, offer an effective and secure, and rapid method of addressing the requirements of animal identification and traceability systems. In this paper, we propose a robust and fast cattle identification through using Gabor filter-based feature extraction method. We extract Gabor features from three different scales of muzzle print images. SVM classifier with its different kernels (Gaussian, Polynomial, Linear and Sigmoid) has been applied to Gabor features. Also, two different levels of fusion are used namely feature fusion and classifier fusion. The experimental results showed that Gaussian-based SVM classifier has achieved the best accuracy among all other kernels and generally our approach is superior than existed works as ours achieves 99.5% identification accuracy. In addition, the identification rate when the fusion is done at the feature level is better than that is done at classification level.

1 Introduction

Animal health and safety of its related products become very crucial for national producers and export markets. This has created a need for source verification, and identification of supply chain of food products. According to [1], beef meat is considered the most consumed meat in the world. So, cattle identification and traceability is currently considered a crucial phase in controlling safety policies of animals, management of food production, and demands of consumers [2]. According to [3], animal traceability process refers to the ability to recognize farm animals and their related products according to their origin in the supply chain to (a) determine ownership, (b) identify parenthood, (c) assure food safety, and (d) ascertain compliance (e.g., for beef export verification, production practice-verification, source-verification, process-verification, and authenticity management). The process of identifying cattle is very important to

A.E. Hassanien et al. (Eds.): AMLTA 2014, CCIS 488, pp. 236–247, 2014.
© Springer International Publishing Switzerland 2014

enable this traceability process and for all entities involved in the food chain including consumers and food industry. Such systems contribute not only to food safety but also to quality assurance. They help to (a) control the spread of animal disease, (b) reduce losses of livestock producers due to disease presence, (c) minimize expected trade loss, and (d) decrease the government cost of control, intervention and eradication of the outbreak diseases [4].

Individual animal identification could be achieved by different methods [4], mechanical, electronic and biometric. As reported in [5] and in [6] mechanical methods are not suitable for large-scale identification programs, could cause animal infections, and not sufficient for traceability purposes. Animal identification through electronic methods [7] make use of external electronic tags (e.g. neck chains or ear tags) which are subject to lose or removal or damage. Biometric animal identification [8] using iris scanning, retinal images and DNA analysis are intrusive for the animals and not cost-effective compared to other approaches (image-processing methods). Machine vision-based solutions [6] can produce accurate results of cattle recognition and do not need to attach any additional elements with or within the animals.

Cattle muzzle print is proven to be a unique feature of each cattle [9]. Consequently, it is concluded that muzzle print is similar to the human's fingerprint. A muzzle pattern could be either lifted on papers or taken as a photo [10]. The lifted on papers images are time-consuming process, requires special skills (controlling the animal and getting the pattern on a paper) and are poor quality. So, in this paper, we will use the muzzle photos and then use Gabor filter to extract features from the collected images of different scales, so overcoming the problem of scale invariance and rotation invariance. These features will be then summed up to overcome the scale invariance problem and increase identification rate.

2   Preliminaries

2.1   Gabor Features

Gabor filter-base method used to extract texture features from gray scale images. It is sensitive to changes in scale and orientation of the texture patterns. Thus, Gabor-filter feature extraction method achieves a relatively small accuracy when the patterns have different scales and orientation[11], [12], [13].

A 2D Gabor function $g(x,y)$ is defined as follows [14]:

$$g(x,y) = \frac{1}{2\pi\sigma_x\sigma_y} exp \left[-\frac{1}{2} \left(\frac{x^2}{\sigma_x^2} + \frac{y^2}{\sigma_y^2} \right) + 2\pi j W x \right] \tag{1}$$

where σ_x and σ_y characterize the spatial extent and frequency bandwidth of the Gabor filter, and W represents the frequency of the filter. Let $g(x,y)$ be the mother generating function of a Gabor filter family. A set of different Gabor functions, $g_{m,n}(x,y) = a^{-2m}g(x',y'))$, can be generated by rotating and scaling $g(x,y)$ to form an almost complete and non-orthogonal basis set, where $\acute{x} = a^{-m}(x\cos\theta_n + y\sin\theta_n)$, $\acute{y} = a^{-m}(-x\sin\theta_n + y\cos\theta_n)$, $a > 1$, $\theta_n = n\pi/K$,

$m = 0, 1, \ldots, S - 1$, and $n = 0, 1, \ldots, K - 1$. The parameter S is the total number of scales, and the parameter K is the total number of orientations. So, S and K represent the total number of generated functions. For, a given image, $I(x, y)$, its Gabor-filtered images is computed as in Equation (2).

$$G_{m,n}(x, y) = \sum_{x_1} \sum_{y_1} I(x_1, y_1) g_{m,n}(x - x_1, y - y_1)) \tag{2}$$

2.2   Feature Fusion

Combining or fusion of many independent sources of information may help to take the most suitable decisions. The combination may be in many levels such as feature or classification level. The goal of feature level fusion is to combine or concatenate the output of two or more independent feature vectors to get one new features vector. Assume $f_1 = [x_1, \ldots, x_r]$, $f_2 = [y_1, \ldots, y_s]$, and $f_3 = [z_1, \ldots, z_t]$ are three feature vectors with three different sizes r, s, and t, respectively. The concatenation of these three feature vectors is calculated by $f_{new} = [x_1, \ldots, x_r, y_1, \ldots, y_s, z_1, \ldots, z_t]$ [15].

Features fusion may lead to a problem of the compatibility of different features, i.e. the features would be in various ranges of numbers. So, it is needed to transform these features into a common domain. To address this problem, normalization techniques such as Z_{Score}, Min-Max, and Decimal Scaling are used [16]. Z_{score} method maps the input scores to distribution with mean of *zero* and standard deviation of 1 [17]. $\acute{f}_i = (f_i - \mu_i)/\sigma_i$ represents Z_{score} feature normalization method, where f_i is the ith feature vector, μ_i and σ_i are the mean and standard deviation of the ith vector, respectively, $\acute{f}_i$ is the ith normalized feature vector.

The fusion of all feature vectors is computed by concatenating the normalized feature vectors. However, concatenation of feature vectors will increase the dimension of the features, thus leading to high computation time and needing more storage space. Thus, dimensionality reduction technique, such as LDA (Linear Discriminant Analysis), is used to reduce a largest set of features and discriminate between classes [16].

2.3   Linear Discriminant Analysis (LDA)

LDA is one of the most famous dimensionality reduction method used in machine learning. LDA attempts to find a linear combination of features which separate two or more classes [18]. The goal of LDA is to find a matrix $W = \max \left| \frac{W^T S_b W}{W^T S_w W} \right|$ that maximizing Fisher's formula. $S_w = \sum_{j=1}^{c} \sum_{i=1}^{N_j} (x_i^j - \mu_j)(x_i^j - \mu_j)^T$ represents a within-class scatter matrix , where $x_i{}^j$ is the ith sample of class j, μ_j is the mean of class j, c is the number of classes, and N_j is the number of samples in class j. $S_b = \sum_{j=1}^{c} (\mu_j - \mu)(\mu_j - \mu)^T$ is a between-class scatter matrix, where μ represents the mean of all classes. The solution of Fisher's formula is a set of eigne vectors (V) and eigne values (λ) of the fisher's formula.

2.4 Support Vector Machine (SVM)

In this paper, we have applied SVM which is one of the classifiers which deals with a problem of high dimensional datasets and gives very good results. SVM tries to find out an optimal hyperplane separating 2-classes basing on training cases [19].

Given a training dataset, $\{x_i, y_i\}$, where $i = 1, 2, 3, \cdots, N$, where N is the number of training samples, x_i is a features vector, and $y_i \in \{-1, +1\}$ is the target label, $y = +1$, for samples belong to class C_1 and $y = -1$ denotes to samples belong to class C_2. Classes C_1 and C_2 are assumed to be linearly separable classes. Geometrically, the SVM modeling algorithm finds an optimal hyperplane or decision surface with the maximal margin to separate two classes and has a maximum distance to the closest points in the training set which are called support vectors, which requires solving the optimization problem in equation 3.

$$max \sum_{i=1}^{n} \alpha_i - \frac{1}{2} \sum_{i,j=1}^{n} \alpha_i \alpha_j y_i y_j K(x_i, x_j). \quad \text{subject to:} \sum_{i=1}^{n} \alpha_i y_i, 0 \leq \alpha_i \leq C \quad (3)$$

where, α_i is the weight assigned to the training sample x_i (if $\alpha_i > 0$, then x_i is called a support vector); C is a regulation parameter used to find a trade-off between the training accuracy and the model complexity so that a superior generalization capability can be achieved.

In case of nonlinear separable classes, each point x in the input or original space is mapped or transformed to a point $z = \phi(x)$ of alternative higher dimensional space, called feature space; which gives a much probability that the mapped points will be linearly separable. The dot product of two points in the feature space $\phi(x).\phi(y)$ can be rewritten as a kernel function $K(x, y)$, where K is a kernel function. If a kernel function must be continuous, symmetric and positive (semi-) definite, so the meaning their kernel matrices have no non-negative Eigne values . Then the optimization problem is convex quadratic of problem, hence the convergence towards the global optimization can be guaranteed and the solution will be unique. There are many types of kernels as follows:

– Linear Kernel : is the simplest kernel function. Linear Kernel is computed as $K(x, y) = xy + c$. Kernel algorithms using a linear kernel are often equivalent to their non-kernel counter parts.
– Polynomial Kernel: is an important family of kernel functions. Ploynomial kernel computed as $K(x, y) = (c + xy)^d$, where d is the degree of the polynomial (if $d = 1$, linear kernel) and c is the intercept constant. Higher order of d, leads to overfitting problem. In overfitting problem, the model may success to fit training data set perfectly with minimum errors, while fitting test or new data will cause a high error.
– Gaussian Kernel: is one of the popular kernel functions. $K(x, y) = exp(\frac{\|x-y\|^2}{2\sigma^2})$ represents Gaussian kernel, where σ plays a major role in the performance of the kernel and it should be carefully tuned to achieve a suitable result.

– Sigmoid (Hyperbolic Tangent) Kernel: Sigmoid Kernel comes from the Neural Networks field, where the bipolar sigmoid function is often used as an activation function for artificial neurons. SVM model using a sigmoid kernel function is equivalent to a two-layer, perceptron neural network. In Sigmoid Kernel, which is computed as $K(x, y) = tanh(\alpha xy + c)$, there are two adjustable parameters, the slope α and the intercept constant c. A common value for α is $1/N$, where N is the data dimension.

Choosing suitable kernel function will make the data easily separable in a feature space despite it is not separable in the original space. However, such choice depends on the problem being addressed- and fine tuning its parameters can easily become a tedious.

2.5 Classifier Fusion

Fusion in classification level may improve the performance of the systems if the classifiers are independent. Fusion of different classifiers may be in abstract, rank or measurement level. Fusion in abstract level considers the simplest fusion method and easiest one to implement. One of the most famous combination methods used in combining classifiers in abstract level is majority voting.

3 Two Proposed Approaches

We have proposed two approaches to identify cattle using muzzle print images. The first one is designed based on feature fusion while the second is designed based on classifiers fusion. The two approaches are summarized in Figure (1).

3.1 Feature Fusion-Based Approach

The Feature fusion-based (FF) approach consists of two main phases: Training and Testing phase.

Training Phase: In this phase the following processes are performed.

1. Collecting all training muzzle print images.
2. Resize the muzzle print images into three different scales $I_{128} = 128 \times 128$, $I_{64} = 64 \times 64$ and $I_{32} = 32 \times 32$.
3. Extracting the features from each resized muzzle print images (I_{128}, I_{64} and I_{32}) using Gabor feature extraction method
4. Representing each image by one feature vector. To reduce the number features in the vector, we used LDA as a dimensionality reduction method,
5. Normalize each feature vector after LDA using Z_{Score} normalization ($\acute{I}_{128}, \acute{I}_{64}$ and $\acute{I}_{32}$),
6. Concatenate the three normalized feature vectors into one new feature vector, i.e., $f_{new} = [\acute{I}_{128} \ \acute{I}_{64} \ \acute{I}_{32}]$

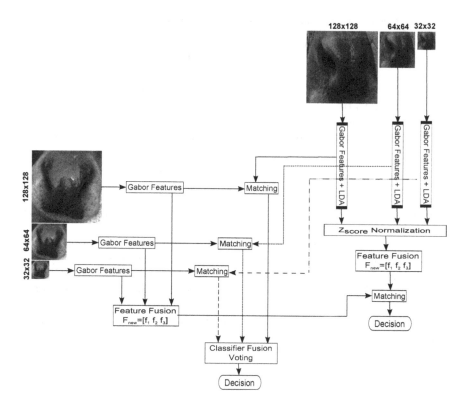

Fig. 1. A block diagram of cattle identification system using muzzle print images

Testing Phase: In this phase, the following operations are performed.

1. Collecting the muzzle print image,
2. Resize the muzzle print images into three different scales: $T_{128} = 128 \times 128$, $T_{64} = 64 \times 64$ and $T_{32} = 32 \times 32$.
3. Extracting the features from each muzzle print images (Testing images) (T_{128}, T_{64} and T_{32}) using Gabor feature extraction method
4. Each feature vector is projected on LDA space.
5. Concatenate the three normalized feature vectors into one new feature vector T_{new}
6. Matching or classifying the testing feature vector T_{new} with training feature vectors f_{new} to identify final decision (i.e. whether the animal is identified or not).

3.2 Classifier Fusion-Based Approach

The Classifier Fusion-based (CF) approach will combine two or more classifiers to identify cattle animals. CF approach, like FF approach, consists of training and testing phases.

Training Phase: In this phase, the system is trained as follows:

1. Collecting all training muzzle print images.
2. Resize the muzzle print images into three different scales $I_{128} = 128 \times 128$, $I_{64} = 64 \times 64$ and $I_{32} = 32 \times 32$.
3. Extracting the features from each muzzle print images (I_{128}, I_{64} and I_{32}) using Gabor feature extraction method
4. Representing each image by one feature vector. To reduce the number features in the vector, we used LDA as a dimensionality reduction method,

Testing Phase: In this phase, the system is tested as follows:

1. Collecting the muzzle print image,
2. Resize the muzzle print images into three different scales $T_{128} = 128 \times 128$, $T_{64} = 64 \times 64$ and $T_{32} = 32 \times 32$.
3. Extracting the features from each muzzle print images (Testing images) (T_{128}, T_{64} and T_{32}) using Gabor feature extraction method
4. Each feature vector is projected on LDA space to reduce its dimensionality.
5. Classifying the testing feature vectors using different scales with training feature vectors to identify final decision in each scale D_1, D_2 and D_3.
6. Combine the output of the three classifiers (decisions) D_1, D_2 and D_3 in abstract level fusion (i.e. voting) to get the final decision (i.e. whether the animal is identified or not).

4   Experimental Results

To evaluate the proposed approach, we have use Matlab platform to implement it and run some experiment. The experiments have been conducted using a PC with the following sepcifications: Intel(R) Core(TM) i5-2400 CPU @ 3.10 GHz, and 4.00 GB RAM, and under windows 32-bit operating system.

The dataset used in the experiments is a muzzle print database image consisting of 217 gray level muzzle print images with size 300×400. These images are collected from 31 cattle animals (7 muzzle print image for each cattle). The muzzle photos are collected in different illumination, rotation, quality levels, and image partiality. Examples of these images are shown in Fig 2.

To test our approaches, we have design three scenarios to test our approach. The first experiment scenario is conducted to understand the effect of changing the number of training data and to evaluate the performance stability over the standardize data (without occlusion nor rotation). In this scenario, testing images are matched using SVM classifier with its different kernels (Gaussian, Linear, Polynomial with different degrees, and Sigmoid). A summary of this scenario is shown in Table 1.

The second experiment scenario is used to prove that our proposed method is robust against rotation. In this scenario, we used four training images. The testing images are rotated and used to identify the cattle. As shown in Fig 3, different orientations are used in our experiment. The results of this experiment is shown in Table 2.

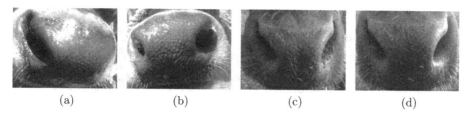

(a)	(b)	(c)	(d)

Fig. 2. A sample of collected muzzle images (a and b belong to one cattle, while c and d belong to another

Table 1. Accuracy results (in %) of our system using different training images

Kernel	No. of Training Images											
	FF						CF					
	6	5	4	3	2	1	6	5	4	3	2	1
Gaussian	100	98.9	98.9	97.9	96.8	96.8	100	100	98.9	97.9	97.9	96.8
Linear	100	98.9	98.9	96.8	96.8	96.8	100	97.9	98.9	96.8	97.9	95.9
Poly (d=3)	100	98.9	98.9	97.9	96.8	96.8	100	98.9	98.9	97.9	97.9	96.8
Poly (d=5)	100	98.9	97.9	97.9	96.8	93.6	96.8	96.8	97.9	96.8	97.9	84.9
Sigmoid	98.9	97.9	97.9	97.9	97.9	96.8	98.9	98.9	97.9	96.8	97.9	96.8

Table 2. Accuracy results (in %) of our system while using rotated images

Kernel	Angles of Rotation (o)													
	FF							CF						
	45	90	135	180	225	270	315	45	90	135	180	225	270	315
Gaussian	100	100	100	100	100	100	100	98.9	100	98.9	100	98.9	100	100
Linear	100	97.9	100	100	100	100	98.9	94.6	97.9	95.7	97.9	89.3	97.9	94.6
Poly (d=3)	100	98.9	100	100	100	100	100	98.9	100	97.9	100	97.9	97.9	98.9
Poly d=5	98.9	98.9	100	100	100	98.9	96.8	96.8	96.8	97.9	98.9	97.9	98.9	89.8
Sigmoid	96.9	96.9	98.9	100	100	98.9	97.9	98.9	98.9	97.9	96.9	97.9	97.9	98.9

In the third scenario, our approaches are tested for images' occlusion shown in Fig 3. In this experiment, the training and testing images are occluded horizontally and vertically in different percentage of its sizes, as shown in Fig 3, and then they are used to identify cattle. The results of this experiment is shown in Table 3.

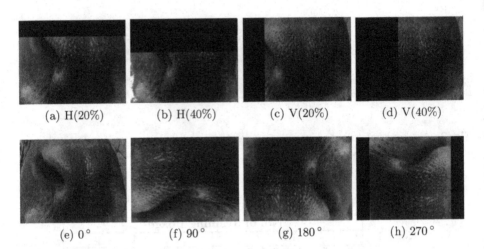

| (a) H(20%) | (b) H(40%) | (c) V(20%) | (d) V(40%) |
| (e) 0° | (f) 90° | (g) 180° | (h) 270° |

Fig. 3. Sample of occluded and rotated muzzle images, top row (a, b, c and d) represents horizontal and vertical occlusion, bottom row (e, f, g and h) represents rotation in different angles

Table 3. Accuracy (in %) of our system while images are occluded

Kernel	Percentage of Occlusion (%)															
	FF								CF							
	H				V				H				V			
	20	40	60	80	20	40	60	80	20	40	60	80	20	40	60	80
Gaussian	100	100	100	94.6	100	100	100	95.7	100	100	95.7	92.5	100	100	95.7	91.4
Linear	100	100	98.9	94.6	100	100	100	98.9	98.3	96.8	89.3	74.2	98.9	96.8	89.9	74.2
Poly (d=3)	100	97.9	97.9	20.4	100	98.9	98.9	49.3	100	97.9	89.3	16.1	100	96.8	91.4	33.3
Poly d=5	97.9	47.3	10.75	3.2	100	45.2	17.2	7.5	92.5	69.9	7.5	0	90.3	67.7	8.6	5.4
Sigmoid	96.8	96.8	97.9	20.4	100	100	98.9	49.3	95.7	94.6	89.3	83.9	96.8	98.9	89.9	33.3

5   Discussion

The performance of FF and CF approaches are evaluated by the percentage of the total number of cattle identifications which were correct(the accuracy). Our discussion will be conducted based on the results gained in the tables above.

From Table 1, the following remarks can be noticed. The accuracy of identifying cattle animals achieve excellent results in both FF and CF approaches. The accuracy is slightly decreased when the number of training images are decreased. Also, all SVM kernels achieved nearly the same accuracy except polynomial function at $d = 5$. In general, CF approach achieves accuracy rate which is slightly lower than FF approach.

From Table 2, a number of points can be noticed. Firstly, FF and CF approaches are robust against images rotation which could take place in different angels. Secondly, Gaussian-based SVM has achieved the best results, while Polynomial kernel with $d <= 5$ has achieved the worst accuracy. Generally, polynomial kernel has achieved accuracy relatively better when its nonlinear has lower degrees such ($d = 2$ or $d = 3$). Thirdly, using rotated images, FF-based approach has achieved accuracy rate better than CF-based approach.

Also, from Table 3, it can remarked that Gaussian kernel-based SVM has achieved the best accuracy, while polynomial-based SVM, with degrees ($d = 5$) or above, has achieved the worst accuracy. Also, it can be seen that polynomial-based SVM has accomplished good results when it is linear or has degrees less than $d <= 4$. Furthermore, FF approach is better than CF approach in the identification rate.

Table 4. Effect of applying LDA on the extracted feature vectors

Image Size	Before LDA		After LDA	
	Length of Feature Vector	Time to Extract Features (Sec)	Length of Feature Vector	Time to Extract Features (Sec)
Original Size (300x400)	240000	30.65 (or Out of Memory)	NA	NA
128x128	32768	1.932	31	0.444
64x64	8192	1.262	31	0.3
32x32	2048	1.063	31	0.255

To show the effect of applying LDA to the performance of our proposed approaches, we have run two experiments, one without applying LDA and another with applying it. The summary of the results obtained from these experiments are shown in Table 4. From this table, it can be noticed that, extracting Gabor features from the original image (with original size) takes more time, leading to out of memory problem. On the other hand, when resizing the muzzle print image into lower scales and applying LDA, the processing time is significantly decreased. This proves the good results obtained by our two proposed approaches.

As a conclusion, form Table 1, 2 and 3, it is noticed that our two approaches achieve a high accuracy rate compared to Awad's system in [5] (93.3%). In addition, from Table 2, we conclude that our two methods achieved excellent accuracy (99.5%) when testing images are rotated or occluded in different angels or percentages respectively. Also, it can be noticed that FF-based approach has achieved accuracy rate better than the one achieved by CF-based approach because the information in feature level are much more than the one in the classification level. Also, the abstract level has only decisions so, it has the minimum information compared with all other classification levels of fusion. Finally, it can be said that our proposed approaches are robust against any distortion in the animal image. This is very important feature when dealing with un-controlled animal while capturing images.

6   Conclusion

In this paper, we have proposed two approaches for identifying cattle animals using muzzle print images. The two approaches make use of Gabor filter to extract robust texture features which are invariant to rotation or occlusion. The features are extracted from three different scales of the images. Two levels of combination or fusion, at feature level of classification level, are then used to increase the animal identification accuracy. The dimensionality problem of the extracted features are addressed by applying LDA which also produced discrimination between different classes and improve the accuracy of our proposed system. Our two proposed approaches make use of SVM classifier with its different kernels function (i.e. Gaussian, Polynomial, Linear, and Sigmoid). The experiment result showed that the two approaches have achieved an excellent accuracy (99.90%). Also, our approaches are tested against any rotation, occlusion, or illumination and they achieved an identification rate of (99.50%). Among these kernel functions, Gaussian-based SVM classifier has achieved the best accuracy in all experiments. In addition, Polynomial-based SVM has achieved a good accuracy but when it is linear or with degrees lower than 5. Also we note that, feature level fusion achieved accuracy better than classifier fusion.

References

1. FAO: World agriculture: Towards 2015/2030. an fao perspective (2003), http://www.fao.org/docrep/005/y4252e/y4252e05b.htm (Online; accessed in April 2014)
2. Bowling, M., Pendell, D., Morris, D., Yoon, Y., Katoh, K., Belk, K., Smith, G.: Review: Identification and traceability of cattle in selected countries outside of north america. The Professional Animal Scientist 24(4), 287–294 (2008)
3. Gonzales Barron, U., Corkery, G., Barry, B., Butler, F., McDonnell, K., Ward, S.: Assessment of retinal recognition technology as a biometric method for sheep identification. Computers and Electronics in Agriculture 60(2), 156–166 (2008)
4. Marchant, J.: Secure animal identification and source verification. JM Communications, UK. Copyright Optibrand Ltd., LLC (2002)
5. Awad, A.I., Zawbaa, H.M., Mahmoud, H.A., Nabi, E.H.H.A., Fayed, R.H., Hassanien, A.E.: A robust cattle identification scheme using muzzle print images. In: 2013 Federated Conference on Computer Science and Information Systems (FedCSIS), pp. 529–534. IEEE (2013)
6. Ahrendt, P., Gregersen, T., Karstoft, H.: Development of a real-time computer vision system for tracking loose-housed pigs. Computers and Electronics in Agriculture 76(2), 169–174 (2011)
7. Voulodimos, A.S., Patrikakis, C.Z., Sideridis, A.B., Ntafis, V.A., Xylouri, E.M.: A complete farm management system based on animal identification using rfid technology. Computers and Electronics in Agriculture 70(2), 380–388 (2010)
8. Allen, A., Golden, B., Taylor, M., Patterson, D., Henriksen, D., Skuce, R.: Evaluation of retinal imaging technology for the biometric identification of bovine animals in northern ireland. Livestock Science 116(1), 42–52 (2008)

9. Baranov, A., Graml, R., Pirchner, F., Schmid, D.: Breed differences and intra-breed genetic variability of dermatoglyphic pattern of cattle. Journal of Animal Breeding and Genetics 110(1-6), 385–392 (1993)
10. Minagawa, H., Fujimura, T., Ichiyanagi, M., Tanaka, K.: Identification of beef cattle by analyzing images of their muzzle patterns lifted on paper. Publications of the Japanese Society of Agricultural Informatics 8, 596–600 (2002)
11. Jain, A.K., Farrokhnia, F.: Unsupervised texture segmentation using gabor filters. In: Conference Proceedings of the IEEE International Conference on Systems, Man and Cybernetics 1990, pp. 14–19. IEEE (1990)
12. Zhang, J., Tan, T., Ma, L.: Invariant texture segmentation via circular gabor filters. In: Proceedings of the 16th International Conference on Pattern Recognition 2002, vol. 2, pp. 901–904. IEEE (2002)
13. Kong, W.K., Zhang, D., Li, W.: Palmprint feature extraction using 2-d gabor filters. Pattern Recognition 36(10), 2339–2347 (2003)
14. Han, J., Ma, K.K.: Rotation-invariant and scale-invariant gabor features for texture image retrieval. Image and Vision Computing 25(9), 1474–1481 (2007)
15. Rattani, A., Kisku, D.R., Bicego, M., Tistarelli, M.: Feature level fusion of face and fingerprint biometrics. In: First IEEE International Conference on Biometrics: Theory, Applications, and Systems, BTAS 2007, pp. 1–6. IEEE (2007)
16. Auckenthaler, R., Carey, M., Lloyd-Thomas, H.: Score normalization for text-independent speaker verification systems. Digital Signal Processing 10(1), 42–54 (2000)
17. Jain, A., Nandakumar, K., Ross, A.: Score normalization in multimodal biometric systems. Pattern Recognition 38(12), 2270–2285 (2005)
18. Scholkopft, B., Mullert, K.R.: Fisher discriminant analysis with kernels (1999)
19. Elhariri, E., El-Bendary, N., Fouad, M.M.M., Platos, J., Hassanien, A.E., Hussein, A.M.M.: Multi-class svm based classification approach for tomato ripeness. In: Abraham, A., Krömer, P., Snášel, V. (eds.) Innovations in Bio-inspired Computing and Applications. AISC, vol. 237, pp. 175–186. Springer, Heidelberg (2014)

Logistic Test Case Regression Model of Petroleum Availability Prediction System

Senan A. Ghallab, Nagwa L. Badr, Abdel Badeeh Salem, and Mohamed Fahmy Tolba

Ain Shams University, Faculty of Computer and Information Sciences, Egypt
senanphd@gmail.com, nagwabadr@cis.asu.edu.eg,
abmsalem@yahool.com

Abstract. Crude oil represents the most important energy source as a life engine. Nowadays, oil industries are trying to find a way to reduce its costs and make a better use of the crude oil prediction process. In addition, multiple computational intelligence techniques are used to apply prediction systems on petroleum field. Uncertainty qualification of petroleum prediction systems is one of the present challenges. Using test cases process on this domain approves more accurate results. The basic idea of this approach is to retesting both historical and test cases statues through regression processes. Validation of test case statues through logistic regression model terminates the controversial about the system efficiency. In this paper, we propose a logistic test case regression model to argue the predicted statues of an intelligent petroleum prediction system. This approach produces results that are closely to measured values which predicted for emphasis petroleum historical data. The petroleum dataset are collected from distinct sources. The historical dataset and previous predicted values of oilfields and reservoirs are exploited.

Keywords: Regression, Prediction, Test Case, Petroleum.

1   Introduction

The petroleum domain contains huge amount of data, geophysics, images and other [1]. On the other hand, according to different definitions of regression test process, regression is a statistical technique to determine the linear relationship between two or more variables, is primarily used for prediction and causal inference [2]. Conceptually, retesting software results after validation and evaluation called regression process. This processes attempts to validate modified software and ensure that no new errors are introduced [3]. Consequently, the prediction results of any system remain controversial; the results supported by experts' knowledge validation. Furthermore, predicted results testing are ubiquitous technique, but many of those concepts still poorly understood. On prediction process, propagate the process of prediction which considers measurements values and exploiting the specific inputs through software testing [4].

A.E. Hassanien et al. (Eds.): AMLTA 2014, CCIS 488, pp. 248–257, 2014.

The test case processes explains the fundamental differences between systems results and testing values. In addition, test case on petroleum prediction domain able to manipulate the faults of predicted systems result. The historical test case performance data might be used to improve long run regression testing performance [5]. There are three types of test case processes which are approved: execution, infection and propagation. In prediction systems, while the impact of system represents execution process, the changes of the data manipulation is the infection process [6]. The precision of predicted test cases statues depends on data validations and approved by using regression test to propagate failures of the prediction system [7]. Eventually, using regression test process on petroleum domain aims engineers to enhance oil industry and emphasis of prediction statues, achieve this goal through the work implementation.

The rest of the paper is organized as follows. Section 2 shows related work. Section 3 explains approach overview. Section 4 shows regression model. Section 5 displays the evaluation of the proposed system. Finally, Section 6, shows the system results and suggestions for future work.

2   Related Work

Most related work on regression testing has used coverage information. More than one related research application concerned of oil prediction. Use the test cases regression to improve the faults of results still demand more researches. It produces a prioritized list of test cases in terms of sequence base on the number of impacted blocks that will be covered by each test. Furthermore, a list of impacted block, which may not be executed by any exist test [8]. Using regression test within static coverage values for prediction purpose is achieved with multiple test cases [9]. The proposed regression test neglects the petroleum software code, previous predicted results and historical values. We interest about retesting the test cases statues and prove regression test of the system results.

A binary matching tool (BMAT), compute the changes between two versions of the program at the basic block granularity. Some researches concerns of forecast the oil production by using data mining technique for improving estimation of oil consumption base of history of data. The case study with ANFIS has less than %5 Error.

The proposed algorithms were implemented on 91 rows of data (monthly oil consumption) from January 2003 to July 2010 for Canada and the rate of accuracy is 95% [10]. Another statistical research, concludes that the horsepower and weight (W) of a vehicle proved to be statistically important in predicting petroleum consumption in vehicles; width and engine displacement of a vehicle have positive impacts on fuel cost [11]. Another related work is considers a data driven approach in modeling uncertainty in spatial predictions, based on Bayesian inference presented as an envelope of p10/p90 credible intervals for reservoirs production prediction [12].

Another related research, focuses on the regression analysis and time series analysis of the Petroleum Product Sales in Masters Energy oil and Gas. it shows that time series analysis explain the analysis of trend and seasonal influence on the data. It was also used to forecast the future product sales of the companies [13].

A Study on Prediction of Output in Oilfield Using Multiple Linear Regression, achieved that shown that the percentage error of predicted value from the actual output is only 4.57%. This validate that this method can be implement to forecast the oilfield output [14].

3   Approach Overview

The major goal of this research is to produce approximated prediction results through using regression of test case statues on petroleum prediction system. Exploiting historical data, predicted results and measured values of oilfields to achieve dazzle prediction results. Previous test case for petroleum prediction system is achieved through Petroleum Prediction System. The previous predicted results extracted, whereas the historical data collected from distinct petroleum sources. Ten wells on Daqing oilfields and other oilfields in Yemen as samples are predicted, shown on table (1). Test case process is applying by build relational rules from comparing, testing the predicted results with measured values of the oilfields. In addition, test cases (Tc) process applied on ten of predicted wells and measurements values (real coverage). Table (1), display ten wells within predicted results as Previous Predicted results (PP) and the empirical values as Coverage petroleum values (CP), as follow:.

Table 1. CP, PP and Tc results

Sample wells	CP (%)	CP statue	PP (%)	PP statue	Tc Statue
1	24.91	M	20.353	M	M
2	12.50	VL	14.6709	VL	VL
3	14.7	VL	13.1073	VL	VL
4	30.24	M	35.7685	L	M
5	37.62	M	26.8873	L	M
6	36.88	M	26.5348	VL	M
7	37.50	M	26.8688	M	M
8	61.13	H	62.6241	M	H
9	32.31	M	32.31	H	M
10	38.74	M	37.4407	M	M

The regression methodology goal is to retest the previous test case statues that are shown on table (1) through logistic regression model processes. Design of logistic regression model will be utilized to efficiently minimize the number of test cases, running time and will be used to predict software failures. Conceptually, there is inherently an unbounded test case on the petroleum prediction system; the number of test cases calculated by An, where A, is number of the detected memberships, whereas n is the number of statue probabilities. The memberships of previous fuzzy petroleum prediction research are depends on five parameters [15]. On the other hand, the output, test case statues for every single membership are: Very High (VH), High (H), Medium (M), Low (L), and Very Low (VL). The number of test cases on the system is $5^5 = 3125$. This huge number of test cases causes conflict on the predicted statues.

4   Regression Model

Systems improvements should result in better coverage of the input domain, efficient test cases, and in spending fewer testing resources. Regression testing is a software engineering activity in which a suite of tests covering the expected behavior of a software system. On another form, it represents re-run of outputs compared against expected results [16].

Usually, the regression test concerns of failures of software's code, multiple projects are interest about retesting the software failures such as PETS project (Prediction of software Error rates based on test and Software maturity results) funded by the European Commission [17].

4.1   Regression Analysis

Regression analysis is a statistical tool for the investigation of relationships between variables. Usually, the investigator seeks to ascertain the causal effect of one variable upon another into systems. Consequently, regression techniques have long been investigate hypothesis, imagine that gathering data on multiple domains [18].

Realistically, the prediction results accuracy of any intelligent system never achieve one hundred percent of precision [19]. Test case processes and regression test aims to validate system failures and focus on real factor which affect of results. In this case, test case process applying on the petroleum prediction system to emphasis the predicted result; regression test reduce the test case number, minimize error time, share petroleum knowledge and declares more precision of the system [20]. Regression analysis for petroleum test cases data which are collected from distinct sources calculated through the following functions:

$$\sum_{i=1}^n X_i \ , \sum_{i=1}^n y_i \ , \sum_{i=1}^n X_i^2 \ , \sum_{i=1}^n y_i^2 \ , \sum_{i=1}^n x_i y_i \tag{1}$$

$$\bar{x} \ , \bar{y} \ , S_{xx} \ = \ \sum_{i=1}^n X_i^2 - \frac{\left[\sum_{i=1}^n X_i\right]^2}{n} \tag{2}$$

$$S_{yy} = \sum_{i=1}^n y_i^2 - \frac{\left(\sum_{i=1}^n y_i\right)^2}{n} \ , S_{xy} \ = \ \sum_{i=1}^n x_i y_i - \frac{\left(\sum_{i=1}^n X_i\right)\left(\sum_{i=1}^n y_i\right)}{n} \tag{3}$$

$$b_0 = \bar{y} - b_1 \bar{x} \ , \quad b_1 \ = \ \frac{S_{xy}}{S_{xx}} \tag{4}$$

$$S_{yy}/x = S_{yy} - \frac{(S_{xy})^2}{S_{xx}}$$

(5)

Where X1-Xn, Y1-Yn are the input data; S is the summation operation; b0, b1 are the slope, intercept estimate of regression curve and n is the number of experimental dataset. To applying the last functions (1-5), using ANOVA function proves the regression value and variance error between values relationships. The regression result achieves that the best results on two detected parameters (Temperatures and Pressure) test cases was close to zero which represents the test case system failures, Table (2) shows the different regression results between values relationships, as follow:.

Table 2. Sample of test cases regression dataset model

Sample	Temperature	Pressure	Test Cases	Regression
1	108.3	67.5	2^5	1
2	62.8	71.5	2^5	0.910
3	57.2	55.6	2^5	0.403
4	85.0	97.0	2^5	0.231
5	85.6	40.4	2^5	0

The regression model shows that the temperature and pressure memberships achieve the best predictable test cases of petroleum prediction system, whereas, three memberships are neglected because just two memberships that effects of the test cases results and the applied regression model produced higher values.

4.2   Regression Approach

In petroleum domain, re-run the predicted results and retesting the test case statues will ensure precision of the prediction systems. Figure (1), shows that regression model methodology, as follow:

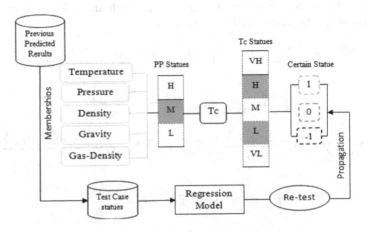

Fig. 1. Regression model methodology

Figure (1), shows the petroleum memberships, statues of every membership, previous predicted, test cases statues and the certain statues values (0, 1,-1).

The Tc which affect on the result is (1), Tc which not affect on the result is (0) and the neglects test cases (-1). The major goal of this model is to retesting the statues values through regression model. Other goals are reduce the number of test cases and minimize the running time. The simplest regression form that shows the relationship between one independent variable (X) and a dependent variable (Y):

$$Y = \beta_0 + \beta_1 X + u \tag{6}$$

Where the value of intercept represents by ($\beta 0$), whereas, the value of slope represents as ($\beta 1$) and (u) is the no predicted variation by the slope and intercept terms. Conceptually, slope and intercept represents through (the neglects test cases (-1)):

$$\text{Slope } (m) = - \frac{n(\Sigma xy) - (\Sigma x)(\Sigma y)}{n(\Sigma x^2) - (\Sigma x^2)} \tag{7}$$

$$\text{Intercept} = \frac{n(\Sigma y) - m(\Sigma x)}{n} \tag{8}$$

Where x, y is the relational cases, whereas, n is the number of case. This work proves the certain test cases statues and then using regression model to emphasis system results, table (3), shows a sample of two petroleum factors data(Temperature (x) and Pressure (y)), as follow:

Table 3. CP, PP and Tc results

Sample	X	Y	X^2	XY
1	108.3	67.5	11728.89	7310.25
2	62.8	71.5	3943.84	4490.2
3	57.2	55.6	3271.84	3180.32
4	85.0	97.0	7225	8245
5	85.6	40.4	7327.36	3458.24
Sum (5)	398.9	332	139057.04	26684.01

Slope (m) =   5 (26684.01)-(398.9)(332) / 5 (139057.04)- 159121.21 = 0.0018

Intercept (c) = 332-0.0018(398.9) / 5 = 66.26

As shown on figure (2), X-axis represents slope (m) and Y axis is the intercept (c); it is interprets how the values scattered (m, c) with separate slope and infrequently limited values between (5,-5).

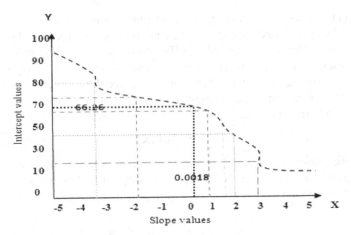

Fig. 2. Slope and intercept curve

The number of test cases of tow petroleum parameters (Temperature and pressure) becomes (25). The logistic regression of the test cases calculated through Logit p(x) function (9) where x represents parameters values, β is the intercept value and X is test case parameter, the result of different test cases regression close to zero, shown on table (4):

$$logit\ p(x) = \ln p(x) / (1 - p(x)) \tag{9}$$

Where $P(x) = 1/ [1+\exp(-(\beta_0 + \beta_1 Xi)]$

As a sample of petroleum factors test cases logistic regression, the first values calculated through:

P(X) = $1/[1+\exp(-(0.0018+7175.958)] = $ **- 1.39,**
Logit $\ln$ $p(x) = (-1.39) / (1-(-1.39)) = $ **0.23.**

Table 4. Logistic regression sample

Tc	Temperature	Pressure	P(X)	Logit P(X)
1	108.3	67.5	-1.39	0.23
2	62.8	71.5	-2.16	0.35
3	57.2	55.6	-.001	0.006
4	85.0	97.0	-2.78	0.27
5	85.6	40.4	-1.72	0.2

As a remedy of logistic regression, the test cases regression efficiency achieves 92% (median average (real Tc (Tc coverage)/100): [0.23*32(1250) / 100], whereas the previous regression efficiency for all test cases 15.62% the results of test cases regression shows different dazzle inferences:

- All results close to zero, which means, test cases prediction failures trends to zero,
- Divide slope and intercept range into three levels (0,-1,1))instead of tow (0,1),
- Reduce the number of test cases aims to achieve more accurate prediction results ,especially in applicable fields,
- Minimizing of regression time error depends on classifying and analyzing inputs and split test cases into several statues,
- The efficiency of regression model achieves 92% and
- Regression analysis results stay on the same range of test cases regression model of petroleum prediction results.

5   Model Validation

Obviously, the number of test cases affecting of regression result and system failures. This work aims to re-run the intelligent petroleum prediction software through test cases regression model. Furthermore, validation of the coverage test cases statues assume that there are three real values (1,0,-1) instead of tow on some previous researches, whereas, the highest value of other test cases logistic regression achieves 90.05% [8].

The evaluation criteria shows that the number of test cases are reduces from 3125 case into 1250 case. The values represents Tc's which affected of the predicted result (1), Tc's which not affected of the predicted result (0) and Tc's which neglected from the system (-1). The criterion of predictive validity for model validation and assessment anticipate.

Consequently, the neglected test cases which are not affected on the prediction result should be rejected to minimize machine time.

The model validations interprets that 3125 test cases, there are 1875 as a neglected test cases. The other test cases, forty four only was incorrectly, the same of 32 predicted test case, only four was incorrectly statuses. The overall regression fitness is 91.95 % of petroleum case study, as shown on table (5), as follow:

Table 5. Model Validation results

	Observed	Predectid		Percent correct
Regression	0	1218	**32**	96.4
	1	44	**4**	87.5
	-1	1875	--	
	Overall fitness: 91.95%			

6   Conclusions and Future Works

The complexity, cost and reliability of the systems are growing rapidly. Proposing a new model to release system faults are critical. This paper elaborated the usage of Logistic Regression Model to predict the test cases statues of intelligent petroleum prediction system. Retest the intelligent petroleum prediction system within test cases statues shows that there are two out of five standard petroleum factors that affect of test cases prediction result, other factors are neglected. The novel logistic regression model updates probabilities of outcome variables into three levels, the test cases prediction failure close to zero. In addition, regression model achieves 92% efficiency; the overall fitness confidence of petroleum system test cases achieves 91.95%. Hopefully on near future researches, increases the results confidence, using regression model to examine several types of applications. In our future works, rough sets-based feature extraction, rule generation and classification will provide more challenging and may allow us to refine our learning algorithms and/or approaches to the incorporating logistic regression to decision-theoretic rough sets for classifications [22].

References

1. Bao, Y., Yang, Y.: A Comparative Study of Multi-step-ahead Prediction for Crude Oil Price with Support Vector Regression. Computational Sciences and Optimization 1, 598–602 (2011)
2. Dongzhi, Z., Guoqing, H., Bokai, X.: Analysis of Multi-factor Influence on Measurement of Water Content in Crude Oil and Its Prediction Model. Chinese Control 2, 430–436 (2008)
3. Spacey, S., Wiesmann, W., Kuhnand, D., Luk, W.: Robust software partitioning with multiple instantiation. INFORMS 24, 500–515 (2012)
4. Mayo, M., Spacey, S.: Predicting Regression Test Failures using Genetic Algorithm-Selected Dynamic Performance Analysis Metrics. Technical report, New Zeland university (2014)
5. Salem, A.M., Rekab, K., Whittaker, J.A.: Prediction of software failures through logistic regression. Information and Software Technology 26, 781–789 (2004)
6. Kordnoori, S., Mostafaei, H.: Grey Markov Model For Prediction The Crude Oil Production And Exportation In Iran. Academic Research 3, 1029–1033 (2011)
7. Anifowose, F.A., Abdulraheem, A.: A Functional Networks-Type-2 Fuzzy Logic Hybrid Model for the Prediction of Porosity and Permeability of Oil and Gas Reservoirs. Computational Intelligence, Modeling and Simulation 37, 193–198 (2009)
8. Rothermel, G., Untch, R.H., Chu, C., Harrold, M.J.: Prioritizing Test Cases For Regression Testing. IEEE Transactions on Software Engineering 27, 929–948 (2002)
9. Kim, J.-M., Porter, A.: A history-based test prioritization technique for regression testing in resource constrained environments. In: ICSE, vol. 86, pp. 119–129 (2002)
10. Mahdavi, Z., Khademi, M.: Prediction of Oil Production with: Data Mining, Neuro-Fuzzy and Linear Regression. Computer Theory and Engineering 4, 446–447 (2012)
11. Abude, F.M.: Modeling of Determinants of Petroleum Consumption of Vehicles in Ghana. Applied Science and Technology 3, 111–115 (2013)

12. Demyanov, V., Kanedvski, M., Pazdnoukhov, A., Christie, M.: Uncertainty quantification with support vector regression prediction models. In: Accuracy Symposium, vol. 50, pp. 133–136 (2010)
13. Daniel, E.C., Chuka, C.E., Victor, O.A., Nwosu Moses, C.: Regression and Time Series Analysis of Petroleum Product Sales in Masters energy oil and Gas. IJETTCS 2, 1285–1303 (2013)
14. Mustafar, I.B., Razali, R.: A Study on Prediction of Output in Oilfield Using Multiple Linear Regression. Applied Science and Technology 1, 107–113 (2011)
15. Ghallab, S.A., Badr, N., Salem, A.B., Tolba, M.F.: A Fuzzy Expert System For Petroleum Prediction. In: WSEAS, Croatia, vol. 2, pp. 77–82 (2013)
16. Srivastava, A., Thiagarajan, J.: Effectively prioritizing tests in development environment. ISSTA, Technical report (2002)
17. Dong, G., Xu, J., Song, Y.: Static Coverage Prediction for Regression Test. IEEE Transaction on Software Engineering 27, 236–248 (2004)
18. Grottke, M., Dussa-Zieger, K.: Prediction of Software Failures Based on Systematic Testing. In: EuroSTAR, vol. 50, pp. 1–12 (2000)
19. Ghallab, S.A., Badr, N., Salem, A.B., Tolba, M.F.: Intelligence Test Case Based-Approach for Crude Oil Prediction System. In: IS 2014. IEEE Press, Poland (2014)
20. Mayo, M., Spacey, S.: Predicting Regression Test Failures using Genetic Algorithm-Selected Dynamic Performance Analysis Metrics. In: Ruhe, G., Zhang, Y. (eds.) SSBSE 2013. LNCS, vol. 8084, pp. 158–171. Springer, Heidelberg (2013)
21. Ghallab, S.A., Badr, N., Salem, A.B., Tolba, M.F.: Integration Web-Based Crude Oil Ontology. In: ICICI, Egypt, vol. 1, pp. 91–96 (2013)
22. Hassanien, A.E., Suraj, Z., Slezak, D., Lingras, P.: Rough computing: Theories, technologies and applications. IGI Publishing Hershey, PA (2008)

Parallel Ward Clustering for Chemical Compounds Using MapReduce

Mohamed G. Malhat, Hamdy M. Mousa, and Ashraf B. El-Sisi

Computer Science dept., Faculty of Computers and Information,
Menofia University, Egypt
m.gmalhat@yahoo.com, {hamdimmm,ashrafelsisi}@hotmail.com

Abstract. The availability of chemical libraries with millions of compounds makes the process of identifying similar chemical compounds more challengeable. Compounds with similar structure are likely to exhibit similar biological activity. So, the identification of these compounds is a key step in the drug discovery process. Hierarchical clustering is developed for that purpose. One of the most popular hierarchical clustering algorithms that are used in many applications in the drug discovery process is ward clustering algorithm. A fundamental problem with the previous implementations of this clustering method is its limitation to handle large data sets within a reasonable time and memory resources. In this paper, MapReduce framework is used to run ward clustering algorithm in parallel manner. The results show considerable reduction in computational time. The parallel ward algorithm saves 17% of time using 3 map instances and saves 58% of time using 6 map instances.

Keywords: Drug Discovery, Hierarchical Clustering, Ward Clustering, MapReduce.

1 Introduction

The drug discovery is the process of making drugs that response to diseases with fewer side effects. It consists of seven steps: disease selection, target hypothesis, lead compound identification, lead optimization, pre-clinical trial, and clinical trial and pharmacogenomic optimization [1]. Chemoinformatics are a new discipline emerging from storing, manipulating, processing, design, creation, organization, management, retrieval, analysis, dissemination, visualization, and use of chemical information. Sometimes it's defined as the application of informatics methods to solve chemical problems [2].

Chemoinformatics are used in lead compound identification and optimization steps. It becomes a critical part of the drug discovery process because it accelerates and improves the drug discovery process and reduces the overall cost [3]. Clustering algorithms is one of the chemoinformatics methods that is used in preliminary analyses of large data sets of medium and high dimensionality as a method of selection, diversity analysis and data reduction [4].

A.E. Hassanien et al. (Eds.): AMLTA 2014, CCIS 488, pp. 258–267, 2014.

The applications of clustering algorithms in drug discovery are compound selection, compound acquisition, High-Throughput Screening (HTS), Quantitative Structure-Activity Relationship (QSAR) analysis and Absorption, Distribution, Metabolism, Elimination, Toxicity (ADMET) prediction [5-12]. Central tasks of most of these applications are the establishment of a relationship between a chemical structure and its biological activity. So, the quality of any clustering algorithm is determined by its ability to group more homogenous compounds in one cluster.

Many clustering algorithms are available for these applications but the most popular clustering algorithm is ward clustering algorithm because of its ability to minimize square-error increase in the produced clusters. But, it is very demanding of computational resources. With the availability of chemical libraries with millions of compounds, ward clustering algorithm fails to handle large chemical data sets [13].

In this age of data size and processing explosion, parallel processing is essential to process a massive volume of data in a timely manner. MapReduce, which has been popularized by Google, is a scalable and fault-tolerant data processing tool that enables to process a massive volume of data in parallel with many low-end computing nodes [14]. In this paper, ward clustering algorithm is modified to run on parallel manner using MapReduce to solve traditional algorithm limitations. The results show that MapReduce saves 17% of time using 3 map instances and 58% using 6 map instances.

The organization of paper is as follow. In section 2, ward clustering algorithm is discussed. In section 3, MapReduce as a parallel data processing mechanism is overviewed. In section 4, the using of MapReduce in ward clustering algorithm is proposed and the result is discussed. Finally in section 5, conclusion and future work are given.

2   Ward Clustering Algorithm

Clustering methods are used in a number of disciplines such as computer science, information technology, information system, engineering, bioinformatics and chemoinformatics. The main using of clustering methods in chemoinformatics is to group similar compounds in a cluster based on a model of similarity measures. After grouping these compounds, the activity of compounds is predicted based on known compounds activity that are in the same cluster [3].

Ward is one of the most used agglomerative hierarchal clustering methods in drug discovery process. It is implemented using stored-matrix algorithm. Stored-matrix contains all pairwise proximities between compounds in the data set to be clustered. Each cluster initially contains an individual compound. As clustering proceeds, each cluster may contain one or more compounds [13]. The stored-matrix algorithm proceeds as follows:

1. Calculate the initial stored-matrix containing the pairwise proximities between all pairs of clusters in the data set.
2. Scan the matrix to find the most similar pair of clusters, and merge them into a new cluster.

3. Update the stored-matrix by inactivating one set of entries of the original pair and updating the other set with the proximities between the new cluster and all other clusters.
4. Repeat steps 2 and 3 until just one cluster remains.

The initial stored-matrix in step 1 is usually calculated using the Euclidean distance measure for numerical chemical representations and Tanimoto coefficient for binary chemical representations [15]. In step 2, the most similar pair of clusters is the one with the smallest proximity whose merger produces the minimum change in square–error. The proximity between clusters in step 3 is calculated using the Lance–Williams formula:

$$d[k,(i,j)] = \alpha_i \, d[k,i] + \alpha_j \, d[k,j] + \beta d[i,j] + \gamma |d[k,i] - d[k,j]| \qquad (1)$$

$$\alpha_i = \frac{N_i + N_k}{N_i + N_j + N_k}, \alpha_j = \frac{N_j + N_k}{N_i + N_j + N_k}, \beta = \frac{-N_k}{N_i + N_j + N_k} \text{ and } \gamma = 0.$$

Where $d[k,(i,j)]$ is the proximity between cluster k and cluster (i, j) formed from merging clusters i and j. The main advantage of ward algorithm is ability to minimize square-error increase in the produced clusters. But Ward requires $O(n^3)$ computation time and $O(n^2)$ space where n is the number of compounds to be clustered [13]. So, it fails to handle large chemical data sets.

The applications of ward in drug discovery are QSAR analysis, compound selection and diversity analysis. In [13], different clustering algorithms were tested on 10 small data sets for which certain properties were known. The results indicated that the ward hierarchical algorithm gave the best overall performance. But Ward algorithm was not well suited to processing large data sets. In [16], ward algorithm is compared with the Max-Min diversity selection method, Kohonen maps and a simple partitioning method to help select diverse yet representative subsets of compounds for further testing. The result shows that ward clustering was the only method that gave results consistently better than random selection of compounds for small data sets. In [17], ward algorithm is used to show that cluster representatives provide a significantly better sampling of activity space than random selection. It shows how clustering can separate actives from in-actives in a small chemical data set, so that a cluster contain in at least one active will contain more than an average number of other actives. In [18], ward algorithm is incorporated with level selection to supports structure browsing and the development of structure-activity relationship in HTS data sets. The clustering was used to produce smaller, more homogeneous subsets from which one representative compound was selected as a screening candidate to determine the optimal clustering level.

Different parallel implementations of ward were proposed to solve computational resource limitations for handling large chemical data sets. In [19], parallel implementation of ward methods for chemical structure clustering was proposed using distributed array processor. In [20], a parallel version of ward algorithm was designed on an n-node hypercube and n-node butterfly. In [21], ward algorithm was performed on parallel using several proximity matrixes based on Parallel Random Access Machine (PRAM).

Recently, MapReduce framework is popularized by Google. It enables a massive volume of data to be processed in parallel manner. MapReduce is one of the most popular parallel processing mechanisms because data is stored locally on the corresponding machine when the job is being executed, great deal for reducing computation time for large data sets, scalability of data, ease of use and fault tolerance when any machine fail other machine is launched automatically. In the next section, MapReduce as a parallel processing mechanism is overviewed.

3   MapReduce Overview

MapReduce is a programming model and an associated implementation for processing and generating large data sets. Users specify a map function that processes a key/value pair to generate a set of intermediate key/value pairs, and a reduce function that merges all intermediate values associated with the same intermediate key. Programs written in this functional style are automatically parallelized and executed on a large cluster of commodity machines. The run-time system takes care of the details of partitioning the input data, scheduling the program's execution across a set of machines, handling machine failures, and managing the required inter-machine communication. This allows programmers without any experience with parallel and distributed systems to easily utilize the resources of a large distributed system [14].

Fig.1 shows MapReduce architecture consists of two phases: map and reduce. In the map phase, the mapper (the algorithm that specifies a map function) takes as input a key/value pair, and it outputs a sequence of key/value pairs. In the reduce phase, the reducer (the algorithm that specifies a reduce function) takes as input all the key/value pairs that have the same key, and it outputs a sequence of key/value pairs which have the same key as the input pairs; these pairs are either the final output, or they become the input of the next MapReduce round [22]. We consider Hadoop, the most popular open-source implementation of MapReduce. Hadoop uses block-level scheduling and a sort-merge technique to implement the functionality for parallel processing. The Hadoop Distributed File System (HDFS) is a distributed file system designed to run on commodity hardware. It is highly fault-tolerant and is designed to be deployed on low-cost hardware. It provides high throughput access to application data and is suitable for applications that have large data sets. An HDFS instance may consist of hundreds or thousands of server machines, each storing part of the file system's data [23]. HDFS has a master/slave architecture as shown in Fig.2. An HDFS cluster consists of a single Name Node, a master server that manages the file system namespace and regulates access to files by clients. In addition, there are a number of Data Nodes manage storage attached to the nodes that they run on. HDFS exposes a file system namespace and allows user data to be stored in files. Internally, a file is split into one or more blocks and these blocks are stored in a set of Data Nodes. The Name Node executes file system namespace operations like opening, closing, and renaming files and directories. It also determines the mapping of blocks to Data Nodes. The Data Nodes are responsible for serving read and write requests from the file system's clients. The Data Nodes also perform block creation, deletion, and replication upon instruction from the Name Node.

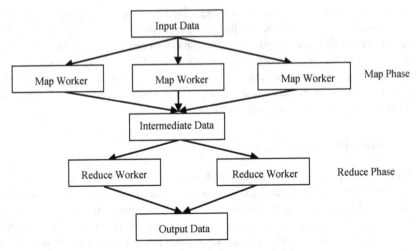

Fig. 1. Architecture of MapReduce

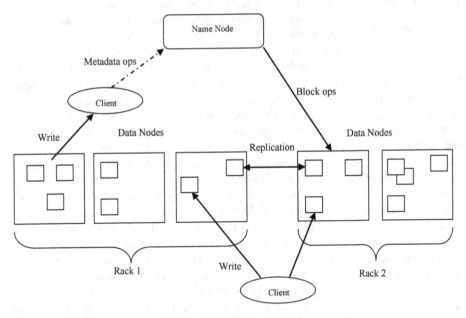

Fig. 2. Architecture of HDFS

4   Implementation and Experimental Results

We take the advantage of MapReduce framework in order to cluster chemical compounds data sets in reasonable time and memory resources. As the stored-matrix is symmetric, with zeros in the main diagonal, only a triangular half of the matrix is stored and computed. The computation of stored-matrix is divided across P map instances as show in Fig.3. Stored-matrix contains proximities between compounds (C_1 ... C_n). Each map instance stores and computes its own stored-matrix. Then each map scans its own matrix to find the local minimum distance and sends it to the reduce instance. The reduce instance collects the local minimum distances and choose the minimum one as the global minimum distance and send the global minimum distance to the corresponding map to merge the two clusters. Then all map instances update its own stored-matrix by computing the distances between the new cluster and all other clusters as shown in Fig.4. This procedure is repeated until all clusters are merged and only one cluster remains that contains all compounds.

All experiments are done on Amazon Elastic MapReduce (EMR). EMR is a web service that enables businesses, researchers, data analysts, and developers to easily and cost-effectively process vast amounts of data. It utilizes a hosted Hadoop framework running on the web-scale infrastructure of Amazon Elastic Compute Cloud (Amazon EC2) and Amazon Simple Storage Service (Amazon S3). EMR enable user to determine number and type of map and reduce instances. In our experiments, parallel ward clustering algorithm is run over four instances (three map instances and one reduce instance) and seven instances (six map instances and one reduce instance) with M1.large processing capabilities for each instance. For sequential ward clustering algorithm only one instance with M1.large processing capabilities is used.

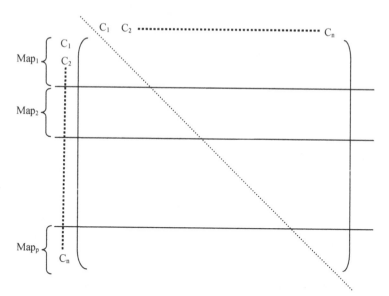

Fig. 3. Parallel Stored-Matrix Computation

Table 1 shows the Processor Architecture, Virtual Central Processing Unit (VCPU), Elastic Compute Unit (ECU), Memory, Instance Storage, Elastic Block Store (EBS) and Network Performance specification of each M1.large instance.

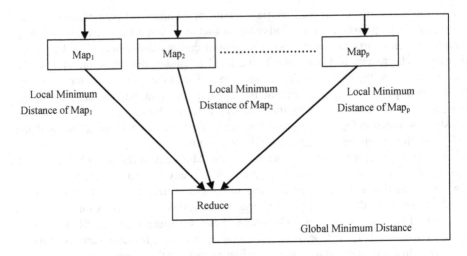

Fig. 4. Parallel Ward Clustering Algorithm

Table 1. Specification of EC2 instance

M1.Large instance specification	Value
Processor Arch	64-bit
VCPU	2
ECU	4
Memory (GB)	7.5
Instance Storage (GB)	2 x 240
EBS-optimized available	Yes
Network Performance	High

The NCI data set is used [24]. Two random subsets are taken from NCI data set. NCI-1 subset contains 500 compounds and NCI-2 subset contains 1,000 compounds. Over this subset, BCUT descriptor found in Chemical Development Kit is used [25]. For each subset, eight different runs time are taken and average time is calculated. Table 2, Table 3 and Table 4 show the average time required in minutes to run sequential ward using 1 instance, parallel ward using 4 instances (3 map instances and 1 reduce instance) and parallel ward using 7 instances (6 map instances and 1 reduce instance) over two subsets of NCI data set.

Table 2. Results of running sequential ward clustering

Subset name	Average (Time in minutes)
NCI-1	29.03
NCI-2	482.98

Table 3. Results of running parallel ward clustering on 3 map instances and 1 reduce instance

Subset name	Average (Time in minutes)
NCI-1	24.15
NCI-2	402.46

Table 4. Results of running parallel ward clustering on 6 map instances and 1 reduce instance

Subset name	Average (Time in minutes)
NCI-1	12.11
NCI-2	201.28

Fig.5 shows time required for sequential and parallel ward clustering; NCI-1 subset takes in average 29.03 minutes in sequential, 24.15 minutes in parallel using 3 map instances, 12.11 minutes in parallel using 6 map instances. NCI-2 subset takes in average 482.98 minutes in sequential, 402.46 minutes in parallel using 3 map instances, 201.28 minutes in parallel using 6 map instances.

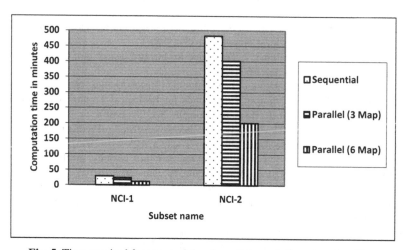

Fig. 5. Time required for sequential and parallel ward clustering algorithm

From these results we conclude that parallel ward clustering saves 17% of time using 3 map instances and saves 58% of time using 6 map instances compared to sequential one. So, as the number of map instances increase, the time required to cluster data set decrease. But this not always true because of overhead time for MapReduce framework to split data across several maps and time needed to collect output.

5   Conclusion and Future Work

In this paper, Hadoop MapReduce framework is used to run ward clustering algorithm in parallel manner. Experimental results show that our approach saves 17% of time using 3 map instances and saves 58% of time using 6 map instances. So, the drug discovery process as whole is improved and accelerated. In future work a different number of map and reduce instances will be used to determine the best number of map instances that can be used with respect to overhead time of MapReduce. Our approach will be modified to be load balancing between maps by distributing stored matrix computation equally between maps to reduce computation time and increase CPU utilizations.

References

1. Engel, T.: Basic Overview of Chemoinformatics. J. Chem. Inf. Model 64, 2267–2277 (2006)
2. Leach, A.R., Gillet, V.J.: An Introduction to Chemoinformatics. Springer (2003)
3. Aggarwal, C., Wang, H.: Managing and Mining Graph Data. Springer (2010)
4. Keseru, G.M., Makra, G.M.: Hit discovery and hit-to-lead approaches. Drug Discovery Today 11, 741–748 (2006)
5. Brown, R.D., Martin, Y.C.: Use of Structure–Activity Data to Compare Structure-Based Clustering Methods and Descriptors for Use in Compound Selection. J. Chem. Inf. Comput. Sci. 36 (1996)
6. Menard, P.R., Mason, J.S.: Rational Screening Set Design and Compound Selection: Cascaded Clustering. J. Chem. Inf. Comput. Sci. 38 (1998)
7. Doman, T.N., et al.: Algorithm5: A Technique for Fuzzy Clustering of Chemical Inventories. J. Chem. Inf. Comput. Sci. 36 (1996)
8. Willett, P., et al.: Implementation of Nonhierarchic Cluster-Analysis Methods in Chemical Information Systems; Selection of Compounds for Biological Testing and Clustering of Substructure Search Output. J. Chem. Inf. Comput. Sci. 26 (1986)
9. Jenkins, J.L., Bender, A., Davies, J.W.: In silico target fishing: Predicting biological targets from chemical structure. Drug Discovery Today 3, 413–421 (2006)
10. Mishra, M., Fei, H., Huan, J.: Computational Prediction of Toxicity. In: IEEE International Conference on Bioinformatics and Biomedicine, pp. 686–691 (2010)
11. Korn, C., Balbach, S.: Compound selection for development – Is salt formation the ultimate answer? Experiences with an extended concept of the "100 mg approach". European Journal of Pharmaceutical Sciences (2013)
12. Gaikwad, V.J.: Application of chemoinformatics for innovative drug discovery. International Journal of Chemical and Application 1, 16–24 (2010)
13. Dean, J., et al.: MapReduce: Simplified data processing on large clusters. Communications of the ACM 51, 107–113 (2008)
14. Zhao, J., et al.: Hadoop MapReduce Framework to Implement Molecular Docking of large-scale virtual screening. In: IEEE Asia-Pacific Services Computing Conference, pp. 350–353 (2012)
15. Wale, N., Watson, I.A., Karypis, G.: Comparison of descriptor spaces for chemical compound retrieval and classification. Knowledge and Information Systems 14, 347–375 (2007)

16. Barnard, J.M., Downs, G.M.: Chemical Similarity Searching. J. Chem. Inf. Comput. Sci. 38, 983–996 (1998)
17. Olson, C.F.: Parallel Algorithms for Hierarchical clustering. Parallel Computing 21, 1313–1325 (1995)
18. Sibson, R.: SLINK: an optimally efficient algorithm for the single link cluster methods. Computer Journal 16, 30–34 (1973)
19. Li, X., Fang, Z.: Parallel clustering algorithms. Parallel Computing 11, 275–290 (1989)
20. Li, Q., Salman, R., et al.: Parallel multitask cross validation for support vector machine using GPU. J. Parallel Distrib. Comput. 73, 293–302 (2013)
21. Upadhyaya, S.R.: Parallel approaches to machine learning - A comprehensive survey. J. Parallel Distrib. Comput 73, 284–292 (2013)
22. Karloff, H.J., Suri, S., Vassilvitskii, S.: A model of computation for mapreduce. In: SODA, pp. 938–948 (2010)
23. White, T.: Hadoop: The Definitive Guide. O'Reilly Media. Inc. (2009)
24. NCI data set, http://cactus.nci.nih.gov/download/nci/ (accessed on May 14, 2014)
25. Chemical Development Kit project, http://sourceforge.net/projects/cdk/ (accessed on May 14, 2014)

Memetic Artificial Bee Colony
for Integer Programming

Ahmed Fouad Ali[1], Aboul Ella Hassanien[2], and Vaclav Snasel[3]

[1] Suez Canal University, Dept. of Computer Science,
Faculty of Computers and Information, Ismailia, Egypt,
Member of Scientific Research Group in Egypt
[2] Cairo University, Faculty of Computers and Information, Cairo, Egypt,
Chair of Scientific Research Group in Egypt
[3] VSB-Technical University of Ostrava, Czech Republic

Abstract. Due to the simplicity of the Artificial Bee Colony (ABC) algorithm, it has been applied to solve a large number of problems. ABC is a stochastic algorithm and it generates trial solutions with random moves, however it suffers from slow convergence. In order to accelerate the convergence of the ABC algorithm, we proposed a new hybrid algorithm, which is called Memetic Artificial Bee Colony for Integer Programming (MABCIP). The proposed algorithm is a hybrid algorithm between the ABC algorithm and a Random Walk with Direction Exploitation (RWDE) as a local search method. MABCIP is tested on 7 benchmark functions and compared with 4 particle swarm optimization algorithms. The numerical results demonstrate that MABCIP is an efficient and robust algorithm.

Keywords: Artificial bee colony, local search algorithm, memetic algorithm, integer programming problems.

1 Introduction

A group can be defined as a structured collection of interacting organisms or members. The global behavior of a group (swarm) of social organisms therefore emerges in a nonlinear manner from the behavior of the individuals in that group. Thus, there exists a tight coupling between individual behavior and the behavior of the entire group. Swarm intelligence (SI) is an innovative distributed intelligence paradigm for solving optimization problems which takes inspiration from the behavior of a group of social organisms. There are many algorithms belong to SI such as Ant Colony Optimization (ACO) [2], Artificial Bee Colony [4], Particle Swarm Optimization (PSO) [5], Bacterial foraging [7], Bat algorithm [14], Bee Colony Optimization (BCO) [12], Wolf search [11], Cat swarm [1], Cuckoo search [13], Firefly algorithm [15], Fish swarm/school [6], etc. Artificial Bee Colony (ABC) algorithm which is proposed by Karaboga [4], is one of the most applied SI algorithm to solve many different types of applications in the past few years. In this paper, we proposed a new hybrid ABC algorithm to

A.E. Hassanien et al. (Eds.): AMLTA 2014, CCIS 488, pp. 268–277, 2014.

solve integer programming problems by combining the ABC algorithm, which is powerful for doing the global search as an exploration process in the algorithm and the random walk with direction exploitation as a local search method which is applied in order to refine the best obtained solution at each iteration. The proposed algorithm is called Memetic Artificial Bee Colony for Integer Programming (MABCIP). The general performance of the MABCIP algorithm is tested on 7 benchmark functions and it is also compared with 4 particle swarm optimization algorithms. The reminder of the paper is organized as fellow. The integer programming problem is defined in Section 2. In section 3, we describe the proposed algorithm with its main components in detail. The numerical experimental results are presented in Section 4. Finally, we conclude the paper in Section 5.

2   Integer Programming Problem Definition

An integer programming problem is a mathematical optimization problem in which all of the variables are restricted to be integers. The unconstrained integer programming problem can be defined as fellow.

$$min f(x), \ \ x \in S \subseteq \mathbb{Z}^n, \tag{1}$$

Where $\mathbb{Z}$ is the set of integer variables, S is a not necessarily bounded set.

3   Memetic ABC for Integer Programming Algorithm

In this section, we highlight the main components of the proposed algorithm. In the MABCIP algorithm, we tried to combine the ABC algorithm with its ability to global search (exploration process) and the random walk with direction exploitation with its ability to local search (exploitation process). Before describing the main steps of the MABCIP algorithm, we present the main steps of the random walk with direction exploitation as a local search algorithm in the following subsection as fellow.

3.1   Local Search Algorithm

The MABCIP algorithm uses a Random Walk with Direction Exploitation (RWDE) as a local search algorithm in order to refine the best obtained solution at each iteration in the the ABC algorithm. Invoking the RWDE algorithm as a local search algorithm in the MABCIP algorithm accelerates the convergence of the algorithm and represents the exploitation process in the MABCIP algorithm. The main steps of the RWDE algorithm are presented in Algorithm 1 as follows.

In Algorithm 1, the initial solution x^0 is given and evaluated by calculating its objective function $f(x^0)$. At the each iteration, the new solution is created by using Equation 2 as follow.

$$x^{(t+1)} = x^{(t)} + \Delta u^{(t)} \tag{2}$$

Algorithm 1. Random walk with direction exploitation

1: Set $t = 0$
2: Set the initial values of a scaler step length Δ
3: Start with an initial solution x^0
4: Evaluate the initial solution $f(x^0)$
5: **repeat**
6:     Set $t = t + 1$
7:     Generate a unit length random vector $u^{(t)}$
8:     Set $x^{(t+1)} = x^{(t)} + \Delta u^{(t)}$
9:     Evaluate the new solution $f(x^{(t+1)})$
10:    **if** $f(x^{(t+1)}) < f(x^{(t)})$ **then**
11:        Set $f(x^{(t+1)}) = f(x^{(t+1)})$
12:    **else**
13:        Set $f(x^{(t+1)} = x^{(t)}$
14:        Set $\Delta = \Delta/2$
15:    **end if**
16: **until** $(t \leq t_{max})$

Where Δ is a prescribed scaler step length and $u^{(t)}$ is unit length random vector. The new solution $x^{(t+1)}$ is evaluated $f(x^{(t+1)})$ and compared with the objective function value of previous solution $f(x^t)$. If the new solution is better than the old solution, then the new solution is accepted and the operation is repeated until the current iteration reaches to the maximum number of iteration t_{max}, otherwise the new solution is rejected and the step length is reduced to $\Delta = \Delta/2$, and the steps 5-16 are repeated.

3.2 MABCIP Algorithm

In the MABCIP algorithm, the initial population is generated randomly, which contains NS solutions x_i, $i = \{1, \ldots, NS\}$, each solution is a D dimensional vector. The number of the NS solutions (food sources) is equal to the number of employed bees. The solutions in the initial population are evaluated by calculating their fitness function as fellow.

$$f(x_i) = \begin{cases} \frac{1}{1+f(x_i)} & \text{if } f(x_i) \geq 0 \\ 1 + abs(f(x_i)) & \text{if } f(x_i) < 0 \end{cases} \qquad (3)$$

The best food source is memorized x_{best} and the probability of each food source is calculated in order to generate a new trail solution v_i by an onlooker bees. The associated probability of each food source P_i is defined as fellow.

$$P_i = \frac{f_i}{\sum_{j=1}^{NS} f_j} \qquad (4)$$

and the trail solution can be generated as $v_{ij} = x_{ij} + \phi_{ij}(x_{ij} - x_{kj})$, $\phi_{ij} \in [-1, 1]$, $k \in \{1, 2, \ldots, NS\}$, $j \in \{1, 2, \ldots, D\}$ and $i \neq k$. The trail solution is evaluated and if it is better than or equal the old solution, then the old solution is replaced with the new solutions, otherwise the old solution is retained. The best solution is memorized and the local search algorithm starts to refine the best found solution so far. If the food source cannot improved for a limited number of cycles, which is called "limit", the food source is considered to be abandoned and replaced with a new food source by scout. The operation is repeated until termination criteria satisfied, i.e the algorithm reaches to *MCN* maximum cycle number.

Algorithm 2. MABCIP algorithm

1: Generate the initial population x_i randomly, $i = \{1, \ldots, NS\}$ {**Initialization**}
2: Evaluate the fitness function $f(x_i)$ of all solutions in the population
3: Keep the best solution x_{best} in the population {**Memorize the best solution** }
4: Set cycle=1
5: **repeat**
6:     Generate a new solution v_i from the old solution x_i, where $v_{ij} = x_{ij} + \phi_{ij}(x_{ij} - x_{kj})$, $\phi_{ij} \in [-1, 1]$, $k \in \{1, 2, \ldots, NS\}$, $j \in \{1, 2, \ldots, D\}$ and $i \neq k$ {**Employed bees**}
7:     Evaluate the fitness function $f(v_{ij})$ for all solution in the population
8:     Keep the best solution between current and candidate solutions {**Greedy selection**}
9:     Calculate the probability P_i, for the solutions x_i, where $P_i = \frac{f_i}{\sum_{j=1}^{NS} f_j}$
10:    Generate the new solutions v_i from the selected solutions depending on its P_i {**Onlooker bees**}
11:    Evaluate the fitness function f_i for all solutions in the population
12:    Keep the best solution between current and candidate solutions {**Greedy selection**}
13:    Apply a local search algorithm as shown in Algorithm 1 on the best solution to obtain a new solution {**Intensification process**}
14:    Determine the abandoned solution if exist, replace it with a new randomly solution x_i {**Scout bee**}
15:    keep the best solution x_{best} found so far in the population
16:    $cycle = cycle + 1$
17: **until** cycle $\leq MCN$

---

4   Numerical Experiments

The efficiency of the MABCIP algorithm is tested in this section by evaluating its general performance on 7 benchmark functions and comparing its results with the other 4 particle swarm optimization algorithms. In the following subsections, we will describe in details the parameter setting of the proposed algorithm and the applied test functions. Also the performance analysis of the proposed algorithm is presented with the comparative results between it and the other algorithms.

4.1   Parameter Setting

Before discussing the proposed MABCIP algorithm, we present the main components which are used in the proposed algorithm, these components are listed as fellow.

- **Population Size.** The number of food sources NS is the number of the population size, the number of food source is equal to the number of employed bees (scouts and employed bees) and the number of onlooker bees.
- **limit.** The maximum number of trails "**limit**", which the employed bees with unimproved solutions become scouts and their solutions are abandoned.
- **Tol.** Is one of the applied termination criteria in the MABCIP algorithm, which is the value of the error goal for each function.
- **Δ.** Is the scaler step length, which is increases or decreases due to the objective function of the current solution.
- **MCN.** Maximum cycle number is the main termination criterion in the standard ABC algorithm.

The parameters setting of the MABCIP algorithm are listed in Table 1.

Table 1. The Parameter setting of the MABCIP algorithm

Parameters	Definitions	value
NS	population size	20
limit	Maximum number of trails	30
Tol	Error goal accuracy	10^{-6}
Δ	Scaler step length	4.0
MCN	Maximum cycle number	2500

4.2   Integer Programming Test Functions

In order to investigate the general performance of the MABCIP algorithm, 7 benchmark functions $f_1 - f_7$ have been selected. The properties of the benchmark functions (function number, dimension of the problem, the range of the problem and the global optimal of the problem) are listed in Table 2 and the functions with their definitions are reported as fellows.

Test problem 1[10]. This problem is defined by

$$F_1(x) = \|x\|_1 = |x_1| + \ldots + |x_n|$$

Test problem 2[10]. This problem is defined by

$$F_2 = x^T x = \begin{bmatrix} x_1 & \cdots & x_n \end{bmatrix} \begin{bmatrix} x_1 \\ \vdots \\ x_n \end{bmatrix}$$

Table 2. The properties of the Integer programming test functions

Function	Dimension	Bound	Optimal
f_1	5	[-100 100]	0
f_2	5	[-100 100]	0
f_3	5	[-100 100]	-737
f_4	2	[-100 100]	0
f_5	4	[-100 100]	0
f_6	2	[-100 100]	-6
f_7	2	[-100 100]	-3833.12

Test problem 3[3]. This problem is defined by

$$F_3 = \begin{bmatrix} 15 & 27 & 36 & 18 & 12 \end{bmatrix} x + x^T \begin{bmatrix} 35 & -20 & -10 & 32 & -10 \\ -20 & 40 & -6 & -31 & 32 \\ -10 & -6 & 11 & -6 & -10 \\ 32 & -31 & -6 & 38 & -20 \\ -10 & 32 & -10 & -20 & 31 \end{bmatrix} x,$$

Test problem 4[3]. This problem is defined by

$$F_4(x) = (9x_1^2 + 2x_2^2 - 11)^2 + (3x_1 + 4x_2^2 - 7)^2$$

Test problem 5[3]. This problem is defined by

$$F_5(x) = (x_1 + 10x_2)^2 + 5(x_3 - x_4)^2 + (x_2 - 2x_3)^4 + 10(x_1 - x_4)^4$$

Test problem 6[9]. This problem is defined by

$$F_6(x) = 2x_1^2 + 3x_2^2 + 4x_1x_2 - 6x_1 - 3x_2$$

Test problem 7[3]. This problem is defined by

$$F_7(x) = -3803.84 - 138.08x_1 - 232.92x_2 + 123.08x_1^2$$
$$+ 203.64x_2^2 + 182.25x_1x_2$$

4.3   Performance Analysis

In order to evaluate the performance of the MABCIP algorithm, we test it on 7 benchmark functions $f_1 - f_7$ and compare it with 4 particle swarm optimization algorithms. We highlight the performance of the MABCIP algorithm in the subsequent sections.

The General Performance of MABCIP with Integer Programming Functions.
Before we compare the MABCIP algorithm with the other 4 comparative algorithms, we present the general performance of it with 4 test functions (randomly picked) as shown in Figure 1. We can observe from Figure 1 that the function values for each test function are rapidly converging as the number of iterations increases.

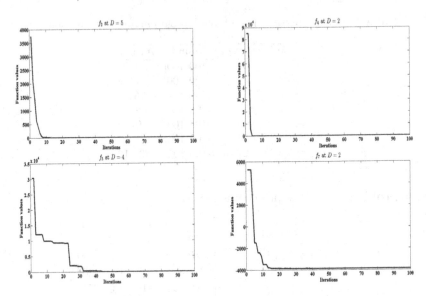

Fig. 1. The general performance of the MABCIP with integer programming functions

4.4   MABCIP and Other Algorithms

The MABCIP algorithm was compared with 4 particle swarm optimization (PSO) [8] as fellow.

- **RWMPSOg.** RWMPSOg is a Random Walk Memetic Particle Swarm Optimization (with global variant), which combines the particle swarm optimization with random walk with direction exploitation.
- **RWMPSOl.** RWMPSOl is a Random Walk Memetic Particle Swarm Optimization (with local variant), which combines the particle swarm optimization with random walk with direction exploitation.
- **PSOg.** PSOg is a standard particle swarm optimization with global variant without local search method.
- **PSOl.** PSOl is a standard particle swarm optimization with local variant without local search method.

The PSO global variant means the best position of the particle (with the best objective function value) in the swarm is assigned to all particles in the swarm at each iteration, while the PSO local variant means each particle in the swarm is assigned to some particles in its neighborhood.

In order to make a fair comparison, we used the same applied termination criteria in the other algorithms, which are the desired accuracy is $Tol = 10^{-6}$ and the maximum number of allowed function evaluations is 25,000. The results of the MABCIP algorithm are taken over 50 runs for each test problem, while the results of the four comparative algorithms are taken from their original papers. The run is consider successful, if the desired error goal is archived with

Table 3. Experimental results for RWMPSOg, RWMPSOl, PSOg, PSOl and MABCIP for $f_1 - f_7$ function

Function	Method	Min	Mean	Max	StD	Suc
f_1	RWMPSOg	17,160	27,176.3	74,699	8656.9	50
	RWMPSOl	24,870	30,923.9	35,265	2405.0	50
	PSOg	14,000	29,435.0	261,110	420,39.1	34
	PSOl	27,400	31,252.0	35,800	1817.8	50
	MABCIP	**8355**	**8955**	**10,055**	**662.38**	50
f_2	RWMPSOg	252	578.5	912	136.5	50
	RWMPSOl	369	773.9	1931	285.5	50
	PSOg	400	606.4	1000	119.0	50
	PSOl	450	830.2	1470	206.0	50
	MABCIP	**240**	**294.4**	**315**	**31.37**	50
f_3	RWMPSOg	**1361**	6490.6	41,593	6912.8	50
	RWMPSOl	5003	9292.6	15,833	2443.7	50
	PSOg	2150	12,681.0	187,000	350,66.8	50
	PSOl	4650	11,320.0	22,650	3802.8	50
	MABCIP	1545	**3215.8**	**4120**	**1201.59**	50
f_4	RWMPSOg	76	215.0	468	97.9	50
	RWMPSOl	**73**	218.7	620	115.3	50
	PSOg	100	369.6	620	113.2	50
	PSOl	120	390.0	920	134.6	50
	MABCIP	115	**197.5**	**230**	**60.58**	50
f_5	RWMPSOg	687	1521.8	2439	360.7	50
	RWMPSOl	**675**	2102.9	3863	689.5	50
	PSOg	680	1499.0	3440	513.0	43
	PSOl	800	2472.4	3880	637.5	50
	MABCIP	1150	**1192.4**	**1225**	**29.75**	50
f_6	RWMPSOg	**40**	**110.9**	238	48.6	50
	RWMPSOl	**40**	112.0	235	48.7	50
	PSOg	80	204.8	350	62.0	50
	PSOl	70	256.0	520	107.5	50
	MABCIP	120	146.2	**175**	**24.07**	50
f_7	RWMPSOg	72	242.7	620	132.2	50
	RWMPSOl	**70**	248.9	573	134.4	50
	PSOg	100	421.2	660	130.4	50
	PSOl	100	466.0	820	165.0	50
	MABCIP	135	**154**	**175**	**15.96**	50

the maximum number of the function evaluations. The minimum (Min), average (Mean), Maximum (Max), Standard deviation (StD) and the success rate (Suc) of the evaluation function for each function are reported in Table 3. The best results of any algorithm are marked in **bold face**. We can observe from the results in Table 3, that the MABCIP algorithm can obtain the desired error goal with 100% rate of success and it is faster than the other algorithm in most cases.

5   Conclusion

A new hybrid ABC algorithm has been proposed in this paper in order to overcome the slow convergence of the ABC algorithm and improve its performance.
The proposed algorithm is a hybrid algorithm between the ABC algorithm and
a Random Walk with Direction Exploitation (RWDE) as a local search method.
The ABC algorithm has a good ability to look for a global optimum solution
by applying a global search with random moves in the search space. The Random Walk with Direction Exploitation (RWDE) method is considered as a local
search method and it is powerful to exploit the promising region in order to
refine the best found solution so far. The proposed algorithm is called Memetic
Artificial Bee Colony for integer programming (MABCIP). The MABCIP algorithm is tested on 7 benchmark functions and compared with 4 particle swarm
optimization algorithms with different search strategies. The numerical results
show that the proposed algorithm is a promising algorithm and can obtain the
optimal solution or near optimal solution faster than the other algorithms.

References

1. Chu, S.-C., Tsai, P.-w., Pan, J.-S.: Cat swarm optimization. In: Yang, Q., Webb, G.
   (eds.) PRICAI 2006. LNCS (LNAI), vol. 4099, pp. 854–858. Springer, Heidelberg
   (2006)
2. Dorigo, M.: Optimization, learning and natural algorithms. Ph. D. Thesis, Politec-
   nico di Milano, Italy (1992)
3. Glankwahmdee, A., Liebman, J.S., Hogg, G.L.: Unconstrained discrete nonlinear
   programming. Engineering Optimization 4, 95–107 (1979)
4. Karaboga, D., Basturk, B.: A powerful and effficient algorithm for numerical function optimization: artificial bee colony (abc) algorithm. Journal of Global Optimization 39(3), 459–471 (2007)
5. Kennedy, J., Eberhart, R.: Particle swarm optimization. In: Proceedings of the
   IEEE International Conference on Neural Networks 1995, vol. 4, pp. 1942–1948.
   IEEE (1995)
6. Li, X.L., Shao, Z.J., Qian, J.X.: Optimizing method based on autonomous animats:
   Fish-swarm algorithm. Xitong Gongcheng Lilun yu Shijian/System Engineering
   Theory and Practice 22(11), 32 (2002)
7. Passino, M.K.: Biomimicry of bacterial foraging for distributed optimization and
   control. IEEE Control Systems 22(3), 52–67 (2002)
8. Petalas, Y.G., Parsopoulos, K.E., Vrahatis, M.N.: Memetic particle swarm optimization. Ann. Oper. Res. 156, 99–127 (2007)
9. Rao, S.S.: Engineering optimization-theory and practice. Wiley, New Delhi (1994)
10. Rudolph, G.: An evolutionary algorithm for integer programming. In: Davidor,
    Y., Männer, R., Schwefel, H.-P. (eds.) PPSN 1994. LNCS, vol. 866, pp. 139–148.
    Springer, Heidelberg (1994)
11. Tang, R., Fong, S., Yang, X.S., Deb, S.: Wolf search algorithm with ephemeral
    memory. In: 2012 Seventh International Conference on Digital Information Man-
    agement Digital Information Management (ICDIM), pp. 165–172 (2012)

12. Teodorovic, D., DellOrco, M.: Bee colony optimizationa cooperative learning approach to complex tranportation problems. In: Advanced, O.R., Methods, A.I. (eds.) Advanced OR and AI Methods in Transportation: Proceedings of 16th MiniEURO Conference and 10th Meeting of EWGT, September 13-16, pp. 51–60. Publishing House of the Polish Operational and System Research, Poznan (2005)
13. Yang, X.S., Deb, S.: Cuckoo search via levy flights. In: World Congress on Nature & Biologically Inspired Computing, NaBIC 2009, pp. 210–214. IEEE (2009)
14. Yang, X.-S.: A new metaheuristic bat-inspired algorithm. In: González, J.R., Pelta, D.A., Cruz, C., Terrazas, G., Krasnogor, N. (eds.) NICSO 2010. SCI, vol. 284, pp. 65–74. Springer, Heidelberg (2010)
15. Firefly, X.S.Y.: algorithm, stochastic test functions and design optimisation. International Journal of Bio-Inspired Computation 2(2), 78–84 (2010)

Automatic Fruit Image Recognition System Based on Shape and Color Features

Hossam M. Zawbaa[1,2,5], Mona Abbass[3,5], Maryam Hazman[3,5],
and Aboul Ella Hassenian[4,5]

[1] Faculty of Mathematics and Computer Science, Babes-Bolyai University, Romania
[2] Faculty of Computers and Information, Beni-Suef University, Egypt
[3] Central Lab. for Agricultural Expert System, Agricultural Research Center, Egypt
[4] Faculty of Computers and Information, Cairo University, Egypt
[5] Scientific Research Group in Egypt (SRGE)
www.egyptscience.net

Abstract. This paper presents an automatic fruit recognition system for classifying and identifying fruit types. The work exploits the fruit shape and color, to identify each image feature. The proposed system includes three phases namely: pre-processing, feature extraction, and classification phases. In the pre-processing phase, fruit images are resized to 90 x 90 pixels in order to reduce their color index. In feature extraction phase, the proposed system uses scale invariant feature transform (SIFT) and shape and color features to generate a feature vector for each image in the dataset. For classification phase, the proposed model applies K-Nearest Neighborhood (K-NN) algorithm classification, and support vector machine (SVM) algorithm of different kinds of fruits. A series of experiments were carried out using the proposed model on a dataset of 178 fruit images. The results of carrying out these experiments demonstrate that the proposed approach is capable of automatically recognize the fruit name with a high degree of accuracy.

Keywords: Fruit classification, Image classification, Features extraction, K-Nearest Neighborhood (K-NN), Support Vector Machine (SVM).

1   Introduction

Nowadays, process automation plays an important role in industries. Many automatic highly efficient methods are developed to use in producing and checking processes. The topic of digital image processing has found many applications in the field of automation [1]. In computer vision and pattern recognition, shape matching is an important problem of which is defined as the establishment of a similarity measure between shapes and its use for shape comparison. A byproduct of recognition task might also be a set of point identical between shapes. Shape matching which is intuitively accurate for humans is a needed job that is not solved yet in its full generality. Its applications include object detection and recognition, image registration, and content based retrieval of images [2].

A.E. Hassenian et al. (Eds.): AMLTA 2014, CCIS 488, pp. 278–290, 2014.

Several image processing methods are applied to analyze the agricultural images for monitoring crop ripeness [3], detecting crop diseases [4], and recognition fruit and vegetables [5]. Fruit recognition and classification systems can be used by many real life applications. Such as a supermarket checkout system where it can be used instead of manual barcodes, and as an educational tool to enhance learning, especially for small children and Down syndrome patients [5,6]. It can assist the plant scientists, where shape and color values of the fruit images that have been computed can assist them do further analysis on variation in morphology of fruit shape in order and can help them understand the genetic and molecular mechanisms of the fruits [6]. Also, it can be used as aiding tool for eye weakness people which can aid them in shopping as a mobile application.

Recognizing different types of vegetables and fruits is a repeated chore in supermarkets, where the cashier has to define each item type which will determine its cost. The barcodes usage mostly ended this packaged products difficulty but when consumers want picking their produce; they will not be able to package it, and thus should be weighted. A popular solution to this difficulty is supplying codes for every type of fruit and vegetable; that has problems precondition that the memorization is sticky, leading to errors in pricing. Another solution is a small book with pictures and codes; the difficulty with this solution is that flipping over the pamphlet is time-consuming [5]. A fruit and vegetable recognition system which automates labeling and computing the price is a good solution for this problem.

Therefore the main goal of this work is to automatic recognize fruit image by classifying it according to its features using machine learning techniques. An image recognition model is proposed which contains three phases: pre-processing, feature extraction, and classification phases. In feature extraction phase, Scale Invariant Feature Transform (SIFT), and shape and color algorithms are used to extract a feature vector for each image. The classification phase uses two algorithms: K-Nearest Neighborhood (K-NN), and support vector machine(SVM). Evaluating the recognition model is done by carrying out a series of experiments. The results of carrying out these experiments demonstrate that the proposed approach is capable of automatically classify the fruit name with a high degree of accuracy.

The rest of this paper is organized as follows. Section 2 introduces some recent research works related to fruit recognition and classification. Section 3 presents the main concepts used for feature extraction and classification algorithms. Section 4 describes the different phases of the proposed model for recognition system: pre-processing, feature extraction, and classification phases. The experimental results are presented in section 5. Finally, section 6 presents the conclusion and future work.

2   Related Work

Many fruit recognition and classification systems were proposed using color and shape features. However, different fruit images may have similar color and shape values. In previous years, many types of image analysis techniques are applied to analyze the agricultural for recognition and classification purposes. Sego and Mirisaee proposed fruit recognition system which recognizes seven fruits. First, they classify fruit images using the KNN algorithm based on mean color values, shape roundness value, area and perimeter values of the fruit. The Euclidean distance is used to measure the distance between the features values of the unknown fruit with the stored features values of every fruit class to find out its nearest fruit class. Their recognition results achieved accuracy up to 90% [6].

Rocha et.al. proposed a fruit and vegetable recognize approach that combines many features and uses most appropriate classifier for each one in order to improve the overall classification accuracy. This approach categorizes fruits and vegetables based on color, texture, and appearance features. Each feature is concatenated and fed independently to its suitable classification algorithm [5]. Another approached uses Artificial Neural Network (ANN), Fourier Descriptors (FD) and Spatial Domain Analysis (SDA) for automatic identifying and sorting fruit. First, the fruit shape recognition is based on the shape boundary and signatures using FD and SDA technique. Detecting the fruit color is done using color information obtained during training process from Artificial Neural Network (ANN). Then, fruit shape recognition and color recognition paths are combined for identifying and sorting fruits purposes. This approached is evaluated using apple, banana and mango images and results achieved accuracy of 99.1% [7].

An approach to identify fruit and vegetable in supermarket is introduced in [8]. First, the image color is described using global measures which are histograms, mean, contrast, homogeneity, energy, variance, correlation, and entropy over the histograms for each color channel. The images are compered based on color coherence vectors (CCVs). While, the border/interior pixel classifier (BIC) is used to classify the image border. The appearance feature is obtained using a vocabulary of parts which is found using K-means and a bottom-up clustering algorithms [8].

3   Preliminaries

3.1   Shape and Color Features

Many fruit recognition and classification systems are developed based on color and shape feature [6]. Color consider as an important feature for image representation due to the color are invariance with respect to image translation, rotation, and scaling [9]. In this paper, color moments that used to describe the images are color mean, color variance, color skewness, and color kurtosis. The Mean, standard deviation, and skewness for a colored image with size $N \times M$ pixels are defined by the equations 1, 2, and 3 [9,10].

$$\overline{x_i} = \sum_{j=1}^{M.N} x_{ij} \frac{N}{.} M \tag{1}$$

$$\delta_i = \sqrt{1\frac{N}{\cdot}M \sum_{j=1}^{M.N}(x_{ij} - \overline{x_i})^2} \tag{2}$$

$$S_i = 3\sqrt{1\frac{N}{\cdot}M \sum_{j=1}^{M.N}(x_{ij} - \overline{x_i})^3} \tag{3}$$

Where x_{ij} is the value of image pixel j of color channel i, $\overline{x_i}$ is the mean for each channel i, δ_i is the standard deviation, and S_i is the skewness for each channel. This paper uses Centroid, Eccentricity, and Euler Number features to describe the shape. The shape centroid (center of gravity) determines the image centroid position which is fixed in relation to the shape. Eccentricity measures the aspect ratio of the length of major axis to the length of minor axis. It is computed by minimum bounding rectangle method or principal axes method. The relation between the number of connecting parts and the number of holes on image shape is described by Euler number image describes. Euler number is compute by subtract the number of holes on a shape from the number of contiguous parts minus [11].

3.2 The Scale Invariant Feature Transform

The Scale Invariant Feature Transform (SIFT) algorithm is an algorithm for image features generation which is invariant to image translation, scaling, rotation and partially invariant to illumination changes and affine projection [12,13,14]. SIFT includes four major stages: scale-space extrema detection, keypoint localization, orientation assignment and keypoint descriptor. The first stage used difference-of-Gaussian function (DOG) to identify the potential interest points, which were invariant to scale and orientation as in equation 4 [15].

$$D(X,Y,\delta) = (G(X,Y,K\delta) - G(X,Y,\delta)) \times I(X,Y)$$
$$= L(X,Y,K\delta) - L(X,Y,\delta) \tag{4}$$

Given a digital image $I(X,Y)$, its scale space representation will be $L(X,Y,)$. $G(X,Y,)$ is the variable-scale Gaussian kernel with the standard deviation δ. In the keypoint localization step, the low contrast points are rejected and the edge response is eliminated. Hessian matrix was used to compute the principal curvatures and eliminate the keypoints that have a ratio between the principal curvatures that are greater than the ratio. An orientation histogram was formed from the gradient orientations of sample points within a region around the keypoint in order to get an orientation assignment. The keypoint descriptors are computed from the local gradient orientation and magnitudes in a certain neighborhood around the identified keypoint. The gradient orientations and magnitudes are combined in a histogram representation from that the descriptor is formed [16].

3.3   K-Nearest Neighborhood

K-Nearest Neighborhood (K-NN) classifier has a faster execution time and is dominant [17]. It works based on minimum distance between the query instance and the training data set to determine the k-nearest neighbors. Computing the distance between two scenarios is computed using distance function (x, y) , where x, y are query instance and the training data set composed of N features as showing in equation 5. Distance measure is computed using one of the two functions (Absolute distance measuring and Euclidean distance measuring) as in the equations 6, and 7.

$$x = \{x_1, x_2, .., x_n\}, y = \{y_1, y_2, .., y_n\} \tag{5}$$

$$d_i(x, y) = \sum_{i=1}^{n} x_i - y_i \tag{6}$$

$$d_i(x, y) = \sum_{i=1}^{n} \sqrt{x_i^2 - y_i^2} \tag{7}$$

After gathering K nearest neighbors, simple majority of these K-nearest neighbors is taken to be the prediction of the query instance. An object is classified by a majority vote of its neighbors [18], with the object being assigned to the class most common amongst its k nearest neighbors.

3.4   Support Vector Machine

The Support Vector Machine (SVM) classifier is a theoretically superior machine learning methodology that used for classification and regression of high-dimensional datasets with great results [3,19,20,21,22]. SVM tries to find an optimal separating hyperplane which effectively separates between classes for solving the classification problem [19]. SVM aims to maximize the margin around a hyperplane that separates a positive class from a negative class [3,20,21,22]. Consider a training dataset with n samples (x_1, y_1), (x_2, y_2), ..., (x_n, y_n). Where a feature vector x_i is in n-dimensional feature space and with labels $y_i \in \{-1, 1\}$ belonging to any of two linearly separable classes C_1 and C_2. The SVM finds an optimal hyperplane with the maximum margin between two classes by solving the optimization problem, as shown in the equations 8 and 9.

$$maximize \sum_{i=1}^{n} \alpha_i - \frac{1}{2} \sum_{i,j=1}^{n} \alpha_i \alpha_j y_i y_j . K(x_i, x_j) \tag{8}$$

$$Subject - to : \sum_{i=1}^{n} \alpha_i y_i, 0 \leq \alpha_i \leq C \tag{9}$$

Where, α_i is the assigned weight to the training sample x_i. when $\alpha_i > 0$, the x_i is a support vector. A regulation parameter C is used to trade-off the training accuracy and the model complexity to achieve a superior generalization capability. K is a kernel function, which is measure the similarity between two samples.

4 The Proposed Fruit Recognition System

This research aims to automatic recognize a fruit image from a collection of images. The proposed fruit recognition system consists of three phases namely: pre-processing, feature extraction, and classification phases.

4.1 Pre-processing Phase

In the pre-processing phase, the proposed model resizes images to 90 x 90 pixels; in order to reduce their color index. Also, the acquired image is in RGB format which is a real color format for an image [23].

4.2 Feature Extraction Phase

This phase is responsible of extracting the characteristics or attributes of an image. Since, the system accuracy are mainly depends upon feature extraction phase, the proposed system has been implemented considering two scenarios of feature extraction. In the first scenario, the shape and color features are used to generate a feature vector for each image in the dataset. The used shape features are Centroid, Eccentricity, and Euler Number. While color mean, color variance, color skewness, and color kurtosis features used for color moments. The second scenario, the scale invariant feature transform (SIFT) is use to generates the interesting points for image as a feature vector for each image in the dataset.

4.3 Classification Phase

Finally the classification phase, the proposed model applied K-Nearest Neighborhood (K-NN) algorithm and Support Vector Machine (SVM) to recognize different kinds of fruits. This phase takes the training dataset feature vectors with their corresponding classes as input, as well as the testing dataset. The output is the name of each fruit image in the testing dataset. The K-Nearest Neighborhood (K-NN) algorithm is worked as in the algorithm 1. While, the algorithm of Support Vector Machine algorithm main steps mentioned in algorithm 2.

5 Experimental Results

A set of experiments has been done to evaluate our proposed model. The built model was implemented using Matlab R2013a on Windows 8.1 operation system. The proposed system was evaluated using around 46 of Orange pictures and around 55 strawberry fruits. Some samples of both training and testing datasets are shown in figure 1.

1: Set k,$1 \le k \le n$
2: Initialize $i = 1$
3: **repeat**
4:     Compute distance from y to x_i
5:     **if** $(i \le k)$ **then**
6:         Include x_i in the set of k-nearest neighbors
7:     **else if** $(x_i$ is closer to y rather than any previous nearest neighbor **then**
8:         Delete farthest in the set of k-nearest neighbors
9:         Include x_i in the set of k-nearest neighbor
10:     **end if**
11:     $i = ++$
12: **until** k- nearest neighbor found
13: Determine the majority of class represented in the set of k-nearest neighbors
14: **if** a tie exists **then**
15:     Compute sum of distance of neighbors in each class which tied
16:     **if** no tie occurs **then**
17:         Classify y in the class of the class of minimum sum
18:     **else**
19:         Classify y in the class of last minimum found
20:     **end if**
21: **else**
22:     Classify y in the majority class
23: **end if**

Algorithm 1. K-Nearest Neighborhood Algorithm

1: Construct N binary SVM
2: Keypoint localization Each SVM separates one class from the rest classes
3: Train the ith SVM with all training samples of the ith class with positive labels, and training samples of other classes with negative labels

Algorithm 2. Support Vector Machine Algorithm

Fig. 1. Examples of training and testing fruit images

As our knowledge there was no specific benchmark data for the fruit types, therefore the used dataset in the experiments has been collected with different transformations (scale change, rotation, illumination, image blur, viewpoint change, and compression) for each fruit category. Each fruit category includes images for different fruit varieties .The dataset was split randomly into two halves, one for training and rest for testing. The results were introduced are the average of these runs. Also, these raw images were used after resize it to 90*90 pixels. We did not make other pre-processing such as: cropping, gray scaling, histogram equalization, etc. in order to assess the robustness of the feature extraction algorithms in the comparison.

5.1   Evaluation Results

As monition before, we used two algorithms for features extraction which are shape and color algorithm and scale invariant feature transform (SIFT) algorithm. So, the selected fruit types are chosen to represent the similarities and differences between shape and color. Apple and orange are similar in shape and different in color. Apple and strawberry are similar in color and different in shape. While, orange and strawberry are different in both shape and color. For each group type, the proposed model has been evaluated considering the following four scenarios:

Scenario 1: features extraction based on shape and color are classified using KNN
Scenario 2: features extraction based on shape and color are classified using SVM
Scenario 3: features extraction based on SIFT are classified using KNN
Scenario 4: features extraction based on SIFT are classified using SVM

The implemented model was run using these four scenarios twice on each group dataset. In the first one, the total dataset was divided into 60% for training and 40% for testing. In the second, the total dataset was divided into 70% for training and 30% for testing.

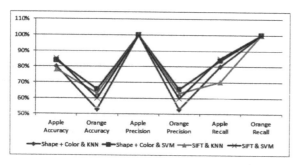

Fig. 2. Results of Apple and Orange for different feature extraction and classifiers (60% training)

First, we classify apple and orange group which are similar in shape and different in color. Figure 2 illustrates for each scenario the accuracy, the precision and the recall for each fruit category when the dataset is divided into 60% training and 40% test. As shown in this figure, classification apple achieves high accuracy than orange. The highest accuracy for classifying apple and orange is achieved when using the SVM classifier (85% for apple, 65.5% for orange). Trying to achieve better accuracy, training set is increasing. The image dataset is divided into 70% for training and 30% for testing. As shown in figure 3, the

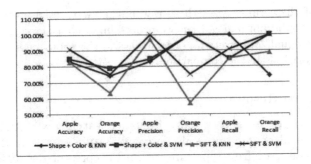

Fig. 3. Results of Apple and Orange for different feature extraction and classifiers (70% training)

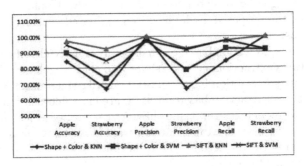

Fig. 4. Results of Apple and Strawberry for different feature extraction and classifiers (60% training)

accuracy is increased when increasing the training set (90.91% for apple, 78.89% for orange).

Classify apple and strawberry group, which are similar in color and different in shape achieves high accuracy than the similar in shape, as shown in Figure 4 and Figure 5. In Figure 4, extracting features based on SIFT and using KNN classifier achieves the highest accuracy (apple 97.37%, and strawberry 92.31%) when training set is 60%. While in Figure 5, the highest accuracy when 70% training is achieved using SIFT as feature extraction: apple (96.97% with SVM classifier), strawberry (85.71% with KNN).

Finally, we classified the distinct fruits in shape and color (orange and strawberry). The results of this group is presented in Figure 6 which shows the accuracy for each scenario, as well as the precision and recall for each fruit category when the dataset is divided into 60% training and 40% test. As shown in this figure, extracting features based on SIFT algorithm achieves high accuracy 100% (for both orange and strawberry) when using the KNN classifier. While, using shape and color to generate images feature achieves the lower accuracy (71.42% orange and 72.72% strawberry) when using KNN classifier. More images in training set leads to achieve high accuracy when using SIFT for feature extraction,

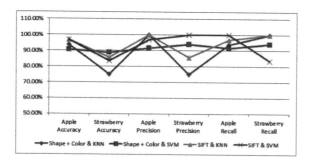

Fig. 5. Results of Apple and Strawberry for different feature extraction and classifiers (70% training)

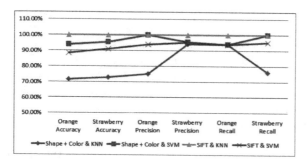

Fig. 6. Results of Orange and Strawberry for different feature extraction and classifiers (60% training)

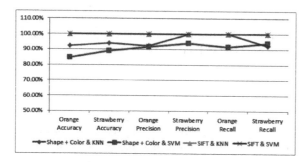

Fig. 7. Results of Orange and Strawberry for different feature extraction and classifiers (70% training)

as shown in Figure 7. Shape and color as a feature extraction achieve accuracy 92.31% orange and 94.12% strawberry when using KNN classifier, and 84.62% orange and 88.89% strawberry when using SVM classifier.

6   Conclusions and Future Work

This paper has presented an approach for automatically recognize fruit images. The proposed model includes three phases: pre-processing, feature extraction, and classification. Feature extraction is done based on two algorithms; the first algorithm shape and color algorithm generates the feature vector from color moments (color mean, color variance, color skewness, and color kurtosis) and shape (Centroid, Eccentricity, and Euler Number). The second algorithm was the Scale Invariant Feature Transform (SIFT). Once the feature vectors are generated for each image, the classifier can be work. The proposed model uses two classifiers, the K-Nearest Neighborhood (K-NN) algorithm classification, and support vector machine (SVM) algorithm.

The proposed model has been evaluated using around 46 of Orange pictures and 55 strawberry fruits. The experimental results show that the classification accuracy depends on the differentiations degree between the fruit types. The accuracy of similarities on shape group archives the lower accuracy among the three groups. It highest achieved is 90.91% for apple and 78.89% for orange when using the SVM classifier. Classify similarities fruits on color achieves 96.97% for apple and 85.71% for strawberry using SIFT as feature extraction. The highest accuracy is achieved with the distinct fruits in shape and color, it archives 100%.

In the future, we intend to experiment using more fruit types and go behind that to recognize the fruit verity. Also, we plan to investigate using other feature extraction and classifiers algorithms in order to increase the obtained accuracy.

Acknowledgment. This work was partially supported by the IPROCOM Marie Curie initial training network, funded through the People Programme (Marie Curie Actions) of the European Union's Seventh Framework Programme FP7/2007-2013/ under REA grant agreement No. 316555. This fund only apply for one author (Hossam M. Zawbaa).

References

1. Rege, S., Memane, R., Phatak, M., Agarwal, P.: 2D Geometric shape and color recognition using digital image processing. International Journal of Advanced Research in Electrical, Electronics and Instrumentation Engineering 2(6), 2479–2481 (2013)
2. Oikonomidis, I., Argyros, A.A.: Deformable 2D shape matching based on shape contexts and dynamic programming. In: Bebis, G., Boyle, R., Parvin, B., Koracin, D., Kuno, Y., Wang, J., Pajarola, R., Lindstrom, P., Hinkenjann, A., Encarnação, M.L., Silva, C.T., Coming, D. (eds.) ISVC 2009, Part II. LNCS, vol. 5876, pp. 460–469. Springer, Heidelberg (2009)
3. Elhariri, E., El-Bendary, N., Fouad, M.M.M., Platos, J., Hassanien, A.E., Hussein, A.M.M.: Multi-class SVM based classification approach for tomato ripeness. In: Abraham, A., Krömer, P., Snášel, V. (eds.) Innovations in Bio-inspired Computing and Applications. AISC, vol. 237, pp. 175–186. Springer, Heidelberg (2014)

4. Camargo, A., Smith, S.: An image-processing based algorithm to automatically identify plant disease visual symptoms. Biosystems Engineering 102(1), 9–21 (2009)
5. Rocha, A., Hauagge, D.C., Wainer, J., Goldenstein, S.: Automatic fruit and vegetable classification from images. Computers and Electronics in Agriculture 70(1), 96–104 (2010)
6. Seng, W.C., Mirisaee, S.H.: A new method for fruits recognition system. In: International Conference on Electrical Engineering and Informatics, ICEEI 2009, Selangor, Malasia, pp. 130–134 (2009)
7. Aibinu, A.M., Salami, M.J.E., Shafie, A.A., Hazali, N., Termidzi, N.: Automatic fruits identification system using hybrid technique. In: Sixth IEEE International Symposium on Electronic Design, Test and Application (DELTA), Queenstown, pp. 217–221 (2011)
8. Rocha, A., Hauagge, D.C., Wainer, J., Goldenstein, S.: Automatic produce classification from images using color, texture and appearance cues. In: XXI Brazilian Symposium on Computer Graphics and Image Processing, SIBGRAPI 2008, Campo Grande, pp. 3–10 (2008)
9. Shahbahrami, A., Borodin, D., Juurlink, B.: Comparison between color and texture features for image retrieval. In: Proceedings 19th Annual Workshop on Circuits, Systems and Signal Processing (ProRisc), Veldhoven, The Netherlands (2008)
10. Soman, S., Ghorpade, M., Sonone, V., Chavan, S.: Content based image retrieval using advanced color and texture features. In: IJCA Proceedings on International Conference in Computational Intelligence (ICCIA), New York, USA, vol. 9, pp. 1–5 (2012)
11. Mingqiang, Y., Kpalma, K., Ronsin, J.: A Survey of shape feature extraction techniques. Pattern Recognition, 43–90 (2008)
12. Lowe, D.G.: Object recognition from local scale-invariant features. In: Proceedings of the Seventh IEEE International Conference on Computer Vision, Corfu, Greece, pp. 1150–1157 (1999)
13. Lowe, D.G.: Local feature view clustering for 3D object recognition. In: Proceedings of the IEEE Computer Society Conference on Computer Vision and Pattern Recognition, CVPR, Kauai, Hawaii, pp. 682–688 (2001)
14. Lowe, D.G.: Distinctive image features from scale-invariant key-points. International Journal of Computer Vision 60(2), 91–110 (2004)
15. Juan, L., Gwun, O.: A comparison of SIFT, PCA-SIFT and SURF. International Journal of Image Processing (IJIP) 3(4), 143–152 (2009)
16. Hamid, N., Yahya, A., Ahmad, R.B., Al-Qershi, O.M.: A Comparison between using SIFT and SURF for characteristic region based image steganography. International Journal of Computer Science Issues 9(3), 111–112 (2012)
17. Teoh, A., Samad, S.A., Hussain, A.: Nearest Neighbourhood Classifiers in Biometric Fusion. International Journal of the Computer, the Internet and Management 12(1), 23–36 (2004)
18. Ho, T.K., Hull, J.J., Srihari, S.N.: Decision combination in multiple classifier systems. IEEE Transactions on Pattern Analysis and Machine Intelligence 16(1), 66–75 (1994)
19. Tzotsos, A., Argialas, D.: Support vector machine approach for object-based image analysis. In: Object-Based Image Analysis. Lecture Notes in Geoinformation and Cartography, pp. 663–677 (2008)

20. Wu, Q., Zhou, D.X.: Analysis of support vector machine classification. Journal of Computational Analysis & Applications 8(2), 99–119 (2006)
21. Zawbaa, H.M., El-Bendary, N., Hassanien, A.E., Abraham, A.: SVM-based Soccer Video Summarization System. In: Third World Congress Nature and Biologically Inspired Computing, Salamanca, Spain, pp. 7–11 (2011)
22. Zawbaa, H.M., El-Bendary, N., Hassanien, A.E., Kim, T.-h.: Machine Learning-Based Soccer Video Summarization System. In: Kim, T.-h., Gelogo, Y. (eds.) MulGraB 2011, Part II. CCIS, vol. 263, pp. 19–28. Springer, Heidelberg (2011)
23. Berns, R.S.: Principles of Color Technology, 3rd edn. Wiley, New York (2000)

Melanoma Classification Based on Ensemble Classification of Dermoscopy Image Features

Gerald Schaefer[1], Bartosz Krawczyk[2], M. Emre Celebi[3],
Hitoshi Iyatomi[4], and Aboul Ella Hassanien[5]

[1] Department of Computer Science, Loughborough University, Loughborough, U.K.
[2] Department of Systems & Computer Networks,
Wroclaw University of Technology, Wroclaw, Poland
[3] Department of Computer Science, Louisiana State University, Shreveport, USA
[4] Department of Applied Informatics, Hosei University, Tokyo, Japan
[5] Faculty of Computers and Information, Cairo University, Egypt

Abstract. Malignant melanoma is the deadliest form of skin cancer and is one of the most rapidly increasing cancers in the world. If diagnosed early, it can be easily cured, and consequently early diagnosis of melanoma is of vital importance. In this paper, we present an effective approach to melanoma classification from dermoscopy images of skin lesions. First, we perform automatic border detection to delineate the lesion from the background skin. We then extract shape features from the border, while colour and texture features are obtained based on a division of the image into clinically significant regions using a Euclidean distance transform. The derived features are then used in a pattern classification stage for which we employ a dedicated ensemble learning approach to address the class imbalance in the training data. In particular, we employ a committee of one-class classifiers for that purpose. One-class classification uses samples from a single distribution to derive a decision boundary, and employing this method on the minority class can significantly boost its recognition rate and hence the sensitivity of our approach. We combine several one-class classifiers using a random subspace approach and a diversity measure to select members of the committee. Experimental results on a large dataset of dermoscopic skin lesion images show our approach to work well, the employed classifier selection stage to be crucial for achieving this performance, and the classifier ensembles to perform statistically better compared to several state-of-the-art ensembles.

Keywords: melanoma, skin cancer, dermoscopy, image features, ensemble classification, one-class classification.

1 Introduction

Malignant melanoma is the deadliest form of skin cancer and one of the most rapidly increasing cancers in the world. For example, in the United States 76,690 cases and 9,480 deaths are predicted for 2013 alone [17]. Early diagnosis of melanoma is particularly important since melanoma can be cured with a simple excision if detected early.

A.E. Hassanien et al. (Eds.): AMLTA 2014, CCIS 488, pp. 291–298, 2014.
© Springer International Publishing Switzerland 2014

Dermoscopy has become one of the most important tools in diagnosing melanoma and other pigmented skin lesions. It is a non-invasive skin imaging technique that involves optical magnification, along with optics that minimise surface reflection, making subsurface structures more easily visible when compared to conventional clinical images [2]. This in turn reduces screening errors and provides greater differentiation between difficult lesions such as pigmented Spitz nevi and small, clinically equivocal lesions [18]. However, it has also been shown that dermoscopy may lower the diagnostic accuracy in the hands of inexperienced dermatologists [3]. Therefore, in order to minimise diagnostic errors that result from the difficulty and subjectivity of visual interpretation, computerised image analysis techniques are highly sought after [11].

Computer-aided approaches to diagnosing melanoma typically proceed in three main stages: border detection, feature extraction and classification [6]. In this paper, we also follow this strategy, but pay particular attention to the final stage, i.e. the pattern classification task. We first perform automatic border detection based on a variant of the JSEG image segmentation algorithm. From the obtained lesion definition, we extract a set of shape features, while colour and texture features are calculated based on a division of the image into clinically significant regions using a Euclidean distance transform. The obtained features are then employed in a pattern classification stage for which, in order to provide improved and more robust performance, we use a multiple classifier system rather than relying on a single classifier. At the same time, we are addressing the present class imbalance which stems from the fact that far fewer malignant samples are available for training compared to benign cases. We do this using an ensemble of one-class classifiers (OCCs) by constructing a committee of support vector data description (SVDD) [20] boundary models. By training the one-class classifier ensemble on the minority class we boost its recognition rate and hence increase sensitivity. Using classifier pruning, based on a diversity measure for OCCs, we discard irrelevant classifiers and improve the quality of the combined model. Experimental results on a large dataset of 564 skin lesion images show our approach to work well, giving a sensitivity of 94.92% coupled with a specificity of 94.02%, and also confirm our approach to outperform several other state-of-the-art ensemble classifiers dedicated to imbalanced classification.

2   Segmentation and Feature Extraction

Automated border detection is typically the first step in the automated analysis of dermoscopy images [7] and is crucial for two main reasons. First, the border structure provides important information for accurate diagnosis, as many clinical features, such as asymmetry, border irregularity, and abrupt border cutoff, are calculated directly from the border. Second, the extraction of other important clinical features such as atypical pigment networks, globules, and blue-white areas, critically depends on the accuracy of border detection.

In our approach, we perform automated border detection using the technique from [5] which in turn is based on the JSEG algorithm [10]. Following a preprocessing step to smooth the image and a colour quantisation process, the image is thresholded to arrive at an approximate outline of the lesion. This is then refined using region growing on a local homogeneity channel and colour-based region merging. Finally, in a post-processing step, background regions and isolated areas are removed and the remaining regions merged to give the final segmentation.

From the segmented lesion area, we then extract a series of features including shape, colour and texture features, where some of the colour and texture features are calculated based on the definition of three significant image regions – lesion, inner and out periphery – which are obtained based on a Euclidean distance transform. In particular, the following descriptors are extracted [6]:

- *Shape features:* lesion area; aspect ratio of lesion; two asymmetry features; compactness; maximum lesion diameter; eccentricity; solidity; equivalent diameter; rectangularity and elongation of the object-oriented bounding box
- *Colour features:* mean and standard deviation of each channel in RGB, rgb, HSV, l1l2l3 and CIEL*u*v* colour spaces; ratios and differences of mean and standard deviation from the different image regions for all colour spaces; two colour asymmetry features each for R, G, and B channels; centroidal distances for each channel of all colour spaces; CIEL*u*v* L_1 and L_2 histogram distances between the different image regions
- *Texture features:* maximum probability, energy, entropy, dissimilarity, contrast, inverse difference, inverse difference moment, and correlation of the normalised gray-level co-occurrence matrix [12] (averages over the four major orientations); ratios and differences of the same co-occurrences features from the different image regions

In total, we compute 11 shape, 354 colour and 72 texture features for each image.

3 Classification

In one-class classification (OCC) [13], training is performed solely on samples from a single class, known as the target concept ω_T. The purpose of OCC is to create a decision surface that encloses all available data samples and thus describes the concept. During the exploitation step, objects unseen during the training phase may appear. These represent data outside the target concept, and are labelled as outliers ω_O. OCC can thus be seen as learning in the absence of counterexamples. The target class should be separated from all possible outliers, and hence the decision boundary should be estimated in all directions in the feature space around the target class.

By labelling the minority class as ω_T, we create a decision boundary around its objects and treat the objects from the majority class as outliers ω_O. We thus obtain a solution that focusses on correct recognition of the minority class and is hence capable to deal with class imbalance.

For a single OCC classifier, it may be difficult to find a good model due to limited training data, high dimensionality of the feature space and/or the properties of the particular classifier, which might result in model overfitting. Multiple classifier systems (MCSs) hence form an attractive perspective for OCC problems, as they allow to train less complex base classifiers, thus reducing the risk of model overfitting.

Multiple classifier systems (MCSs) can improve upon the performance of their base classifiers since they are able to exploit their individual strengths while eliminating their weaknesses. Classifier fusion algorithms can in general be divided into methods that make decisions on the basis of outputs (labels) of individual classifiers, and approaches that construct new discriminant functions based on continuous outputs (supports) of individual classifiers. In this paper, we focus on the latter, as it has been shown that this type of fuser often offers superior performance [23].

Assume that we have R classifiers $\{\Psi^{(1)}, \Psi^{(2)}, ..., \Psi^{(R)}\}$. For a given object $x \in \mathcal{X}$, each individual classifier decides for class $i \in \mathcal{M} = \{1, ..., M\}$ based on the values of discriminants. Let $F^{(l)}(i, x)$ denote a function that is assigned to class i for a given value of x, and that is used by the l-th classifier $\Psi^{(l)}$. A combined classifier Ψ can then be derived by [23]

$$\Psi(x) = i \quad \text{if} \quad \hat{F}(i, x) = \max_{k \in M} \hat{F}(k, x), \tag{1}$$

where

$$\hat{F}(i, x) = \sum_{l=1}^{R} w^{(l)} F^{(l)}(i, x) \quad \text{and} \quad \sum_{i=1}^{R} w^{(l)} = 1. \tag{2}$$

One-class boundary methods are based on computing the distance between object x and the description (decision boundary) that encloses the target class ω_T. Therefore, in order to apply a fusion method based on discriminants, we require the support function of object x for a given class. For this, we use a heuristic solution which maps a distance into the discriminant,

$$\hat{F}(\omega_T, x) = \frac{1}{c_1} exp(-d(x, \omega_T)/c_2), \tag{3}$$

and models a Gaussian distribution around the classifier, where $d(x, \omega_T)$ is a Euclidean distance metric between the considered object and a decision boundary, c_1 and c_2 are a normalisation constant respectively a scale parameter and are fitted to the target class distribution.

After this mapping, we can apply one of several possibilities to fuse the outputs of the base OCC models [19]. With R one-class classifiers in the pool, we use the mean of the estimated probabilities

$$y_{mp}(x) = \frac{1}{R} \sum_{k} (F_k(\omega_T, x)), \tag{4}$$

which assumes that the outlier object distribution is independent of x and thus uniform in the area around the target concept.

In order to choose the most valuable individual models for the ensemble, careful classifier selection must be conducted. One of the most popular criteria for this task is ensemble diversity, which aims at choosing predictors that are as different from each other as possible. This is motivated by the fact that adding similar classifiers to the committee does not improve its quality but only increases its complexity. On the other hand, diverse base classifiers might be mutually supplementary and hence allow to exploit different areas of competence.

Let us assume that the highest ensemble diversity for a given object $x_j \in X$ is given by $R/2$ of the ensemble votes with the same value (ω_T or ω_O) and the remaining $R - R/2$ with the other value. Denoting by $r(x_j)$ the number of one-class classifiers that correctly recognise object x_j, and assuming there are N objects in the training set, we use the entropy [14]

$$E_{oc}(\Pi^r) = \frac{1}{N} \sum_{j=1}^{N} \frac{1}{R - R/2} \min\{r(x_j), R - r(x_j)\} \tag{5}$$

as a diversity measure, where Π^r is the considered pool of classifiers. This measure may take values from the interval $[0, 1]$, where 0 corresponds to identical ensembles and 1 to the highest possible diversity respectively.

4   Experimental Results

In our experiments, we use a dataset of 564 skin lesion images obtained from two dermoscopy atlases. The samples stem from three university hospitals (University of Graz, University of Naples and University of Florence) [2] and from the Sydney Melanoma Unit [16]. All images are true-colour images with a typical resolution of 768×512 pixels. Of the 564 cases, 88 were melanoma while the remaining 476 were benign, justifying our dedicated approach to address class imbalance.

As base classifiers, we use support vector data description (SVDD) [20] classifiers with RBF kernels.The classifier pool consisted of 15 models built on the basis of a random subspace approach with each subspace consisting of 60% of the original features. After the diversity-based classifier selection, the OCC ensemble consisted of 6 to 8 individual classifiers (depending on the fold of CV).

In order to put the obtained results into context, we have also evaluated a single SVDD and a combined pool of SVDD classifiers without diversity-based classifier selection. In addition, we implemented several classifier ensembles that are dedicated to imbalanced classification, namely SMOTEBagging, SMOTE-Boost, IIvotes, and EasyEnsemble, all with support vector machines (SVMs) [21] (with a Gaussian RBF kernel and classifier tuning) as base classifiers.

SMOTEBagging [22] and SMOTEBoost [9] use, as the names imply, SMOTE [8] to introduce new objects into each of the bagging/boosting iterations separately. IIvotes [4] is an approach which fuses a rule-based ensemble with a SPIDER preprocessing scheme so as to be more robust with respect to atypical data distributions and to automatically find an optimal number of bags. EasyEnsemble [15] uses bagging as the main concept and, employing AdaBoost for each of the bags, can be viewed as an ensemble of ensembles.

A combined 5x2 CV F test [1], repeated ten times, was carried out to assess the statistical significance of the obtained results. A classifier is assumed as

Table 1. Classification results for all tested algorithms

	sensitivity	specificity	accuracy
SVDD	89.49	90.07	89.66
SVDD ensemble	85.81	90.93	90.14
SMOTEBagging	92.54	93.06	92.98
SMOTEBoost	91.85	92.89	92.73
IIVotes	93.05	93.56	93.48
EasyEnsemble	91.85	92.89	92.73
OCC ensemble	94.92	94.02	94.16

Table 2. Statistical significance results: + signifies that the algorithm in this row statistically outperforms the algorithm in this column, − indicates statistical inferiority.

	SVDD	SVDD ensemble	SMOTEBagging	SMOTEBoost	IIVotes	EasyEnsemble	OCC ensemble
SVDD		+	−	−	−	−	−
SVDD ensemble	−		−	−	−	−	−
SMOTEBagging	+	+		+	−	+	−
SMOTEBoost	+	+	−		−		−
IIVotes	+	+	+	+		+	−
EasyEnsemble	+	+	−		−		−
OCC ensemble	+	+	+	+	+	+	

statistically significantly better compared to another one if one of the following is true:

- its sensitivity is statistically significantly better and its overall accuracy is not statistically significantly worse;
- its overall accuracy is statistically significantly better and its sensitivity is not statistically significantly worse.

The results of our experimental comparison are given in Table 1, which lists sensitivity (i.e. the probability that a case identified as malignant is indeed malignant), specificity (i.e. the probability that a case identified as benign is indeed benign) and overall classification accuracy (i.e. the percentage of correctly classified patterns) for each approach. In addition, we provide the results of the statistical significance test in Table 2.

Looking at the results, we can first of all notice that a single SVDD model outperforms the unpruned (i.e., consisting of all models) SVDD ensemble. This highlights the difficulties in combining OCC predictors.

Improved performance is achieved through application of ensemble techniques that are dedicated to deal with class imbalance. All four of the implemented approaches, i.e. SMOTEBagging, SMOTEBoost, IIvotes and EasyEnsemble, achieve both better sensitivity and better specificity. This confirms that appropriate ensemble classifiers typically lead to better classification.

Finally, looking at the results achieved by our proposed multiple classifier system, we can see that it clearly provides the best overall performance. The achieved sensitivity of 94.92% and specificity of 94.02% are both highest among all methods, while Table 2 shows that our method also statistically outperforms all other approaches. This demonstrates that our carefully crafted one-class classifier ensemble provides a powerful method for classifying skin lesion attributes.

5 Conclusions

In this paper, we have proposed an effective method for the automated identification of melanoma from dermoscopy skin lesion images. We first segment the area of the lesion using an approach based on thresholding, region growing and region merging. Based on the lesion border, we then extract a set of shape features, while colour and texture features are derived based on the definition of three clinically important image areas. Finally, the extracted features are analysed in a pattern classification stage. For this, we employ a carefully crafted ensemble classifier that addresses the encountered class imbalance by combining several one-class classifiers and employing an ensemble diversity measure for effective classifier selection. Based on a dataset of 564 skin lesion images, our approach is shown to work very well, giving a sensitivity of 94.92% coupled with a specificity of 94.02%, while we further demonstrate it to give statistically better classification performance compared to several state-of-the-art ensemble classifiers dedicated to imbalanced classification.

References

1. Alpaydin, E.: Combined 5 x 2 CV F test for comparing supervised classification learning algorithms. Neural Computation 11(8), 1885–1892 (1999)
2. Argenziano, G., Soyer, H.P., De Giorgi, V.: Dermoscopy: A Tutorial. EDRA Medical Publishing & New Media (2002)
3. Binder, M., Schwarz, M., Winkler, A., Steiner, A., Kaider, A., Wolff, K., Pehamberger, H.: Epiluminescence microscopy. a useful tool for the diagnosis of pigmented skin lesions for formally trained dermatologists. Archives of Dermatology 131(3), 286–291 (1995)
4. Błaszczyński, J., Deckert, M., Stefanowski, J., Wilk, S.: Integrating selective preprocessing of imbalanced data with Ivotes ensemble. In: Szczuka, M., Kryszkiewicz, M., Ramanna, S., Jensen, R., Hu, Q. (eds.) RSCTC 2010. LNCS (LNAI), vol. 6086, pp. 148–157. Springer, Heidelberg (2010)
5. Celebi, M.E., Aslandogan, Y.A., Stoecker, W.V., Iyatomi, H., Oka, H., Chen, X.: Unsupervised border detection in dermoscopy images. Skin Research and Technology 13(4), 454–462 (2007)
6. Celebi, M.E., Kingravi, H., Uddin, B., Iyatomi, H., Aslandogan, A., Stoecker, W.V., Moss, R.H.: A methodological approach to the classification of dermoscopy images. Computerized Medical Imaging and Graphics 31(6), 362–373 (2007)

7. Celebi, M., Iyatomi, H., Schaefer, G., Stoecker, W.: Lesion border detection in dermoscopy images. Computerized Medical Imaging and Graphics 33(2), 148–153 (2009)
8. Chawla, N.V., Bowyer, K.W., Hall, L.O., Kegelmeyer, W.P.: SMOTE: Synthetic minority over-sampling technique. Journal of Artificial Intelligence Research 16, 321–357 (2002)
9. Chawla, N.V., Lazarevic, A., Hall, L.O., Bowyer, K.W.: SMOTEBoost: improving prediction of the minority class in boosting. In: Lavrač, N., Gamberger, D., Todorovski, L., Blockeel, H. (eds.) PKDD 2003. LNCS (LNAI), vol. 2838, pp. 107–119. Springer, Heidelberg (2003)
10. Deng, Y., Manjunath, B.S.: Unsupervised segmentation of color-texture regions in images and video. IEEE Trans. Pattern Analysis and Machine Intelligence 23(8), 800–810 (2001)
11. Fleming, M.G., Steger, C., Zhang, J., Gao, J., Cognetta, A.B., Pollak, I., Dyer, C.R.: Techniques for a structural analysis of dermatoscopic imagery. Computerized Medical Imaging and Graphics 22(5), 375–389 (1998)
12. Haralick, R.M.: Statistical and structural approaches to texture. Proceedings of the IEEE 67(5), 786–804 (1979)
13. Koch, M.W., Moya, M.M., Hostetler, L.D., Fogler, R.J.: Cueing, feature discovery, and one-class learning for synthetic aperture radar automatic target recognition. Neural Networks 8(7-8), 1081–1102 (1995)
14. Kuncheva, L.I., Whitaker, C.J.: Ten measures of diversity in classifier ensembles: limits for two classifiers. In: IEE Workshop on Intelligent Sensor Processing, pp. 10/1–10/6 (2001)
15. Liu, X., Wu, J., Zhou, Z.: Exploratory undersampling for class-imbalance learning. IEEE Trans. Systems, Man and Cybernetics - Part B: Cybernetics 39(2), 539–550 (2009)
16. Menzies, S.W., Crotty, K.A., Ingwar, C., McCarth, W.H.: An atlas of surface microscopy of pigmented skin lesions: dermoscopy, 2nd edn. McGraw-Hill (2003)
17. Siegel, R., Naishadham, D., Jemal, A.: Cancer statistics, 2013. CA: A Cancer Journal for Clinicians 63(1), 11–30 (2013)
18. Steiner, K., Binder, M., Schemper, M., Wolff, K., Pehamberger, H.: Statistical evaluation of epiluminescence dermoscopy criteria for melanocytic pigmented lesions. Journal of the American Academy of Dermatology 29(4), 581–588 (1993)
19. Tax, D.M.J., Duin, R.P.W.: Combining one-class classifiers. In: Kittler, J., Roli, F. (eds.) MCS 2001. LNCS, vol. 2096, pp. 299–308. Springer, Heidelberg (2001)
20. Tax, D.M.J., Duin, R.P.W.: Support vector data description. Machine Learning 54(1), 45–66 (2004)
21. Vapnik, V.N.: Statistical Learning Theory. John Wiley & Sons (1998)
22. Wang, S., Yao, X.: Diversity analysis on imbalanced data sets by using ensemble models. In: IEEE Symposium on Computational Intelligence and Data Mining, pp. 324–331 (2009)
23. Wozniak, M., Zmyslony, M.: Designing combining classifier with trained fuser – analytical and experimental evaluation. Neural Network World 20(7), 925–934 (2010)

Classification of HEp-2 Cell Images Using Compact Multi-Scale Texture Information and Margin Distribution Based Bagging

Gerald Schaefer[1], Niraj P. Doshi[1], Qinghua Hu[2], and Aboul Ella Hassanien[3]

[1] Department of Computer Science, Loughborough University, U.K.
[2] School of Computer Science and Technology, Tianjin University, China
[3] Faculty of Computers and Information, Cairo University, Egypt

Abstract. Indirect immunofluorescence imaging is commonly employed for screening of antinuclear antibodies based on HEp-2 cells which is used for diagnosing autoimmune diseases and other important pathological conditions involving the immune system. For this purpose, observed HEp-2 cells are categorised into homogeneous, fine speckled, coarse speckled, nucleolar, cytoplasmic, and centromere cells. Typically, this categorisation is performed manually by an expert and is hence both time consuming and subjective.

In this paper, we present a method for automatically classifiying HEp-2 cells using multi-scale texture information in conjunction with an ensemble classification system. We extract multi-dimensional local binary pattern (MD-LBP) texture features of the cell area, which we then compactify using principal component analysis (PCA). PCA-projected features of reduced dimensionality are then employed as input for the subsequent classification stage. For classification, we use a margin distribution based bagging pruning (MAD-Bagging) classifier ensemble. We evaluate our algorithm on the ICPR 2012 HEp-2 contest benchmark dataset, and demonstrate it to give excellent performance, superior to all algorithms that were entered in the competition.

Keywords: Indirect immunofluorescence imaging, HEp-2 cell classification, texture, multi-dimensional LBP, ensemble classification, bagging.

1 Introduction

Indirect immunofluorescence (IIF) imaging is used to identify antinuclear antibodies in HEp-2 cells which founds the basis for diagnosis of diseases such as systemic rheumatic disease, systemic sclerosis and mellitus (type-I) diabetes [8]. In IIF, cultured HEp-2 cells are observed under a fluorescence microscope and then categorised based on fluorescence intensity and on the type of staining patterns.

This classification of HEp-2 cells is crucial for diagnosis, since different patterns yield information for different autoimmune diseases. At the same time, since performed manually by an expert, it is a laborous and time consuming

A.E. Hassanien et al. (Eds.): AMLTA 2014, CCIS 488, pp. 299–308, 2014.

task. A computer-aided (CAD) approach would hence not only speed up the task but also lead to objective, reproducible results. HEp-2 cells are generally categorised into six groups: homogeneous, fine speckled, coarse speckled, nucleolar, cytoplasmic, and centromere cells, which are also the classes we consider in this paper. Example images are shown in Fig. 1.

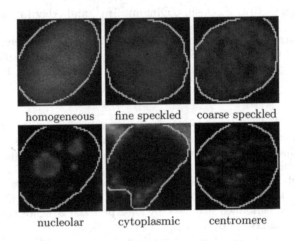

homogeneous   fine speckled   coarse speckled

nucleolar   cytoplasmic   centromere

Fig. 1. Sample HEp-2 cell images (with manually defined borders in white) from the ICPR 2012 contest dataset

In this contribution, we utilise multi-resolution texture information for categorising HEp-2 cell images. In particular, we employ multi-dimensional local binary pattern (MD-LBP) features to characterise the cell area. MD-LBP [22] is a multi-scale extension of LBP [19] that also preserves the relationships between the scales in form of a multi-dimensional histogram. To reduce the dimensionality of the feature data, we employ principal component analysis (PCA). The PCA-compactified MD-LBP features then form the input for a classification stage for which we utilise a margin distribution based bagging pruning (MAD-Bagging) [26] classifier ensemble. We evaluate our algorithm on the ICPR 2012 contest dataset [10], and show that it provides very good performance, superior to all algorithms entered in the competition.

2   Related Work

Automated classification of HEp-2 cell images has recently received increased attention, in particular with the running of a competition at ICPR 2012 [10]. A number of approaches were presented at the contest, of which we summarise a select few in the following.

In [5], images are contrast normalised and statistical texture features based on the grey level co-occurrence matrix (GLCM) [13] as well as frequency domain texture features based on the discrete cosine transform (DCT) [23] are extracted.

To improve classification performance, a two-step feature selection method is employed where the first step is based on a minimum redundancy maximum relevance algorithm to select a candidate feature set, while a final feature set is obtained using a sequential forward selection method. A support vector machine (SVM) [25] is used for classification.

In [16], DCT coefficient features, local binary pattern (LBP) [19] and Gabor texture descriptors [18] as well as various global appearance statistical features (area, perimeter, average intensity and standard deviation of the cell region as well as the ratio of cell and background) are utilised. A multiclass boosting SVM [12] is employed for classification with different SVMs merged into a classifier and boosted using a modified AdaBoost.M1 algorithm [17].

Shape and texture features are combined in [24]. The cell images are thresholded at different intensity levels and shape based descriptors (perimeter, eccentricity, etc.) and intensity based features (average intensity, standard deviation, etc.) are extracted at each level. Following this, gradient magnitude features are calculated after smoothing the image using Gaussian kernels with different parameters. Finally, GLCM texture features are also calculated. The obtained features are fed to a Random Forest classifier [2].

In [9], shape features based on the Hessian matrix are employed, where the eigenvalues of the Hessian and the related eigenvector orientations are used for shape characterisation. Edge features are extracted using an adaptive robust structure tensor and histogram of oriented gradients (ARST-HOG) [20] approach. Finally, texture information, based on LBP, is also utilised. Classification is performed using regression trees as base classifiers and a ShareBoost algorithm [21] for classifier fusion.

In [11], along with GLCM and HOG [4] features, region-of-interest (ROI)-based descriptors are used which include shape features (eccentricity, perimeter, etc.) and intensity based features (derived from intensity percentiles). For classification, an SVM is chosen as the best performing algorithm.

3   HEp-2 Cell Multi-Scale Texture Features

Previous approaches to automatically classifying HEp-2 cells are typically based on several features, while texture features are employed in the majority of algorithms. In this paper, we utilise a single, relatively simple type of texture feature based on local binary patterns (LBP) that we show to yield very good HEp-2 cell recognition.

LBP [19] describes the local neighbourhood of a pixel by thresholding neighbouring pixels g_p with the centre pixel value g_c. The resulting sequence of 0s and 1s is then known as the local binary pattern, formally expressed as

$$\text{LBP} = \sum_{p=1}^{8} s(g_p - g_c)2^{p-1}, \tag{1}$$

where

$$s(x) = \begin{cases} 1 \text{ for } x \geq 0 \\ 0 \text{ for } x < 0 \end{cases}, \tag{2}$$

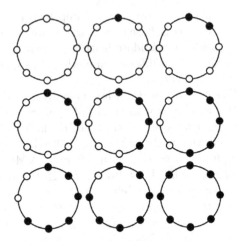

Fig. 2. Uniform LBP patterns

and a histogram of these patterns is generated to summarise texture information of an image or a region of interest. LBP is inherently invariant to monotonic intensity transformations and hence more robust than other techniques.

LBP patterns are typically obtained from a circular neighbourhood where locations in the neighbourhood that do not fall exactly at the centre of a pixel are obtained through interpolation. If a texture is rotated, essentially the patterns (that is, the 0s and 1s around the centre pixel) rotate with respect to the centre. Rotation invariance can hence be obtained by mapping all possible rotated patterns to the same descriptor. Furthermore, certain patterns are fundamental properties of texture and may account for the majority of LBP patterns. To address this, only uniform patterns can be utilised where a uniformity measure is defined by the number of transitions from 0 to 1 or vice versa in the LBP code.

Based on 8 neighbouring pixels, 9 different rotation invariant uniform patterns (with maximal two transitions) can be defined (see Fig. 2), while the remaining patterns are accumulated in a single bin, thus giving a histogram of 10 bins. This yields a powerful texture descriptor that was shown to work well for texture classification, especially when obtained at multiple scales [19,6].

In conventional multi-scale LBP, the histograms for each scale are simply concatenated to form a one-dimensional feature vector. This leads to a loss of information regarding the relationships between patterns across different scales and additional ambiguity. Multi-dimensional LBP (MD-LBP) [22] addresses this by preserving the joint distribution of LBP codes at different scales in form of a multi-dimensional histogram of LBP values. To do so, for each pixel LBP codes at different scales are obtained, while the combination of these codes identifies the histogram bin that is incremented.

MD-LBP leads to clearly improved texture classification, in particular on more challenging datasets [22]. However, the generation of multi-dimensional histograms also leads to rather large feature lengths, and hence increased computational requirements. For example, the feature length of a 3-dimensional MD-LBP histogram is $10^3 = 1000$, while conventional multi-scale LBP has a feature length of 30. Hence, to employ a more compact texture representation, we perform dimensionality reduction based on principal component analysis (PCA), leading to the derivation of MD-LBP-PCA features [7].

To characterise HEp-2 images, we use the green channel of IIF images resized to 64×64 pixels from which we extract rotation invariant uniform LBP texture information – in form of MD-LBP-PCA features based on three scales with radii $\{1, 3, 5\}$ and compactified to a feature length of 50 – from the cell area to be used in the subsequent classification stage.

4 HEp-2 Cell Classification Using MAD-Bagging

Based on the texture features derived above, we then perform classification for which we employ an ensemble learning approach. The idea of ensemble classifiers [15] is to exploit the strengths and local competencies of a pool of classifiers, while at the same time reducing their individual weaknesses. Consequently, an appropriately constructed combination of several predictors can lead to better and more robust classification performance compared to any single classifier

In particular, we utilise a margin distribution based bagging pruning (MAD-Bagging) [26] classifier ensemble. Bagging is a well known, simple yet effective technique for generating an ensemble of classifiers [1]. A collection of base classifiers is trained on bootstrap replicates of the training set and the outputs of all trained base classifiers are combined using simple voting. In general, the error of bagging decreases as base classifiers aggregated in the ensemble increase. Eventually, the error asymptotically approaches a constant level when the size of the ensemble becomes very large.

It is well accepted that generalisation performance of base classifiers and the diversity among base classifiers greatly influence the performance of the ensemble. Bagging uses bootstrapping to generate diverse training sets which then leads to diverse classifiers. On the other hand, selecting only a subset of candidate base classifiers may lead to a significant improvement of the final ensemble. Algorithms were developed to select diverse and accurate base classifiers to yield compact and powerful sub-ensembles [27].

MAD-Bagging utilises the margin of the ensemble as an optimisation objective and derives an L_1 regularised squared loss function. By solving the resulting optimisation problem, a sparse weight vector for the candidate base classifiers can be obtained. Then, only base classifiers with non-zero weights are included in the final ensemble, while the others are discarded.

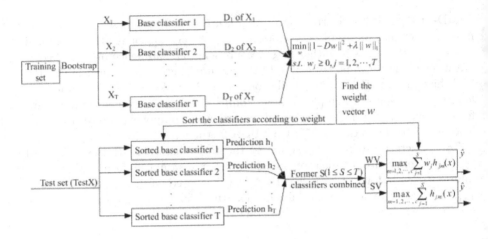

Fig. 3. MAD-Bagging framework

Assuming a set of samples $X = \{(x_i, y_i)\}_{i=1}^{N}$ with $y_i \in \{-1, +1\}$, a base classifier h_j performs a mapping from X to $\{-1, +1\}$. The voted ensemble $f(x)$ is of the form

$$f(x) = \sum_{j=1}^{T} w_j h_j(x),\tag{3}$$

where w_j is the (non-negative) weight assigned to base classifier h_j, $\sum_{j=1}^{T} w_j = 1$, and T is the number of the candidate base classifiers available. An error occurs for x_i if and only if the output of the voting classifier and the label y_i do not have the same sign. Hence, $y_i h_i(x)$ is the difference between the weights assigned to the correct label and the weights assigned to the incorrect label.

$y_i h_i(x)$ is considered the sample margin r_i with respect to the voting classifier f. In order to obtain a good margin for each sample, a loss function $\sum_i C(y_i h_i(x))$ is designed in MAD-Bagging, leading to solving for

$$\min \sum_i C(y_i h_i(x)) + \lambda ||w||_1 \tag{4}$$

subject to $W_j \geq 0$. This objective function leads to an optimal margin distribution over the training samples.

The overall procedure of MAD-Bagging is illustrated in Fig. 3.

5 Experimental Results

For evaluation, we use the ICPR 2012 HEp-2 classification contest dataset [10] which is based on 28 HEp-2 images acquired by means of a fluorescence microscope under 40-fold magnification coupled with a 50W mercury vapour lamp. Images were taken with a SLIM system digital camera, and stored in 24-bit

Table 1. HEp-2 cell classification results on ICPR12 training data

algorithm (evaluation)	accuracy
Cataldo et al. [5] (10CV)	86.96
Li et al. [16] (5CV)	98.34
Strandmark et al. [24] (LOOCV)	97.40
Ersoy et al. [9] (5CV)	92.80
Ghosh and Chaudhary [11] (10CV)	91.13
Proposed (10CV)	92.49

true-colour format with a resolution of 1388 × 1038 pixels. Cells were manually segmented and annotated by a specialist to obtain a ground truth for the competition.

The training dataset provided to contestants comprises 721 samples of individual cells, extracted from part of the captured images. There are 150 homogeneous, 94 fine speckled, 109 coarse speckled, 102 nucleolar, 58 cytoplasmic, and 208 centromere cells (an example of each class is given in Fig. 1). The testing dataset, extracted from different images, contains 734 cells in total of which 172 are homogeneous, 114 fine speckled, 101 coarse speckled, 139 nucleolar, 51 cytoplasmic, and 149 centromere cells.

For MAD-Bagging, 100 bags were used, and support vector machines [25], in particular one-against-one multi-class SVMs [14], for which the parameters were optimised [3], served as base classifiers.

We first evaluate the performance on the training dataset, by performing 10-fold cross validation (10CV), where the dataset is split into 10 partitions and training is performed on all but one partition while testing is conducted on the remaining one, reporting the average classification accuracy over all 10 folds. The results are reported in Table 1, which also lists the classification accuracies on the same dataset of the methods discussed in Section 2[1].

From Table 1, we can see that our MD-LBP based HEp-2 cell classification approach affords good performance, giving a classification accuracy of about 92.5%.

Table 2. Classification results on ICPR12 test data

method	accuracy [%]
Cataldo et al. [5]	48.50
Li et al. [16]	64.17
Strandmark et al. [24]	47.82
Ersoy et al. [9]	49.18
Ghosh and Chaudhary [11]	59.81
top ICPR contest entry (Nosaka and Fukui, unpublished)	68.66
Proposed	70.44

---

[1] Note, that not all papers use 10CV as evaluation method, and that hence the cited numbers are only partially comparable.

Table 3. Confusion matrix for proposed method

actual \ predicted	homogeneous	fine speckled	coarse speckled	nucleolar	cytoplasmic	centromere	accuracy
homogeneous	128	32	12	1	3	4	71.11
fine speckled	20	61	1	0	0	32	53.51
coarse speckled	4	15	72	0	0	10	71.29
nucleolar	8	1	2	78	30	20	56.12
cytoplasmic	0	0	3	0	48	0	94.12
centromere	0	5	1	13	0	130	87.25

As the ICPR contest revealed [10], while several of the 28 submitted entries obtained high classification performance (95+%) on the training dataset, accuracy on the test dataset was significantly lower, suggesting that the test data is much more challenging. The best approach, by Nosaka and Fukui, was reported to give a classification accuracy of 68.66%, while about half of the submitted approaches reached less than 50% [10] including some of those discussed in Section 2 (most submitted approaches were not published). Interestingly, also a medical doctor, a specialist with 12 years experience in immunology, did not fare much better with a correct recognition rate of 73.30%.

In Table 2, we report the results obtained on the test dataset. As we can see from there, impressively our method outperforms all competition entries on the test dataset with a classification accuracy of 70.44%.

A more detailed analysis is provided in Table 3 in form of a confusion matrix. From there we can observe that cytoplasmic and centomere cells are well identified. The performance for homogeneous and coarse speckled cells is somewhat worse, and the poorest performance is achieved for fine speckled and nucleolar cells. The highest degree of confusion occurs between fine speckled and homogeneous cells.

6   Conclusions

In this paper, we have presented an effective approach to the automated classification of HEp-2 cell images obtained through indirect immunofluorescence imaging. Our method is based on multi-dimensional local binary pattern (MD-LBP) descriptors of the cell area, which are then compactified using principal component analyis. For classification, we employ a margin distribution based bagging pruning (MAD-Bagging) classifier ensemble. Based on the ICPR 2012 competition dataset, our approach is shown to deliver very good classification performance and to outperform all entries submitted to the contest.

References

1. Breiman, L.: Bagging predictors. Machine Learning 24(2), 123–140 (1996)
2. Breiman, L.: Random forests. Machine Learning 45(1), 5–32 (2001)
3. Chang, C.C., Lin, C.J.: libSVM: A library for support vector machines. ACM Transactions on Intelligent Systems and Technology 2, 27:1–27:27 (2011)
4. Dalal, N., Triggs, B.: Histograms of oriented gradients for human detection. In: IEEE Int. Conference Computer Vision and Pattern Recognition, vol. 1, pp. 886–893 (2005)
5. Di Cataldo, S., Bottino, A., Ficarra, E., Macii, E.: Applying textural features to the classification of HEp-2 cell patterns in IIF images. In: 21st Int. Conference on Pattern Recognition, pp. 3349–3352 (2012)
6. Doshi, N.P., Schaefer, G.: A comparative analysis of local binary pattern texture classification. In: Visual Communications and Image Processing (2012)
7. Doshi, N.P., Schaefer, G.: Texture classification using compact multi-dimensional local binary pattern descriptors. In: 2nd Int. Conference on Informatics, Electronics and Vision (2013)
8. Egerer, K., Roggenbuck, D., Hiemann, R., Weyer, M.G., Büttner, T., Radau, B., Krause, R., Lehmann, B., Feist, E., Burmester, G.R.: Automated evaluation of autoantibodies on human epithelial-2 cells as an approach to standardize cell-based immunofluorescence tests. Arthritis Research and Therapy 12(2), R40 (2010)
9. Ersoy, I., Bunyak, F., Peng, J., Palaniappan, K.: HEp-2 cell classification in IIF images using ShareBoost. In: 21st Int. Conference on Pattern Recognition, pp. 3362–3365 (2012)
10. Foggia, P., Percannella, G., Soda, P., Vento, M.: Benchmarking HEp-2 cells classification methods. IEEE Trans. Medical Imaging 32(10), 1878–1889 (2013)
11. Ghosh, S., Chaudhary, V.: Feature analysis for automatic classification of HEp-2 florescence patterns: Computer-aided diagnosis of auto-immune diseases. In: 21st Int. Conference on Pattern Recognition, pp. 174–177 (2012)
12. Gonen, M., Tanugur, A., Alpaydm, E.: Multiclass posterior probability support vector machines. IEEE Trans. Neural Networks 19(1), 130–139 (2008)
13. Haralick, R.M.: Statistical and structural approaches to texture. Proceedings of the IEEE 67(5), 786–804 (1979)
14. Hsu, C.W., Lin, C.J.: A comparison of methods for multiclass support vector machines. IEEE Trans. Neural Networks 13(2), 415–425 (2002)
15. Kuncheva, L.I.: Combining pattern classifiers: Methods and algorithms. Wiley-Interscience, New Jersey (2004)
16. Li, K., Yin, J., Lu, Z., Kong, X., Zhang, R., Liu, W.: Multiclass boosting SVM using different texture features in HEp-2 cell staining pattern classification. In: 21st Int. Conference on Pattern Recognition, pp. 170–173 (2012)
17. Li, X., Wang, L., Sung, E.: Adaboost with SVM-based component classifiers. Engineering Applications of Artificial Intelligence 21(5), 785–795 (2008)
18. Manjunath, B.S., Ma, W.Y.: Texture features for browsing and retrieval of image data. IEEE Trans. Pattern Analysis and Machine Intelligence 18(8), 837–842 (1996)
19. Ojala, T., Pietikainen, M., Maenpaa, T.: Multiresolution gray-scale and rotation invariant texture classification with local binary patterns. IEEE Trans. Pattern Analysis and Machine Intelligence 24, 971–987 (2002)

20. Palaniappan, K., Bunyak, F., Kumar, P., Ersoy, I., Jaeger, S., Ganguli, K., Haridas, A., Fraser, J., Rao, R., Seetharaman, G.: Efficient feature extraction and likelihood fusion for vehicle tracking in low frame rate airborne video. In: 13th Int. Conference on Information Fusion (2010)
21. Peng, J., Barbu, C., Seetharaman, G., Fan, W., Wu, X., Palaniappan, K.: Share-Boost: boosting for multi-view learning with performance guarantees. In: Gunopulos, D., Hofmann, T., Malerba, D., Vazirgiannis, M. (eds.) ECML PKDD 2011, Part II. LNCS (LNAI), vol. 6912, pp. 597–612. Springer, Heidelberg (2011)
22. Schaefer, G., Doshi, N.P.: Multi-dimensional local binary pattern descriptors for improved texture analysis. In: 21st Int. Conference on Pattern Recognition, pp. 2500–2503 (2012)
23. Sorwar, G., Abraham, A., Dooley, L.S.: Texture classification based on DCT and soft computing. In: 10th IEEE International Conference on Fuzzy Systems, pp. 545–548 (2001)
24. Strandmark, P., Ulen, J., Kahl, F.: HEp-2 staining pattern classification. In: 21st Int. Conference on Pattern Recognition, pp. 33–36 (2012)
25. Vapnik, V.N.: Statistical Learning Theory. John Wiley & Sons (1998)
26. Xie, Z., Xua, Y., Hu, Q., Zhu, P.: Margin distribution based bagging pruning. Neurocomputing 85, 11–19 (2012)
27. Zhou, Z.H., Wu, J.X., Tang, W.: Ensembling neural networks: many could be better than all. Artificial Intelligence 137(1-2), 239–263 (2002)

Directional Stationary Wavelet-Based Representation for Human Action Classification

M.N. Al-Berry[1], M.A.-M. Salem[1], Hala M. Ebeid[1], A.S. Hussein[2], and Mohamed Fahmy Tolba[1]

[1] Scientific Computing Department, Faculty of Computer and Information Sciences, Ain Shams University, Egypt
{maryam_nabil,salem}@cis.asu.edu.eg,
hala_mousher@hotmail.com,
fahmytolba@gmail.com
[2] Faculty of Computer Studies, Arab Open University, Kuwait
ashrafh@acm.org

Abstract. This paper proposes a directional wavelet-based representation of natural human actions in realistic videos. This task is very important for human action recognition, which has become one of the most important fields in computer vision. Its importance comes from the large number of applications that employ human action classification and recognition. The proposed method utilizes the 3D Stationary Wavelet Analysis to encode the directional spatio-temporal characteristics of the motion available in video sequences. It was tested using the Weizmann dataset, and produced promising preliminary results (92.47 % classification accuracy) when compared to existing state–of–the–art methods.

Keywords: Human Action Classification, Stationary Wavelet Analysis, Global motion representation, Motion History Images, Motion Energy Images.

1 Introduction

Human action and activity recognition has become one of the most important areas in computer vision. This importance emanates from a wide spectrum of applications, such as intelligent surveillance [1], content-based video retrieval, behavioral biometrics, medical studies, robotics, security, animation, and human-computer interaction (HCI) [2]. The field of automatic human action recognition is still non-trivial [3, 4]. Its complexity results from a number of challenges such as: varying performance of actions, change in illumination, and dynamic or cluttered environments [4, 5, 6].

Poppe [5], defined vision-based human action recognition as: "The process of labeling image sequences with action labels". Following Weinland et al. [7], an action is a sequence of movements generated by a performer during the performance of a task, and an action label is a name, such that an average human agent can understand and perform the named action.

Different methods have been proposed for segmenting, representing, and classifying actions. These methods can be classified into different taxonomies [7, 8, 9]. One

A.E. Hassanien et al. (Eds.): AMLTA 2014, CCIS 488, pp. 309–320, 2014.

of the famous methods that have been used for holistic motion representation is the Motion History Image (MHI) [10, 11, 12]. Motion History Images are temporal templates that are simple but robust in motion representation, and they are used for action recognition by several research groups [12].

In this paper, a stationary wavelet-based directional action representation is proposed. The proposed representation is based on the 3D Stationary Wavelet Transform (SWT) that has been proposed and used in [13] for spatio-temporal motion detection. The 3D SWT succeeded in motion detection in the presence of illumination variations in both indoor and outdoor scenarios while having reasonable complexity. In the proposed action representation, the 3D SWT is used to encode the action into 3 directional wavelet-based templates. Hu invariant moments [14] have been used for describing the templates obtained using the proposed method in combination with different classifiers using benchmark datasets. The preliminary results obtained using simple features and classifiers show that the proposed representation results are comparable to state-of-the-art methods. The ultimate goal of this work is to use the 3D SWT output for all processing in a visual surveillance framework, i.e., to compute once and use the output in all processing steps (motion detection, tracking, action recognition).

The rest of the paper is organized as follows: Section 2 provides a short review of related work. Section 3 describes the proposed method in detail. Section 4 demonstrates the experimental results. And finally section 5 concludes the paper and highlights some future directions.

2   Related Work

Nowadays, there are many applications that require human motion analysis and recognition, including intelligent surveillance, content-based video retrieval, behavioural biometrics, medical studies, robotics, security, animation, and human-computer interaction (HCI) [3], [9],[15]-[17]. A common task in these applications is automatic human action and activity recognition, which is now one of the most promising fields of computer vision.

There are a lot of papers concerned with the field of action recognition. Some of them provide a very good and detailed review of the field [3], [5], [7], [9], [16]. For example, in [16] a comprehensive survey for activity recognition in video surveillance is provided. The survey describes simple and complex activities along with various applications, provides a categorization of various techniques, describes various datasets, and gives future directions to work on.

Actions in images can be represented using a global or local representation. The global representation encodes the whole motion into a single representation; while local representation represents the motion using a number of independent spots [5]. Both representations have been used and reported in the literature with different performances and applications.

For action classification, a direct classification can be used where the observed template is compared to action class prototypes, or discriminative classifiers can be used to learn a function that discriminates between two or more classes [5]. Other approaches for recognizing actions and activities are described in [9].

In the direction of local representation local interest points are extracted from space-time or 3D volumes. Chen [18], proposed the MoSIFT to detect interest points and describe local features for action recognition. It encodes both local appearance and motion as histograms of gradients in space and histograms of optical flow. In [17], Yan and Luo proposed new action descriptor based on the space-time interest points (STIPs). It was called the histogram of interest point locations (HIPLs). HIPL reorganizes STIPs and reflects the spatial location information, and can be viewed as a useful supplement to the bag-of-interest-point (BIP) feature. They used a combined AdaBoost and sparse representation classifier for classifying actions in benchmark datasets. Their descriptor resulted in good performance but it captured only the spatial information of interest points, without inclusion of the temporal domain. Bregonzio et al. [19] proposed a spatio-temporal technique for action representation in which only the global distribution information of interest points is utilized. Holistic features from clouds of interest points accumulated over multiple temporal scales are extracted. The proposed spatio-temporal distribution representation contains complementary information to the conventional Bag of Words representation. Based on Multiple Kernel Learning the features are fused. In [20], Rapantzikos et al., explored the ability of the 3D wavelet transform to efficiently locate and represent dynamic events while keeping the computational complexity low. They proposed a framework for representing human actions as spatiotemporal salient regions in the 3D wavelet domain. They represented a video sequence as a solid in the three-dimensional Euclidean space, with time being the third dimension, and applied a multiscale 3D wavelet transform to decompose the volume into subbands and use the resulting coefficients to compute saliency. The efficiency of their method was proven by comparison against a well established technique on a public video dataset consisting of six actions. In [21] they used saliency for feature point detection in videos and incorporate color and motion apart from the intensity. Their method used a multiscale volumetric representation of the video and involved spatiotemporal operations at the voxel level.

Sharma et al. [22], investigated the efficacy of directional information of wavelet multi-resolution decomposition for histogram-based classification of human gestures represented by spatio-temporal templates. They used global templates that collapse temporal component into gesture representation. These templates were modified to be invariant to translation, rotation and scale. Histograms of wavelet coefficients at different scales are compared to establish the significance of available information for classification. Their experiments showed that the available information in high pass or low pass decompositions by itself was not sufficient to provide significant accuracy. One of the most successful global representations is the Motion History Images (MHI), proposed by Davies and Bobick [23]. MHI is a global view-based approach that is simple but efficient in representing actions and used in many applications. Ahad et al. [12], provide an overview on the techniques and applications that are based on MHI. They also present various variants that were proposed to enhance the basic MHI, and direct researchers to some future directions. In [10], Davies used MHI for real-time action recognition and categorization. The holistic motion was encoded into a single template, and for recognition, higher order moments were computed and statistically matched to trained models. Babu and Ramakrishnan, [11] used the encoded motion information available in a compressed MPEG stream to construct the Motion History Image (MHI) and the corresponding Motion Flow History (MFH). Different sets of features were extracted and used to train a number of classifiers, then the performance of each feature set with respect to various classifiers were analysed. Their results showed that the K- nearest neighbour (KNN), neural network, and support vector machine (SVM) classifiers give the highest classification accuracy. MHI has been also used by Shao et al [24], in a framework that detects various types of exercises and counts the cycle of the exercise in an indoor environment. A shape-based feature descriptor is extracted from the MHI and Motion Energy Image (MEI) and is used for recognition with high accuracy. The idea of motion history images was extended to 3D history volumes by Weinland et al [25].

From the above discussion, it can be concluded that the problem of human action recognition is still not totally solved and more research need to be elaborated to overcome some of the challenges that face it. For example, global template-based approaches require accurate extraction of objects' silhouettes, a process that can be hard in the presence of varying illumination. In this research, we use the 3D SWT, which proved to be robust to different types of illumination variations, to obtain a directional spatio-temporal representation of human actions.

3   Proposed Directional Wavelet-Based Templates

This section provides a description of the proposed stationary wavelet-based action representation method. The proposed representation is motivated by the motion history images [23] and based on the 3D SWT proposed in [13]. First, the directional wavelet energy images are proposed in section A. Section B proposes the directional wavelet-based history images. Finally, section C describes the features used in classification.

3.1   Proposed Directional Wavelet Energy Images

The first proposal is to build a directional Wavelet-based Energy Image (WEI) using the 3D SWT proposed in [13], where the video sequence is represented as a 3D volume of frames with time being the third dimension. The video sequence is divided into blocks of 8 frames, and a 3 level SWT is applied on the block. The coefficients of three sub-bands (ADD, DAD, DDD) are thresholded to obtain foreground images. The foregrounds obtained at the eighth layer of 3 different scales are fused into three sub-band foreground images ($O^d(x,y,t)$) (d = 1, 2, 3). These sub-band foreground images encode the directional motion energy during the processed 8 frames at 3 different scales, and thus can be used to represent the action in the duration of these 8 frames. This is illustrated in Fig. 1.

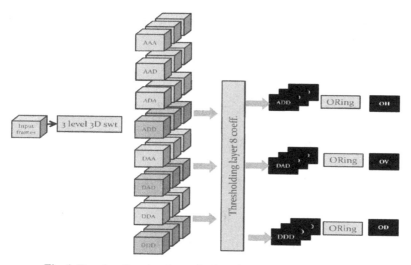

Fig. 1. Forming directional wavelet-based templates using 3D SWT

3.2   Proposed Directional Wavelet-Based History Images

Motion history image is a way of representing the motion sequence in one gray scale view-based template. The motion history image $H\tau(x,y,t)$ is computed using an update function $\psi(x,y,t)$ as follows [12]:

$$H_\tau(x, y,t)=\begin{cases} \tau & if\, \psi(x, y,t)=1 \\ \max(0, H_\tau(x, y,t)-\delta) & Otherwise \end{cases} \quad (1)$$

(x,y) and t are the location and time and $\psi(x,y,t)$ indicates the presence of motion in the current video image, τ indicates the temporal extent of the movement and δ is the decay parameter.

In the proposed method, when multiple blocks of frames are processed using the 3D SWT, the resulting foreground images can be combined into three directional Wavelet–based History Image (DWH_τ) template as follows:

$$DWH_\tau^d(x, y,t)=\begin{cases} \tau & if\, O^d(x, y,t)=1 \\ \max(0, DWH_\tau^d(x, y,t-1)-\delta) & Otherwise \end{cases} \quad (2)$$

(x,y) and t are the location and time, d is the sub-band and $O(x,y,t)$ is the binary motion mask obtained by the motion detector at this sub-band, i.e., it signals the presence of motion in this direction.

3.3   Feature Extraction

For describing the obtained templates, the seven Hu invariant moments [14] are computed. The Hu moments are known to result in good discrimination of shapes while being translation, scale, mirroring, and rotation invariant [26]. The set of seven invariant moments is defined as follows [26]:

$$\Phi_I = \eta_{20} + \eta_{02} \tag{1}$$

$$\Phi_2 = (\eta_{20} - \eta_{02})^2 + 4\eta_{11}^2 \tag{2}$$

$$\Phi_3 = (\eta_{30} - 3\eta_{12})^2 + (3\eta_{21} - \eta_{03})^2 \tag{3}$$

$$\Phi_4 = (\eta_{30} + \eta_{12})^2 + (3\eta_{21} + \eta_{03})^2 \tag{4}$$

$$\Phi_5 = (\eta_{30} - 3\eta_{12})(\eta_{30} + \eta_{12})[(\eta_{30} + \eta_{12})^2 - 3(\eta_{21} + \eta_{03})^2] \\ + (3\eta_{21} - \eta_{03})(\eta_{21} + \eta_{03})[3(\eta_{30} + \eta_{12})^2 - (\eta_{21} + \eta_{03})^2] \tag{5}$$

$$\Phi_6 = (\eta_{20} - \eta_{02})[(\eta_{30} + \eta_{12})^2 - (\eta_{21} + \eta_{03})^2] \\ + 4\eta_{11}(\eta_{30} + \eta_{12})(\eta_{21} + \eta_{03}) \tag{6}$$

$$\Phi_7 = (3\eta_{21} - \eta_{03})(\eta_{30} + \eta_{12})[((\eta_{30} + \eta_{12})^2 - 3(\eta_{21} - \eta_{03})^2] \\ + (3\eta_{12} - \eta_{30})(\eta_{21} + \eta_{03})[3(\eta_{30} + \eta_{12})^2 - (\eta_{21} + \eta_{03})^2] \tag{7}$$

where, η_{pq} is the normalized central moments defined as

$$\eta_{pq} = \frac{\mu_{pq}}{\mu_{00}^{\gamma}} \tag{10}$$

μ_{pq} is the central moment of order $(p+q)$, $p = 0,1,2,...$ and $q = 0,1,2,...,$

$$\gamma = \frac{p+q}{2} + 1 \tag{11}$$

or $p+q = 2,3, ...$

4   Results and Discussion

The classification results obtained using the proposed action representations are presented in this section. Section A describes the used dataset, while section B presents the experimental results.

4.1 Description of Dataset

The proposed representation was tested using the Weizmann dataset [27]. The dataset contains 93 video sequences for 10 actions, bend, jack, jump, pjump, run, side, skip, wave1, wave2, and walk. They are performed by 9 persons in front of a static background. Sample frames from the dataset are shown in Fig. 2.

Fig. 2. Four sample frames of the "jack" action from Weizmann dataset [27]

4.2 Experimental Setup

The human action recognition system consists of a feature extraction step and a classification step. In the experiments a feature space is constructed by using the seven Hu invariant moments. The quadratic discriminant analysis classifier is used to find out the true class of the test patterns. The classifier assumes a Gaussian mixture model and does not use prior probabilities or costs for fitting.

The human action recognition experiments are performed on the Weizmann database. Through the experiments, the images of 10 actions have been checked; each action performed by 9 different persons. The correct classification rate is used for evaluation. The correct classification rate (CCR) is the percentage of correctly classified samples of the dataset.

4.3 Experiments and Results

First, the 3D SWT is used to build a directional Wavelet-based Energy Image (WEI). After that, the motion history image is used to represent the motion sequence in one gray scale view-based template. The coefficients of three sub-bands (ADD, DAD, DDD) are used to check which one achieves a better performance. Fig. 3 shows the directional wavelet energy (first row), and history (second row) templates obtained for the "jack" action.

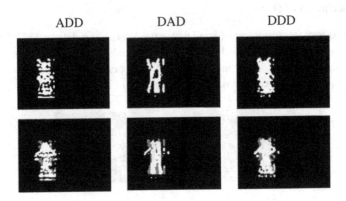

Fig. 3. Directional wavelet energy (first row), and directional wavelet history (second row) of jack action

Then, the seven Hu invariant moments are computed for both the directional wavelet energy and the history templates. Finally, the feature vector that contained the seven Hu moments of the corresponding directional template or history templates is passed to a quadratic discriminant analysis classifier. Confusion matrices obtained from classifying Directional Wavelet Energy Images (WEI) are shown in Fig. 4. The ADD band gave the highest correct classification rate 90.32%, while using the DDD band resulted in high confusion between different action classes.

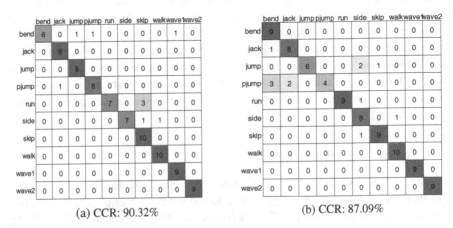

(a) CCR: 90.32%

	bend	jack	jump	pjump	run	side	skip	walk	wave1	wave2
bend	6	0	1	1	0	0	0	0	1	0
jack	0	9	0	0	0	0	0	0	0	0
jump	0	0	9	0	0	0	0	0	0	0
pjump	0	1	0	8	0	0	0	0	0	0
run	0	0	0	0	7	0	3	0	0	0
side	0	0	0	0	0	7	1	1	0	0
skip	0	0	0	0	0	0	10	0	0	0
walk	0	0	0	0	0	0	0	10	0	0
wave1	0	0	0	0	0	0	0	0	9	0
wave2	0	0	0	0	0	0	0	0	0	9

(b) CCR: 87.09%

	bend	jack	jump	pjump	run	side	skip	walk	wave1	wave2
bend	9	0	0	0	0	0	0	0	0	0
jack	1	8	0	0	0	0	0	0	0	0
jump	0	0	6	0	0	2	1	0	0	0
pjump	3	2	0	4	0	0	0	0	0	0
run	0	0	0	0	9	1	0	0	0	0
side	0	0	0	0	0	8	0	1	0	0
skip	0	0	0	0	0	1	9	0	0	0
walk	0	0	0	0	0	0	0	10	0	0
wave1	0	0	0	0	0	0	0	0	9	0
wave2	0	0	0	0	0	0	0	0	0	9

Fig. 4. Confusion matrices obtained using different directional wavelet energy sub-bands. (a) ADD band , (b) DAD band, and (c) DDD band.

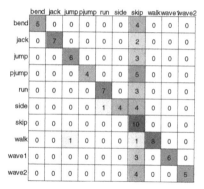

(c) CCR: 66.7%

Fig. 4. (*Continued*)

The classification results obtained using the proposed directional wavelet history templates are illustrated in Fig. 5. Again the DDD band didn't contain enough information to discriminate between different action classes, while in this case the DAD band recorded the highest classification rate.

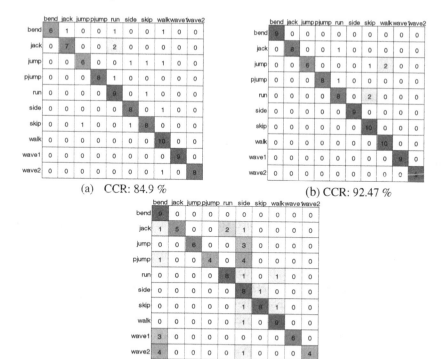

(a) CCR: 84.9 %   (b) CCR: 92.47 %

(c) CCR: 72.04 %

Fig. 5. Confusion matrices obtained using different directional history sub-bands. (a) ADD band , (b) DAD band, and (c) DDD band.

The performance of the proposed method is compared with some of the state–of–the–art methods. The selected methods are Scovanner et al. [28], Ballan et al. [29], Kong et al. [30], and Schindler and Gool [31]. One can observe from fig. 6 that the proposed method showed a promising accuracy compared to the reference methods that have more complex features and classification techniques.

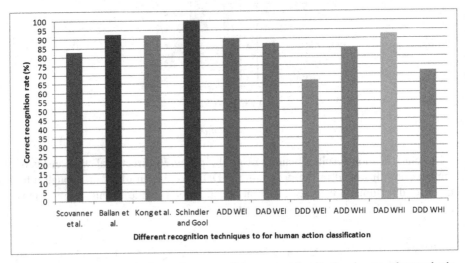

Fig. 6. Comparison between the performance of the proposed method and some other methods

5   Conclusions and Future Work

Automatic human action and activity classification are very promising and are hot topics in computer vision. Their applications range from intelligent surveillance systems and intelligent vehicles, to Human Computer Interaction. Many techniques have been proposed for human action recognition, but the field is still open due to the variability in action performance, and varying environments and illumination conditions among other difficulties and challenges.

This paper proposes a set of directional stationary wavelet-based action representations. The proposed directional representations are evaluated individually to test the efficiency of the directional information contained in the templates. Hu invariant moments have been used for action description in combination with discriminant analysis classification using the Weizmann benchmark datasets. Preliminary results and a comparison with the state–of–the–art methods show that the information in some of the proposed directional representations can be by themselves efficient and promising for action classification. A correct classification rate of 92.47% was achieved in the best case using the DAD sub-band history image.

Future work may include testing the effect of combining the features extracted from directional templates on classification results, using other features to describe the proposed representation, and validating the accuracy of the proposed method using other datasets that contain more challenges.

References

1. Kim, I.S., Choi, H.S., Yi, K.M., Choi, J.Y., Kong, S.G.: Intelligent Visual Surveillance: a survey. Int. J. Control, Automation, and Systems 8, 926–939 (2010)
2. Pantic, M., Nijholt, A., Pentland, A., Huanag, T.S.: Human- Centered Intelligent Human-Computer Interaction (HCI2): How far are we from attaining it? Int. J. Autonomous and Adaptive Communications Systems 1(2), 168–187 (2008)
3. Moeslund, T.B., Hilton, A., Krüger, V.: A survey of advances in vision-based human motion capture and analysis. Comput. Vis. Image Und. 104, 90–126 (2006)
4. Thi, T.H., Cheng, L., Zhang, J., Wang, L., Satoh, S.: Structured learning of local features for human action classification and localization. Image Vision Comput. 30, 1–14 (2012)
5. Poppe, R.: A survey on vision-based human action recognition. Image Vision Comput. 28, 976–990 (2010)
6. Cristani, M., Raghavendra, R., Del Bue, A., Murino, V.: Human Behavior Analysis in Video Surveillance: Social Signal Processing Perspective. Neurocomputing 100, 86–97 (2013)
7. Weinland, D., Ranford, R., Boyer, E.: A survey of vision-based methods for action representation, segmentation and recognition. Comput. Vis. Image Und. 115, 224–241 (2011)
8. Pantic, M., Nijholt, A., Pentland, A., Huanag, T.: Machine Understanding of Human Behavior. In: ACM Int. Conf. Multimodal Interface (2006)
9. Turaga, P., Chellappa, R., Subrahamanian, V., Udrea, O.: Machine Recognition of Human Activities: A Survey. IEEE T. Circ. Syst. Vid. 18(11), 1473–1487 (2008)
10. Davis, J.W.: Representing and Recognizing Human Motion: From Motion Templates to Movement Categories. In: Digital Human Modeling Workshop, IROS 2001 (2001)
11. Babu, R.V., Ramakrishnan, K.R.: Recognition of human actions using motion history information extrcted from the compressed video. Image Vision Comput. 22, 597–607 (2004)
12. Ahad, M., Tan, J., Kim, H., Ishikawa, S.: Motion history image: its variants and applications. Mach. Vision Appl. 23, 255–281 (2012)
13. Al-Berry, M.N., Salem, M.A.-M., Hussein, A.S., Tolba, M.F.: Spatio-Temporal Motion Detection for Intelligent Surveillance Applications. Int. J. Computational Methods 11(1), 1 (2014)
14. Hu, M.-K.: Visual Pattern Recognition by Moment Invariants. IEEE T. Inform. Theory 8(2), 179–187 (1962)
15. Aggrawal, J., Cai, Q.: Human Motion Analysis: A Review. Comput. Vis. Image Und. 73(3), 428–440 (1999)
16. Vishwakarma, S., Agrawal, A.: A survey on activity recognition and behavior understanding in video surveillance. The Visual Computer 29(10), 983–1009 (2013)
17. Yan, X., Luo, Y.: Recognizing human actions using a new descriptor based on spatial–temporal interest points and weighted-output classifier. Neurocomputing 87, 51–61 (2012)
18. Chen, M.Y.: MoSIFT: Resognizing Human Actions in Surveillance Videos, School of Cegie Mellon Universityomputer Science at Research Showcase . Carnegie Mellon University (2009)
19. Bregonzio, M., Xiang, T., Gong, S.: Fusing appearance and distribution information of interest points for action recognition. Pattern Recogn. 45, 1220–1234 (2012)
20. Rapantzikos, K., Tsapatsoulis, N., Avrithis, Y., Kollias, S.: Spatiotemporal saliency for event detection and representation in the 3D Wavelet Domain: Potential in human action recognition. Signal Processing: Image Communication 24, 557–571 (2009)

21. Rapantzikos, K., Avrithis, Y., Kollias, S.: Spatiotemporal saliency for event detec-tion and representation in the 3D Wavelet Domain: Potential in human action recognition. In: 6th ACM International Conference on Image and Video Retrieval, pp. 294–301 (2007)
22. Sharma, A., Kumar, D.K., Kumar, S., McLachlan, N.: Wavelet Directional Histograms for Classification of Human Gestures Represented by Spatio-Temporal Templates. In: 10th International Multimedia Modeling Conference MMM 2004, pp. 57–63 (2004)
23. Davies, J., Bobick, A.F.: The representation and recognition of human movements using temporal templates. In: IEEE Computer Society Conference on Computer Vision and Pattern Recognition, pp. 928–934 (1997)
24. Shao, L., Ji, L., Liu, Y., Zhang, J.: Human action segmentation and recognition via motion and shape analysis. Pattern Recogn. Lett. 33, 438–445 (2012)
25. Weinland, D., Ronfard, R., Boyer, E.: Free Viewpoint Action Recognition using Motion History Volumes. Comput. Vis. Image Und. 104(2), 249–257 (2006)
26. Gonzalez, R.C., Woods, R.E.: Digital Image Processing, 3rd edn. Printice Hall (2008)
27. Blank, M., Gorelick, L., Shechtman, E., Irani, M., Basri, R.: Actions as Space-time Shapes. In: 10th International Conference on Computer Vision, pp. 1395–1402 (2005)
28. Scovanner, P., Ali, S., Shah, M.: A 3-dimensional sift descriptor and its application to action recognition. In: 15th ACM International Conference on Multimedia, pp. 357–360 (2007)
29. Ballan, L., Bertini, M., Del Bimbo, A., Seidenari, L., Serra, G.: Recognizing Human Actions by fusing Spatio-temporal Appearance and Motion Descriptors. In: 16th IEEE International Conference on Image Processing, ICIP 2009, pp. 3569–3572 (2009)
30. Kong, Y., Zhang, X., Hu, W., Jia, Y.: Adaptive learning codebook for action recog-nition. Pattern Recogn. Lett. 32, 1178–1186 (2011)
31. Schindler, K., Gool, L.V.: Action snippets: how many frames does human action recognition require. In: IEEE Conf. Computer Vision and Pattern Recognition, CVPR 2008, pp. 1–8 (2008)

Learning Hierarchical Features Using Sparse Self-organizing Map Coding for Image Classification

Saleh Aly

Department of Electrical Engineering, Faculty of Engineering
Aswan University, Aswan, Egypt
Saleh@aswu.edu.eg

Abstract. Image descriptor is a critical issue for most image classification problems. Low-level image descriptors based on Gabor filters, SIFT and HOG features have exhibited good image representation for many applications. However, these descriptors are not appropriate for image classification and need to be converted into appropriate representations. This process can be performed by applying two operations of coding and pooling. In coding operation, an appropriate codebook is learned and better adapted from training data, while pooling operation summarizes the coded features over larger regions. Several methods of coding and pooling schemes have been proposed in the literature. In this paper, self-organizing map (SOM) is employed to learn a topologically adapted codebook instead of the well-known k-means algorithm. In addition, a new non-negative sparse coding technique for SOM is also proposed and tested. The new sparse coding method utilizes Non-negative Least Squares (NNLS) optimization to best reconstruct every input pattern using K-best-matching codewords. Experimental results using Caltech-101 database show the effectiveness of the proposed method compared with other state-of-the-art methods.

Keywords: Image Classification, Sparse Coding, Self-organizing Map.

1 Introduction

The human visual system can efficiently detect and identify objects within cluttered scenes [6]. However, for artificial systems this is still a difficult problem due to the high in-class variability of many object types. Unsupervised feature learning can fill this gap due to its learning capability to efficiently represent images. Recently, state-of-the-art image classification systems contain two parts: bag-of-features (BoF) [11] and spatial pyramid matching (SPM) [15]. These methods either employ local image patches or handcrafted descriptors such as SIFT [16] or HOG [12], and encode these features into an over-completed representation using various algorithms such as k-means [7] or sparse coding [19]. After coding, global image representations are fed into linear [18,19] or non-linear [15]

A.E. Hassanien et al. (Eds.): AMLTA 2014, CCIS 488, pp. 321–330, 2014.

classifiers. However, linear classifier is most preferable due to its computation efficiency.

Although k-means is a simple algorithm to learn the dictionary from training data, it is sensitive to initialization and data outliers. However, self-organizing map (SOM) learning algorithm overcomes these two problems through its efficient neighborhood update strategy. The main contribution in this paper is to analyze the performance of the topological dictionary learning method using self-organizing map learning algorithm as a variant for the classical non-topological feature learning methods (e.g. k-means). Recently, a new modification in SOM is proposed in [14] to accurately approximate the input pattern to be represented by a linear mixture of neurons which is computed by a least-squares fitting procedure where the coefficients in the linear mixture of models are constrained to nonnegative values. Based on this method , we propose a new coding scheme which use only a few best-matching neurons (K-neurons) instead of using all neurons to reconstruct input patterns.

In this work, we also examine the combination of the learned topographic dictionary using Dot-product SOM with various encoding schemes. The dictionary learned by SOM has the advantage that neuron centers near to each other extract similar features while far neurons extract different features. The topographic order of the neurons helps to find a good sparse representation for each input pattern. Three various encoding schemes which map the input pattern to a feature vector are investigated. The first one is the hard encoding scheme which use the index of the best matching neuron as a feature. Secondly, soft threshold encoding is employed by thresholding the similarity between the input pattern and all neuron centers. Third, non-negative sparse coding which tries to explain the input pattern as a linear combination of a small number of neuron centers is employed to find the sparse non-negative feature codes.

The paper is organized as follows, section 2 describes the works related to the proposed method. Section 3 explains the proposed system and the details of the learning algorithm used for topological dictionary learning. Section 4 explains the details of various encoding schemes and pooling operations. Section 5 shows the experimental setup and results using Caltech-101 image databases. Finally, conclusions are described in section 6.

2   Related Works

Designing a good image features is very important for the success of many category-level image classification problems. Many methods first extract low-level descriptors such as scale-invariant feature transform (SIFT) [16] and histogram of oriented gradients (HOG) [12] at interest point locations. In this paper, we consider the problem of learning good image representations from these features which is suitable for classification using linear support vector machine classifier. Extracting such high-level features involves two alternative operations of coding and pooling. Successful methods employ SIFT features to learn good image representations are: spatial pyramid matching [15], sparse coding [19], Locality-constrained Linear Coding (LLC) [18] and soft-thresholding k-means [8]. Theses methods either apply k-means or sparse coding methods to learn dictionary from SIFT features.

Dictionary learning algorithms have been proposed to find a set of basis vectors that quantize/reconstruct local image patches or low-level features (SIFT) [8,9]. Feature learning algorithms are often employ k-means for dictionary learning and constructing heigher level image representations. In [15], k-means used to train a dictionary from low-level descriptors (SIFT) that are then used to define an encoding of the descriptor into a new feature space. Recently, new powerful algorithms have proposed to generate better dictionary using sparse coding [19]. Although dictionary produced by sparse coding is more efficient than that generated by k-means algorithm, the heavy computation of this algorithm make it not scalable for large data sets. In addition, k-means algorithm suffers from sensitivity to initialization and data outliers. SOM can be considered as another powerful substitution for k-means to learn an efficient dictionary. Due to its neighborhood update strategy, SOM overcomes the initialization and outliers problems exists in k-means algorithm.

Various encoding methods have been proposed to map the original data to a high-dimensional space that emphasis locality [18] or sparsity [17, 19]. The relationship between encoding and dictionary learning have been studied in [8], which emphasis that selecting the appropriate encoding scheme is much important than dictionary learning. The problem of feature encoding has become a topic of growing interest in recent years. In this work, we investigate the performance of three different encoding schemes: hard quantization, soft thresholding and non-negative sparse coding. In addition, the computation of SOM is very fast and does not require solving any special optimization problem.

3   Proposed System

In the proposed system, low-level features of the input image are extracted using dense-SIFT algorithm [16]. The 128-dimensional SIFT features are computed at regular grid locations of the input image. As shown in Fig. 1, the collected local feature vectors are used to learn higher level dictionary using self-organizing map. The trained SOM is then employed to encode each SIFT local feature. The encoded features are spatially pooled at three different spatial levels to efficiently represent the input image. The concatenated feature vector is finally classified using support vector machine classifier with linear kernel.

3.1   Self-organizing Map

The SOM [14] is an unsupervised learning algorithm which learns the distribution of a set of patterns without any class information. A pattern is projected from an input space to a position in the map (information is coded as the location of an activated node). SOM is previously employed to learn a set of topographic hierarchical features and can be used as an invariant feature extractor to solve image recognition problem [2, 3]. Recently, an efficient learning algorithm [1]

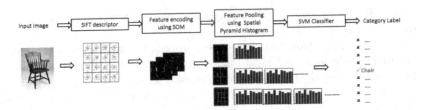

Fig. 1. Proposed image classification system using self-organizing map

based on SOM is proposed to learn invariant local feature representation. The SOM is unlike most clustering techniques (k-means) in that it provides a topological ordering of the features. Similarity in input patterns is preserved in the output of the process. The topological preservation of the SOM process makes it especially useful to find an efficient sparse code representation for each input pattern.

3.2 Dot-Product SOM Learning

Most dictionary learning algorithms [10] employ dot-product similarity measure instead of Euclidean distance for its computational efficiency. A practical computation of the SOM is based on their dot products. For Euclidean vectors this method is particularly advantageous, if there are plenty of zero elements in the vectors, because they are skipped in the evaluation of similarities. However, the neuron centers m_j, for their comparison with the input x_i must be kept normalized to constant length all the time.

Batch learning of SOM is preferred than sequential learning because it is faster and safer [14] (it is not affected by data outliers). However for huge image data, batch learning is not practical. A modified stochastic mini-batch learning algorithm is proposed in [1] to learn neuron centers using randomly selected patterns.

Assume training data $X^{N \times D} = \{x_i | x_i \in R^D, i = 1, ..., N\}$ where N is the number of samples and d is the data dimension, m_j is neuron center, C is the number of neurons in the map, n is the mini-batch size, α is the learning rate, and σ is the neighborhood radius. For each neuron, there is an associated mean m_j where $j = 1, .., C$.

In the mini-batch learning algorithm [1], the training data is divided into N/n divisions, i.e. $\{X^b, b = 1, ..., N/n\}$. For each sample in the b mini-batch, the competition step using dot-product similarity is applied to find the best matching unit. A subset of the batch is assigned for each neuron according to the competition results. After applying all samples in each mini-batch, mean vectors (neuron centers) for all neurons are updated. Neuron centers are updated in similar way as batch learning of SOM to preserve the topological structure of the data. Moreover, the radius and learning rates are decreased monotonically at each step of the mini-batch training. The following three main steps are repeated for each mini-batch.

Step 1: Neuron Competition: Each neuron in the SOM network is associated with a mean vector. The winner neuron c can be found by maximizing the dot-product similarities between every input sample $x(t)$ and all neurons in the SOM map.

$$c = \underset{j}{\operatorname{argmax}} \left(x(t)^T m_j(t) \right) \tag{1}$$

Step 2: Network Update: When a batch of samples $X^b(t)$ is randomly selected from the data set X, all winning neurons are computed for each sample using Eq.(1). Means vectors of all neurons $(j = 1, ..., C)$ are updated as follows:

$$m_j(t+1) = \alpha(t)m_j(t) + (1 - \alpha(t)) \frac{\sum_t h_{cj} x(t)}{\sum_t h_{cj}} \tag{2}$$

where $\alpha(t)$ is the learning rate (a monotonically linear decreasing function of t) and $h_{cj}(t)$ is the neighborhood function of the jth neuron given the winning neuron c. According to Eq. (2), the winning neuron and all its spatial neighbors in the grid are modified. Updating the neighbors of the winner neuron is the main difference between k-means and SOM. Neighbor updating helps neuron centers to be more robust to initialization and data outliers. The rates of the modifications at different nodes depend on the mathematical form of the function h_{cj}. A common choice for the neighborhood d $h_{cj}(t)$.

$$h_{cj} = e^{(-\|c-j\|^2)/2\sigma^2(t)} \tag{3}$$

Where j and c represent the index of the j-neuron and the c-winner neuron in the SOM map and their dimensionality depend on the dimension of the SOM map (i.e for 2D maps this value represent a vector of the neuron row and column position in the map). The amount of variations for each neighbor neurons depend on its position with respect to the winner neuron (i.e. near neurons receive large amount of variations than far neurons). $\sigma(t)$ is the neighborhood radius functions where the following formula is used to calculate its value at each iteration t.

$$\sigma(t) = \sigma_i + \frac{t}{T}(\sigma_f - \sigma_i) \tag{4}$$

Where T, σ_i and σ_f are the number of iterations, the initial and the final radius values respectively. The neighborhood radius function $\sigma(t)$ is a monotonically decreasing function of t. The mathematical form of $\sigma(t)$ is given in Eq. (4), The radius value is fairly large at the beginning of training σ_i and gradually decrease to small value σ_f at the end of training. The topological order is developed at the beginning, and the final convergence to nearly optimal values at the end of training. The neighborhood function has very important role in self-organization. However in the application of image representation, the objective is to decrease the total quantization error of the training samples. To achieve this goal, the radius should have relatively moderate value in the beginning (i.e. the initial value is set to be 1) and the final radius value should not be zero (i.e. it can be selected equal 0.01) in order to preserve the topological order of the map.

Step 3: Neuron centers normalization: Since the computation of the SOM is in practice carried out by the mini-batch algorithm, the mapping of all of the input items onto the respective winner nodes (i.e., the associated lists) is made using Eq. 2. The only modification is the normalization of the neuron centers to constant length shall be made after each iteration cycle.

$$m_j(t) = \frac{m_j(t)}{\|m_j(t)\|_2^2}, j = 1, ..., C \qquad (5)$$

4   Feature Encoding and Pooling

In the encoding step, the low-level descriptors (e.g. SIFT or HOG) are mapped to feature vectors. Following similar notations as [4], let I denote an input image which is represented by a set of SIFT descriptors $x_i, i = 1, ..., N$ extracted at N locations. The image is divided into P regions ($1 \times 1, 2 \times 2, 4 \times 4$) of three level spatial pyramid [15]. Let f and g denote the encoding and pooling operation respectively. The vector z representing the whole image obtained by applying coding and pooling operations sequentially. In this work, dot-product SOM learning algorithm is employed to learn C centroids ($m_j, j = 1, ..., C$) from training data. Given the learned centroids, three encoding schemes include hard, soft and non-negative sparse coding are considered for feature mapping f. Since max-pooling is more efficient than average-pooling when considering linear support vector machine for classification [4, 19], we employ max-pooling operation in this work.

4.1   Hard Encoding

Hard coding is the classical encoding for vector quantization methods and SOM which is originally used with bag-of-features [11]. The feature mapping f is the standard 1-of-C, hard assignment coding:

$$f_k(x) = \begin{cases} 1 & \text{if } k = \text{argmax}_j \left(x^T m_j \right) \\ 0 & \text{otherwise} \end{cases} \qquad (6)$$

This is the extremal sparse representation that has been used frequently in computer vision.

4.2   Soft-Threshold Encoding

The second choice for feature mapping is a non-linear mapping [8] which is softer than the previous encoding while also keeping some sparsity.

$$f_k(x) = \max\{0, x^T m_j - \alpha\} \qquad (7)$$

Where α is a fixed threshold controls the degree of sparseness. This mapping function outputs 0 for any feature f_k where its similarity to the centroid

m_j is below fixed threshold α. In practice, this simple thresholding function is appropriate to cases where an input patch is closer to more than one centroid. Soft coding has been shown to improve performance over hard coding [4, 8, 10]. However, soft encoding still tries to match each input pattern to one centroid competitively.

4.3   Non-Negative Sparse Coding

Sparse representation tries to explain the input pattern as a linear combination of a small number of centroids [17]. Recently, Kohonon [14] employed non-negative least squares method to find the linear combination which best approximate each input vector from all neurons of the SOM. In this paper, a modified version of this approximation is proposed, we only use K-best neurons instead of using all neurons to approximate each input pattern. Using K nearest neighbor centroids helps to increase sparseness and reduce ambiguity in representation.

$$f_i = \underset{f}{\operatorname{argmin}} \| M f_i^T - x_i \|_2^2 \quad \text{subject to} \quad f_i \geq 0 \tag{8}$$

Where M is the dictionary comprises the best matching K neuron centers of the input pattern x_i, and f_i is the optimized non-negative sparse codes obtained by solving Eq.8.

4.4   Feature Pooling

Each of the above encoding schemes give a new sparse representation for each input pattern. A pooling operator takes the codes that are located within P regions and summarize them to a single vector of fixed length. The representation for the global image is obtained by concatenating the representations of each region. There are two popular pooling methods used for representation, namely average and max pooling. Average pooling is mostly performed well in combination with histogram intersection non-linear SVM [15], however, max pooling give better performance when combine with linear SVM [19]. Linear SVM is computationally efficient and fast in both training and testing compared with non-linear kernels such as histogram intersection or Chi-square kernels. In this work, we consider only max pooling which is proved to be more suitable in combination with linear classifier. Max pooling computes the maximum of each region:

$$g_{p,j} = \max_{i \in N_p} f_{i,j}, \text{for} \quad p = 1, ..., P \tag{9}$$

Where N_p denotes the set of locations/indices within region p. The vector z represents the whole image is obtained by sequentially coding and pooling over all regions, and then concatenating all features.

$$h_m = g(f(x_i)_{i \in N_p}), p = 1, ..., P \tag{10}$$

$$z^T = [h_1^T....h_P^T] \tag{11}$$

By applying the proposed feature extraction method on a set of labeled data set, these features vectors are employed to train a multi-class linear support vector machine classifier [5].

5   Experimental Results

In this section, the performance of SOM using various encoding schemes is examined using the challenged Caltech-101 dataset.

5.1   Caltech-101 Dataset

The Caltech-101 dataset [13] contains 9144 images in 101 classes including animals, vehicles, flowers , etc, with significant variations in shape, viewpoint and illumination conditions. The number of images per category varies from 31 to 800. Most images have medium resolution (i.e. around 300×300 pixels). Caltech-101 is probably one of the most diverse available databases for image classification. Most images have a small clutter, and objects are centered and occupy most of the image. As suggested by the original creator of the dataset and by other researchers, the whole dataset is partitioned into 30 training images per class and testing on the rest of images (using no more than 50 images per class).

5.2   Comparison with State-of-the-Art Methods

In this experiment, the performance is meaured using average accuracy over 102 classes (i.e. 101 classes and another "background" class) with 30 training images per class. Dot-product SOM learning algorithm is employed to learn a codebook of 512 neurons. The number of the K best matching units used in non-negative sparse coding is chosen empirically to be 5. Classification is performed using linear spatial pyramid matching of $4 \times 4, 2 \times 2$ and 1×1 subregions and linear support vector machine. All images were resized to be no longer that 300×300 pixels with preserved aspect ratio. Results are compared with other state-of-the-art methods, some methods are based on a non-topological dictionary learning method like k-means algorithm [8, 15] and others used dictionary learned by sparse coding criteria [18, 19].

The obtained results shown in Table (5.2) indicate that the proposed sparse SOM method has a comparable performance with other dictionary learning methods based on sparse coding optimization. However, calculating sparse SOM codes does not require solving any special optimization technique as compared with other sparse coding methods. As expected, the proposed sparse SOM representation performs better than hard and soft encoding because of the richer information obtained from the non-negative reconstruction of each input pattern. The Locality-constrained Linear Coding (LLC) [18] method has better performance than the proposed method because it use multi-scale SIFT features and big dictionary size (2048 basis). The proposed method gives almost similar results compared with sparse coding k-means [8] using only half dictionary size. We note that it is often possible to achieve better performance simply by using large dictionary size. This is can be easily achieved with sparse SOM and because solving non-negative least square optimization does not depends on the dictionary size rather it depends on the number of K best matching units.

Table 1. Classification rate for Caltech database

Algorithm	Dictionary Size	Classification rate(%)
spatial pyramid matching (SPM) [15]	200	64.60
Linear SPM using sparse coding [19]	1024	73.20
Locality-constrained Linear Coding (LLC) [18]	2048	73.44
soft-threshold k-means [8]	1024	63.20
Sparse coding k-means [8]	1024	71.9
Hard SOM	512	65.34
soft threshold SOM	512	66.14
Non-negative Sparse coding SOM	512	70.4

6   Conclusion and Future Works

In this paper, a topological dictionary learning method using SOM is introduced to address the problem of image representation. Low-level feature are computed using dense SIFT, while high-level features are learned and approximated with neuron centers of the self-organizing map (SOM). Dictionary learning using SOM shows better robustness to initialization and outliers compared with a non-topological feature learning method (k-means). The proposed sparse SOM not only perform better than other encoding schemes like hard and soft coding but also it does not require solving any special optimization technique compared with sparse coding based methods. In future work, different problems related to image classifications such as face recognition and handwritten digit recognition will be examined.

References

1. Aly, S.: Learning invariant local image descriptor using convolutional mahalanobis self-organizing map. Neurocomputing 142, 239–247 (2014)
2. Aly, S., Shimada, A., Tsuruta, N., Taniguchi, R.I.: Robust face recognition using multiple self-organized gabor features and local similarity matching. In: 2010 20th International Conference on Pattern Recognition (ICPR), pp. 2909–2912. IEEE (2010)

3. Aly, S., Tsuruta, N., Taniguchi, R.I., Shimada, A.: Visual feature extraction using variable map-dimension hypercolumn model. In: IEEE International Joint Conference on Neural Networks, IJCNN 2008 (IEEE World Congress on Computational Intelligence), pp. 845–851. IEEE (2008)

4. Boureau, Y.L., Bach, F., LeCun, Y., Ponce, J.: Learning mid-level features for recognition. In: 2010 IEEE Conference on Computer Vision and Pattern Recognition (CVPR), pp. 2559–2566. IEEE (2010)

5. Chang, C.C., Lin, C.J.: Libsvm: a library for support vector machines. ACM Transactions on Intelligent Systems and Technology (TIST) 2(3), 27 (2011)

6. Cireşan, D., Meier, U., Masci, J., Schmidhuber, J.: Multi-column deep neural network for traffic sign classification. Neural Networks 32, 333–338 (2012)

7. Coates, A., Karpathy, A., Ng, A.Y.: Emergence of object-selective features in unsupervised feature learning. In: NIPS, vol. 25, pp. 2690–2698 (2012)

8. Coates, A., Ng, A.Y.: The importance of encoding versus training with sparse coding and vector quantization. In: Proceedings of the 28th International Conference on Machine Learning (ICML 2011), pp. 921–928 (2011)

9. Coates, A., Ng, A.Y.: Learning feature representations with k-means. In: Montavon, G., Orr, G.B., Müller, K.-R. (eds.) NN: Tricks of the Trade, 2nd edn. LNCS, vol. 7700, pp. 561–580. Springer, Heidelberg (2012)

10. Coates, A., Ng, A.Y., Lee, H.: An analysis of single-layer networks in unsupervised feature learning. In: International Conference on Artificial Intelligence and Statistics, pp. 215–223 (2011)

11. Csurka, G., Dance, C., Fan, L., Willamowski, J., Bray, C.: Visual categorization with bags of keypoints. In: Workshop on Statistical Learning in Computer Vision, ECCV, vol. 1, pp. 1–2 (2004)

12. Dalal, N., Triggs, B.: Histograms of oriented gradients for human detection. In: IEEE Computer Society Conference on Computer Vision and Pattern Recognition, CVPR 2005, vol. 1, pp. 886–893. IEEE (2005)

13. Fei-Fei, L., Fergus, R., Perona, P.: Learning generative visual models from few training examples: An incremental bayesian approach tested on 101 object categories. Computer Vision and Image Understanding 106(1), 59–70 (2007)

14. Kohonen, T.: Essentials of the self-organizing map. Neural Networks 37(0), 52–65 (2013)

15. Lazebnik, S., Schmid, C., Ponce, J.: Beyond bags of features: Spatial pyramid matching for recognizing natural scene categories. In: 2006 IEEE Computer Society Conference on Computer Vision and Pattern Recognition, vol. 2, pp. 2169–2178. IEEE (2006)

16. Lowe, D.G.: Distinctive image features from scale-invariant keypoints. International Journal of Computer Vision 60(2), 91–110 (2004)

17. Olshausen, B.A., Field, D.J.: Sparse coding with an overcomplete basis set: A strategy employed by v1? Vision Research 37(23), 3311–3325 (1997)

18. Wang, J., Yang, J., Yu, K., Lv, F., Huang, T., Gong, Y.: Locality-constrained linear coding for image classification. In: 2010 IEEE Conference on Computer Vision and Pattern Recognition (CVPR), pp. 3360–3367. IEEE (2010)

19. Yang, J., Yu, K., Gong, Y., Huang, T.: Linear spatial pyramid matching using sparse coding for image classification. In: IEEE Conference on Computer Vision and Pattern Recognition, CVPR 2009, pp. 1794–1801. IEEE (2009)

PSilhOuette: Towards an Optimal Number of Clusters Using a Nested Particle Swarm Approach for Liver CT Image Segmentation

Abder-Rahman Ali[1], Micael S. Couceiro[2,3], and Aboul Ella Hassenian[1,4]

[1] Scientific Research Group in Egypt (SRGE), Egypt
[2] Artificial Perception for Intelligent Systems and Robotics (AP4ISR),
Institute of Systems and Robotics (ISR), University of Coimbra, Pinhal de Marrocos,
Polo II, 3030-290, Coimbra, Portugal
[3] Ingeniarius, Lda., Rua da Vacaria, n.37, 3050-381, Mealhada, Portugal
[4] Faculty of Computers and Information, Computer Science Department,
Cairo University, Egypt
abder-rahman.a.ali@ieee.org, micaelcouceiro@isr.uc.pt, micael@isec.pt,
aboitcairo@gmail.com

Abstract. This paper proposes a nested particle swarm optimization (PSO) method to find the optimal number of clusters for segmenting a grayscale image. The proposed approach, herein denoted as PSilhOuette, comprises two hierarchically divided PSOs to solve two dependent problems: i) to find the most adequate number of clusters considering the silhouette index as a measure of similarity; and ii) to segment the image using the Fuzzy C-Means (FCM) approach with the number of clusters previously retrieved. Experimental results show that parent particles converge towards maximizing the silhouette value while, at the same time, child particles strive to minimize the FCM objective function.

Keywords: segmentation, fuzzy c-means, CT.

1 Introduction

Image segmentation is the process of subdividing the image into its constituent parts, and is considered one of the most difficult tasks in image processing [1]. Medical images, in their raw form, are represented by arrays of numbers in the computer, with the numbers indicating the values of relevant physical quantities that show contrast between different types of body tissue. Segmentation results in medical images make it possible for shape analysis, detecting volume change, and making a precise radiation therapy treatment plan. However, despite the intensive research, segmentation remains a challenging problem due to the diverse image content, cluttered objects, image noise, non-uniform image texture, etc [2].

Spiral computed tomography (CT) has rapidly gained acceptance as the preferred CT technique for routine liver evaluation because it provides image acquisition at peak enhancement of the liver parenchyma during a single breath hold [3].

A.E. Hassanien et al. (Eds.): AMLTA 2014, CCIS 488, pp. 331–343, 2014.
© Springer International Publishing Switzerland 2014

Researchers have been working towards providing a diagnostic support of liver diseases, liver volume measurements, and 3D liver volume rendering, without the need of any manual process and visual inspection, which requires mental work and huge time consuming processes. Image segmentation has been one of the many image processing methods employed on that particular task. Nevertheless, still many challenges remain before one can provide a fully autonomous image segmentation method of liver CT images. The physical attributes of the liver, combined with the limitations inherent to CT technology, provide low-level contrast and blurry edged images. Also, other organs in the vicinities, like the spleen and stomach, share similar gray levels, making it even harder to clearly identify the liver [4].

In this paper, we propose a nested Particle Swarm Optimization (PSO) approach, hierarchically divided into two problems: *i)* finding the most adequate number of clusters considering the silhouette index as a measure of similarity; *ii)* segment the image using the Fuzzy C-Means (FCM) approach with the number of clusters previously retrieved. In other words, the overall process will comprise two PSO-based approaches divided into different levels. Within the higher-level PSO, herein denoted as PSO-NC, particles will have the main objective of finding the most adequate number of clusters (NC). An approach that used Fuzzy C-Means with silhouette was presented in [19], however, silhouette in that study was used as a measure of normalized dissimilarity, such that average silhouette width was used as a threshold that is utilized to decide whether or not to recompute a new set of membership functions with their corresponding new cluster centers, or to stop the process. Thus, the approach proposed in [19] does not change the number of clusters used with an image, in contrary to what we propose in this paper.

The paper is organized as follows: Section 2 describes Fuzzy C-Means Clustering; Sections 3 discusses the optimum number of clusters issue; Section 4 presents the proposed approach; Results and evaluation are shown in Section 5; and the paper is concluded in Section 6.

2   Fuzzy C-Means Clustering

The automatic selection of a robust optimal method remains a challenge in segmentation of medical imaging. Most of the existing image segmentation techniques comprise on threshold, regional, edge detection and clustering methods. Within the class of clustering methods, the Fuzzy C-Means (FCM) clustering algorithm is perhaps the most widely used in image segmentation, due to its overall performance [5,20].

Clustering is the process of partitioning a data set into different classes, such that the data in each class share same common features according to a defined distance measure (*i.e.* Euclidean distance). The standard crisp C-Means clustering scheme is very popular in the field of pattern recognition [6]. However, this scheme uses hard partitioning, in which each data point belongs to exactly one class. Fuzzy C-Means (FCM) is a generalization of the standard crisp c-means scheme, in which a data point can belong to all classes with different degrees of membership [7].

FCM [16-18] is an unsupervised clustering algorithm. It divides n vectors into c fuzzy partitions, calculates the clustering center to each group, and minimizes the non-similarity index value function. With fuzzy partitions, elements of the membership matrix are allowed to have values between 0 and 1. After normalizing, the combined membership of a dataset would be as follows [8]:

$$\sum_{i=1}^{c} \mathbf{u}_{ij} = 1, \forall_j = 1, 2, , ..., n \tag{1}$$

Let $X = \{x_1, ..., x_b, ..., x_n\}$ be a set of n objects, and $V = \{v_1, ..., v_b, ..., v_c\}$ be a set of c centroids in a p-dimensional feature space. The Fuzzy C-Means partitions X into c clusters by minimizing the following objective function [9]:

$$J = \sum_{j=1}^{n} \sum_{i=1}^{c} (\mathbf{u}_{ij})^m \left\| \mathbf{x}_j - \mathbf{v}_i \right\|^2 \tag{2}$$

where $1 \leq m \leq \infty$ is the *fuzzifier*, $\mathbf{v}_i$ is the ith centroid corresponding to cluster β_i, $\mathbf{u}_{ij} \in [0, 1]$ is the fuzzy membership of the pattern $\mathbf{x}_j$ to cluster β_i, and $\|.\|$ is the distance norm such that,

$$\mathbf{v}_i = \frac{1}{n_i} \sum_{j=1}^{n} (\mathbf{u}_{ij})^m x_j \quad where \quad n_i = \sum_{j=1}^{n} (\mathbf{u}_{ij})^m \tag{3}$$

and,

$$\mathbf{u}_{ij} = \frac{1}{\sum_{k=1}^{c} \left(\frac{d_{ij}}{d_{kj}} \right)^{\frac{2}{m-1}}} \quad where \quad d_{ij}^2 = \left\| \mathbf{x}_j - \mathbf{v}_i \right\|^2 \tag{4}$$

FCM starts by randomly choosing c objects as centroids (means) of the c clusters. Memberships are calculated based on the relative distance (*i.e.* Euclidean distance) of the object $\mathbf{x}_j$ to the centroids using Eq. (4). After the memberships of all objects have been found, the centroids of the clusters are calculated using Eq. (3) [9].. The process stops when $max_{ij} \left\| u_{ij}^{(k+1)} - u_{ij}^{(k)} \right\| < \varepsilon$, where ε is a termination criterion between 0 and 1, whereas k are the iteration steps. This process converges to a local minimum of J [22].

3   Optimum Number of Clusters

Most clustering algorithms require the user to specify a priori the number of clusters [10]. However, the cluster number remains an open challenge to answer, varying from a given dataset to another. Choosing a too large cluster number causes the separation of similar points. On the other hand, choosing a too small cluster number causes the grouping of dissimilar points into the same cluster. The partitions of interest are those composed of compact and well separated clusters that best meet our expectations [11].

3.1 Silhouette Based Cluster Validity

The notion of silhouette as a measure of clustering quality was first introduced by Rousseeuw [12]. The measure is evaluated at each data point, and the clustering quality of one cluster is measured by taking the average silhouette over the member points. Silhouettes can be evaluated on fuzzy partitions by carrying out a defuzzification of the membership matrix. However, it is possible for different fuzzy partitions on the same dataset to evaluate to the same silhouette values, provided that they are converted to the same crisp partition by means of defuzzification. Such fuzzy partitions can be obtained on one dataset by slightly changing the fuzzifier m when performing FCM clustering [11].

3.2 Generalized Intra-Inter Silhouettes

A natural generalization of the construction of silhouettes to fuzzy partitions is given in [22]. The rationale behind this generalization is based on a distance view of the clustering problem and the problem of cluster validity. A crisp clustering of the data points is essentially a clustering of the associated pairwise distances into intra-distances and inter-distances. This clustering can be modeled by associating two scores to each distance, *i.e.* intra-score and inter-score that only assume the values 0 and 1. A distance becomes intra-distance, hence intra-score of 1 and inter-score of 0, if the end points are assigned to the same cluster and inter-distance otherwise. Evaluating fuzzy logical connectives on point (pixel) membership values computes the intra- and inter-scores, either crisp or fuzzy, as shown in the definition below [11]:

Definition: Let $d_j k = d(x_j, x_k)$ denote the distance between the data points x_j and x_k; $1 \leq j \neq k \leq n$. Let $\mathbf{u}_i$ denote a cluster, and $\mathbf{u}_{ij}$ be the membership of $\mathbf{x}_j$ to cluster $\mathbf{u}_i$; $1 \leq i \leq c$. The intra-score for d_{jk} with respect to cluster $\mathbf{u}_i$ is defined as [11]:

$$intra_i(d_{jk}) = (\mathbf{u}_{ij} \wedge \mathbf{u}_{ik}) \tag{5}$$

The inter-score for d_{jk} with respect to cluster $\mathbf{u}_r$ and $\mathbf{u}_s$; $1 \leq r < s \leq c$, is defined as [11]:

$$inter_{rs}(d_{jk}) = (\mathbf{u}_{rj} \wedge \mathbf{u}_{sk}) \vee (\mathbf{u}_{sj} \wedge \mathbf{u}_{rk}) \tag{6}$$

The compactness distance a_j and the separation distance b_j of each data point $\mathbf{x}_j$ are defined as weighted means of the associated pairwise distances using [11]:

$$a_j = min \left\{ \frac{\sum_{k=1}^{n} intra_i \left(j, k\right) . d_{jk}}{\sum_{k=1}^{n} inter_{rs} \left(j, k\right)} ; \sum_{k=1}^{n} intra_i(j, k) > 0, 1 \le i \le c \right\} \quad (7)$$

$$b_j = min \left\{ \frac{\sum_{k=1}^{n} inter_{rs} \left(j, k\right) . d_{jk}}{\sum_{k=1}^{n} inter_{rs} \left(j, k\right)} ; \sum_{k=1}^{n} inter_{rs}(j, k) > 0, 1 \le r < s \le c \right\} \quad (8)$$

Here, the silhouette s_j of $\mathbf{x}_j$ can be computed using [11]:

$$s_j = \frac{b_j - a_j}{max\{a_j, b_j\}} \quad (9)$$

Eq.9 evaluates to $[-1, +1]$. The difference $(b_j - a_j)$ indicates good clustering of $\mathbf{x}_j$ if positive, and poor clustering if negative. Values around zero either emerge from mis-clustering or the point being in overlapping regions. The average generalized silhouette *gSil* computed using *Eq.10* over all the data set X having n points, is a measure of the dataset clustering quality [11].

$$\mathbf{Sil}(x) = \frac{\sum_{j=1}^{n} s_j}{n} \quad (10)$$

gSil returns a vector of silhouette values, one value for each data point (pixel). If one point has a silhouette value near *1*, then its clustering is very good *i.e.* the clustering algorithm has successfully grouped the point with similar points in the same cluster. In contrast, if the silhouette is near *-1*, then the clustering of the point is very bad. A silhouette value of *0* indicates an intermediate case.

Therefore, each silhouette is considered a measure of the clustering quality of the associated point; the higher the silhouette value, the better the clustering of the point is.

4   PSilhOuette Approach

The *Particle Swarm Optimization (PSO)* algorithm has been one of the most widely used stochastic optimization methods since it was first proposed by Kennedy and Eberhart in 1995 [23]. The PSO algorithm is a biologically inspired technique derived from the collective behavior of birds flocks. The stochastic optimization ability of the algorithm is enhanced due to its cooperative simplistic mechanism, wherein each particle presents itself as a possible solution of the problem (*e.g.* the best cluster centers or the most ideal number of clusters of a given segmented image). These particles travel through the search space to find an optimal solution, by interacting and sharing information with other particles, namely their individual best solution (local best) and computing the global best.

In each step t of the PSO, a given fitness function (Eq.(2)) is used to evaluate the particles success. To model the swarm, each particle n moves in a multidimensional space according to the position $\mathbf{x}_n[t]$, and velocity $\mathbf{v}_n[t]$, which are highly dependent on local best $\mathbf{x}_n^{\sim}[t]$ and global best $\mathbf{g}_n^{\sim}[t]$ information:

$$\mathbf{v}_n[t+1] = w\mathbf{v}_n[t] + \rho_1\mathbf{r}_1\left(\mathbf{g}_n^{\sim} - \mathbf{x}_n\right) + \rho_2 r\mathbf{r}\left(\mathbf{x}_n^{\sim} - \mathbf{x}_n[t]\right) \tag{11}$$

$$\mathbf{x}_n[t+1] = \mathbf{x}_n[t] + \mathbf{v}_n[t+1] \tag{12}$$

Coefficients ρ_1 and ρ_2 are assigned weights, which control the inertial influence of the *globally best* and the *locally best*, respectively, when the new velocity is determined. Typically, ρ_1 and ρ_2 are constant integer values, which represent *cognitive* and *social* components with $\rho_1 + \rho_2 < 2$ [24]. However, different results can be obtained by assigning different influences for each component.

The parameter w will weigh the inertial property of particles on determining a new velocity, $0 < w < 1$. With a small w, particles ignore their previous activities, thus ignoring the system dynamics and being susceptible to get stuck in local solutions (*i.e.* exploitation behavior). On the other hand, with a large w, particles will present a more diversified behavior, which allows exploration of new solutions and improves the long-term performance (*i.e.* exploration behavior). However, if the exploration level is too high, then the algorithm may take too much time to find the global solution.

The parameters r_1 and r_2 are random vectors with each component generally a uniform random number between 0 and 1. The intent is to multiply a new random component per velocity dimension, rather than multiplying the same component with each particles velocity dimension.

In brief, the PSO can be used to solve many different problems. In this work, we propose a nested PSO approach hierarchically divided into two problems: *i)* finding the most adequate number of clusters considering the silhouette index as a measure of similarity; *ii)* segment the image using the Fuzzy C-Means (FCM) approach with the number of clusters previously retrieved. In other words, the overall process will comprise two PSO-based approaches divided into different levels. Within the higher-level PSO, herein denoted as PSO-NC, particles will have the main objective of finding the most adequate number of clusters (NC). Each PSO-NC particle, on the other hand, will be the parent of a whole lower-level swarm, herein denoted as PSO-FCM. In the PSO-FCM, each children particle will have the main objective of segmented the image based on the number of clusters provided by the parent swarm. The flowchart depicted in Fig. 1 summarizes the proposed approach, denoted *PSilhOuette*.

Given that this is a hierarchical and inter-dependent process, we will start by presenting the problem addressed to the lower-level swarm of particles; the PSO-FCM.

4.1   PSO-FCM

Here, the velocity and position dimensions of each particle corresponds to the total number of desired cluster centers of the image, *i.e.* $dim\ v_n[t] = dim\ x_n[t] = C$.

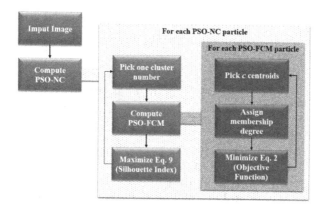

Fig. 1. Optimization of the average silhouette value over the number of trials

In other words, each particles position is represented as a C-dimension vector. Moreover, each particle moves in a multidimensional space according to position $x_n[t]$ from the discrete time system (1), wherein $x_n[t] \in \mathbb{R}^C \wedge 0...0^T \leq x_n(L-1) \times 1...1^T$.

As particles move in that multidimensional space C, they share their own solution to their teammates (*i.e.* other particles inside the same swarm) using the fitness function presented in (2). To analyze the general state of the swarm, the fitness of all particles is evaluated and the collective and individual best positions of each of the particles are updated.

It is noteworthy that, in this problem, the dimension of the space, C, directly depends on the solution retrieved by the parent particle of the PSO-NC.

The algorithm for this approach is as follows:

PSO-FCM()
1 **Initialize** w, ρ_1, ρ_2 / * *internal coefficient, global and local weights* * /
2 **Initialize** n / * *initial number of particles within the population* * /
3 **Initialize** Δv / * *maximum number of levels a particle can travel*
4 *between iterations* * /
5 **Initialize** I_τ / * *total number of iterations* * /
6 **Initialize** $[0...0]^T \leq x_n[0] \leq (L-1) \times [1...1]^T$ / * *randomly initialize*
7 *the cluster centroids, i.e, position of particles* * /
8 **Initialize** $x_n^\sim, g_n^\sim$ *based on* $x_n[0]$ / * *initial local best and global best*
9 *positions* * /
10 **Initialize** $J_n^{best}, J_\tau^{best}$ *based on* $x_n^\sim, g_n^\sim$ / * *initial local best and global*
11 *best solution* * /
12 **for** t *until* I_τ / * *main loop* * /
13   **do for** *each particle* n
14   **do**
15            $v_n[t+1] = w v_n[t] + \rho_1 r_1 (g_n^\sim - x_n[t]) + \rho_2 r_2 (x_n^\sim - x_n[t]),$
16            $|v_n[t+1]| \leq \Delta v$

$17 \qquad \mathbf{x}_n[t+1] = \mathbf{x}_n[t] + \mathbf{v}_n[t+1], [0 \ldots 0]^\tau \le \mathbf{x}_n[t+1] \le (L-1)\times$

$18 \qquad [1 \ldots 1]^\tau$

$19 \qquad / * \; compute \; \mathbf{J}_n[t+1]^n \; based \; on \; the \; vector \; of \; clusters \; defined$

$20 \qquad by \; \mathbf{x}_n[t+1] \; * /$

$21 \qquad \mathbf{J}_n[t+1] = \sum_{j=1}^{n} \sum_{i=1}^{c} (\mathbf{u}_{ij})^m \, \|\mathbf{X}_j - (\mathbf{x_n}[t+1])_i\|^2$

$22 \qquad (\mathbf{x_n}[t+1])_i = \frac{1}{n_i} \sum_{j=1}^{n} (\mathbf{u}_{ij})^m \, \mathbf{X}_j \quad where \quad n_i = \sum_{j=1}^{n} (\mathbf{u}_{ij})^m$

$23 \qquad \mathbf{u}_{ij} = \frac{1}{\sum_{k=1}^{c} \left(\frac{d_{ij}}{d_{kj}}\right)^{\frac{2}{m-1}}} \; where \; d_{ij}^2 = \|\mathbf{X}_j - (\mathbf{x_n}[t+1])_i\|^2$

24

$25 \qquad \textbf{if } \mathbf{J}_n[t+1] < J_n^{best} \; / * particle \; n \; has \; improved \; * /$

$26 \qquad \textbf{then } J_n^{best} = \mathbf{J}_n[t+1]$

$27 \qquad \mathbf{x}_n^{\sim} = \mathbf{x}_n[t+1]$

$28 \qquad \textbf{if } \mathbf{J}_n[t+1] < J_\tau^{best} \; / * swarm \; has \; improved \; * /$

$29 \qquad \textbf{then } J_\tau^{best} = \mathbf{J}_n[t+1]$

$30 \qquad \mathbf{g}_n^{\sim} = \mathbf{x}_n[t+1]$

4.2 PSO-NC

In our previous work [27], this problem was solved by iteratively increasing the number of clusters. This may be a thorough process if one does not have a priori information about the number of clusters from segmented images of a given dataset usually have. Therefore, this problem will once again benefit from PSOs low CPU processing requirements, in which the velocity and position dimensions of each particle corresponds to the dimension of a single scalar value, i.e. $dim \; v_n[t] = dim \; x_n[t] = 1$. In other words, each particles position is represented as a single scalar value, moving in a one-dimensional space (i.e. a line) according to position $x_n[t]$ from the discrete time system (1), wherein $x_n \; t \in \mathbb{N} \, \Lambda \, 2 \le x_n \le L$.

As particles move in that one-dimensional space, they share their own solution to their teammates (i.e. other particles inside the same swarm) using the fitness function presented in Eq.(2). To analyze the general state of the higher-level swarm, the fitness of all parent particles is evaluated and the collective and individual best positions of each of the parent particles are updated.

It is noteworthy that, in this problem, the fitness of each parent particle depends on the overall best solution of each of its children particles from the lower-level PSO-FCM.

For calculating the average silhouette value, 30 trials were being run on each image, and the average of the average silhouette values of the 30 trials was taken as a representation of the average silhouette value. As for the number of clusters, the median of the optimum number of clusters for the 30 trials was being taken to represent the optimum number of clusters. Wherever the median resulted in a double value, this was converted to an integer value.

5 Results and Discussion

5.1 Abdominal CT Data Collection

CT scanning is a diagnostic imaging procedure that uses X-rays in order to present cross-sectional images (slices) of the body. The proposed system will be applied on a complex dataset. The dataset is divided into seven categories, depending on the tumor type: Benign (Cyst (CY), Hemangioma (HG), Hepatic Adenoma (HA), and Focal Nodular Hyperplasia (FNH)); or Malignant (hepatocellular carcinoma (HCC), Cholangiocarcinoma (CC), and Metastases (MS)). Each of these categories have more than 15-patients; each patient has more than one hundred slices, and more than one phase of CT scan (arterial, delayed, portal venous, non-contrast). The dataset includes a diagnosis report for each patient. All images are in JPEG format, selected from a DICOM file, and have dimensions of 630630, with horizontal and vertical resolution of 72 DPI, and bit depth of 24 bits [13]. All CT images were captured from Radiopaedia[1].

5.2 PSilhOuette Performance

The proposed approach was tested on *10* liver CT images. The range of clusters c being tested was *2-10*. Based on [25], when $c = 3$, the liver, marrow, and the spleen merge to form a cluster, and the other regions merge to form a single cluster. When $5 \leq c \leq 7$, the segmentation results improve because the blood vessels and stomach are visualized. However, the liver and spleen are classified into the same cluster. When $c = 8$, the tumors are visualized and almost all of the regions are correctly identified. When $c - 9$, the same segmentation as when $c = 8$ is obtained. Increasing the number of clusters above the aforementioned values may result in poor segmentation since the actual number of tissues present in the data is less than the specified number of clusters.

However, the optimum number of clusters for a given dataset is usually not known as a priori. It is thus advantageous that if this number can be determined based on the given dataset [26].

As one may observe, the PSilhOuette, although stochastic, results in a stable final average silhouette value regardless of the image, as the STD is small (near 0) in most situations. On the other hand, the CPU processing time significantly varies along trials. This phenomenon is better explained in Fig. 2, in which the evolution of the best average silhouette value over each iteration of the algorithm is represented. Note that the best silhouette value corresponds to the one retrieved by the best particle, *i.e.*, the particle presenting the number of clusters that result in a higher average silhouette value from the remaining swarm.

Fig. 2 also clearly depicts that, although the progression of the average silhouette value is mostly monotonically increasing, one can observe some stagnation in certain situations (*i.e.*, when the average silhouette value does not improve

---

[1] http://radiopaedia.org/search?q=CTscope=all

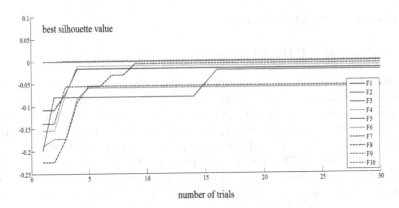

Fig. 2. Optimization of the average silhouette value over the number of trials

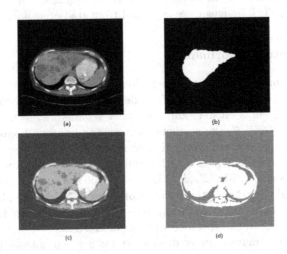

Fig. 3. (a) original image; (b) ground truth; (c) PSO-FCM result using 4-clusters (recommended); (d) result of PSO-FCM using 2-clusters

over a number of iterations). Under these situations, particles within PSilhOuette are unable to converge, and the CPU processing time decreases since most number of clusters found by them were already evaluated. Put it differently, as particles are unable to provide new alternative solutions, the overall swarm does not improve and there is no need of computing a new average silhouette value.

Tab. 1 depicts the performance of PSilhOuette using the provided dataset. The performance is essentially measured by considering the average (AVG) and standard deviation (STD) values of both the average silhouette index and the CPU processing time of all trials. It can be observed that the optimal number of clusters found by the proposed PSilhOuette method varies between 2 and 4 clusters.

Table 1. Performance of PSilhOuette (AVG±STD)

Image	Average Silhouette	CPU Time	Clusters
1	-0.056±0.09	1066±1087	2
2	-0.117±0.12	1255±1189	2
3	0.000±0.00	1099±1113	4
4	-0.101±0.08	1112±1298	2
5	0.000±0.00	1007±1092	4
6	-0.078±0.11	1226±1260	2
7	-0.068±0.09	1219±1418	2
8	-0.085±0.08	1112±1169	2
9	-0.001±0.01	700±466	3
10	0.000±0.00	1191±1149	3

Table 2. Average silhouette values for different images using different numbers of clusters

Image	c=2	c=3	c=4	c=5	c=6	c=7	c=8	c=9	c=10
1	-0.056					-0.3167			
2	-0.117						-0.3429		
3	1.67×10^{-14}		0.000						
4	-0.101			-0.2065					
5	1.54×10^{-14}		0.000						
6	-0.078								-0.2972
7	-0.068	-0.055							
8	-0.085				-0.2664				
9	9.09×10^{-4}	-0.001							
10	1.37×10^{-14}	0.000							

As can be seen from Tab.2, the proposed approach shows better average silhouette results for all images, except for image 7.

Fig. 3 shows the segmentation result of image *3* (from the tables), using the recommended number of clusters (*i.e.* 4), as compared to the result using 2-clusters.

5.3 Evaluation

The evaluation will be carried out using the average silhouette value. Where the optimum number of clusters was *larger* than *2*, this will be compared with PSO-FCM using 2-clusters, especially that in some literature [14,15], 2-clusters seems to be chosen in the experiments. And, where the optimum number of clusters was *equal* to 2, this will be compared with PSO-FCM using a random number of clusters larger than *2* chosen from the range *3* to *10* clusters.

6   Conclusion

Choosing the most adequate number of clusters is of vital importance in Fuzzy C-Means clustering. Therefore, it is necessary to find a way to retrieve the optimal number of clusters that can represent the relevant information within a given image . This paper proposed a particle swarm optimization approach, denoted as PSilhOuette, which simultaneously solves two optimization problems in clustered fashion: maximize the average silhouette, and minimize the Fuzzy C-Means objective function. The herein proposed methodology was evaluated on a liver CT dataset, depicting a higher clustering quality. In the future, the authors will benchmark the proposed approach over other state-of-the-art alternatives.

References

1. Annadurai, S., Shanmugalakshmi, R.: Fundamentals of Digital Image Processing. Dorling Kindersley, Delhi (2006)
2. Huang, X., Tsechpenakis, G.: Medical Image Segmentation. In: Hristidis, V. (ed.) Information Discovery on Electronic Health Records, ch. 10, Chapman & Hall (2009)
3. Leeuwen, M., Noorddzij, J., Feldberg, M., Hennipman, A., Doornewaaard, H.: Focal Liver Lesions: Characterization with Triphasic Spiral CT. Radiology 201, 327–336 (1996)
4. Mharib, A., Ramli, A., Mashohor, S., Mahmood, R.: Survey on Liver CT Image Segmentation Methods. Artificial Intelligence Review 37(2), 83–95 (2012)
5. Oliveira, J., Pedrycz, W.: Advances in Fuzzy Clustering and its Applications. John Wiley Sons Ltd., England (2007)
6. Duda, R., Hart, P., Stork, D.: Pattern Classification, 2nd edn. John Wiley, Chichester (2001)
7. Bezdek, J.: Fuzzy Mathematics in Pattern Classification, Ph.D. thesis, Applied Mathematic Center, Cornell University, Ithaca, NY (1973)
8. Cao, B.-y., Wang, G.-j., Guo, S.-z., Chen, S.-l. (eds.): Fuzzy Information and Engineering 2010. AISC, vol. 78. Springer, Heidelberg (2010)
9. Maji, P., Pal, S.: Maximum Class Separability for Rough-Fuzzy C-Means Based Brain MR Image Segmentation. T. Rough Sets 9, 114–134 (2008)
10. Halkidi, M., Batistakis, Y., Vazirgiannis, M.: On Clustering Validation Techniques. Journal of Intelligent Information Systems 17, 107–145 (2001)
11. Rawashdeh, M., Ralescu, A.: Center-Wise Intra-Inter Silhouettes. In: Hüllermeier, E., Link, S., Fober, T., Seeger, B. (eds.) SUM 2012. LNCS (LNAI), vol. 7520, pp. 406–419. Springer, Heidelberg (2012)
12. Rousseeuw, P.: Silhouettes: A Graphical Aid to the Interpretation and Validation of Cluster Analysis. Computational and Applied Mathematics 20, 53–65 (1987)
13. Anter, A., Azar, A., Hassanien, A., El-Bendary, N., ElSoud, M.: Automatic Computer Aided Segmentation for Liver and Hepatic Lesions Using Hybrid Segmentations Techniques. In: IEEE Proceedings of Federated Conference on Computer Science and Information Systems 2013, pp. 193–198 (2013)
14. Aguiar, R., Sales, R.: Dependence Analysis of the Market Index Using Fuzzy C-Means Algorithm. International Proceedings of Economics Development & Research 1, 362–365 (2011)

15. Wu, K.: Analysis of Parameter Selections for Fuzzy C-Means. Pattern Recognition 45, 407–415 (2012)
16. Bush, B.: Fuzzy Clustering Techniques: Fuzzy C-Means and Fuzzy Min-Max Clustering Neural Networks (retrieved), http://benjaminjamesbush.com/fuzzyclustering/fuzzyclustering.docx (accessed on March 18, 2014)
17. Bezdek, J., Ehrlich, R., Full, W.: FCM: The Fuzzy C-Means Clustering Algorithm. Computers and Geosciences 10, 191–203 (1984)
18. Abdel-Dayem, A., El-Sakka, M.: Fuzzy C-Means Clustering for Segmenting Carotid Artery Ultrasound Images. In: Kamel, M.S., Campilho, A. (eds.) ICIAR 2007. LNCS, vol. 4633, pp. 935–948. Springer, Heidelberg (2007)
19. Kannan, S.R.: A new segmentation system for brain MR images based on fuzzy techniques. Applied Soft Computing 8(4), 1599–1606 (2008)
20. Zhou, H., Schaefer, G., Sadka, A.H., Celebi, M.E.: Anisotropic mean shift based fuzzy c-means segmentation of dermoscopy images. IEEE Journal of Selected Topics in Signal Processing 3(1), 26–34 (2009)
21. Rawashdeh, M., Ralescu, A.: Crisp and Fuzzy Cluster Validity: Generalized Intra-Inter Silhouette Index. In: 2012 Annual Meeting of the North American Fuzzy Information Processing Society (NAFIPS), pp. 1–6. IEEE (2012)
22. Lupaşcu, C.A., Tegolo, D.: Stable Automatic Unsupervised Segmentation of Retinal Vessels Using Self-organizing Maps and a Modified Fuzzy C-Means Clustering. In: Petrosino, A. (ed.) WILF 2011. LNCS (LNAI), vol. 6857, pp. 244–252. Springer, Heidelberg (2011)
23. Kennedy, J., Eberhart, R.: A New Optimizer Using Particle Swarm Theory. In: Proceedings of the IEEE Sixth International Symposium on Micro Machine and Human Science, Nagoya, Japan, pp. 39–43 (1995)
24. Couceiro, M.S., Martins, F.M., Rocha, R.P., Ferreira, N.M.: Analysis and Parameter Adjustment of the RDPSO - Towards an Understanding of Robotic Network Dynamic Partitioning based on Darwin's Theory. International Mathematical Forum 7(32), 1587–1601 (2012)
25. Wong, K.: Medical Image Segmentation: Methods and Applications in Functional Imaging. In: Handbook of Biomedical Image Analysis Topics in Biomedical Engineering International Book Series, pp. 111–182 (2005)
26. Wong, K.-P., Dagan, F., Meikle, S., et al.: Segmentation of Dynamic PET Images Using Cluster Analysis. IEEE Trans. Nucl. Sci. 49, 200–207 (2002)
27. Ali, A.-R., Couceiro, M., Hassanien, A.E., Tolba, M.F., Snasel, V.: Fuzzy C-Means Based Liver CT Image Segmentation with Optimum Number of Clusters. In: Abraham, A., Körner, P., Snášel, V. (eds.) Proceedings of the Fourth Intern. Conf. on Innov. in Bio-Inspired Comput. and Appl. IBICA 2014. AISC, vol. 303, pp. 131–139. Springer, Heidelberg (2014)

Part V
Rough/ Fuzzy Sets and Applications

Soft Rough Sets for Heart Valve Disease Diagnosis

H. Hannah Inbarani[1], S. Senthil Kumar[1],
Ahmad Taher Azar[2], and Aboul Ella Hassanien[3,4]

[1] Department of Computer science, Periyar University, Salem-636011
[2] Faculty of Computers and Information, Benha University, Benha, Egypt
[3] Faculty of Computer and Information, Cairo University, Cairo, Egypt
[4] Scientifc Research Group in Egypt (SRGE)
{hhinba,pkssenthilmca}@gmail.com, ahmad.azar@fci.bu.edu.eg
www.egyptscience.net

Abstract. Heart murmur (systolic and diastolic) is the result of different cardiac valve disorders. The auscultation of the heart is still the first basic analysis tool used to calculate the functional state of the heart, as well as the first pointer used to submit the patient to a cardiologist. In order to develop the diagnosis capabilities of auscultation, pattern recognition algorithms are currently being technologically advanced to assist the physician at primary care centers for adult and pediatric residents. A basic task for the diagnosis from the phonocardiogram is to detect the events (heart sounds and murmurs) present in the cardiac cycle. Four common murmurs were considered including aortic stenosis, aortic regurgitation, mitral stenosis, and mitral regurgitation. In this work modified soft rough set is used as a classifier in the classification of three heart valve data sets. Four types of classification approaches were compared to evaluate the discriminatory power of the classification such as (Decision table, MultiLayer Perceptron (MLP), Back Propagation Network (BPN) and Navie Bayes). The best results were achieved by soft rough sets. The favorable results demonstrate the effectiveness of the proposed approach for heart sounds' classification.

Keywords: Heart murmur, Soft rough set, Classification, phonocardiogram, auscultation.

1 Introduction

Heart sound is a biomedical signal with valuable diagnostic information about structural abnormality of the heart valves and associated great vessels. However, making diagnosis based on the sounds heard by means of a stethoscope is a difficult skill. Signal processing techniques can be employed to process the phono cardiographic signals (PCG) towards improving the accuracy of diagnosis. Phonocardiography is the registration of sound vibrations of heart and blood flow that has a capability of presenting various heart valve disorders. Heart murmurs are often the first sign of pathological alterations of heart valves, and they are usually found for the period of auscultation in major health care. Heart murmurs are an important feature to identify cardiac disorders [1]. Main objective of this study is on

A.E. Hassanien et al. (Eds.): AMLTA 2014, CCIS 488, pp. 347–356, 2014.

pathological murmurs, which require medical care. Murmurs are high rate and noise like sounds formed from the turbulence on the blood moving from a narrow cardiac valves or reflow from the atrio-ventricular valves. Mitral regurgitation, aortic stenosis (diastolic) aortic regurgitation, mitral stenosis (systolic) are amongst the most common murmurs [2-7]. Rough set theory is a new mathematical tool to handle uncertainty and incomplete information. Polish mathematician Pawlak. Z initially proposed it [8-11]. The theory consists of finite sets, equivalence relations and cardinality concepts. A principal goal of rough set theoretic analysis is to synthesize or construct approximations (upper and lower) offsets concepts from the acquired data [12]. Rough set theory clarifies set-theoretic characteristics of the classes over combinatorial patterns of the attributes. Data discretization is the process of transforming data containing a quantitative attribute so that the attribute in question is replaced by a qualitative attribute. Data attributes are either numeric or categorical. While categorical attributes are discrete, numerical attributes are either discrete or continuous. This work adopts class attribute contingency co-efficient discretization approach. First, the continuous variables are discretized to reduce the effect of distribution imbalance. Many practical applications of classification involve a large volume of data and/or a large number of features/attributes. Since these datasets are usually collected for reasons other than mining the data (e.g. classification), there may be some redundant or irrelevant features. Hence it is necessary to remove irrelevant attributes and only the relevant attributes are used for deriving new knowledge. In order to solve such a problem in medical diagnosis, attribute reduction, also called feature subset selection, is usually employed as a preprocessing step to select part of the attributes and focus the learning algorithm on relevant information [13-16]. Attribute reduction can facilitate data visualization and data understanding, as well as decrease the storage space needed to store the data. In recent years, rough set theory has been widely discussed and used in attribute reduction. Rough set has strong ability in data processing and can extract useful rules from them. The main aim of Feature Selection is to determine a minimal feature subset from a problem domain. Feature Selection is used to improve the classification accuracy and reduce the computational time of classification algorithms. In this work, rough entropy [13] based feature selection approach is applied for selecting relevant features. Rough set theory and soft set theory [17-21] are two different tools to deal with uncertainty. Apparently there is no direct connection between these two theories, however efforts similar. Modified Soft Rough (MSR) sets satisfy all the basic properties of rough sets and soft sets [17-21]. In some situations, equivalence relation cannot be defined which is the basic requirement in rough set theory. In these situations, MSR-sets can help us to find approximations of subsets. The proposed work consists of two parts: Initially in the pre-processing stage, redundant data are removed and rules are derived from reduced data set. In this paper, modified soft rough set based classification is applied for generating decision rules from the reduced data set. Heart valve diagnosis is regarded as an important yet complicated task that needs to be executed accurately and efficiently. The automation of this system would be extremely advantageous. Appropriate computer-based information and/or decision support systems can aid in achieving clinical tests at a reduced cost. Efficient and accurate implementation of automated system needs a comparative study of various techniques available. This paper aims to analyze the different predictive/ descriptive data mining techniques proposed in recent years for the diagnosis of heart diseases. The rest of the paper is structured as follows: Section 2 describes over all methodology of the proposed work and section 3 explains the proposed algorithm and supervised rough entropy based feature selection. It converses how MSR-Sets classification can be used for generating classification rules from a set of perceived samples of the given data set. Section 4 describes the experimental results of the proposed work for the detection of heart valve diseases and comparative study is made with decision table, Naive Bayes classifier, Back propagation neural network and multilayer perceptron classifier algorithms. Finally conclusion is given in Section 5.

2   Methodology

The methodology adopted in this work is given in Fig. 1.

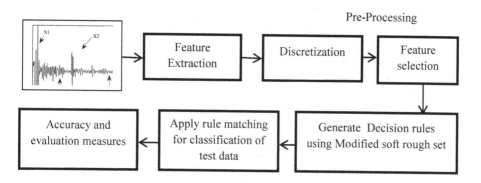

Fig. 1. Overall Methodology

Heart valve disease diagnosis requires a sequence of processes to be performed, data acquisition, feature extraction, preprocessing and classification. Each process is discussed separately for heart murmur classification in this work. In the first step, characteristics of heart murmur: time, intensity, pitch, location and duration based features are extracted from the phonocardiographic (PCG) signals. Extraction of localized features depends on precise knowledge about the timing of the heart cycles. In the second step, Data discretization process is applied. The goal of discretization is to partition the range into a small number of intervals that have good class coherence, which is usually measured by an evaluation function. One solution to this problem is to partition numeric variables into a number of sub-ranges and treat each such sub-range as a category. This process of partitioning continuous variables into categories is usually termed discretization. If feature information is irrelevant or redundant or the data is noisy and unreliable, then knowledge discovery during training is more difficult. Feature selection is the process of identifying and removing as much of the irrelevant and redundant information as possible and informative attributes are selected by using rough entropy algorithm. In the third step, modified soft rough set based classification [17] approach is applied for generating rules from the trained data and rule matching is applied for test data to compute the decision class based on reliability analysis. In this study, the proposed approach is applied for heart valve disease diagnosis.

3   Proposed Work

In this work, Rough entropy based feature selection algorithm is applied for feature selection. Data feature indicates the importance of feature in information systems, offering the powerful reference to the decision. It can be seen from this method that the completeness for the minimal reduction of the below method is proved obviously. That is, none of the features in the subset B can be eliminated again without decreasing its discriminating capability, the positive region, the feature core and the calculation complexity of this algorithm. Hence, the reduction process is shown in Fig. 2 [13]:

Rough entropy based feature selection algorithm
Input: A decision information system S = (U, C, D, F).
Output: A relative reduction set of S.
Step 1:   Calculating $Pos_C(D)$ and $U - Pos_C(D)$ based on radix sorting and hash to obtain U/R
Step 2:   Calculating $E_C(D)$ and $E_{C-\{a\}}(D)$, where a $\in$ C.
Step 3:   Calculating Core(C) and let B = Core(C).
Step 4:   If B = $\varnothing$ then go to (6).
Step 5:   If $E_C(D) = E_C(D)$ and $E_B(D)$ then go to (8).
Step 6:   Making H = {a $\in$ C - B l max {significance (a)}}.
i)   If l H l= 1 then selecting a $\in$ H else selecting a with the minimum
ii)   l U/ (B $\cup$ {a}) l = min{l U/(B $\cup$ {a}) ll a$\in$ H}.
iii)   B = B $\cup$ {a}.
Step 7:   If $E_B(D)/ = E(D)$ then go to (6) else
a.   B = B - Core(C).
b.   t =l B l.
c.   For (i = 1; i = t; i + +)
a_i $\in$ B.
B = B - a
If $E_{Core}(C) \cup B(D)/ = E_C(D)$ then B = B $\cup$ ai
B = B $\cup$ Core(C).
Step 8:   The B is a minimum relative reduction of condition features set C.

Fig. 2. Rough entropy based feature selection

MSR based classification algorithm [17] is presented in Fig.3 In this approach, lower and upper soft rough approximations of the given data set based on Decision class X are constructed. In the second step, AND operation is applied to combine the soft sets. In the third step, deterministic rules are generated based on lower soft rough approximation. In the fourth step, non-deterministic rules are generated based on upper soft rough approximation and support of each non-deterministic rule is computed using step 5 of the algorithm.

Algorithm: MSR based Classification
Input: Given murmur Dataset with conditional attributes and the Decision attributes aortic regurgitation, aortic stenosis, mitral regurgitation, and mitral stenosis. **Output:** Generated Decision Rules **Step1:** Construct MSR approximation space for the given murmur dataset **Step 2:** Apply AND operation for all conditional attributes. **Step 3:** Generate deterministic rules using $$\underline{X}\phi = \{X\epsilon U: \phi(x) \neq \phi(y), for\ all\ y\epsilon x^c\}$$ **Step 4:** Generate non-deterministic rules by using $$\bar{X}\phi = \{X\epsilon U: \phi(x) = \phi(y), for\ some\ y\epsilon x\}$$ **Step 5:** Compute the support value for each non-deterministic rule $$support = \frac{support(A \wedge B)}{support(A)}$$

Fig. 3. MSR based classification algorithm

4   Experimental Results

4.1   Data Set Description

The prediction of heart valve diseases using MSR based classification is applied for three different data sets of heart sound signals. The first data set "AR_MS diastolic murmur" consists of diastolic diseases where it contains 38 instances of Aortic Regurgitation (AR) cases and 38 instances of Mitral Stenosis (MS) cases. The second data set "AS_MR systolic Murmur" contains diastolic diseases where it contains 41 instances of Aortic Stenosis (AS) cases and 43 instances of Mitral Regurgitation (MR) cases. The third data set "Healthy and unhealthy" contains 70 instances, where 39 instances represent healthy patients and the other 31 represents unhealthy, murmur diseased patients [5]. The proposed algorithm is applied to training data and the generated classification rules are matched with test data to determine exact class. The attributes in these data sets are all numerical. 80% of the data is chosen as the training set and 20% as testing data. Comparative analysis of the proposed approach is made with other classification approaches like Decision table, Naive Bayes and MLP.

4.2   Accuracy Measures

Precision is the average probability of relevant retrieval. Recall is the average probability of complete retrieval. F-Measure is a measure that combines precision and recall is the harmonic mean of precision and recall.

Precision   = True positive / (True positive + False negative)
Recall   = True positive / (True positive + False positive)
F-Measure = (2*Precision*Recall) / (Precision + Recall)

Accuarcy = (True positive + True negative)/ (True positive + True negative + False Negative + False Positive)

A Precision, Recall, F-measure and Accuracy Evaluation collect and report a suite of descriptive statistics for binary classification tasks. The basis of a precision recall evaluation is a matrix of counts of actual and predicted classifications. Tables 2, 3 and 4 depict the attributes of heart valve diseases data set.

4.3   Analysis

In this paper, the features are reduced by the supervised rough entropy based feature selection method. The classification is initially performed on the reduced data set. The supervised rough entropy method and classification of the reduced data shows that the method selects useful features which are of comparable quality. Table 1 represents the number of attributes reduced using rough entropy approach.

Table 1. Reduced attribute set for heart valve datasets

Datasets	Total attributes	Number of Reduced attributes	Reduced Attributes
Diastolic	100	3	(96,99,100)
Systolic	100	4	(31,34,99,100)
Healthy and unhealthy	100	7	(1,37,40,89,97,99,100)

Performances of Classification algorithms are presented in Table 2 and Fig 4. The proposed method (Modified Soft Rough set based classification) provides high accuracy of 100% for heart valve diseases (diastolic murmur) data. Other classifier algorithms classifier algorithms provide accuracy of 61%, 66%, 60% and 93.1% for heart valve diseases (diastolic murmur) data set.

Table 2. Performance analysis of the classification algorithms for diastolic murmur data

Measures	MSR	Naïve Bayes	MLP	Decision Table	BPN
Precision	1.000	0.629	0.722	0.569	0.94
Recall	1.000	0.6	0.667	0.6	0.933
F-Measure	1.000	0.604	0.667	0.536	0.933
Accuracy	1.000	0.61	0.66	0.60	0.931

Table 3 and Fig 5 represent the accuracy of the proposed method (Modified Soft Rough set based classification) which is able to classify 100% correctly for systolic murmur data. Other classifier algorithms Naïve Bayes, Multi-layer perceptron, Decision table and Back propagation neural network provide an accuracy of 82.30%, 70.50%, 76.40% and 81.30% for systolic Murmur data set respectively.

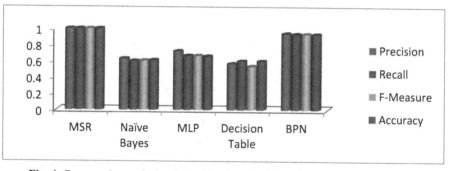

Fig. 4. Comparative analysis of classification algorithms for diastolic murmur data

Table 3. Performance analysis of the classification algorithms for systolic murmur data

Measures	MSR	Naïve Bayes	MLP	Decision Table	BPN
Precision	1.000	0.827	0.729	0.843	0.85
Recall	1.000	0.824	0.706	0.765	0.812
F-Measure	1.000	0.822	0.693	0.755	0.796
Accuracy	1.000	0.823	0.705	0.764	0.813

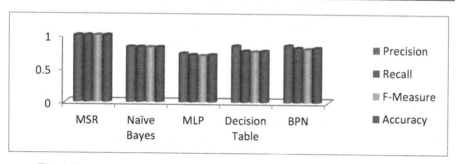

Fig. 5. Comparative analysis of classification algorithms for systolic murmur data

Table 4 and Fig 6 demonstrates the accuracy of proposed method (MSR based classification) as 100% and a comparative analysis is made with Back propagation neural network, Naïve Bayes, Multi-layer perceptron and Decision table algorithms. These methods provide accuracies of 80.1%, 85.8%, 78.6% and 79.8% for healthy and unhealthy data respectively.

Table 4. Performance analysis of the classification algorithms for healthy and unhealthy data

Measures	MSR	Naïve Bayes	MLP	Decision Table	BPN
Precision	1.000	0.905	0.775	0.835	0.816
Recall	1.000	0.857	0.786	0.786	0.803
F-Measure	1.000	0.863	0.776	0.735	0.797
Accuracy	1.000	0.858	0.786	0.798	0.801

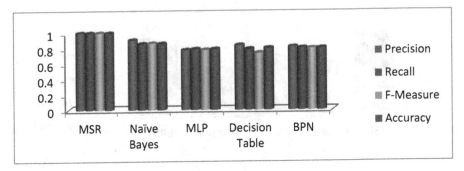

Fig. 6. Comparative analysis of classification algorithms for healthy and unhealthy data

Different heart valve diseases distinguish between pathological and physiological murmurs. A feature set making use of heart sound characteristics from several signal domains has been derived. The derived feature set aims at facilitating the classification step by providing relevant information to the classifier. Since diastolic murmurs and systolic murmurs are mostly pathological, in many other studies, the data is either provided from teaching tapes or from specially selected heart cycles of very high quality and with typical morphology. The results have shown that the measures suggested in this work can provide an effective and accurate delineation of heart murmurs and correlate closely to clinical descriptions documented in the medical care. The proposed system provides the possibility to correctly classify heart murmurs. Quantitative description of the important heart murmur measures such as precision, recall and f-measure can be obtained from the analysis. This accurate information can be used to minimize human error, alleviate differences in individual opinions and to aid the classification of heart murmurs. A more extensive comparative study would be required to evaluate the proposed algorithm against other algorithms introduced for heart murmur classification on the same heart valve database. In this study, normal heart sounds were differentiated from four types of murmurs. Last three tables and figures demonstrate precision, recall, F-measure and accuracy values of classification approaches. It is interesting to note that an increase in overall classification accuracy is recorded for the proposed method. In this work, the system which is able to differentiate between healthy and unhealthy heart defects using intelligent techniques. The intelligent diagnostic system not only helps in accurate detection, it is also useful for the physician who is in charge to help people saving lives of many cases of abnormality. This technique can be applied for high-volume screening of people suspected of having a heart disease.

5   Conclusion

From the above results, we can easily conclude that the MSR-algorithm is an effective method for detection of heart valve diseases. The proposed classification method is applied to reduced data set thus reducing the number of rules while leading to significantly to improved classification accuracy. Comparisons of the proposed approach with familiar, Decision Table, Naive Bayes classifier, BPN and MLP algorithms are made using performance metrics. The healthier results show that MSR-based classification is suitable for the diagnosis of heart valve diseases.

Acknowledgment. The authors would like to highly appreciate and gratefully acknowledge, Ilias Maglogiannisa from University of Central Greece, Department of Computer Science and Biomedical Informatics, Lamia, Greece, for supporting us and providing with heart sound data set.

References

1. Sanei, S., Ghodsi, M., Hassani, H.: An adaptive singular spectrum analysis approach to murmur detection from heart sounds. Medical Engineering & Physics 33(3), 362–367 (2011)
2. Hamdy, A., Hefny, H., El-Bendary, N., Khodeir, A., Hassanien, A.E.: Cardiac disorders detection approach based on local transfer function Classifier. In: Federated Conference on Computer Science and Information Systems, Kraków, Poland, September 8-11, pp. 55–61 (2013)
3. Amiri, A.M., Armano, G.: Heart Sound Analysis for Diagnosis of Heart Diseases in Newborns. APCBEE Procedia 7(2013), 109–116 (2013)
4. Safara, F., Doraisamy, S., Azman, A., Jantan, A., Ranga, S.: Diagnosis of Heart Valve Disorders through Trapezoidal Features and Hybrid Classifier. International Journal of Bioscience, Biochemistry and Bioinformatics 3(6), 662–665 (2013)
5. Elbedwehy, M.N., Zawbaa, H.M., Ghali, N., Hassanien, H.E.: Detection of Heart Disease using Binary Particle Swarm Optimization. In: Proceeding of 2012 Federated Conference on Computer Science and Information Systems (FedCSIS), Wroclaw, September 9-12, pp. 177–182 (2012)
6. Maglogiannis, I., Loukis, E., Zafiropoulos, E., Stasis, A.: Support Vectors Machine-based identification of heart valve diseases using heart sounds. Computer Methods and Programs in Biomedicine 95(1), 47–61 (2009)
7. Salama, M.A., Hassanien, A.E., Platos, J., Fahmy, A.A., Snasel, V.: Rough Sets-Based Identification of Heart Valve Diseases Using Heart Sounds. In: Corchado, E., Snášel, V., Abraham, A., Woźniak, M., Graña, M., Cho, S.-B. (eds.) HAIS 2012, Part III. LNCS, vol. 7208, pp. 667–676. Springer, Heidelberg (2012)
8. Pawlak, Z.: Rough sets Rough sets. International Journal of Parallel Programming 11(5), 341–356 (1982)
9. Pawlak, Z., Skowron, A.: Rough sets: some extensions. Information Science 177(1), 28–40 (2007)
10. Pawlak, Z., Skowron, A.: Rough sets and Boolean reasoning. Information Science 177(1), 41–73 (2007)
11. Pawlak, Z.: Rough classification. Int. J. Hum.-Comput. Stud. 51(2), 369–383 (1999)
12. Khoo, L.P., Tor, S.B., Zhai, Y.L.: A Rough-Set-Based approach for Classification and Rule Induction. International Journal Advanced Manufacturing Technology 15(6), 438–444 (1999)
13. Sun, L., Xu, J., Xue, Z., Zhang, L.: Rough Entropy-based Feature Selection and Its Application. Journal of Information & Computational Science 8(9), 1525–1532 (2011)
14. Inbarani, H.H., Azar, A.T., Jothi, G.: Supervised hybrid feature selection based on PSO and rough sets for medical diagnosis. Computer Methods and Programs in Biomedicine 113(1), 175–185 (2014)

15. Inbarani, H.H., Banu, P.K.N., Azar, A.T.: Feature selection using swarm-based relative reduct technique for fetal heart rate. Neural Computing and Applications (2014), doi:10.1007/s00521-014-1552-x

16. Azar, A.T., Banu, P.K.N., Inbarani, H.H.: PSORR - An Unsupervised Feature Selection Technique for Fetal Heart Rate. In: 5th International Conference on Modelling, Identification and Control (ICMIC 2013), Egypt, August 31-September 1-2 (2013)

17. Senthilkumar, S., Hannah Inbarani, H., Udhayakumar, S.: Modified Soft Rough set for Multiclass Classification. In: Krishnan, G.S.S., Anitha, R., Lekshmi, R.S., Senthil Kumar, M., Bonato, A., Graña, M. (eds.) Computational Intelligence, Cyber Security and Computational Models. AISC, vol. 246, pp. 379–384. Springer, Heidelberg (2014)

18. Feng, F., Liu, X., Leoreanu-Fotea, V., Young Jun, Y.B.: Soft sets and soft rough sets. Information Sciences 181(6), 1125–1137 (2011)

19. Shabir, S., Ali, M.I., Shaheen, T.: Another approach to soft rough sets. Knowledge-Based Systems 40(2013), 72–80 (2013)

20. Udhaya Kumar, S., Hannah Inbarani, H., Senthilkumar, S.: Bijective soft set based classification of Medical data. In: International Conference on Pattern Recognition, Informatics and Medical Engineering (PRIME), pp. 517–521 (2013)

21. Udhaya Kumar, S., Hannah Inbarani, H., Senthil Kumar, S.: Improved Bijective-Soft-Set-Based Classification for Gene Expression Data. In: Krishnan, G.S.S., Anitha, R., Lekshmi, R.S., Senthil Kumar, M., Bonato, A., Graña, M. (eds.) Computational Intelligence, Cyber Security and Computational Models. AISC, vol. 246, pp. 127–132. Springer, Heidelberg (2014)

A Novel Hybrid Perceptron Neural Network Algorithm for Classifying Breast MRI Tumors

Amal M. ElNawasany[1], Ahmed Fouad Ali[1,2], and Mohamed E. Waheed[1]

[1] Suez Canal University, Faculty of Computers and Information, Ismailia, Egypt
[2] Scientific Research Group in Egypt (SRGE)
www.egyptscience.net

Abstract. Breast cancer today is the leading cause of death amongst cancer patients inflicting women around the world. Breast cancer is the most common cancer in women worldwide. It is also the principle cause of death from cancer among women globally. Early detection of this disease can greatly enhance the chances of long-term survival of breast cancer victims. Classification of cancer data helps widely in detection of the disease and it can be achieved using many techniques such as Perceptron which is an Artificial Neural Network (ANN) classification technique. In this paper, we proposed a new hybrid algorithm by combining the perceptron algorithm and the feature extraction algorithm after applying the Scale Invariant Feature Transform (SIFT) algorithm in order to classify magnetic resonance imaging (MRI) breast cancer images. The proposed algorithm is called breast MRI cancer classifier (BMRICC) and it has been tested tested on 281 MRI breast images (138 abnormal and 143 normal). The numerical results of the general performance of the BMRICC algorithm and the comparasion results between it and other 5 benchmark classifiers show that, the BMRICC algorithm is a promising algorithm and its performance is better than the other algorithms.

Keywords: Artificial Neural Network (ANN), perceptron algorithm, magnetic resonance imaging (MRI), breast cancer.

1 Introduction

To date, 1.38 million new breast cancer cases have been diagnosed, which is 23% of total new cancer cases in the world [18]. Breast cancer is the most common cancer worldwide, and the second leading cause of cancer death. One in nine women in the UK and USA will develop the disease in their lifetimes [1]. The observed annual incidence of breast cancer globally is about one million cases, with more than half occurring in the Western world: 200,000 cases in the United States and 320,000 cases in Europe [1]. As the tumor gets bigger, the center of it gets further and further away from the blood vessels in the area where it is growing. So the center of the tumor gets less and less of the oxygen and the other nutrients all cells need to survive. Without oxygen and nutrients, the cell will die. So it needs to grow its own blood supply to survive. This is called angiogenesis

A.E. Hassanien et al. (Eds.): AMLTA 2014, CCIS 488, pp. 357–366, 2014.
© Springer International Publishing Switzerland 2014

[19]. The cancer cells may be able to stimulate normal cells to produce angiogenic factors to help produce new blood vessels. The cancer can't grow much bigger than a pin head before it needs to develop its own blood supply [23]. A tumor or cancer is unable to grow beyond 1-2 mm in size without the development of a new blood supply. Tumor endothelial cells divide much more rapidly than normal endothelial cells; up to 50 times as fast, as in breast cancer [12]. Once a cancer can stimulate blood vessel growth, it can grow bigger and grow more quickly. It will stimulate the growth of hundreds of new capillaries from the nearby blood vessels to bring it nutrients and oxygen [23]. MRI is commonly used for breast screening to explore the small details between breast tissues [9], [10], [15], [18]. Breast MRI uses magnets and radio waves to produce detailed 3-dimensional images of the breast tissue. Before the test, patient may need to have a contrast solution (dye) injected into his arm through an intravenous line. The solution will help any potentially cancerous breast tissue show up more clearly. Notable advantages of MRI compared with conventional imaging techniques enhanced detection of recurrence, better evaluation of lesions in the augmented breast, and improved screening of high-risk women [7], [18]. The tumor needs to support its center with blood, the mean values of the ratios of tumor to normal blood flow and blood volume are significantly higher than those for benign or normal tissue [3], [14]. On a breast MRI, the contrast tends to become more concentrated in areas of cancer growth. Image classification algorithms are used to detect and diagnose tumor masses in medical images. There are many classification algorithms that can be used to classify data extracted from images. Also there are many feature extraction algorithms that used to extract features from images. In this paper, we apply the feature extraction algorithm based on the information that the tissue around tumor tends to be darker and the contrast tends to become more concentrated in areas of cancer growth so the cancer area is more intensity than the surrounding tissue in MRI. Also, we use SIFT to detect the most descriptive points in the images then extract feature that care with darkness and intensity of the images in the same time. A feature matrix of size 8×60 is extracted from regions around a set of points detected by SIFT [13], [16] for each image. The perceptron algorithm is applied to the feature matrix of all images in order to classify the obtained data.

The reminder of the paper is organized as fellow. Section 2 describes the SIFT algorithm with its main steps. The perceptron algorithm is presented in Section 3. In Section 4, we explain the proposed algorithm. The numerical experimental results are presented in Section 5. Finally, The conclusion of the paper is presented in Section 6.

2   Scale Invariant Feature Transform (SIFT) Algorithm

The SIFT algorithm [22] is a popular feature extraction technique that mainly used in object recognition [13], [16] as it detect the most descriptive points inside the image. The SIFT algorithm has four major phases as fellow.

- **Scale space extrema detection.** Extrema Detection phase examines the image to detect a set of candidate points by comparing the pixel's intensities in different scales. The points that are different from their surroundings are potential candidates for image features and called extreme. Choose all extrema within $3 \times 3 \times 3$ neighborhood [14]. A point is selected to be a candidate point if it has an intensity that is larger than or less than its 26 neighbors in a 3×3 window [11].

 – **Key point Localization.** The next phase, Key point Detection or Key point Localization, the final key points are selected from the extrema after discarding the candidate points that are lie along an edge of the image or points have low contrast [22].

 – **Orientation Assignment.** The third phase, Orientation Assignment, one or more orientations are assigned to each key point. This phase adds a set of points that are misses in the first two phases. These missed key points are not an extremum but have a significant scale. The algorithm now has identified a final set of key points.

 – **Key point Descriptor.** The last phase, Key point Descriptor Generation, SIFT algorithm computes a set of descriptors around the region for each key point identified so far.

3   The Perceptron Algorithm

The perceptron algorithm is a learning procedure, which is used to obtain the weights of a perceptron that separate two classes. Form the obtained weights, the equation of the separating hyperplane can be derived. The perceptron training algorithm does not assume any a priori knowledge about the specific classification problem being solved. The initial weights are generated randomly. The Input samples are repeatedly presented to the perceptron and the performance of the perceptron observed. If the performance on a given input sample is satisfactory (the current network output is the same as the desired output for a given sample), then the weights are not changed in this step. Otherwise, the weights must be changed in such a way as to reduce system error. The operation is repeated until termination criterion satisfied. The main termination criterion of the perceptron algorithm is reaching to the correct classification of all samples if the input data is linearly separable or if the data is nonseparable, the user may be satisfied with a misclassification data with low rate. The main steps of the perceptron algorithm are presented in Algorithm 1.

4   The BMRICC Algorithm

In this section, we present the BMRICC algorithm and its main steps, which are listed as fellow.

4.1   Input Images

We use 281 breast cancer MRI images (138 abnormal and 143 normal) from the cancer genome atlas for breast cancer TCGA-BRCA data collection from the

---

Algorithm 1. The perceptron algorithm

1: Set the initial value of the learning rate η.
2: Start with a randomly chosen weight vector w_0.
3: Set $k = 1$
4: **repeat**
5:    Let i_j be a misclassified input vector.
6:    Let $x_k = class(i_j).i_j$, implying that $w_{k-l}.x_k < 0$.
7:    Update the weight vector to $w_k = w_{k-l} + \eta x_k$.
8:    Set $k = k + 1$.
9: **until** Termination criteria satisfied.

---

cancer imaging archive (TCIA) the Frederick national laboratory for cancer research [24]. The images are in Digital Imaging and Communications in Medicine (DICOM) format. We need some preprocessing steps to deal with images. The collected images are diagnosed to normal and abnormal by specialist radiologists.

4.2   Preprocessing Steps

In order to start working with the MRI images, some preprocessing processes have been made as fellow.

Convert DICOM Format to Joint Photographic Experts Group (JPEG).
The DICOM images is a medical images format that is read by special programs. DICOM is a standard for handling, storing, printing, and transmitting information in medical imaging. It includes a file format definition and a network communications protocol [17]. We use software called "RaniAnt DICOM Viewer" version 1.9.14 to read the collected DICOM images then convert them to joint photographic experts group (JPEG) format, readable format, to be able to deal with.

Select the Affected Side. The radiologist examines the images and identifies the normal and abnormal sides (right or left) then we collect the affected images to classify them.

Convert 3D Images to 2D. SIFT algorithm deals with 2D images only so we convert all the images to 2D.

4.3   Feature Extraction

We use the SIFT algorithm in order to detect as many descriptive points as possible within each image and extract features from all of the points then shorten the matrix to 8 rows and 60 column using different statistics. For each seed point p delivered by SIFT, a region is constructed around the seed point, R (e.g., 40×40 pixels) and the following features are extracted from R, the discrete cosine transform (DR) [19] of R and the approximation coefficients matrix (WR) of R (computed using the wavelet decomposition of R):

1. The mean gray-level which is an indication of darkness (3 features).
2. The standard deviation that indicate to intensity variation (3 features).
3. The mean, median, standard deviation, covariance, range, and the maximum of the descriptor vector of p (6 features).
4. A set of texture features from the gray level co-occurrence matrix (GLCM) in directions 0, 45, 90 and 135 (48 features) [16].

The matrix for each image is of size 8×60. The final matrix for 281 images is 2248×60 that is used to be classified by the perceptron algorithm. The effect of the SIFT algorithm with the MRI images before and after applying the SIFT algorithm is shown in Figure 1.

4.4 BMRICC Algorithm

The BMRICC algorithm starts with the input MRI images and the preprocessing steps have been applied in order to convert the image type from DICOM format to JPEG format and from 3D to 2D. The Sift algorithm is applied on the 2D to detect the key point and extract the feature from the detected key points. The perceptron algorithm is applied with specific number of iterations as a termination criterion in order to classify the output data. The main steps of the BMRICC algorithm are presented in Algorithm 2.

Algorithm 2. The BMRICC algorithm

INPUT: Get the breast MRI images.
OUTPUT: Classified data and their accuracies

1: Convert images from DICOM format to JPEG format.
2: Convert images from 3D to 2D in order to apply SIFT algorithm. {**Preprocessing steps**}
3: Apply SIFT algorithm to detect key points.
4: Extract feature from the detected key points.
5: **repeat**
6:    Apply the perceptron algorithm as shown in Algorithm 1.
7: **until** Termination criteria satisfied

---

5 Numerical Experiments

The BMRICC algorithm is tested on 281 MRI images (138 abnormal and 143 normal) and compared with 5 benchmark classifier algorithms. Before discussing the numerical results of the general performance of the proposed algorithm, we present the parameter setting, which have been applied on the BMRICC algorithm.

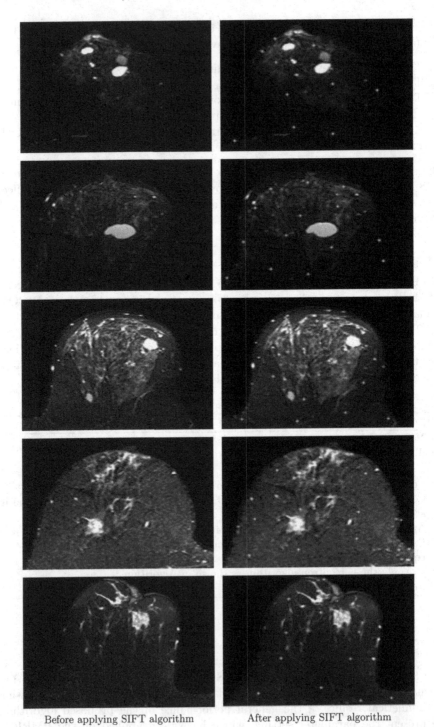

Before applying SIFT algorithm   After applying SIFT algorithm

Fig. 1. Results of applying SIFT algorithm on breast MRI images

5.1   Parameter Setting

The parameters setting of the BMRICC algorithm are listed in Table 1 as fellow.

Table 1. The Parameter setting of the BMRICC algorithm

Parameters	Definitions	value
η	Learning rate	0.9
Max_{iter}	Maximum number of iterations	400
M	Size of training data	800
N	Size of tested data	1448

5.2   The BMRICC Algorithm and other Algorithms

The BMRICC algorithm have been compared with 5 benchmark classifier algorithms. The description of the these classifier algorithms are reported as fellow.

- **Naive Bayes (NB) classifier.** NB is a probabilistic classifier based on the Bayes theorem [5]. Rather than predictions, the Nave Bayes classifier produces probability estimates. For each class value they estimate the probability that a given instance belongs to that class. Requiring a small amount of training data to estimate the parameters necessary for classification is the advantage of the Naive Bayes classifier [8].
- **Support Vector Machine (SVM).** SVM is introduced by Vapnik [21]. The basic concept in SVM is the hyper plane classifier, or linear separability. SVM projects the input data into a kernel space, then it builds a linear model in this kernel space. A classification SVM model attempts to separate the target classes with the widest possible margin. In WEKA application the SVM classifier is called SMO [20].
- **K-Nearest Neighbors (KNN).** The K nearest neighbors is a simple non-parametric instance-based learning algorithm that stores the training data set and classifies new unclassified cases based on its similarity to its neighbors by measuring the distance between them.
- **Random Forest Tree (RFT).** Random forests are an ensemble learning method for classification (and regression) that operate by constructing a multitude of decision trees at training time and outputting the class that is the mode of the classes output by individual trees [11].
- **Discriminant analysis.** Discriminant analysis is a classification method. Linear discriminant analysis is also known as the Fisher discriminant, named for its inventor, Sir R. A. Fisher [6]. For each class to be identified it calculate a different function of the attributes. The class function yielding the highest score represents the predicted class. It assumes that different classes generate data based on different Gaussian distributions. In the training (create) a classifier, the fitting function estimates the parameters of a Gaussian distribution for each class. To predict the classes of new data, the trained classifier finds the class with the smallest misclassification cost [2].

Comparison between NB, SVM, KNN, RFT and BMRICC. The comparison results between the BMRICC algorithm and the other algorithms are reported in Table 2. The results are taken over 30 runs for each algorithm. The minimum, maximum and the average accuracy for each algorithm is reported in Table 2. Also the comparison results of the 6 algorithms are plotted in Figure 2.

Table 2. The comparison results of the BMRICC and the other algorithms

Algorithm	Minimum	Maximum	average
BMRICC	76.80	86.74	83.37
NB	65.43	76.54	71.36
KNN	62.96	79.01	72.72
Discriminant	70.37	80.25	76.30
SVM	76.24	81.77	77.07
RFT	76.54	86.42	81.48

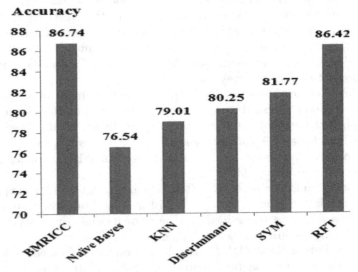

Fig. 2. Comparisons between BMRICC and other algorithms

We can conclude from Table 2 and Figure 2, that the BMRICC algorithm is promising algorithm and can classify the MRI images with higher accuracy than the other algorithms.

6   Conclusion and Future Work

In this paper, we proposed a new hybrid algorithm which is called breast MRI cancer classifier (BMRICC) algorithm in order to classify the breast tumor MRI images . The BMRICC algorithm combines the percepteron algorithm with the feature extraction algorithm that uses the SIFT algorithm to extract feature from the MRI breast images. The proposed algorithm has been tested on 281 MRI images and compared with 5 benchmark classification algorithms. The experimental results show that the proposed algorithm is a promising algorithm and can classify the MRI images with higher accuracy than the other classification algorithms. AS part of our future work, we will try to improve the performance of the BMRICC algorithm in order to increase its accuracy by applying other neural network algorithms and replace the termination criterion of the algorithm by another self adaptive termination instead of using the specific number of iterations. In our future work, to increase the efficiency of the proposed computer aided classification of the breast MRI tumors process, an intensity adjustment process will provide more challenging and may allow us to refine our segmentation algorithm using pulse coupled neural network hybrid with wavelet theory [4].

References

1. Abdulkareem, I.: A Review on Aetio-Pathogenesis of Breast Cancer. J. Genet. Syndr. Gene. Ther. 4, 142 (2013), doi:10.4172/2157-7412.1000142
2. Altman, N.S.: An introduction to kernel and nearest-neighbor nonparametric regression. The American Statistician 46(3), 175–185 (1992)
3. DeLille, J.: Breast cancer: regional bloodflow and blood volume measured with magnetic susceptibility-based MR imaging - initial results. Radiology 223, 558–565 (2002)
4. Own, H.S., Hassanien, A.E.: Rough Wavelet Hybrid Image Classification Scheme. Journal of Convergence Information Technology (JCIT) 3(4), 65–75 (2008)
5. Dumitru, D.: Prediction of recurrent events in breast cancer using the Naive Bayesian classification. Annals of University of Craiova, Math. Comp. Sci. Ser. 36(2), 92–96 (2009) ISSN: 1223-6934
6. Fisher, R.A.: The Use of Multiple Measurements in Taxonomic Problems. Annals of Eugenics 7, 179–188 (1936)
7. Goscin, P.C., Berman, G.C., Clark, R.A.: Magnetic Resonance Imaging of the Breast. Cancer Control 8(5) (2001)
8. Han, J., Kamber, M.: Data Mining Concepts and Techniques. Morgan Kauffman Publishers, USA (2006)
9. Hassanien, A.E., Moftah, H.M., Azar, A.T., Shoman, M.: MRI breast cancer diagnosis hybrid approach using adaptive ant-based segmentation and multilayer perceptron neural networks classifier. Appl. Soft Comput. 14, 62–71 (2014)
10. Hassanien, A.E.: Fuzzy rough sets hybrid scheme for breast cancer detection. Image Vision Comput. 25(2), 172–183 (2007)
11. Leo, B.: Random Forests. Machine Learning 45(1), 5–32 (2001), doi:10.1023/A:1010933404324
12. Leibowitz, B.: Dawning of the age of angiogenesis. american boards of internal medicine and subspecialties of medical oncology and hematology (April 2004)
13. Lowe, D.: Object recognition from local scale-invariant features. In: Proceeding of the IEEE International Conference on Computer Vision, vol. 2, pp. 1150–1157 (1999)

14. Lowe, D.: Distinctive image features from scale-invariant keypoints. International Journal of Computer Vision 60(2), 91–110 (2004)
15. Moftah, H.M., Azar, A.T., Al-Shammari, E.T., Ghali, N.I., Hassanien, A.E., Shoman, M.: Adaptive k-means clustering algorithm for MR breast image segmentation. Neural Computing and Applications 24(7-8), 1917–1928 (2014)
16. Othman, A., Tizhoosh, H.: Image Classification using Evolving Fuzzy Inference Systems. In: IEEE International Conference on Fuzzy Systems, pp. 1435–1438. IEEE (2013)
17. Pianykh, O.: Digital Imaging and Communications in Medicine (DICOM). A Practical Introduction and Survival Guide. Springer (2008)
18. Al-Faris, A.Q., Ngah, U.K., Isa, N.A.M., Shuaib, I.L.: Breast MRI Tumor Segmentation using Modified Automatic Seeded Region Growing Based on Particle Swarm Optimization Image Clustering. In: Snášel, V., Krömer, P., Köppen, M., Schaefer, G. (eds.) Soft Computing in Industrial Applications. AISC, vol. 223, pp. 49–60. Springer, Heidelberg (2014)
19. Rouhi, P.: Role Of Angiogenesi. In: Cancer Invasion And Metastasis. Karolinska Institutet (2013)
20. Salama, G.I., Abdelhalim, M.B., Zeid, M.A.: Breast Cancer Diagnosis on Three Different Datasets Using Multi-Classifiers. International Journal of Computer and Information Technology (2277 - 0764) 01(01) (September 2012)
21. Vapnik, V.N.: The Nature of Statistical Learning Theory, 1st edn. Springer, New York (1995)
22. Vinukonda, P.: A Study of The Scale-Invariant Feature Transform on A Parallel Pipeline, B.TECH., JNTU University (May 2011)
23. Woollams, C.: Everything You Need To Know To Help You Beat Cancer, 4th ed., CANCERactive, a UK registered Charity (March 2013)
24. http://cancerimagingarchive.net/gettingaccess.html

Rough Set Based Feature Selection
for Egyptian Neonatal Jaundice

P.K. Nizar Banu[1], H. Hannah Inbarani[2],
Ahmad Taher Azar[3], Hala S. Own[4], and Aboul Ella Hassanien[5]

[1] Department of Computer Applications, B. S. Abdur Rahman University, Chennai, India
[2] Department of Computer Science, Periyar University, Salem, India
[3] Faculty of Computers and Information, Benha University, Benha, Egypt
[4] National Research Institute of Astronomy and Geophysics, Helwan, Egypt
[5] Faculty of Computers and Information, Cairo University, Cairo, Egypt
{nizarbanu,hhinba,halaown,aboitcairo}@gmail.com,
ahmad_t_azar@ieee.org

Abstract. This paper analyses rough set based feature selection methods for early intervention and prevention of neurological dysfunction and kernicterus that are the major causes of neonatal jaundice. Newborn babies develop some degree of jaundice which requires high medical attention. Improper prediction of diseases may lead to choose unsuitable type of treatment. Traditional rough set based feature selection methods and tolerance rough set based feature selection methods for supervised and unsupervised approach is applied for Egyptian neonatal jaundice dataset. Features responsible for prediction of Egyptian neonatal jaundice is analyzed using supervised quick reduct, supervised entropy based reduct and Unsupervised Tolerance Rough Set based Quick Reduct (U-TRS-QR). Results obtained demonstrate features selected by U-TRS-QR are highly accurate and will be helpful for physicians for early diagnosis.

Keywords: Neonatal Jaundice, Rough Sets, U-TRS-QR, Quick Reduct, Entropy.

1   Introduction

Neonatal jaundice is when a baby has high level of bilirubin in the blood. Bilirubin is a yellow pigment produced during normal breakdown of red blood cells. Normally, bilirubin is processed by the liver and then passed through the intestinal tract. Newborns still-developing lever may not be mature enough to remove bilirubin. Babies born before 38 weeks of gestation, babies who are not getting enough breast milk and babies whose blood type is not compatible with their mothers are at the highest risk for developing newborn jaundice. Physiological jaundice and Pathological jaundice are the two basic types of jaundice. Physiologic jaundice usually occurs on second or third day of life and disappears up to seventh day. Pathologic jaundice is a result of different causes such as hemolysis, infections and group incompatibilities between mother and child [1]. Increased bilirubin cause infant's skin and whiteness of the eyes to look yellow. Kernicterus, or bilirubin

A.E. Hassanien et al. (Eds.): AMLTA 2014, CCIS 488, pp. 367–378, 2014.

encephalopathy, is a condition caused by bilirubin toxicity to the basal ganglia and various brainstem nuclei. In the acute phase, severely jaundiced infants become lethargic, hypotonic and suck poorly. If the hyperbilirubinemia is not treated, the infant becomes hypertonic and may develop a fever and a high-pitched cry. The hypertonia is manifested by backward arching of the neck and trunk. Surviving infants usually develop a severe form of athetoid cerebral palsy, hearing loss, dental dysplasia, paralysis of upward gaze and less often intellectual and other handicaps [2]. In [3], Artificial Neural Networks is applied for the diagnosis of neonatal jaundice. Weighted rough set framework for intervention and prevention of Egyptian neonatal jaundice is presented in [4].

Medical datasets have accumulated large quantities of information about patients and their medical conditions. Relationships and patterns within these data could provide new medical knowledge [5-8].Advanced and intelligent techniques have been used in medical data analysis such as neural network, Bayesian classifier, genetic algorithms [9-11], fuzzy theory and rough set. Each technique contributes a distinct methodology for addressing problems in its domain [4].

Rough set theory [12-14], is a new intelligent technique that has been applied to the medical domain and is used for the discovery of data dependencies, evaluates the importance of attributes, discovers the patterns of data, reduces all redundant objects and attributes and finds the minimum subset of attributes. Unsupervised PSO based Relative Reduct is applied for fetal heart rate for finding the most informative features in [15]. Unsupervised feature selection using tolerance rough set based quick reduct for mammogram images is presented in [16]. Unsupervised Quick Reduct proposed in [17], Unsupervised Relative Reduct proposed in [18], U-TRS-RelRed proposed in [19], is applied for gene expression dataset in [20]. In [21], US-PSO-RR and USRR methods are applied for fetal heart rate and proved US-PSO-RR is better in terms of both classification and clustering accuracies.

In [4], weighted rough set framework is applied for real time Egyptian neonatal jaundice dataset. First, a weighted attribute reduction algorithm is applied to find the reduct set. Finally, a set of diagnosis rules are extracted based on weighted MLEM2 algorithm. Authors of [4], proved weighted rough set performs better than weighted SVM and decision tree.

In general, some features in the dataset may not be relevant to describe the problem and degrades classification performance. Finding significant features is very important to improve classification performances and to predict and prevent disease in early stage. In this paper we have applied two supervised feature selection methods and one unsupervised feature selection method for Egyptian neonatal jaundice's decision system and information system respectively.

This paper is organized as follows: section 2 discusses rough set based feature selection methods. Egyptian neonatal jaundice dataset is described and experimental analysis is presented in section 3. Conclusion is presented in section 4.

2   Rough Set Based Feature Selection

Feature selection finds a subset of original features which provides the most useful information by preserving the significant details present in a given dataset [15]. Feature selection can be categorized as supervised and unsupervised. Supervised feature selection requires class (decision) attribute to find reduct set, whereas unsupervised feature selection does not require class attribute. Egyptian neonatal jaundice dataset has 16 conditional attributes and 1 decision attribute with 3 classes. We have applied supervised feature selection methods: quick reduct and entropy based reduct and unsupervised feature selection method: U-TRS-QR for the Egyptian neonatal jaundice dataset. Algorithms are discussed below.

2.1   Quick Reduct

Quick reduct algorithm finds a reduct without exhaustively generating all possible subsets. It starts with an empty set and adds one feature which results in greatest increase in the rough set dependency until it produces maximum possible value for the dataset [22]. Quick Reduct algorithm is shown in Figure1.

Let $I = (U, A)$ be an information system, where U is a nonempty set of finite objects and A is a nonempty finite set of attributes such that $a\colon U \rightarrow V_a$ for every $a \in A$. V_a is the set of values that attribute a may take. With any $P \subseteq A$ there is an associated equivalence relation $IND(P)$.

$$IND(P) = \left\{ (x, y) \in U^2 \mid \forall_a \in P, a(x) = a(y) \right\} \qquad (1)$$

Lower Approximation refers to the set of cases that definitely belongs to a given class. Upper approximation refers to the set of cases that possibly belongs to a given class. Let $X \subseteq U$, the P-lower approximation $\underline{P}X$ and P-upper approximation $\overline{P}X$ of a set X can be defined as

$$\underline{P}X = \left\{ x \mid [x]_p \subseteq X \right\} \qquad (2)$$

$$\overline{P}X = \left\{ x \mid [x]_p \cap X \neq \varnothing \right\} \qquad (3)$$

Positive region of the partition U/Q with respect to P, is the set of all elements of U that can be uniquely classified as blocks of the partition, U/Q by means of P, can be defined as

$$POS_p(Q) = \bigcup_{X \in U/Q} \underline{P}X \qquad (4)$$

Using this positive region, rough set degree of dependency of a set of attributes Q on a set of attributes P is defined as

$$\gamma_P(Q) = \frac{|POS_P(Q)|}{|U|} \qquad (5)$$

2.2   Entropy Based Reduct (EBR)

Entropy-based reduct developed in [23], is based on the entropy heuristic employed by machine learning techniques such as C4.5 [24]. EBR is concerned with examining a dataset and determining those attributes that provide the most gain in information. Entropy of attribute A with respect to the conclusion C is defined as

$$H(C \mid A) = -\sum_{j=1}^{m} p(a_j) \sum_{i=1}^{n} p(c_i \mid a_j) \log_2 p(c_i \mid a_j) \qquad (6)$$

Entropy of every attribute is computed. Search for the best feature subset is stopped when the resulting subset entropy is the lowest. Any subset with entropy 0 will also have a corresponding rough set dependency of 1. Entropy based Reduct algorithm is shown in Figure 2.

Algorithm: Quick Reduct
Quick Reduct (C,D)
C, Set of all conditional features
D, Set of decision features
$R \leftarrow \{\ \}$
do
$T \leftarrow R$
$\forall x \in (C - R)$
$if\ \gamma_{R \cup \{x\}}(D) > \gamma_T(D)$
$T \leftarrow R \cup \{x\}$
$R \leftarrow T$
$until\ \gamma_R(D) = \gamma_c(D)$
return R

Fig. 1. Quick Reduct

Algorithm: Entropy Based Reduct
EBR(C, D)
C, set of all conditional features
D, set of decision features
$R \leftarrow \{\ \}$
do
$T \leftarrow R$
$\forall x \in (C - R)$
$if H(D \mid R \cup \{x\}) < H(D \mid T)$
$T \leftarrow R \cup \{x\}$
$R \leftarrow T$
$until H(D \mid R) = H(D \mid C)$
return R

Fig. 2. Entropy Based Reduct

2.3   Unsupervised Tolerance Rough Set based Quick Reduct (U-TRS-QR)

Tolerance based rough set theory [25-26] is proposed as an extension of original theory; which is defined as a measure of similarity of feature values. Lower and Upper approximations are also based on these similarity measures. Transitivity constraint is relaxed by introducing some degree of flexibility in tolerance rough sets. Due to the additional flexibility of employing similarity measures, some features may belong to more than one tolerance class. A similarity measure is defined for each attribute in Tolerance Rough Set approach. A standard measure given in [27] is

$$SIM_a(x, y) = 1 - \frac{|a(x) - a(y)|}{a_{max} - a_{min}} \tag{7}$$

Where 'a' is the attribute under consideration, and a_{max} and a_{min} denote the maximum and minimum values for the feature taken. For a subset of features P, similarity can be achieved as follows

$$(x, y) \in SIM_{P,\tau} iff \prod_{a \in P} SIM_a(x, y) \geq \tau \tag{8}$$

$$(x, y) \in SIM_{P,\tau} iff \frac{\sum_{a \in P} SIM_a(x, y)}{|P|} \geq \tau \tag{9}$$

Lower $\underline{P_\tau}X$ and Upper $\overline{P_\tau}X$ approximations are defined in a similar way to traditional rough set theory.

$$\underline{P_\tau}X = \left\{ x \mid SIM_{p,\tau}(x) \subseteq X \right\} \tag{10}$$

$$\overline{P_\tau}X = \left\{ x \mid SIM_{p,\tau}(x) \cap X \neq \varnothing \right\} \tag{11}$$

Positive region and dependency functions are as follows

$$POS_{P,\tau}(y) = \bigcup_{X \in U | Y \in U} \underline{P_\tau}X \, ; \, \gamma_{P,\tau}(y) = \frac{\left| POS_{P,\tau}(y) \right|}{|U|} \tag{12}$$

Let $I = (U, A)$ be an information system. Where U is the universe with a non-empty set of finite objects, A is a non-empty finite set of condition attributes. $\forall a \in A$, There is a corresponding function $f_a : U \to V_a$, where V_a is the set of values of a. Here, $\tau \in [0,1]$ is a similarity threshold. τ determines the required level of similarity for inclusion within tolerance classes. With the help of the above mentioned definitions, feature reduction methods use the tolerance based degree of similarity, $\gamma_{P,\tau}(y)$, to estimate the significance of subsets [20]. We have fixed $\tau = 0.8$, for our experimentation. U-TRS-QR algorithm is shown in Figure 3.

Algorithm: U-TRS-QR
U-TRS-QR (C) **C**, Set of all conditional features; **U**, Universe – Set of all objects **τ**, Threshold; **R**, Reduct Set $R \leftarrow \{ \ \} ; \gamma_{Best}^{\tau} = 0$ do $T \leftarrow R$ $\gamma_{Prev}^{\tau} = \gamma_{Best}^{\tau}$ $\forall x \in (C - R)$ $\forall y \in (C)$ $\gamma_{R \cup \{x\}, \tau}(y) = \dfrac{\left| POS_{R \cup \{x\}}(y) \right|}{|U|}$ $if \ \overline{\gamma_{R \cup \{x\}, \tau}}(y) \forall y \in C > \overline{\gamma T, \tau}(y), \forall y \in C$ $T \leftarrow R \cup \{x\}$ $\gamma_{Best}^{\tau} = \overline{\gamma T, \tau}(y)$ $R \leftarrow T$ $until \gamma_{Best}^{\tau} == \gamma_{Prev}^{\tau}$ return R

Fig. 3. U-TRS-QR

3   Experimental Analysis and Discussion

This section discusses on jaundice dataset, experiments carried out, features selected and performance measures of rough set based feature selection algorithms.

3.1   Dataset Description

We have taken the dataset used in [4]. This is a real time dataset collected from the newborns during January to December 2007 in Neonatal Intensive Care Unit in Cairo, Egypt. This dataset consists of 16 features with 3 classes. Dataset with its description is shown in Table 1. Original dataset consists of null values. For our experimentation, we filled null values with average of the corresponding attribute.

Quick Reduct, Entropy based Reduct and U-TRS-QR are applied for Neonatal jaundice dataset for selecting more dominating features for predicting the risk of neonatal jaundice and extreme hyperbilirubinemia. First, the experiment is conducted for all 808 records, and then the dataset is split into 100 records sequentially which has combination of three classes. Participation of direct hyperbilirubinemia is less in the split because it has only 3% of records. Dataset of 808 records is shown in Table 2.

Table 1. Neonatal jaundice data and description

S. No	Attribute Name	Description
1	Sex	Male or Female
2	Age / day	Postnatal age in days on admission
3	Gest. Age	Gestational age (F= full term, N=near term, P=preterm)
4	Wt/g	Weight in grams on admission
5	Onset of J at day	Postnatal age of patient on the day in which onset of jaundice was occurred
6	Days of adm.	Days of admission in hospital
7	Peak of T bil	Peak of total bilirubin level
8	bil peak at day	Postnatal age of patient on the day in which total bilirubin peak was recorded
9	T bil d of presentation	Total bilirubin level on the day of presentation
10	D bil d of presentation	Direct bilirubin level on the day of presentation
11	T bil 24h later	Total bilirubin level after 24 hours of presentation
12	D bil 24h later	Direct bilirubin level after 24 hours of presentation
13	T bil after 2day	Total bilirubin level after 2 days of presentation
14	D bil after 2day	Direct bilirubin level after 2 days of presentation
15	T bil before disc	Total bilirubin level before discharge from hospital or death
16	D bil before disc	Direct bilirubin level before discharge from hospital or death
17	Pattern (Class Attribute)	1. Patient with indirect hyperbilirubinemia 2. Patients with indirect hyperbilirubinemia then changed into direct hyperbilirubinemia 3. Patents with direct hyperbilirubinemia

Table 2. Jaundice dataset and its distribution

S. No	Class Name	Class Size	Class Distribution
1	Indirect hyperbilirubinemia	737	91%
2	Changed from indirect to direct hyperbilirubinemia	41	13%
3	Direct hyperbilirubinemia	25	3%

Quick reduct algorithm finds the dependency of each attribute, the attribute with highest dependency is chosen and its all possible combinations are generated. This process is carried out until the dependency of reduct equals the consistency of dataset. It is 1, if the dataset is consistent. Entropy based reduct first evaluates the entropy of each individual attribute. Attribute with lowest entropy is selected first, and then entropy of all subsets containing this attribute is computed. Subset which gives the lowest entropy is chosen as the entropy based reduct.

As stated in [20], U-TRS-QR deals with conditional attributes alone and the similarity of each attribute is calculated using Eq. 7. Features that result in the highest increase in the tolerance rough set similarity metric are selected. Feature with highest similarity measure is taken and all possible combinations of the selected feature are constructed. Similarities of the selected features with different combinations are computed. This process is carried out until the selected subset produces the maximum similarity for the dataset. The algorithm terminates after evaluating the similarity of the selected feature subset with other subsets and returns the best feature subset.

3.2   Features Selected by the Rough Set Based Algorithms

To avoid informal decision regarding the type of treatment, diseases must be predicted correctly by the physicians. For prediction, features play a major role. Medical datasets may have more number of features, considering all the features for diagnosis may take more time for diagnosis. Hence features suitable for prediction of disease are necessary. Egyptian neonatal jaundice dataset has 16 features with 808 samples. Feature selection methods tries to find subset of original feature set by preserving the original characteristics of all the features. Out of 16 features in the dataset considered for experimentation, 5 features are selected by quick reduct, 7 by entropy based reduct and 12 by U-TRS-QR as shown in Tables 3, 4 and 5. Class attribute is not considered for U-TRS-QR method since it is unsupervised feature selection method.

Final reduct set, used to generate list of rules for classification by [4] is shown in Table 6. U-TRS-QR method selects all the features shown in [4], hence features of weighted rough set framework becomes a subset of U-TRS-QR method and this method does not rely on decision attribute, its F-measure is also high.

Table 3. Features selected by quick reduct

S. No	Features
1	Sex
2	Age / day
3	Days of adm.
4	T bil d of presentation
5	D bil before disc

Table 4. Features selected by entropy based reduct

S. No	Features
1	Sex
2	Gest. Age
3	Wt/g
4	Days of adm.
5	Peak of T bil
6	bil peak at day
7	D bil before disc

To evaluate the performance measures of reduct sets, J48 decision tree classifier is used. For all the three feature selection algorithms 66% split is used for training and the remaining 44% is used for testing. It is observed, U-TRS-QR performs better than other feature selection methods used for comparison. Advantage of using U-TRS-QR is, it does not depend on class attribute alone. It finds dependencies among conditional attributes rather than the class attribute, hence it is easy to intervene and prevent neurological dysfunction due to neonatal jaundice. Table 7. shows the classification % of the algorithms for 44% of test data.

Table 5. Features selected by U-TRS-QR

S. No	Features
1	Age / day
2	Wt/g
3	Days of adm.
4	Peak of T bil
6	T bil d of presentation
7	D bil d of presentation
8	T bil 24h later
9	D bil 24h later
10	T bil after 2day
11	D bil after 2day
12	T bil before disc
13	D bil before disc

Table 6. Reduct Set of Weighted Rough Set Framework

S. No	Features
1	D bil d of presentation
2	D bil 24h later
3	T bil after 2day
4	T bil before disc
5	D bil before disc

Table 7. Classification Results for 44% of 808 dataset

S. No	Method	Correctly Classified Instances	Incorrectly Classified Instances	% of Correct Classification
1	Quick Reduct	261	14	94.91
2	Entropy Based Reduct	266	9	96.73
3	Weighted Rough Set Framework	262	13	95.27
4	U-TRS-QR	263	12	95.63
5	Original data set with all attributes	266	9	96.73

Confusion matrix of all four feature selection methods for 44% of Egyptian neonatal dataset is shown in Table 8.

Table 8. Confusion Matrix for 44% of 808 dataset

Predicted / Actual	Quick Reduct			Entropy Based Reduct			Weighted Rough Set Framework			U-TRS-QR			Original data set		
	A	B	C	A	B	C	A	B	C	A	B	C	A	B	C
A	246	1	0	246	1	0	246	1	0	246	0	1	245	1	1
B	1	11	5	1	14	2	4	9	4	3	10	4	2	13	2
C	0	7	4	0	5	6	0	4	7	0	4	7	0	3	8

Table 9, shows the performance measures for 44% of 808 records with 16 features. As illustrated in Table 8, performance on Class A reported by F-Measure is same for U-TRS-QR, and original dataset but quick reduct achieves best F-measure. For Class B, entropy based reduct achieves best F-Measure which is closer to original dataset. For Class C, weighted rough set frame work gives better results based on F-Measure.

Table 9. Performance measures of Egyptian neonatal jaundice dataset

S. No	Method	Class	TP Rate	FP Rate	Precision	Recall	F-Measure
1	Quick Reduct	A	0.9960	0.0360	0.9960	0.9960	0.9960
		B	0.6470	0.0310	0.5790	0.6470	0.6110
		C	0.3640	0.0190	0.4440	0.3640	0.4000
2	Entropy Based Reduct	A	0.9960	0.0360	0.9960	0.9960	0.9960
		B	0.8240	0.0230	0.7000	0.8240	0.7570
		C	0.5450	0.008	0.7500	0.5450	0.6320
3	Weighted Rough Set Framework	A	0.9960	0.1430	0.9840	0.9960	0.9900
		B	0.5290	0.0190	0.6430	0.5290	0.5810
		C	0.6360	0.0150	0.6360	0.6360	0.6360
4	U-TRS-QR	A	0.9960	0.1070	0.9880	0.9960	0.9920
		B	0.5880	0.0160	0.7140	0.5880	0.6450
		C	0.6360	0.0190	0.5830	0.6360	0.6090
5	Original data set with all attributes	A	0.9920	0.0710	0.9920	0.9920	0.9920
		B	0.7650	0.0160	0.7650	0.7650	0.7650
		C	0.7270	0.0110	0.7270	0.7210	0.7270

To better understand the working procedure of the feature selection methods discussed in this paper, we trained 66% of data and tested for the whole dataset. U-TRS-QR is an unsupervised method which attains good results compared to other supervised methods. Table 10 reports the classification performance and confusion matrix for the whole dataset. U-TRS-QR method achieves best results for Class A, B and C. This is because unsupervised method does not depend only on decision attribute which forces to place the records in any one of the classes without knowing how dependent they are among the conditional attributes.

Experts suggest that they require the following 7 features: Gest. Age, wt/g, T bil d of presentation, D bil d of presentation, T bil after 2 day, D bil after 2 day. Though quick reduct and entropy based reduct gives best results for the 44% of dataset, it does not select more mandatory features as U-TRS-QR. As shown in [4], weighted rough set framework also selects only 2 out of 7 mandatory features and it is to be noted that it is a supervised feature selection method. It looks for the dependency with class attribute alone.

U-TRS-QR, an unsupervised feature selection method proposed in this paper for jaundice dataset, selects 12 features which comprises 6 mandatory features namely, wt/g, T bil d of presentation, D bil d of presentation, T bil after 2 day, D bil after 2 day. This method shows how one feature is dependent on the other feature; as a result other 6 features selected by the proposed method insist physicians these also have impact on neonatal jaundice and may assist them in drug development and for early diagnosis of disease. Out of 737 samples, 733 are correctly predicted as Class A, 36 out of 46 as Class B and 19 out of 25 as Class C as illustrated in Table 10.

Table 10. Confusion Matrix and Performance measures for full dataset

Predicted / Actual	A	B	C	Feature Selection Methods	TP Rate	FP Rate	Precis-ion	Recall	F-Measure
A	730	7	0	Quick Reduct	0.9910	0.0700	0.9930	0.9910	0.9920
B	4	36	6		0.7830	0.0280	0.6320	0.7830	0.6990
C	1	14	10		0.4000	0.0080	0.6250	0.4000	0.4880
A	730	7	0	Entropy Based Reduct	0.9910	0.0700	0.9930	0.9910	0.9920
B	4	30	12		0.6520	0.0250	0.6120	0.6520	0.6320
C	1	12	12		0.4800	0.0150	0.5000	0.4800	0.4900
A	732	5	0	Weighted Rough Set Framework	0.9930	0.1130	0.9890	0.9930	0.9910
B	8	32	6		0.6960	0.0160	0.7270	0.6960	0.7110
C	0	7	18		0.7200	0.0080	0.7500	0.7200	0.7350
A	733	4	0	U-TRS-QR	0.9950	0.1130	0.9890	0.9950	0.9920
B	8	36	2		0.7830	0.0130	0.7830	0.7830	0.7830
C	0	6	19		0.7600	0.0030	0.9050	0.7600	0.8260
A	732	5	0	Original data set	0.9930	0.1130	0.9890	0.9930	0.9910
B	8	36	2		0.7830	0.0140	0.7660	0.7830	0.7740
C	0	6	19		0.7600	0.0030	0.9050	0.7600	0.8260

4   Conclusion

Feature selection methods discussed in this paper include both traditional rough set and tolerance rough set. Tolerance rough set has a relaxation of similarity rather than finding dependency alone. It is clearly illustrated in the experiments carried out. Advantage of using tolerance rough set and unsupervised approach for Egyptian neonatal jaundice dataset is proved to be better in terms of performance measures.

It is proved that U-TRS-QR, an unsupervised approach selects highly informative features as expected by the medical experts without depending on class attribute. Association of conditional attributes and its subset is focused for selecting the important features as a subset by preserving the characteristics of original features.

According to the experts, Gest. Age is also an important feature which defines full term, near term or preterm baby. In future, readers can focus on feature selection method which selects all 7 mandatory features or by tuning the necessary parameters discussed in the algorithms.

References

1. Sohani, M., Kermani, K.K.: A Neuro-Fuzzy Approach to Diagnosis of Neonatal Jaundice. Bio Inspired Models of Network, Computing and Information Systems, pp. 101–104 (2006)
2. http://pediatrics.aappublications.org/cgi/content/full/108/3/763
3. Shrivastava, S.: Diagnosis of Neonatal Jaundice using Artificial Neural Networks. International Indexed & Refereed Research Journal 4, 43–44 (2013)
4. Own, H.S., Abraham, A.: A new weighted rough set framework based classification for Egyptian NeoNatal Jaundice. Applied Soft Computing 12, 999–1005 (2012)
5. Zhong, N., Skowron, A.: Rough sets based knowledge discovery process. International Journal of Applied Mathematics and Computer Science 11(3), 603–619 (2001)
6. Carlin, S., Komorowski, J., Ohrn, A.: Rough set analysis of medical datasets: a case of patients with suspected acute appendicitis. In: Proceedings of Workshop on Intelligent Data Analysis in Medicine and Pharmacology (ECAI 1998), Brighton, UK, pp. 18–28 (1998)
7. Cios, K., Pedrycz, W., Swiniarski, R.: Data Mining Methods for Knowledge Discovery. Kluwer Academic, Boston (1998)
8. Lavrajc, N., Keravnou, E., Zupan, B.: Intelligent Data Analysis in Medicine and Pharmacology. Kluwer Academic, Boston (1997)
9. Beligiannis, G., Hatzilygeroudis, I., Koutsojannis, C., Prentzas, J.: A GA driven intelligent system for medical diagnosis. In: Gabrys, B., Howlett, R.J., Jain, L.C. (eds.) KES 2006, Part I. LNCS (LNAI), vol. 4251, pp. 968–975. Springer, Heidelberg (2006)
10. Adamopoulos, A.V., Anninos, P.A., Likothanassis, S.D., Beligiannis, G.N., Skarlas, L.V., Demiris, E.N., Papadopoulos, D.: Evolutionary self-adaptive multimodel prediction algorithms of the fetal magnetocardiogram. In: Proceedings of the 14th International Conference on Digital Signal Processing (DSP 2002), pp. 1149–1152 (2002)
11. Beligiannis, G.N., Skarlas, L.V., Likothanassis, S.D., Perdikouri, K.G.: Nonlinear model structure identification of complex biomedical data using a genetic programming-based technique. IEEE Transactions on Instrumentation and Measurement 54, 2184–2190 (2005)
12. Pawlak, Z.: Rough sets. International Journal of Computer and Information Science, 341–356 (1982)
13. Pawlak, Z.: Rough Sets Theoretical Aspect of Reasoning about Data. Kluwer Academic, Boston (1991)
14. Pawlak, Z., Grzymala-Busse, J., Slowinski, R., Ziarko, W.: Rough sets. Communications of the ACM 38(11), 89–95 (1995)
15. Azar, A.T., Nizar Banu, P.K., Hannah Inbarani, H.: PSORR— An unsupervised feature selection technique for fetal heart rate. In: Proceedings of the 5th International Conference on Modelling, Identification and Control (ICMIC 2013), Cairo, Egypt, August 31-September 2, pp. 60–65 (2013)

16. Aroquiaraj, I., Thangavel, K.: Unsupervised feature selection in digital mammogram image using tolerance rough set based quick reduct. In: Proceedings of the Fourth International Conference on Computation Intelligence and Communication Networks, pp. 436–440 (2012)

17. Velayutham, C., Thangavel, K.: Unsupervised quick reduct algorithm using rough set theory. Journal of Electronic Science and Technology 9(3), 193–201 (2011)

18. Velayutham, C., Thangavel, K.: Unsupervised feature selection using rough set. In: Proceedings on International Conference-Emerging Trends in Computing, pp. 307–314 (2011)

19. Hannah Inbarani, H., Nizar Banu, P.K.: Unsupervised feature selection using tolerance rough set based relative reduct. In: Proceedings of the International Conference on Advances in Engineering, Science and Management, pp. 326–331 (2012)

20. Nizar Banu, P.K., Hannah Inbarani, H.: A Comparative analysis of rough set based intelligent Techniques for Unsupervised Gene Selection. International Journal of System Dynamics Applications 2(4), 33–46 (2013)

21. Hannah Inbarani, H., Nizar Banu, P.K., Azar, A.T.: Feature Selection using swarm-based relative reduct technique for fetal heart rate. Neural Computing Applications, 1–14 (2014)

22. Jenson, R., Shen, Q.: Rough Computing, Theories, Technologies and applications. In: Hassanien, A.E., et al. (eds.) Roughset based feature selection – A review. Information Science Reference, pp. 70–103 (2008)

23. Jensen, R., Shen, Q.: Fuzzy-Rough attribute reduction with application to web categorization. Fuzzy Sets and Systems 141(3), 469–485 (2004)

24. Quinlan, J.R.: C4.5: Programs for Machine Learning. Kaufmann Series in Machine Learning. Kaufmann Publishers, San Maeto (1993)

25. Skowron, A., Stepaniuk, J.: Tolerance approximation spaces. Fundamenta Informaticae 27(2), 245–253 (1996)

26. Slowinski, R., Vanderpooten, D.: Similarity relation as a basis for rough approximations. In: Wang, P. (ed.) Advances in Machine Intelligence and Soft Computing, vol. IV, pp. 17–33. Duke Univ. Press (1997)

27. Stepaniuk, J.: Similarity based rough sets and learning. In: Tsumoto, S., Kobayashi, S., Yokomori, T., Tanaka, H. (eds.) Proceedings of the Fourth International Workshop on Rough Sets, Fuzzy Sets and Machine Discovery, Tokyo, Japan, pp. 18–22 (1996)

Sparse ICA Based on Infinite Norm
for fMRI Analysis

Liang Chen[1], Shigang Feng[1], Weishi Zhang[1],
Aboul Ella Hassanien[2], and Hongbo Liu[1]

[1] School of Information, Dalian Maritime University, Dalian 116026, China
[2] Information Technology Department, Cairo University, Giza, Egypt
a.hassanien@fci-cu.edu.eg, lhb@dlmu@dlmu.edu.cn

Abstract. Functional MRI (fMRI) is a functional neuroimaging technique that measures the brain activity by detecting the associated changes in blood flow. Independent component analysis (ICA) provides a feasible approach to analyze the collected data sets. In this paper, we introduce a novel criterion via infinity norm to achieve the sparse solution. The experimental result has been shown that the approach can be successfully applied in fMRI data. In memory-imagine cognitive experiment, the activated regions for different tasks are different in brain. But some regions are activated in each runs, which suggests that these brain regions may play an important role in cognition functions of memory-imagine.

Keywords: Sparse, ICA, Infinite Norm, fMRI.

1   Introduction

Functional magnetic resonance imaging (fMRI) is a functional neuroimaging technique based on blood oxygen level dependent (BOLD) signal. Previous studies [5,7] show that the task related signal and noise signal caused by physiological or logical reasons are independent. So the fMRI signal could be seen as a group of time series constructed by several statistical independent components. Therefore we could make use of independent component analysis (ICA) method for fMRI data analysis, which is helpful to obtian brain's activity stimuli factors and further to determine the brain stimuli reaction areas.

Currently ICA algorithm is available to separate mixed functional magnetic resonance imaging (fMRI) signals [2,4,12,18]. Mckeown [12] and Calhoun [2,3] use it to extract the initialized characteristics of the collected fMRI signals. Duann [4] use ICA to obtain the initial features of fMRI signals, to detect the changes of hemodynamic response related to fMRI experiment tasks. Zarzoso and Comon proposed RobustICA [17,18], which is faster than FastICA algorithm, and improves the signal containing bad points and the robustness under the condition of pseudo local extremum points. Sparsity is a new promising method of independent component independence [11,14,16]. Witten and Tibshirani proposed an approach of sparse principal components and canonical correlation analysis which utilize the property of penalized matrix decomposition [15].

A.E. Hassanien et al. (Eds.): AMLTA 2014, CCIS 488, pp. 379–388, 2014.

In this paper, we will propose a new ICA algorithm based on infinity norm which could help us to obtain the coefficient distribution with sparse characteristic.

2 Methodology

Before introducing the Sparse ICA, we briefly review the classical assumption of blind source separate (BSS) problem in fMRI data.

The BSS problem in fMRI data, could be described as an adaptive linear optimization problem [1, 13]. Considering the observed fMRI data, we assume that the fMRI data $X = (x_1, x_2, ..., x_n)^T$ is a group linear mixed data which is mixed by independent components (ICs) $S = (s_1, s_2, ..., s_m)^T$ via a mixture matrix A of size $n \times m$. It could be described as (1), where n is time points and m is the number of ICs . For the sake of simplicity, we assume that $n = m$ in this paper.

$$X = AS \tag{1}$$

The centralization process could be used as the preprocessing method in ICA. In fMRI time series, there is an interesting statistical property, the double zero means property. Without loss of generality, we assumed that the $x_i = (x_{i1}, x_{i2}, ..., x_{iV})$. V represents the number of voxels of brain. Then the fMRI data could be described as (2). If the each column of matrix X minus mean value of corresponding column, both of rows and columns are zero means in sense of probability. This could help us to focus on both variates of BOLD signal and traditional ICA assumption.

$$X = \begin{pmatrix} x_{11} & x_{12} & \cdots & x_{1V} \\ \vdots & \vdots & & \vdots \\ x_{n1} & x_{n2} & \cdots & x_{nV} \end{pmatrix} \tag{2}$$

Data whitening is another preprocessing in ICA. We whiten X to have (3). This preprocessing could lead the mean square of data to unitary. It can simplify and speed up ICA algorithms. In order to simplify the symbolic representation, we still use the X to represent the whitened matrix Z in the rest of the parts. B is the whitening matrix which could calculated by singular value decomposition (SVD) method.

$$Z = BX \tag{3}$$

In order to separate the source signal from fMRI data, we project X onto a weight matrix $W = (w_1, w_2, ..., w_n)^T$ to produce the output $Y = W^T X$. Then the blind source separation problem is transformed into the optimization of W, which could help Y to turn into sparsest extraction. It is easy to prove that $Y = W^T X$ and (1) are equivalent.

The sparse property could be general as a penalty function shown as (4), where $S(y_i)$ can be calculate by $-e^{-y_i^2}$, $log(1 + y_i^2)$ or $|y_i|$, where y_i denotes voxel i.

$$Sparse(Y) = -\sum_i S(y_i) \tag{4}$$

If we define $S(y_i)$ as a $p-$norm to formulate a penalty function, we could have (5), where $p > 0$.

$$Sparse(Y) = -E\left\{ \lim_{p\to\infty} \left(\sum_i |y_i|^p\right)^{\frac{1}{p}} \right\} \tag{5}$$

If p verges to infinity, $\left(\sum_i |y_i|^p\right)^{\frac{1}{p}}$ converge on the maximum value of $|y_i|$, which could be described as (6).

$$\lim_{p\to\infty} \left(\sum_i |y_i|^p\right)^{\frac{1}{p}} = \max_i\{|y_i|\} \tag{6}$$

Then the objective function of sparse ICA could be define as (7).

$$\begin{aligned} F(Y) &= E\left\{ \lim_{p\to\infty} \left(\sum_i |y_i|^p\right)^{\frac{1}{p}} \right\}^2 \\ &= E\left\{ \max_i\{|y_i|^2\} \right\} \end{aligned} \tag{7}$$

While applying on fMRI data, the source signal $|y_i|^2$ could be replaced by $(w_i^T x)^2$ where w_i denotes the i-th column of unmixing matrix W, X the mixture signal of fMRI data. Then we could have an objective function of unmixing matrix W, which is defined as (8).

$$\begin{aligned} F(W) &= E\left\{ \max_i\{|w_i^T X|^2\} \right\} \\ &= \int \max_i\{|w_i^T X|^2\} p(X)dX \end{aligned} \tag{8}$$

The main difficulty in sparse ICA is the incontinuity of (8), because we need to compute $\partial F(W)/\partial W$ to formulate the optimization criterion. So, we will utilize (6) to realize the goal. We assume that $c = arg\max_i\{|w_i^T X|^2\}$. Then we have (9).

$$(w_c^T X)^2 = \max_i \{|w_i^T X|^2\}$$

$$= \lim_{p \to \infty} \left(\sum_i |w_i^T X|^2 \right)^{\frac{1}{p}} \tag{9}$$

And then, let $Q = \sum_i (w_i^T X)^{2p}$, (9) can be rewriten as (10).

$$(w_c^T X)^2 = \lim_{p \to \infty} Q^{\frac{1}{p}} \tag{10}$$

Therefore, the optimization criterion of sparse ICA could be defined as (11). On the other hand, it is easy to find that $\partial Q^{\frac{1}{p}} / \partial w_i$ could be induced by (12).

$$\frac{\partial F(W)}{\partial w_i} = \int \lim_{p \to \infty} \frac{\partial Q^{\frac{1}{p}}}{\partial w_i} p(X) dX \tag{11}$$

$$\frac{\partial Q^{\frac{1}{p}}}{\partial w_i} = \frac{1}{p} Q^{\frac{1-p}{p}} \cdot \frac{\partial Q}{\partial w_i} = 2Q^{\frac{1}{p}} \cdot \frac{(w_i^T X)^{2p-1}}{Q} X \tag{12}$$

While $p \to \infty$, $(w_i^T X)^{2p} / Q$ equals to 1 or 0. If and only if $i = c$, the algebraic expression will equal to 1.

To sum up, the optimization criterion of sparse ICA could be calculated by (13).

$$\frac{\partial F(W)}{\partial w_i} = \begin{cases} \int 2 |w_i^T X| X p(X) dX & \text{iff } i = c \\ 0 & \text{others} \end{cases} \tag{13}$$

Another problem is the evaluation of $p(X)$ while applying sparse ICA algorithm on fMRI data. We will utilize the unbiased estimate to solve this problem.

$$\frac{\partial F(W)}{\partial w_i} = \begin{cases} \frac{\sum 2 |w_i^T X| X}{n} & \text{iff } i = c \\ 0 & \text{others} \end{cases} \tag{14}$$

The optimization criteria of discrete data could be shown as (14), where n equals to the time point number of fMRI time series. Finally, our sparse ICA method is illustrated in Algorithm 1.

3   Data Acquisition

3.1   Subjects and Parameter

A total of 21 Chinese college students (10 male and 11 female, age at 19-26 year old) recruited from the local population of graduates or undergraduates at Dalian Maritime University participated in this study. Informed consent was obtained before participation.

The iterations of Algorithm 1 is set to 500. $m = n$ equals to the time point of fMRI scanner sequence. Just in memory-imagine experiment, the number of time points is 84.

Algorithm 1. Sparse ICA based on Infinite Norm

Require:

  fMRI data set $X_{n \times V}$

Ensure: independent components $Y_{m \times V}$

1: Do zero mean process for X

2: $X = X - mean(X)$

3: Do whitening process for X

4: $X = BX$

5: **for** $sweep = 1$ to $iterine$ **do**

6: $\quad \alpha = 1e - 3 * sqrt(1/sweep)$

7: $\quad Y = WX$

8: $\quad$ Compute $c = \arg\max\limits_{i} \left[\left(w_j{}^T X \right)^2 \right]$

9: $\quad W(sweep + 1) = W(sweep) - \alpha * \frac{\partial F(W)}{\partial W}$, where W is the matrix that the c-th
   row can be calculated by (14) and the others are zero vectors

10: **end for**

11: **return** $Y_{m \times V}$

3.2   Functional Experiment

The design of functional experiment has been adopted memory imagine, where subjects exposed to challenging mental arithmetic presented on a computer screen, to which they have to respond using a two-button mouse in time. In each trial of the experiment condition, initially there appear a group of pictures which is easily interpreted as some normal subjects, such as hill, car, tiger and so on, which of them appears 2 second for each *stimuli*, and the *stimuli* group has been marked 'memory' condition and 'imagine' condition. For the experimental condition, subjects have to decide whether the *stimuli* in 'imagine' condition has been also shown in 'memory' condition or not within 1 second. The *stimuli* in each condition may be different in 'memory' condition and 'imagine' condition, which included pictures, Chinese characters and English words. In the rest condition, the user interface is displayed with a small white attention dot and without deciding tasks being shown.

3.3   Functional Imaging Data Acquisition

Subjects were scanned on a 3.0T Siemens Magnetom Vision Scanner, employing a block design with four runs. In four runs, four experimental blocks were presented in a constant order with four rest blocks was preceding each experimental block. See Fig. 1.

4   Result

The sparse ICA result will show for each task with group ICA toolbox in Matlab. We will show the first four ICs according to the similarity of ICs and task time series. Task time series will be calculated by canonical hemodynamic response

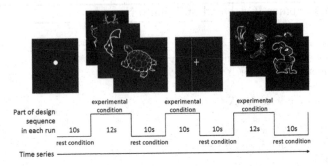

Fig. 1. Examples of the stimuli used for the tasks and experimental design. Each 'memory' condition lasted for 12s, each 'imagine' condition lasted for 10s and each rest condition lasted for 10s.

function (canonical HRF) [5,6,8–10]. The mathematical expression of canonical HRF has been shown by (15), which is constructed by two gamma functions, where p_1 and p_2 are equal to 6 and 16 in default. Its waveform has been shown as Fig. 2. The task time series are constructed by block design sequence and canonical HRF convolution. All the activity position and intensity of the first four ICs for each run are shown in Tables 1-4, respectively.

$$h(t) = \frac{t^{p_1-1} - e^{-t}}{\Gamma(p_1)} - \frac{t^{p_2-1} - e^{-t}}{p_1 \Gamma(p_2)} \tag{15}$$

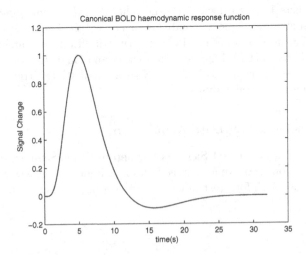

Fig. 2. Canonical BOLD haemodynamic response function

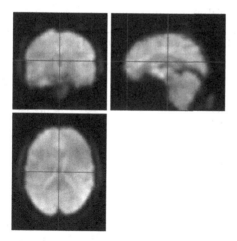

Fig. 3. The first four sparse ICA result in the first run

Table 1. Run 1 analysis of brain areas

intensity	coordinate	number	Brodmann Area	Level3
1242.3107	(45,-60,45)	24	BA 40	Inferior Parietal Lobule
1230.3107	(42,-57,48)	4	BA 7	Superior Parietal Lobule
1175.3107	(57,-51,33)	9	BA 40	Supramarginal Gyrus
1211.0197	(51,-54,45)	22	BA 40	Inferior Parietal Lobule
1216.0197	(42,-57,48)	3	BA 7	Superior Parietal Lobule

Table 2. Run 2 analysis of brain areas

intensity	coordinate	number	Brodmann Area	Level3
1149.4036	(51,-66,30)	7	BA 39	Angular Gyrus
1133.4036	(45,-69,36)	2	BA 39	Precuneus
1131.4036	(39,-81,6)	2	BA 19	Superior Frontal Gyrus
1121.4036	(-33,-51,60)	1	BA 7	Cuneus
1124.8304	(-27,-84,27)	1	BA 19	Cuneus
1220.8304	(48,-60,45)	17	BA 40	Inferior Parietal Lobule
1119.8304	(-30,-84,24)	1	BA 19	Superior Occipital Gyrus
1137.5903	(-27,-84,27)	2	BA 19	Cuneus
1226.5903	(51,-60,42)	15	BA 39	Inferior Parietal Lobule

Table 3. Run 3 analysis of brain areas

intensity	coordinate	number	Brodmann Area	Level3
1186.5418	(-45,-48,51)	16	BA 40	Inferior Parietal Lobule
1176.5418	(48,-66,12)	5	BA 39	Middle Temporal Gyrus
1190.5418	(-9,-63,51)	15	BA 7	Precuneus
1200.5418	(-9,-63,54)	13	BA 7	Superior Parietal Lobule
1168.5418	(48,-60,27)	1	BA 39	Superior Temporal Gyrus
1183.5418	(54,-54,33)	5	BA 40	Supramarginal Gyrus
1194.5393	(-45,-45,51)	14	BA 40	Inferior Parietal Lobule
1158.5393	(48,-63,27)	2	BA 39	Middle Temporal Gyrus
1194.5393	(39,-36,60)	2	BA 2	Postcentral Gyrus
1177.5393	(-9,-63,51)	11	BA 7	Precuneus
1198.5393	(-9,-63,54)	8	BA 7	Superior Parietal Lobule
1155.8189	(48,-63,27)	1	BA 39	Middle Temporal Gyrus
1190.8189	(-9,-63,54)	4	BA 7	Superior Parietal Lobule
1151.8189	(48,-60,27)	1	BA 39	Superior Temporal Gyrus
1185.6925	(-45,-48,51)	16	BA 40	Inferior Parietal Lobule
1148.6925	(48,-63,27)	4	BA 39	Middle Temporal Gyrus
1201.8189	(39,-36,60)	1	BA 2	Postcentral Gyrus

Table 4. Run 4 analysis of brain areas

intensity	coordinate	number	Brodmann Area	Level3
1261.5606	(42,-54,48)	21	BA 40	Inferior Parietal Lobule
1162.5606	(60,-54,-3)	1	BA 37	Inferior Temporal Gyrus
1192.5606	(-51,-69,9)	6	BA 19	Middle Occipital Gyrus
1340.5606	(60,-54,0)	5	BA 21	Middle Temporal Gyrus
1225.5606	(39,-57,51)	7	BA 7	Superior Parietal Lobule
1141.9049	(-36,-75,33)	1	BA 39	Angular Gyrus
1201.9049	(-15,-87,27)	11	BA 19	Cuneus
1156.9049	(39,-66,42)	3	BA 39	Inferior Parietal Lobule
1185.9049	(-24,-90,21)	3	BA 18	Middle Temporal Gyrus
1165.9049	(33,-75,39)	8	BA 19	Precuneus
1197.9049	(-36,-78,30)	4	BA 19	Superior Occipital Gyrus
1156.9049	(-12,-57,60)	2	BA 7	Superior Parietal Lobule
1187.8719	(-18,-90,24)	11	BA 19	Cuneus
1156.8719	(42,-66,39)	6	BA 39	Inferior Parietal Lobule
1148.8719	(-12,-57,60)	2	BA 7	Superior Parietal Lobule
1181.4896	(-24,-90,21)	5	BA 18	Middle Occipital Gyrus
1161.4896	(33,-75,39)	7	BA 19	Precuneus
1174.4896	(-36,-78,30)	4	BA 19	Superior Occipital Gyrus
1155.4896	(-12,-57,60)	2	BA 7	Superior Parietal Lobule

5 Discussion

In this paper, we have analysis fMRI data of different runs with sparse ICA algorithm. The result shows that the main activity regions include inferior parietal lobule (BA40), somatosensory association cortex (BA7), associative visual cortex (BA19), angular gyrus (BA39) and primary somatosensory cortex(BA2).
BA40 and BA7 are the most active regions in all of these four functional experiment tasks. BA7 is believed to play a role in visual-motor coordination, and BA40 is believed to both involve in reading and phonology.
Based on the analysis of experimental results in four groups of tasks can be seen that no matter how the content of the task is, the inferior parietal lobule (BA40), somatosensory association cortex (BA7) is closely related to the memory-imagine experiment thought experiments in working memory task. Among the analysis of experimental results also contains many other areas of the brain, such as associative visual cortex (BA19), presents gyrus (BA39) and primary somatosensory cortex (BA2), and other brain regions also has obvious activation conditions. It also suggests that the functions of memory in the brain imagination are not a certain specific brain regions involved in activity.

6 Conclusion

In this paper, we proposed a sparse ICA approach via infinity norm, and further discusses the application of the algorithm in fMRI data analysis. Application of infinite norm as sparse sex measurement criteria can guarantee independent component coefficient sparsity. FMRI experiment shows that the algorithm can be successfully applied to fMRI data, activating ferrite locality. In memory to imagine cognitive experiment, the brain activated regions for different tasks are different. But there are some regions, such as inferior parietal lobule (BA40), somatosensory association cortex (BA7), associative visual cortex (BA19), presents gyrus (BA39) and primary somatosensory cortex (BA2), and other brain regions, show the activated state in each runs. So that these brain regions play an important role in memory-imagine cognition functions. On the other hand, while analyzing each group different brain regions are activated. This suggests that the brain in implement any function or make any reaction and activation, are not specify one or several brain regions involved, but many brain regions realize the brain function.

Acknowledgment. This work is supported by the National Natural Science Foundation of China (Grant No.61105117, 61173035), the Fundamental Research Funds for the Central Universities (Grant No. 2011JC025, 3132013325), and the Program for New Century Excellent Talents in University (NCET-11-0861). The authors would like to thank Chao Yang, Dongmei Li and Zhe Yin for program technical support.

References

1. Bell, A.K., Sejnowski, T.J.: A non-linear information maximization algorithm that performs blind separation. IEEE Transactions on Knowledge and Data Engineering, 467–474 (1995)
2. Calhoun, V., Adali, T., Pearlson, G., Pekar, J.: A method for making group inferences from functional mri data using independent component analysis. Human Brain Mapping 14(3), 140–151 (2001)
3. Calhoun, V., Adali, T., Pearlson, G., Pekar, J.: Spatial and temporal independent component analysis of functional mri data containing a pair of task-related waveforms. Human Brain Mapping 13(1), 43–53 (2001)
4. Duann, J.R., Jung, T.P., Kuo, W.J., Yeh, T.C., Makeig, S., Hsieh, J.C., Sejnowski, T.J.: Single-trial variability in event-related bold signals. Neuroimage 15(4), 823–835 (2002)
5. Friston, K.J., Ashburner, J.T., Kiebel, S.J., Nichols, T.E., Penny, W.D.: Statistical parametric mapping: The analysis of functional brain images: The analysis of functional brain images. Academic Press (2011)
6. Friston, K.J., Fletcher, P., Josephs, O., Holmes, A., Rugg, M.D., Turner, R.: Event-related fMRI: characterizing differential responses. Neuroimage 7(1), 30–40 (1998)
7. Friston, K.J., Holmes, A.P., Worsley, K.J., Poline, J.P., Frith, C.D., Frackowiak, R.S.: Statistical parametric maps in functional imaging: a general linear approach. Human Brain Mapping 2(4), 189–210 (1994)
8. Friston, K.J., Kahan, J., Razi, A., Stephan, K.E., Sporns, O.: On nodes and modes in resting state fMRI. NeuroImage (2014)
9. Henson, R., Friston, K.: Convolution models for fMRI. Statistical Parametrical Mapping, pp. 178–192 (2007)
10. Li, B., Friston, K.J., Liu, J., Liu, Y., Zhang, G., Cao, F., Su, L., Yao, S., Lu, H., Hu, D.: Impaired frontal-basal ganglia connectivity in adolescents with internet addiction. Scientific Reports 4 (2014)
11. Mairal, J., Bach, F., Ponce, J., Sapiro, G.: Online learning for matrix factorization and sparse coding. Journal of Machine Learning Research 11, 19–60 (2010)
12. McKeown, M.J., Sejnowski, T.J.: Independent component analysis of fmri data: examining the assumptions. Human Brain Mapping 6(5-6), 368–372 (1998)
13. Oja, E., Plumbley, M.: Blind separation of positive sources by globally convergent gradient search. Neural Computation 16(9), 1811–1825 (2004)
14. Pascual-Montano, A., Carazo, J.M., Kochi, K., Lehmann, D., Pascual-Marqui, R.D.: Nonsmooth nonnegative matrix factorization (nsNMF). IEEE Transactions on Pattern Analysis and Machine Intelligence 28(3), 403–415 (2006)
15. Witten, D., Tibshirani, R., Hastie, T.: A penalized matrix decomposition, with applications to sparse principal components and canonical correlation analysis. Biostatistics 10(3), 515–534 (2009)
16. Yang, C.F., Ye, M., Zhao, J.: Document clustering based on nonnegative sparse matrix factorization. In: Wang, L., Chen, K., S. Ong, Y. (eds.) ICNC 2005. LNCS, vol. 3611, pp. 557–563. Springer, Heidelberg (2005)
17. Zarzoso, V., Comon, P.: Comparative speed analysis of fastICA. In: Davies, M.E., James, C.J., Abdallah, S.A., Plumbley, M.D. (eds.) ICA 2007. LNCS, vol. 4666, pp. 293–300. Springer, Heidelberg (2007)
18. Zarzoso, V., Comon, P.: Robust independent component analysis by iterative maximization of the kurtosis contrast with algebraic optimal step size. IEEE Transactions on Neural Networks 21(2), 248–261 (2010)

A Framework for Ovarian Cancer Diagnosis Based on Amino Acids Using Fuzzy-Rough Sets with SVM

Farid A. Badria[1], Nora Shoaip[2], Mohammed Elmogy[3], A. M. Riad[2],
and Hosam Zaghloul[4]

[1] Pharmacognosy Dept., Faculty of Pharmacy, Mansoura University, Egypt
faridbadria@yahoo.com
[2] Information Systems Dept., Faculty of Computers and Information,
Mansoura University, Egypt
norashoaip@yahoo.com, amriad2000@mans.edu.eg
[3] Information Technology Dept., Faculty of Computers and Information,
Mansoura University, Egypt
melmogy@mans.edu.eg
[4] Dept. of Clinical Pathology, Faculty of Medicine, Mansoura University, Egypt
Hosam_z@yahoo.com

Abstract. Ovarian cancer is diagnosed in nearly a quarter of a million women globally each year. It has the highest mortality rate of all cancers for women. The prognosis for ovarian cancer patients is poor, particularly about 20% of ovarian cancers are found at an early stage. CA-125 test is used as a tumor marker. A high level of CA-125 could be a sign of ovarian cancer or other conditions. We use a machine learning technique to explore better knowledge and most important factors useful for detecting early-stage ovarian cancer by evaluating the significance of data between the amino acids and the ovarian cancer. Therefore, we propose a Fuzzy Rough with Support Vector Machine (SVM) classification model to mine suitable rules. In pre-processing stage, we use Fuzzy Rough set theory for feature selection. In post-processing stage, we use SVM to merit of dealing with real and complex data, performing quick learning and having good classification performance.

Keywords: Ovarian cancer, Amino acids, Rough set, Fuzzy-Rough model, SVM classification.

1 Introduction

Medical diagnosis contains high degree of difficulty that faces two main problems. The first problem of the medical diagnosis is a classification process. It must analyze many factors in difficult circumstances, such as diagnosis disparity and limited observation. The second problem is the uncertainty of the processed data which affects the diagnosis process.

Recently, many techniques have been developed to deal with uncertain, incomplete, and even inconsistent data. Rough set theory [1,46] is used to analysis vague and

A.E. Hassanien et al. (Eds.): AMLTA 2014, CCIS 488, pp. 389–400, 2014.
© Springer International Publishing Switzerland 2014

uncertain data. In practice, Rough set classifies the discrete attributes with great accuracy. This cannot be done well with real-valued, continuous attributes. Therefore, it lead to the creation of hybrid systems to integrate the Rough set theory with other machine learning technique, such as Fuzzy set. Both of Rough and Fuzzy sets have different ways for dealing with uncertainty. These techniques are complementary to each other and the combination between them can provide improved solutions for dealing with continuous attributes. Dubois and Prade [2] proposed the concept of fuzzy rough sets. Fuzzy rough sets offer a high degree of flexibility in enabling the vagueness and imprecision present in real-valued data to be modeled effectively. Fuzzy Rough set [3-6] is a generalization of the lower and upper approximation of the rough set to have greater flexibility in handling uncertainty. As compared to the original Rough sets, this hybrid technique has two important differences. First of all, the approximation has been translated from a crisp Rough set to a Fuzzy Rough set. The second is the indiscernibility relation has been extended to a fuzzy equivalence relation.

There are several classification methods widely-used in medical data based on symptoms and health conditions [7-10]. Naïve Bayesian classifier is considered as one of the simplest and rapid algorithms. An important disadvantage of Naïve Bayesian is that it has strong feature independence assumptions. Decision tree classifiers provide good visualization of data. This visualization allows users to readily understand the overall structure of data but decision tree may be too complex when a data set contains many attributes. Neural Network (NN) consists of computational nodes that emulate the functions of the neurons in the brain. NN has good robustness and self-learning, but for massive data it cannot get good effect. Its classification performance is very sensitive to the parameters selected and its training or learning process is very slow and computationally very expensive. Fuzzy-Rough Nearest Neighbour combines fuzzy-rough approximations with the ideas of the classical fuzzy K-nearest neighbour (FNN) approach. The rationale behind the algorithm, that presents the lower and the upper approximation of a decision class, is calculated by means of the nearest neighbours of a test object. It provides good clues to predict the membership of the test object to that class.

On the other hand, Vapnik [11] developed the Support Vector Machine (SVM) for binary classification. The goal is to find a decision boundary between two classes that is maximally far from any point in the training data. When we cannot find a linear separator, data points are projected via kernel techniques into a higher-dimensional space where the data points effectively become separable. SVM has many advantages [12] such as, it has a simple geometric interpretation and it gives a sparse solution. The limitation of the SVM approach lies in choice of the kernel parameters for obtaining good results, which practically means that an extensive search must be conducted on the parameter space before results can be trusted. It generates black box models [13] that do not have an understandable form to extract rules directly for a domain expert.

This paper is organized as follows. Section 2 is a quick review on the prior works of the medical diagnosis based on machine learning techniques. Section 3 discusses the proposed Fuzzy Rough with SVM classification model. In section 4, the experimental results are introduced to illustrate how to apply the proposed model into practical case for detecting early-stage ovarian cancer. Finally, the conclusion and the future work are presented in section 5.

2   Related Work

Medical data usually contain high degrees of uncertainty and they are considered as a classification processes. So, improving the diagnosis process is an emergent task and poses a greater challenge for many researchers to cared about and tried to find the best machine learning technique to accomplish the job. Among these techniques, SVM has become one of the most popular techniques for classification. Therefore, Marchiori and Sebag [14] proposed a method to improve the classification of SVM with recursive feature selection (SVM-RFE). The resulting classifiers are applied to four publically available gene expression data sets from leukemia, ovarian, lymphoma, and colon cancer data, respectively. The results indicated that their proposed approach significantly improves the predictive performance of the baseline SVM classifier, the stability, and the robustness on all tested data sets.

Zhou et al. [15] profiled relative metabolite levels in sera from 44 women diagnosed with serous papillary ovarian cancer (stages I-IV) and 50 healthy women. The profiles were input to a customized functional SVM algorithm for diagnostic classification. 100 % accuracy performance was evaluated through a 64-30 split validation test and with a stringent series of leave-one-out cross-validations.

Srivastava and Bhambhu [16] presented an SVM learning method which is applied on different types of data sets (Diabetes data, Heart data, Satellite data and Shuttle data). Their experimental results showed that the choice of the kernel function and the best value of parameters for particular kernel are the critical parameters for a given amount of data. Sasirekha and Kumar [17] used SVM classification for the heart beat time series. The test result showed that the proposed approach works well in detecting different attacks and achieved high performance 96.63%. Kumari and Chitra [18] proposed a SVM model with radial basis function kernel for diabetes disease classification. The performance parameters are the classification accuracy 78%, sensitivity 80%, and specificity 76.5% of the SVM and RBF. Thus, SVM can be successfully used for diagnosing diabetes disease.

When the input space dimension is large, they will lead to a longer time training of classification algorithms and affect the practicality of performance. Our survey of literature shows that many researchers used feature selection algorithms to improve classification accuracy on different medical data sets. For example, Bhatia et al. [19] used Genetic algorithm for feature selection and enhanced the performance of the SVM to a great extent and a high accuracy. Their system's accuracy was 72.55% which is obtained by using only 6 out of 13 features against an accuracy of only 61.93% by using all the features.

Akay [20] observed that SVM with feature selection using F-score obtains promising results in classifying the potential breast cancer patients. The results showed that the highest classification accuracy (99.51%) is obtained for the SVM model that contains five features, and this is very promising compared to their previously reported results. Tsai et al. [21] used linear regression and variance analysis (ANOVA) for data pre-processing for the purpose of solve the problems such as large number of ovarian cancer gene chip variables. Then, information database is classified by SVM and the comparison of different results of Kernel function is conducted. They discovered that the SVM had considerably fine effect in classification and when different kernel function are used the results will change too.

Srivastava et al. [22] introduced a Rough-SVM classification method with heart data set, which makes great use of the advantages of SVM's greater generalization performance and Rough Set theory in effectively dealing with attributes reduction. The accuracy of the Classification is increased by 4.29% than general SVM. Chen et al. [23] and LI and Ye-LI [24] proposed hybrid models between SVM and Rough Set (RS). The introduction of RS can reduce the number of data entered to the SVM. It is greatly reduced and the system speed is improved. Gandhi and Prajapati [25] used F-score method and K-means clustering for feature selection and the performance of the SVM classifier is empirically evaluated on the reduced feature subset of Pima Indian diabetes dataset and achieved accuracy of 98 %.

Núñez et al. [26] proposed a SVM model with prototype method. Their model combined prototype vectors for each class with support vectors to define an ellipsoid in the input space which are then mapped to if-then rules. This approach does not scale well for a large number of patterns.

Barakat and Diederich [27] handled the SVM rule-extraction task in three basic steps: training, propositional rule-extraction, and rule quality evaluation that achieve high accuracy and fidelity. Barakat and Bradley[28] used the area under the receiver operating characteristics (ROC) curve to assess the quality of rules extracted from an SVM model.

Martens et al. [29] provided an overview of recently proposed SVM rule extraction techniques, complemented with the pedagogical ANN rule extraction techniques which are also suitable for SVMs. Another effort for extracted correct and valid rules from the medical point of view and consistent with clinical knowledge of diabetes risk factors by Barakat and Bradley [30] handled rule-extraction as a learning task. The decision trees and rule sets produced by C5 offer an explanation of the concepts learned by the SVM.

In this paper we tried to find a tumor marker for ovarian cancer by evaluating the significance of data between the amino acids and the ovarian cancer. The dataset collected from Mansoura Cancer Center. This dataset is a complex nature, it contain several attributes all of them real value data type which, posed a greater challenge to find the best hybrid model to accomplish the job. So, we propose a Fuzzy Rough with Support Vector Machine (SVM) classification model. We choose Fuzzy Rough as feature selection algorithm which can deal with complex data without making any transformation for dealing with real data such as discretization to void loss of data. Fuzzy Rough reduced the condition attributes to improve classification accuracy. The classification task was performed using SVM which, become one of the most popular techniques for classification. This model has merit of dealing with real and complex data, performing quick learning and having good classification performance.

3   The Proposed Fuzzy-Rough with SVM Classification Modle

Mining suitable rules to explore most important factors affecting medical diagnosis from complex, real medical data with quick learning and good classification performance is considered as a great challenge. Therefore, this paper proposes a Fuzzy Rough with SVM classification model. It integrates Fuzzy Rough and SVM in four phases, as shown in Fig. 1. The first phase is to prepare the input data and construct the information table. The second phase is to reduce features to get the minimum important ones and exclude the redundant attribute for saving training time. The third phase is the classification stage which is based on SVM. The last phase is to extract the suitable rules to get understandable form for helping physician in diagnosing ovarian cancer in early stages. In the following subsections, we will discuss the main stages of our proposed technique in more detail.

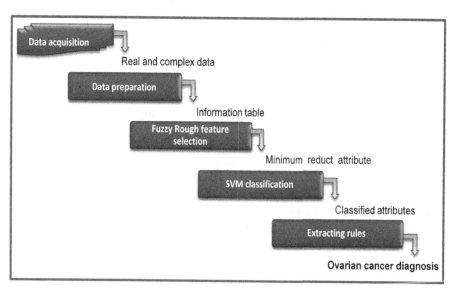

Fig. 1. The bolock diagram of the proposed Fuzzy Rough feature selection based on SVM classification model

3.1   Data Preparation and Information Table Construction

In this stage, we prepare the medical data to be presented in a suitable form to be processed in our proposed system. Data set should be prepared to remove the redundancy, represent the non-numerical attributes in a suitable numerical form, check for the missing values, and rescale the real values for decision attribute. This is done in order to construct the information table   for extracting knowledge in the form of a two-dimensional table. For dealing with real data, we don't have to use discretization algorithms. To void loss of data, we assign membership degrees (fuzzification) as preparing to use Fuzzy Rough feature selection.

3.2   Reducing Features

Feature selection is an important stage in building classification systems. It is advantageous to limit the number of input features to the classifier to have a good prediction and less computationally intensive model [31-36]. We reduced the condition attributes by using the Fuzzy-Rough QuickReduct algorithm [4]. It uses the dependency function to choose which important attributes to add to the current reduct. This algorithm terminates when the addition of any remaining attribute does not increase the dependency. Fuzzy-Rough sets have achieved more accurate results than Pawlak Rough sets. They translated crisp Rough set to a Fuzzy set and extended the indiscernibility relation to a fuzzy equivalence relation. So, we don't have to make any transformation for dealing with real data such as discretization. On the other hand, SVM [22] is unable to simplify the input space dimension in classification process, or to determine what kind of data is redundant or useful. The SVM become powerless when dealing with uncertain information. Therefore, we used Fuzzy Rough as pre-processing stage to get minimum reduct. Then, the result is supplied as input to SVM to get a balance between training data and SVM performance.

3.3   SVM Classification

The classification task was performed using SVM. To evaluate the performance of the SVM models, the data set is partitioned into two numbers of subsets. One subset is used for training and the other subset is used for testing SVM. The SVM classification error is recorded. This is repeated on different training test partition models until the model with the smallest classification error will be adopted. We tested the performance of SVM with different parameter values for the polynomial and RBF kernels to choose the SVM kernel, which has good classification.

The proposed model is evaluated using three parameters: accuracy of the classifier, sensitivity, and specificity. They are statistical measures [18, 20, 25] that describe how well the classifier discriminates between a case with positive and with negative class and they can be visualized and studied using the Receiver Operating Characteristic (ROC) curve.

$$\text{Accuracy} = \frac{TP + TN}{TP + TN + FP + FN} \tag{1}$$

$$\text{Sensitivity (TP Rate)} = \frac{Tp}{TP + FN} \tag{2}$$

$$\text{Specificity} \quad = \frac{TN}{TN + FP} \tag{3}$$

Where TP (True Positive) is the number of examples correctly classified to that class. The TN (True Negative) is the number of examples correctly rejected from that class. The FP (False Positive) is the number of examples incorrectly rejected from that class. Finally, the FN (False Negative) is the number of examples incorrectly classified to that class.

3.4   Extracting Suitable Rules

In order to scoring success in medical diagnosis, our proposal have to produce its result in an understandable form (if then rule). On the other hand, we faced a big problem in our model that the SVM algorithm is considered as a black box model. It has not the ability to explain the result in an understandable form for a domain expert [26]. So, we have to extract rules from the information provided by SVM using boundaries (ellipsoids or hyper-rectangles) which are defined in the input space. They are considered as a combination of prototype vectors and support vectors. Prototype is computed through a clustering algorithm; whereas rule is define for each region.

4   Experimental Results

Since the CA-125 is not a good tumor marker for detecting the early-stage ovarian cancer, we tried to determine the effect of the amino acids in predicting the ovarian cancer. We applied our proposed model on two different studies. The first one is between amino acids and ovarian cancer. The second is between amino acids and CA-125 tumor marker. These two studies are conducted to explore better knowledge and the most important factors useful for detecting early-stage of the ovarian cancer. Waikato Environment for Knowledge analysis (Weka) version 3.7.2 [37-39] has been used to carry out experiments.

- **Samples Collection**

The current investigation is a case-control based study, which was approved by the ethical institutional review board at Mansoura University that complies with acceptable international standards. Written informed consent for participation was obtained from each participant. Case patients were recruited from the population of patients with diagnosed ovarian cancer who were evaluated and treated at Mansoura Cancer Center. The inclusion criteria were as follows: pathologically confirmed diagnosis of ovarian cancer, CA-125 (1.9-16.3 U/ml), a diagnostic cut-off value, and Egyptian residency. From February 2011 through December 2013, statistical analyses indicated that the eligible patients who were not recruited did not differ from the recruited patients in terms of demographical, epidemiological, or clinical factors (retrieved from patients' medical records). The control subjects were healthy and recruited from the diagnostic biochemical lab, AutoLab of Mansoura institution, and were matched by age, sex, and ethnicity to the case subjects. The eligibility criteria for controls were the same as those for patients, except for having a cancer diagnosis. A short structured questionnaire was used to screen for potential controls on the basis of the eligibility criteria. Analysis of the answers received on the short questionnaire indicated that 80% of those questioned agreed to participate in clinical research.

- **Samples Analysis**

CA-125 analysis blood samples (5 mL) were taken, centrifuged and the serum separated and stored at – 80 C until analyzed. Serum samples were assayed. For CA-125 by enzyme-linked immunosorbent assay with commercial kits (Abbott, North

Chicago, IL). For amino acid, analysis is done by high performance liquid chromatography (HPLC) amino acids analyzer, LC 3000 eppendorf (Germany).

Table 1. Description of the ovarian cancer attributes

Amino Acids (Condition Attribute)	Class
1-aspartic, 2-threonine, 3-serine, 4-glutamine, 5-glycine, 6-alanine, 7-cystine, 8-valine, 9-methionine, 10-isoleucine, 11-leucine, 12-tyrosine, 13-phenylalanine, 14-histidine, 15-lysine, 16-Arginine, 17-Proline	ovarian cancer
*All of them is Numeric (real) data type *no missing data	Nominal {negative, positive}

4.1   The Relation between The Amino Acids and The Ovarian Cancer

Data is obtained from 135 female patients. They include 22 negative cases and 113 positive cases. The class target or decision attribute of the ovarian cancer takes the values "negative" or "positive" that means negative or positive test for ovarian cancer, respectively. Table 1 shows the description of the used attributes.

After preparing data set and construct the information table. Fuzzy-Rough reduction is calculated for all amino acids attributes. We found that attributes 2, 3, 13, and 14 are the most important amino acids that have main affect on ovarian cancer. Reduced amino acids data set is split in to training data set (66%) and test data set (33%). We selected polynomial kernel as appropriate SVM kernel function which is suitable for nonlinear data with high complexity. We gain an excellent correctly classified instances (100%) and precision equal to 1, as show in Table 2.

Table 2. The comparison of the performance of the proposed model with other models

Classification technique		Accuracy	Sensitivity Avg (TP Rate)	Specificity Avg (FP Rate)
Fuzzy-Rough with SVM	Polynomial	100 %	1	0
	RBF	97.82 %	0.978	0.145
SVM	Polynomial	95.65 %	0.957	0.29
	RBF	86.95 %	0.87	0.87
FuzzyRoughNN		100 %	1	0
Fuzzy-Rough with J48		97.82 %	0.978	0.145
Fuzzy-Rough with NaiveBayes		95.65 %	0.957	0.29

The proposed Fuzzy-Rough with SVM (Polynomial kernel) classification model achieved classification accuracy greater than SVM (Polynomial kernel) by 4.347%. Our proposal model has achieved the highest accuracy as compared to other technique such as, J48 (97.82%), NaïveBayes (95.65%) and Equal with Fuzzy-Rough-NN classification technique (100%). Therefore, it is interesting to find two classification techniques achieved perfect accuracy that ensures the success of our medical rules.

We extracted rules from the information provided by SVM. We found that the phenylalanine amino acid (no. 13) can be used as an excellent tumor marker with classification error equal to 0.0%. Table 3 shows the most important rules for detecting early-stage ovarian cancer.

Table 3. The resulting rules for detecting early-stage ovarian cancer based on amino acids

	13-phenylalanine	Class	Correctly classified
Rules	'(33.9-63.1]'	Positive	
	'(63.1-inf)'	Negative	100 %
	'(-inf-33.9]'	Negative	

4.2   The Relation between The Amino Acids and CA-125

In this experiment, we want to extract the relationship between the amino acids and the tumor marker (CA-125). The class target or decision attribute is CA-125 which is a real data type. We rescale this attribute to takes the values "normal" inside the range [1.9-16.3 U/ml]. There are 53 cases in the normal range and there are 60 abnormal cases. After using Fuzzy-Rough reduction, we find that the selected attributes (1, 2, 3, 5, 6, 7, and 9) are the minimal set of amino acids that have main affect on the tumor marker CA-125. For classification process it is necessary to validate model with cross validation. hence 3-fold, 5-fold and 10-fold cross validation techniques are used on amino acids dataset. comparative results show that SVM with polynomial kernel works better for 5-fold cross validation. We gain correctly classified instances of 60.177%. By using different classification techniques on the reduced ovarian cancer data set, our proposal model has achieved the highest accuracy as compared to other techniques, such as Fuzzy-Rough-NN (55.75%), J48   (48.67 %), and NaïveBayes (53.98%).

5   Conclusion

In this paper, we proposed a Fuzzy-Rough with SVM classification model to diagnose ovarian cancer based on amino acids. We combined Fuzzy-Rough as pre-processing stage and SVM as post-processing stage in four phases to explore better knowledge and extract most important factors. We did two experimental studies. The first is between amino acids and the ovarian cancer. We found that there's a strong relation between amino acids and ovarian cancer. We found that the phenylalanine amino acid (no. 13) is a very good tumor marker for ovarian cancer which can be considered as the most important factor for detecting early-stage ovarian cancer. The second study is between amino acids and the CA-125 tumor marker that achieved moderate accuracy of 60.177 %. Therefore, in future work, we hope to apply our hybrid model with larger datasets with additional attributes to explore main factors that affect the CA-125.

References

1. Pawlak, Z.: Rough sets. International Journal of Computer & Information Sciences 11(5), 341–356 (1982)
2. Dubois, D., Prade, H.: Rough fuzzy sets and fuzzy rough sets. Int. J. General Syst. 17(2-3), 191–209 (1990)
3. Jensen, R., Shen, Q.: Semantics-Preserving Dimensionality Reduction: Rough and Fuzzy-Rough-Based Approaches. IEEE Transactions on Knowledge and Data Engineering 16(12), 1457–1471 (2004)
4. Jensen, R.: Combining Rough and Fuzzy Sets for Feature Selection. University of Edinburg (2005), http://books.google.com.eg/books?id=84ESSQAACAAJ (retrieved)
5. Shen, Q., Jensen, R.: Rough Sets, their Extensions and Applications. International Journal of Automation and Computing 4, 217–228 (2007)
6. Chen, H., Wang, M., Qian, F., Jiang, Q., Yu, F., Luo, Q.: Research on Combined Rough Sets with Fuzzy Sets. In: Yu, F., Luo, Q. (eds.) ISIP, vol. 1, pp. 163–167. IEEE Computer Society (2008)
7. Yoo, I., Alafaireet, P., Marinov, M., Pena-Hernandez, K., Gopidi, R., Chang, J., Hua, L.: Data Mining in Healthcare and Biomedicine: A Survey of the Literature. J. Medical Systems 1, 2431–2448 (2012)
8. Chen, H.-L., Liu, D.-Y., Yang, B., Liu, J., Wang, G., Wang, S.-J.: An Adaptive Fuzzy k-Nearest Neighbor Method Based on Parallel Particle Swarm Optimization for Bankruptcy Prediction. In: Huang, J.Z., Cao, L., Srivastava, J. (eds.) PAKDD 2011, Part I. LNCS (LNAI), vol. 6634, pp. 249–264. Springer, Heidelberg (2011)
9. Lee, M., Chang, T.: Comparison of Support Vector Machine and Back Propagation Neural Network in Evaluating the Enterprise Financial Distress, CoRR, abs/1007.5133 (2010)
10. Abdul Khaleel, M., Pradham, S.K., Dash, G.N.: A Survey of Data Mining Techniques on Medical Data for Finding Locally Frequent Diseases. International Journal of Advanced Research in Computer Science and Software Engineering (IJARCSSE) 3(8), 149–153 (2013)
11. Meyer, D.: Support Vector Machines. The Interface to libsvm in package e1071. Online-Documentation of the package e1071 for Support Vector Machines (2014)
12. Sewell, M.: Support Vector Machines (SVMs), http://www.svms.org (retrieved)
13. Suganya, G., Dhivya, D.: Extracting Diagnostic Rules from Support Vector Machine. Journal of Computer Applications (JCA) 4, 95–98 (2011)
14. Marchiori, E., Sebag, M.: Bayesian learning with local support vector machines for cancer classification with gene expression data. In: Rothlauf, F., Branke, J., Cagnoni, S., Corne, D.W., Drechsler, R., Jin, Y., Machado, P., Marchiori, E., Romero, J., Smith, G.D., Squillero, G. (eds.) EvoWorkshops 2005. LNCS, vol. 3449, pp. 74–83. Springer, Heidelberg (2005)
15. Zhou, M., Guan, W., Walker, L.D., Mezencey, R., Benigno, B.B., Gray, A., Fernández, F.M., McDonald, J.F.: Rapid Mass Spectrometric Metabolic Profiling of Blood Sera Detects Ovarian Cancer with High Accuracy. American Association for Cancer Research (aacrjournal) 1 (2010), doi:10.1158/1055-9965.EPI-10-0126
16. Srivastava, D.K., Bhambhu, L.: Data Classification Using support vector machine. Journal of Theoretical and Applied Information Technology 12(1), 1–7 (2010)
17. Sasirekha, A., Kumar, P.G.: Support Vector Machine for Classification of Heartbeat Time Series Data. International Journal of Emerging Science and Engineering (IJESE) 1, 38–41 (2013)

18. Kumari, V.A., Chitra, R.: Classification Of Diabetes Disease Using Support Vector Machine. International Journal of Engineering Research and Applications 3, 1797–1801 (2013)
19. Bhatia, S., Prakash, P., Pillai, G.N.: SVM Based Decision Support System for Heart Disease Classification with Integer-Coded Genetic Algorithm to Select Critical Features. In: Proceedings of World Congress on Engineering and Computer Science, WCECS (2008)
20. Akay, M.F.: Support vector machines combined with feature selection for breast cancer diagnosis. Expert Syst. Appl. 36(2), 3240–3247 (2009)
21. Tsai, M.-H., Wang, S.-H., Wu, K.-C., Chen, J.-M., Chiu, S.-H.: Human Ovarian carcinoma microarray data analysis based on Support Vector Machines with different kernel functions. In: International Conference on Environment Science and Engineering IPCBEE, vol. 8, pp. 138–142 (2011)
22. Srivastava, D.K., Patnaik, K.S., Bhambhu, L.: Data Classification: A Rough - SVM Approach. Contemporary Engineering Sciences 3(2), 77–86 (2010)
23. Chen, R.-C., Cheng, K.-F., Hsieh, C.-F.: Using Rough Set and Support Vector Machine for Network Intrusion Detection. International Journal of Network Security & Its Applications (IJNSA) 1(1), 1–13 (2009)
24. Li, Y.-B., Ye, L.: Survey on Uncertainty Support Vector Machine and Its Application in Fault Diagnosis. IEEE 9, 561–565 (2010)
25. Gandhi, K.K., Prajapati, N.B.: Diabetes prediction using feature selection and classification. International Journal of Advance Engineering and Research Development (IJAERD) 1(5), 1–7 (2014)
26. Núñez, H., Angulo, C., Catala, A.: Rule-extraction from Support Vector Machines. In: Proc. of European Symposium on Artificial Neural Networks (ESANN), Burges, pp. 107–112 (2002)
27. Barakat, N., Diederich, J.: Eclectic Rule-Extraction from Support Vector Machines. International Journal of Computational Intelligence 2(1), 59–62 (2005)
28. Barakat, N., Bradley, P.A.: Rule Extraction from Support Vector Machines: Measuring the Explanation Capability Using the Area under the ROC Curve. IEEE, 0-7695-2521 (2006)
29. Martens, D., Huysmans, J., Setiono, R., Vanthienen, J., Baesens, B.: Rule Extraction from Support Vector Machines: An Overview of Issues and Application in Credit Scoring. In: Diederich, J. (ed.) Rule Extraction from Support Vector Machines. SCI, vol. 80, pp. 33–63. Springer, Heidelberg (2008)
30. Barakat, N., Bradley, P.A.: Rule extraction from support vector machines: A review. Neurocomputing 74(1), 178–190 (2010)
31. Huang, C.-L., Liao, H.-L., Chen, M.-C.: Prediction model building and feature selection with support vector machines in breast cancer diagnosis. Expert Systems with Applications 34, 578–587 (2008)
32. Xu, F.F., Miao, D.Q., Wei, L.: Fuzzy-rough Attribute Reduction via Mutual Information with an Application to Cancer Classification. Computer. Math. Appl. 57(6), 1010–1017 (2009)
33. Devi, S.N., Rajagopalan, S.P.: A study on Feature Selection Techniques in Bio-Informatics. International Journal of Advanced Computer Science and Applications (IJACSA) 2(1), 138–144 (2011)
34. Zhao, J., Zhang, Z.: Fuzzy Rough Neural Network and Its Application to Feature Selection. Academic Journal 13(4), 270–275 (2011)
35. Hong, J.: An Improved Prediction Model based on Fuzzy-rough Set Neural Network. International Journal of Computer Theory and Engineering 3(1), 158–162 (2011)

36. Gangwal, C., Bhaumik, R.N.: Intuitionistic Fuzzy Rough Relation in Some Medical Applications. International Journal of Advanced Research in Computer Engineering Technology 1, 28–32 (2012)
37. Markov, Z., Russell, I.: An Introduction to the WEKA Data Mining System, Workshop #5 at the 39th ACM Technical Symposium on Computer Science Education, Portland, Oregon (March 2008), http://www.cs.ccsu.edu/~markov/weka-tutorial.pdf (retrieved)
38. Jensen, R.: Fuzzy-rough data mining with Weka, http://users.aber.ac.uk/rkj/Weka.pdf (retrieved)
39. Jensen, R., Parthaláin, N.M., Shen, Q.: Tutorial: Fuzzy-rough data mining (using the Weka data mining suite) (2014), http://users.aber.ac.uk/rkj/wcci-tutorial-2014 (retrieved)
40. Hua, J.: A Knowledge Acquisition Model of Inconsistent Medical Data Based on Rough Sets Theory. In: 2007 Workshop on Intelligent Information Technology Applications, vol. 1, pp. 176–180. IEEE Computer Society (2008)
41. Tripathy, B.K., Acharjya, D.P., Cynthya, V.: A Framework for Intelligent Medical Diagnosis using Rough Set with Formal Concept Analysis. International Journal of Artificial Intelligence & Applications (IJAIA) 2(2), 45–66 (2011)
42. Durairaj, M., Meena, K.: A Hybrid Prediction System Using Rough Sets and Artificial Neural Networks. International Journal of Innovative Technology Creative Engineering 1(7), 16–23 (2011)
43. An, L., Tong, L.: A rough neural expert system for medical diagnosis. In: International Conference on Services Systems and Services Management, vol. 2, pp. 1130–1135. IEEE Computer Society (2005)
44. Ding, S., Chen, J., Xu, X., Li, J.: Rough Neural Networks: A Review. Journal of Computational Information Systems 7, 2338–2346 (2011)
45. Lavanya, D., Rani, K.U.: A Hybrid Approach to Improve Classification with Cascading of Data Mining Tasks. International Journal of Application or Innovation in Engineering Management (IJAIEM) 2, 345–350 (2013)
46. Hassanien, A.E., Suraj, Z., Slezak, D., Lingras, P.: Rough computing: Theories, technologies and applications. IGI Publishing Hershey, PA (2008)

MRI Brain Tumor Segmentation System Based on Hybrid Clustering Techniques

Eman A. Abdel Maksoud[1], Mohammed Elmogy[2], and Rashid Mokhtar Al-Awadi[3]

[1]Information System Dept., Faculty of Computers and Information,
Mansoura University, Egypt
eng.eman.te@gmail.com
[2] Information Technology Dept., Faculty of Computers and Information,
Mansoura University, Egypt
melmogy@mans.edu.eg
[3] Communication Dept., Faculty of Engineering, Mansoura University, Egypt
actt_egypt@yahoo.com

Abstract. In this paper, we developed a medical image segmentation system based on hybrid clustering techniques to provide an accurate detection of brain tumor with minimal execution time. Two hybrid techniques have been proposed in our proposed medical image segmentation system. The first hybrid technique is based on k-means and fuzzy c-means (KFCM) while the second is based on k-means and particle swarm optimization (KPSO). We compared the two proposed techniques with k-means; fuzzy c-means, expectation maximization, mean shift, and particle swarm optimization using three different benchmark brain data sets. The results clarify the effectiveness of our second proposed technique.

Keywords: Medical image segmentation, K-means, Fuzzy C-means, Expectation Maximization, Mean shift, Particle swarm optimization.

1 Introduction

Image segmentation is a fundamental task in image processing and computer vision disciplines. It refers to the process of partitioning a digital image into multiple non-overlapping regions [1].There are many image segmentation techniques, such as edge base, clustering and region based segmentation techniques [2]. Although of the variety of image segmentation techniques, the selection of an appropriate technique for a special type of images is a difficult problem. Not all techniques are suitable for all types of images [3]. The main problem in segmentation algorithms is the difficulty of balancing the over-segmentation and under-segmentation. On the other hand, medical image segmentation is considered as an active research area. It is a quite challenging problem due to images with poor contrasts, noise, and missing or diffuses boundaries [4]. The magnitude resonance images (MRI) scan is comfortable for diagnosis. It is not affect the human body because it doesn't use any radiation. It is based on the magnetic field and radio waves [5]. On the other hand, a brain tumor can be defined as an

A.E. Hassanien et al. (Eds.): AMLTA 2014, CCIS 488, pp. 401–412, 2014.
© Springer International Publishing Switzerland 2014

abnormal growth of the cells in the brain. Brain tumors are of two types: primary and secondary. Primary tumors are classified as benign and malignant [6]. Benign tumors can be removed. They usually have a border or an edge. Malignant tumors are more serious. They grow rapidly in crowd and invade the nearby healthy tissue. The physician gives the treatment for the strokes rather than the treatment for the tumor. So, detection of the tumor is important for that treatment. The lifetime of the person who affected by the brain tumor will increase if it is detected early. Therefore, an efficient medical image segmentation technique should be developed with advantages of minimum user interaction, fast computation, accurate, and robust segmentation results to help physicians in diagnosing accurately. The most widely used techniques of image segmentation are clustering techniques. Clustering is an unsupervised learning technique which needs the user to determine the number of clusters in advance to classify pixels [7]. Therefore, the cluster is a collection of similar pixels and is dissimilar to the pixels belonging to other clusters [8].   Clustering techniques can perform clustering in one of two ways, either by partitioning or by grouping pixels [9].   In this paper we focused on clustering techniques to detect the brain tumor. We made our experiments by using the most famous five clustering techniques: k means, fuzzy c means, expectation maximization, mean shift, and particle swarm optimization. We applied these techniques on three different data sets which have 254 abnormal MRI brain images. The MRI images were pre-processed at first to enhance the quality of the processed images. We integrated two different image clustering techniques in our proposed medical system in clustering step of the framework to have advantages of these clustering techniques and overcoming the limitations of them in two proposed hybrid techniques (k means with fuzzy c means and k means with particle swarm optimization). Then, extraction of the tumor is done automatically without user interaction by using thresholding and level set methods to contour the tumor area. The last stage of our proposed medical segmentation system framework is the validation stage by comparing the results with the ground truth.

This paper is organized as follows. In Section 2, the current scientific research in medical image segmentation is introduced.   Section 3 presents the materials and methods used in this work. It describes the image data sets used in this work. It also shows the proposed medical image segmentation system based on our proposed hybrid clustering techniques. Section 4 depicts the experimental results obtained from the evaluation of the two proposed techniques using three types of data sets and discusses the main questions derived from them.   Finally, conclusion and future work are drawn in Section 5.

2   Related Work

Medical image segmentation is considered as a hot research topic. Several researchers have suggested various methodologies and techniques for image segmentation. For example, Jumb et al. [10] presented color image segmentation using k-means clustering and Otsu's adaptive thresholding.They started by converting RGB image to HSV color model and extracted value channel. Then, they applied Otsu's multi-thresholding on the value depending on the separation factor (SF). After that, they applied k-means clustering. Finally, they used morphological processing. The main disadvantages of their method are preprocessing and image enhancement stage is missing and segmentation is not fully automatic. It depends on the user notice to apply k-means or not. Sivasangareswari and Kumar [11] proposed a brain tumor segmentation using fuzzy c-means clustering with local information and Kernel metric. At first, they filtered brain images by median filter. Second, they made clustering by using fuzzy c-means technique. Their modified FCM (fuzzy c-means) depends on space distance of all neighbor pixels. They did feature extraction using algorithms of a gray level co-occurrence matrix (GLCM). Classification by SVM was also used. The limitation of their work is that they didn't make skull removal. The thing that increases the amount of used memory and increase the processing time. Abdul-Nasir et al. [12] presented a color image segmentation technique to detect the malaria parasites in red blood cells. They applied partial contrast stretching technique on color malaria images and extract color components from enhanced images. Then, they used k-means clustering technique. Finally, they used median filter and seeded region growing area extraction algorithms.   The limitation of this method is that there is no best model all times. Therefore, HSI color model gives best results in recall but C-Y color model gives best results in precision.Joseph et al. [13] presented brain tumor MRI segmentation .They started by the preprocessing stage. It converts the RGB input image to grey scale. They used a median filter. The preprocessed image is supplied for k-means clustering algorithm then followed by morphological filtering if there are no clustered regions. The main disadvantage is they didn't make skull removing in preprocessing step.Wang et al. [14] implemented an adaptive particle swarm optimization algorithm with mutation operation based on k-means. They combined the particle swarm optimization and k-means in case of local search and global search. The mutation was processed to accelerate a poor particle in population. The main disadvantage is that they didn't care about reducing the data set size by using feature extractions, which can reduce the number of iterations and the execution time. In our proposed medical image segmentation system based on the proposed hybrid clustering techniques,the main objective is to accurately detect the brain tumor in minimal execution time. We put into account the accuracy and minimum execution time in each stage. In the preprocessing stage, we applied the median filter to enhance entire image quality and removed the skull from the processed image. This stage reduces both the processing time and the used amount of memory. In segmentation stage, all advantages of k-means, fuzzy c-means, and particle swarm optimization are preserved; while their main problems have been solved by the proposed hybrid techniques. The over segmentation and under segmentation problems were solved as shown in the experimental results and the iterations and computation time were reduced. The user interaction is eliminated. The thresholding is applied to present a clear brain tumor clustering. Finally, the level set stage is applied to present the contoured tumor area on the original image.

3   MRI Brain Segmentation System

There are many medical image segmentation systems using k-means technique in detecting mass tumor in brain [15]. K-means technique is fast and simple, but it suffers from incomplete detection of tumor mainly if it is malignant tumor. On the other hand, other systems used fuzzy c-means technique because it retains more information from the original image. It can detect malignant tumor cells accurately [16], but these techniques are sensitive to noise and outliers and they take long execution time. Besides these systems, there is other systems used particle swarm optimization to segment tumor from brain images [17]. It may reach the optimal solution or near the optimal solution. It takes more computation time especially in color image segmentation. In our proposed medical image segmentation system, we get benefits from the last three techniques. As shown in Fig.1, the proposed medical image segmentation system consists of four main stages: pre-processing, clustering, tumor extraction and contouring, and validation. The main idea of doing the integration is to reduce number of iterations done by fuzzy c-means clustering technique. Of course, it minimizes execution time and gives qualitative results with KFCM (K means integrated with fuzzy c means) clustering. It also reduces computation time of particle swarm optimization to reach to the optimal clustering in KPSO (k means integrated with particle swarm optimization) clustering. The main stages of the proposed system will be discussed in more detail in the sequent subsections.

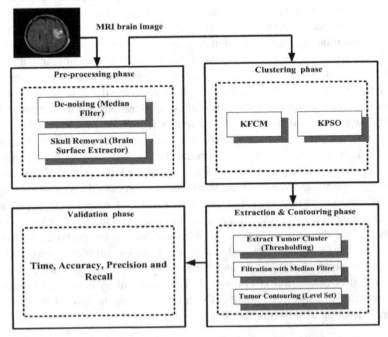

Fig. 1. The framework of the proposed medical image segmentation system

3.1   Pre-processing Stage

This stage is implemented by applying a series of initial processing procedures on the image before any special purposes processing. The main purpose of this stage is to improve the image quality and removes the noise. Since, the brain images are more sensitive than other medical images; they should be of minimum noise and maximum quality. Therefore, this stage consists of de-noising and skull removal sub-stages. De-noising is important for medical images to be sharp, clear, and free of noise and artifacts. MRI images are normally corrupted by Gaussian and Poisson noise [18]. In this paper, we used median filter which is a nonlinear filter. It is often used in image processing to reduce salt and pepper noises [19]. It works by moving pixel by pixel through the image, replacing each value with the median value of neighboring pixels. The pattern of neighbors is called the "window", which slides pixel by pixel over the entire image. The median is calculated by first sorting all the pixel values from the window into numerical order, after that replacing the pixel being considered with the middle (median) pixel value. Median filtering is better than linear filtering for removing noise in the presence of edges [20]. On the other hand, image background doesn't usually contain any useful information but increasing processing time. So, removing background, skull, scalp, eyes, and all structures, which are not in interest, decrease the amount of used memory and increase the processing speed. Skull removed is done by using BSE (brain surface extractor) algorithm. It is used only with MRI images [21]. It filters the image to remove irregularities, detects edges in the image, and performs morphological erosions and brain isolation. It also performs surface cleanup and image masking.

3.2   Clustering Stage

By de-noising the MRI images and removing skulls, the images are fed to one of the proposed techniques: KFCM or KPSO. In case of supplying the image to the first technique KFCM we initialize cluster numbers K, max iterations, and termination parameter. The cluster centers are calculated by:

$$mu = (1: k)*m/(k+1) \qquad (1)$$

Where mu are the initial means that can be calculated due to K the number of clusters and m=max (MRI image) +1. Then, assign each point to the nearest cluster center based on minimum distance and re-compute the new cluster centers. It repeats until some convergence criterion is met. Then, the resulting image can be clustered by initializing number of centroids (centroid of the cluster is the mean of all points in this cluster) equal to the number of k. This will reduce iterations and processing time. If initializing number of centroids differs from K number, it may increase time in some cases. Then, calculating the distance and updating membership and means values with determining the condition of closing. The output of the technique is the clustering image, execution time, and iteration. On the other hand, the free noise MRI images are fed to the second proposed approach (KPSO) by initializing cluster numbers k, population of particles, inertial weight value, and number of iterations. The algorithm follows the same steps of KFCM till determining the clusters means due to initial k. Each particle is updated by two best values. The first is the personal best (*pbest*) which is the best solution or fitness that has achieved so far by that particle. The second value is global best (*gbest*) which is the best value obtained so far by any particle in the neighborhood of that particle. Each particle modifies its position using the current positions, the current velocities, the distance between the current position and *pbest*, and the distance between the current position and the *gbest*. After that, the particle updates its velocity and positions until termination parameter which is the number of iterations. The output of the algorithm is a clustering image with optimal number of clusters, optimal clusters centroids, and computation time.

3.3   Extraction and Contouring Stage

In this stage, we used two segmentation methods: thresholding and active contour level set. Thresholding segmentation is intensity based segmentation. It is one of the important, simple, and popular segmentation techniques [22]. Thresholding segmentation technique converts a multilevel image into a binary image. It is used to extract or separate the objects in an image from the background. The segmenting image obtained from thresholding has the advantages of smaller storage space, fast processing speed, and ease manipulation [23]. The output of this stage is a segmented image with dark background and lighted object which is the brain tumor. On the other hand, the active contours have been used for image segmentation and boundary detection since the first introduction of snakes by Kass et al. [24]. The main idea is to start with initial boundary. Shapes are presented as closed curves, i.e. contours. It iteratively modifies them by applying shrink/expansion according to the constraints. An advantage of the active contours is that they partition an image into regions with continuous boundaries. So, we used level set to contour the boundary of tumor area or shape continuously after thresholding. Level set method is demonstrated in details by Lee et al. [25]. By using level set after thresholding, it gives user the resulting segmenting image of the original image with contoured tumor areas.

3.4   Validation Stage

In validation stage, the resulting segmenting images with the two proposed clustering techniques were compared to the ground truth as illustrated in experimental results. The results were evaluated by performance matrix which contains the precision and recall. Precision is the correct segmentation refers to the percentage of true positive, the number of pixels that belong to a cluster and are segmented into that cluster. Recall or sensitivity is defined as the number of true positives divided by the total number of elements that actually belong to the positive cluster [26]. The performance matrix will be illustrated in details in experimental results.

3.5   Experimental Results

In order to check the performance of our two hybrid proposed clustering techniques in our proposed medical image segmentation system, we used three benchmark data sets. The first is the Digital Imaging and Communications in Medicine (DICOM) data set. DICOM consists of 21 images which contain brain tumors. It has no ground truth images for the contained images. The second data set is Brain Web data set. It contains simulated brain MRI data based on normal and multiple sclerosis (MS). This dataset consists of 152 images. The last data set is BRATS database from Multimodal Brain Tumor Segmentation. The data set consists of multi-contrast MRI scans and has ground truth images. This data set contains 81 images.

In this section, we show the results of our two hybrid proposed clustering techniques obtained using real MRI brain images from the three different databases. This work was implemented using MATLAB 7.12.0 (R2011a). We run our experiments on a core i5/2.4 GHZ computer with 8 GB RAM and a NVEDIA/ (1 GB VRAM) VGA card. Table 1 and Table 2 demonstrate the results of applying the main four stages of our framework including our proposed techniques KFCM and KPSO on the three image data sets. Table 3, shows that EM (Expectation Maximization) like KM (K means) in accuracy but it takes longer time (T in seconds) than K Means. On the other hand, the mean shift clustering technique (MS) need to enter the parameters of bandwidth, threshold and output number of clusters K and time (T in seconds). It takes less processing time but it does not give accurate results as in DS2 when K=3.We observed that without skull removal, the processing time on all techniques was increased in DS1. On the contrary, when removing skull as in DS2 or using images with removed skull like in DS3, the processing time is reduced as shown in Table 3. In table 4, KFCM seems like FCM (Fuzzy C Means) in accuracy but KFCM take less processing time T than FCM with less iteration. From Table 5, we observed that KPSO seems like PSO (Particle Swarm Optimization) in accuracy but KPSO take less processing time than PSO. Table 6, 7 describe the performance matrix of K-means and expectation maximization. The results prove that expectation maximization may be like K-means in accuracy in the last two data sets (DS2 and DS3), but in first data set (DS1), KM accuracy is 85.7% where EM accuracy is 66.6%. From Table 6 and 8, we can observe that MS technique also seems to be the same as KM technique in performance matrix except in the second data set (DS2). Table 9 and 10 ensure that KFCM is more accurate than FCM and it is very clear in the results of first data set where KFCM accuracy is 90.50% but FCM accuracy is 85.7%. Table 11, 12 describe the performance matrix comparisons between particle swarm optimization PSO and the integration between k means and particle swarm optimization KPSO. The results prove that they are the same in accuracy, but PSO takes long time compared to KPSO in Table 5.The results showed that FCM takes longest execution time (T in seconds) in clustering, and then EM. After that, PSO takes the third level in execution time less than FCM and EM. KFCM is in the fourth level and KPSO in the fifth level. KM is in the sixth level and MS is in the last level.

Table 1. The main stages of the proposed framework by using KFCM applied on three benchmark data sets

DS	Original mri	BSE	Median filter	KFCM	Threshold	Level Set	Truth/Normal
Ds1		NO skull removal					No truth or normal images
DS2							
DS3		Already skull removed					

Table 2. The main stages of the proposed framework by using KPSO applied on three benchmark data sets

DS	Original mri	BSE	Median filter	KPSO	Threshold	Level Set	Truth/Normal
Ds1		NO skull removal					No truth or normal images
DS2							
DS3		Already skull removed					

Table 3. The comparison between KM (K mean), EM (Expectation Maximization) and MS (Mean Shift) clustering algorithms

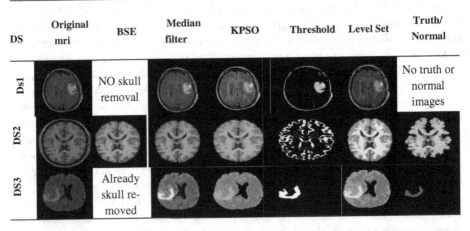

DS	KM			EM			MS		
		K	T(s)		K	T(s)		K	T(s)
Ds1		9	7.5		9	34.47		4	0.35
Ds2		4	1.76		4	8.00		3	0.29
Ds3		12	4.34		12	32.06		5	0.47

Table 4. The comparison between FCM (Fuzzy C means) and our proposed technique (KFCM)

DS	FCM	Iteration No	T(s)	KFCM	Iteration No	T(s)
DS1		51	59.52		8	12.87
DS2		19	15.92		4	5.18
DS3		14	6.89		3	3.46

Table 5. The comparison between PSO (Particle Swarm Optimization) and our proposed technique (KPSO)

DS	PSO	Iteration No	T(s)	KPSO	Iteration No	T(s)
DS1		7	34.58		9	9.23
DS2		25	58.75		25	8.42
DS3		4	6.26		4	6.01

Table 6. The performance metrics of KM

Data Sets	TP	TN	FP	FN	Accuracy	Precision	Recall
DS1	85.7	0	0	14.3	85.7	100	85.7
DS2	96.7	0	0	3.3	96.7	100	96.7
DS3	95.06	0	0	4.94	95.06	100	95.06

Table 7. The performance metrics of EM

Data Sets	TP	TN	FP	FN	Accuracy	Precision	Recall
DS1	66.6	0	0	33.4	66.6	100	66.6
DS2	95.4	0	0	4.6	95.4	100	95.4
DS3	95.06	0	0	4.94	95.06	100	95.06

Table 8. The performance matrices of MS

Data Sets	TP	TN	FP	FN	Accuracy	Precision	Recall
DS1	85.7	0	0	14.3	85.7	100	85.7
DS2	96.05	0	0	3.95	96.05	100	96.05
DS3	95.06	0	0	4.94	95.06	100	95.06

Table 9. The performance matrices of FCM

Data Sets	TP	TN	FP	FN	Accuracy	Precision	Recall
DS1	85.7	0	0	14.3	85.7	100	85.7
DS2	100	0	0	0	100	100	100
DS3	100	0	0	0	100	100	100

Table 10. The performance matrices of KFCM

Data Sets	TP	TN	FP	FN	Accuracy	Precision	Recall
DS1	90.5	0	0	9.5	90.5	100	90.5
DS2	100	0	0	0	100	100	100
DS3	100	0	0	0	100	100	100

Table 11. The performance matrices of PSO

Data Sets	TP	TN	FP	FN	Accuracy	Precision	Recall
DS1	95	0	0	5	95	100	95
DS2	100	0	0	0	100	100	100
DS3	100	0	0	0	100	100	100

Table 12. The performance matrices of KPSO

Data Sets	TP	TN	FP	FN	Accuracy	Precision	Recall
DS1	95	0	0	5	95	100	95
DS2	100	0	0	0	100	100	100
DS3	100	0	0	0	100	100	100

4   Conclusion and Future Work

Image segmentation plays an important role in medical image. In this paper, we proposed a brain image segmentation system based on two different clustering techniques. The first is integration between fuzzy c means with k means which is called KFCM. The second is integration between particle swarm optimization and k means which is called KPSO. We applied the two clustering techniques on the three different data sets to detect the brain tumor. From experiments, we proved the effectiveness of our techniques in segmenting the brain tumor by comparing it with five state-of-the-art algorithms: K-means, Expectation Maximization, Mean Shift, Fuzzy C means, and particle swarm optimization. The result of KPSO is very near to KFCM in accuracy and time but in first data set, the KPSO accuracy is 95% and KFCM is 90.5% and KPSO time is less than KFCM time. In future work, the 3D evaluation of the brain tumor detection using 3D slicer will be carried out. As well as to increase the efficiency of the segmentation process, an intensity adjustment process will provide more challenging and may allow us to refine our segmentation techniques to the MRI brain tumor segmentation, refer to [27].

References

1. Bai, X., Wang, W.: Saliency-SVM: An automatic approach for image segmentation. J. Neur. Comput. 136, 243–255 (2014)
2. Patil, D.D., Deore, S.G.: Medical Image Segmentation: A Review. Int. J. Comp. Sci. Mob. Comput. 2(1), 22–27 (2013)
3. Dass, R., Priyanka, D.S.: Image segmentation techniques. Int. J. Electron. Commun. Technol. 3(1), 66–70 (2012)
4. Fazli, S., Ghiri, S.F.: A Novel Fuzzy C-Means Clustering with Hybrid Local and Non Local Spatial Information for Brain Magnetic Resonance Image Segmentation. J. Appl. Eng. 2(4), 40–46 (2014)
5. Patel, J., Doshi, K.: A Study of Segmentation Methods forDetection of Tumor in Brain MRI. Advance in Electronic and Electric Engineering 4(3), 279–284 (2014)
6. Leela, G.A., Kumari, H.M.V.: Morphological Approach for the Detection of Brain Tumour and Cancer Cells. J. Electron. Comput. Eng. Res. 2(1), 7–12 (2014)
7. Neshat, M., Yazdi, S.F., Yazdani, D., Sargolzaei, M.: A New Cooperative Algorithm Based on PSO and K-Means for Data Clustering. J. Comput. Sci. 8(2), 188–194 (2012)
8. Madhulatha, T.S.: An overview on clustering methods. IOSR. J. Eng. 2(4), 719–725 (2012)
9. Acharya, J., Gadhiya, S., Raviya, K.: Segmentation Techniques For Image Analysis: A review. Int. J. Comput. Sci. Mang. Res. 2(1), 1218–1221 (2013)
10. Jumb, V., Sohani, M., Shrivas, A.: Color Image Segmentation Using K-Means Clustering and Otsu's Adaptive Thresholding. Int. J. Innov. Technol. Explor. Eng. 3(9), 72–76 (2014)
11. Kumar, K.S., Sivasangareswari, P.: Fuzzy C-Means Clustering with Local Information and KernelMetric for Image segmentation. Int. J. Adv. Res. Comput. Sci. Technol. 2(1), 95–99 (2014)
12. Abdul-Nasir, A.S., Mashor, M.Y., Mohamed, Z.: Colour Image Segmentation Approach for Detection of Malaria Parasites Using Various Colour Models and k-Means Clustering. J. WSEAS Transactions. Biol. Biomed. 10(1), 41–55 (2013)
13. Joseph, R.P., Singh, C.S., Manikandan, M.: brain tumor MRI image segmentation and detection in image processing. Int. J. Res. Eng. Technol. 3(1), 1–5 (2014)
14. Wang, X., Guo, Y., Liu, G.: Self-adaptive Particle Swarm Optimization Algorithm with Mutation Operation based on K-means. Advanced materials research. In: 2nd International Conference on Computer Science and Electronics Engineering, pp. 2194–2198. Atlantis Press, Paris (2013)
15. Mohan, P., Al, V., Shyamala, B.R., Kavitha, B.C.: Intelligent Based Brain Tumor Detection Using ACO. Int. J. Innov. Res. Comput. Commun. Eng. 1(9), 2143–2150 (2013)
16. Anandgaonkar, G., Sable, G.: Brain Tumor Detection and Identification from T1 Post Contrast MR Images Using Cluster Based Segmentation. Int. J. Sci. Res 3(4), 814–817 (2014)
17. Arulraj, M., Nakib, A., Cooren, Y., Siarry, P.: Multicriteria Image Thresholding Based on Multiobjective Particle Swarm Optimization. J. Applied Mathe. Sci. 8(3), 131–137 (2014)
18. Rodrigues, I., Sanches, J., Dias, J.: Denoising of Medical Images corrupted by Poisson Noise. In: 15th IEEE International Conference on Image Processing ICIP, pp. 1756–1759. IEEE Press, San Diego (2008)
19. Abinaya, K.S., Pandiselvi, T.: Brain tissue segmentation from magnitude resonance image using particle swarm optimization Algorithm. Int. J. Comput. Sci. Mob. Comput. 3(3), 404–408 (2014)

20. Kumar, S.S., Jeyakumar, A.E., Vijeyakumar, K.N., Joel, N.K.: An adaptive threshold intensity range filter for removal of random value impulse noise in digital images. J. Theoretical Appl. Info. Technol. 59(1), 103–112 (2014)
21. Medical Medica limage processing analysis and visualization, http://mipav.cit.nih.gov/pubwiki/index.php/ Extract_Brain:_Extract_Brain_Surface_BSE
22. Narkhede, H.P.: Review of Image Segmentation Techniques. Int. J. Sci. Modern. Eng 1(8), 54–61 (2013)
23. Saini, R., Dutta, M.: Image Segmentation for Uneven Lighting Images using Adaptive Thresholding and Dynamic Window based on Incremental Window Growing Approach. Int. J. Compu. App. 56(13), 31–36 (2012)
24. Kass, M., Witkin, A., Terzopoulos, D.: Snakes: Active contour Models. Int. J. Compu. Vison 1(4), 321–331 (1988)
25. Lee, G.P.: Robust image segmentation using active contours: level set approaches. PhD, North Carolina State University (2005)
26. Dakua, P.S.: Use of chaos concept in medical image segmentation. J. Comput. Meth. Biomech. Biomed. Eng. Imag. Visua. 1(1), 28–36 (2013)
27. Own, H.S., Hassanien, A.E.: Rough Wavelet Hybrid Image Classification Scheme. Journal of Convergence Information Technology JCIT 3(4), 65–75 (2008)

Part VI
Fuzzy Multi Criteria Decision Making

A Generalized Polygon Fuzzy Number for Fuzzy Multi Criteria Decision Making

Samah Bekheet, Ammar Mohammed, and Hesham A. Hefny

Department of Computer & Information Sciences,
Institute of Statistical Studies and Research, Cairo University, Giza, Egypt
samahbekheet@yahoo.com, ammar@eu.ed.eg, hehefny@ieee.org

Abstract. Allowing various forms of fuzzy numbers to be adopted in fuzzy multi criteria decision making (FMCDM) problems adds more flexibility to decision makers to represent their own opinions to handle uncertainty. For most cases uncertain numbers of the forms of: interval, triangle, or trapezoidal are used. In this paper, polygon fuzzy numbers (PFNs) are introduced so as to allow decision makers to adopt other forms of numbers such as: pentagon, hexagon, heptagon, octagon, etc, to provide more flexibility to represent uncertainty. A case study is given to illustrate the way of manipulation of the proposed PFN.

Keywords: Polygon Fuzzy Number, Multi Criteria Decision Making, Generalized Fuzzy Number.

1 Introduction

Decisions in real world applications are often made under the presence of conflicting, uncertain, incomplete and imprecise information. Decision making is the procedure to find the best alternatives among a set of feasible alternatives and also ranking them as their priorities. Fuzzy multi Criteria Decision making (FMCDM) provides a powerful approach for drawing rational decisions under uncertainty given in the form of linguistic values (e.g. excellent, very good, good, and bad).

To enable the decision makers to express their own opinions. Such linguistic values need fuzzy tools to evaluate their calculations [1]. Examples of FMCDM tools are T-Norm Based, Gaussian fuzzy numbers, Interval fuzzy numbers, Interval type two fuzzy number, Triangle fuzzy numbers and Trapezoidal fuzzy numbers [2]. Interval, Triangle and Trapezoidal fuzzy numbers are more popular due to their conveniences of the arithmetic operations such as: addition, subtraction, multiplication, division, reciprocal, geometric mean, etc. Such operations enable the decision makers to determine the rank of criteria (alternatives) powerfully [3].

Several researchers consider Trapezoidal fuzzy numbers as Generalized fuzzy numbers (GFNs) [4,5,6]. This mainly due to the fact that other popular forms of specific fuzzy numbers including: triangles, intervals, or even singleton can be obtained as special cases of Trapezoidal fuzzy numbers. In this paper, we introduce Polygon Fuzzy Number (PFN) as the actual form of GFN. The proposed form of PFN provides

A.E. Hassanien et al. (Eds.): AMLTA 2014, CCIS 488, pp. 415–423, 2014.

higher flexibility to decision makers to express their own linguistic values rather than other form of fuzzy numbers. Using PFNs, decision makers can freely express their own linguistic values using fuzzy numbers of various shapes e.g. triangle, trapezoidal, pentagon, hexagon, heptagon, octagon and etc. This provides a powerful tool for solving various decision making problems which are based on different views of decision makers [7, 8, 9, 13].

The rest of the paper is organized as follows: Section 2 defines the problem. Section 3 introduces the proposed model and the required definitions of the proposed model. Section 4 introduces a numerical example. Finally section 5 presents the conclusion.

2   Problem Definition

A Trapezoidal fuzzy number with its four vertices is considered a generalized number for other forms including: triangles, intervals and also singletons (i.e. crisp numbers) [4,5,6]. However, it is intuitively clear that allowing more vertices to the fuzzy number adds more flexibility to the decision maker to represent his own opinion to deal with the considered FMCDM problem. Therefore, the ability to introduce generalized piece-wise membership function with n-vertices as a fuzzy number with its own arithmetic operations represents the typical unification of all other forms of fuzzy numbers. Adopting such new forms of generalized fuzzy numbers should considerably enhance modeling and solving FMCDM problems.

3   Polygon Fuzzy Number

3.1   Basic Definitions

A polygon fuzzy number (PFN) is defined as a convex and normal polygon fuzzy set, where Polygon fuzzy sets are firstly addressed in the context of fuzzy interpolative reasoning [10]. A polygon fuzzy set A has n characteristic points $(a_0, a_1, ..., a_{n-1})$ as shown in fig. 1. The core of the fuzzy set, at which the membership equals one, is represented by the interval $[a_{\lfloor (n-1/2) \rfloor}, a_{\lceil (n-1/2) \rceil}]$. There are $\lfloor (n-1)/2 \rfloor +1$ membership levels including bottom and top levels. Thus the cardinality of the level set of a polygon fuzzy set is denoted by V as given in (1), It is clear that V represents the number of α-cuts of the polygon fuzzy sets, namely: $\alpha_0= 0, ...,$ $\alpha_{\lfloor (n-1)/2 \rfloor +1}=1$.

$$V = \lfloor (n-1)/2 \rfloor +1 \qquad (1)$$

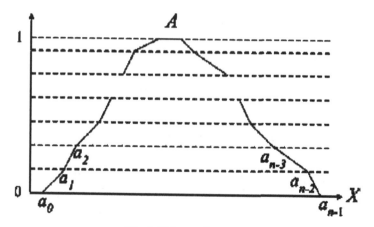

Fig. 1. Polygon fuzzy set

3.2   Ranking of Polygon Fuzzy Number

The centre of area of a fuzzy number is considered the most popular ranking method [11, 12]. Computing the centroid of a general polygon fuzzy number can be obtained by the following theorem.

3.2.1 The Centroid Theorem
The centroid x (center of area) of the polygon which characterized by the n-points $(x_0, x_1,.., x_{n-1})$ with area A and membership level $y_i = \mu A(x_i)$, can be computed using the following formula:

$$\bar{x} = \frac{\sum_{i=1}^{n} \overline{X}_i A_i}{\sum_{i=1}^{n} A_i} \tag{2}$$

$$\overline{X}_i = \frac{1}{3}[(x_i + x_{i-1}) + (\frac{x_i y_i + x_{i-1} y_{i-1}}{y_i + y_{i-1}})] \tag{3}$$

$$A_i = \frac{(x_i - x_{i-1})(y_i + y_{i-1})}{2} \tag{4}$$

3.2.2 The Theorem Proof
The polygon fuzzy number A with its n vertices sees Figure 2, can be divided into (n-1) sub-polygons.

Each sub-polygon can generally be represented as a trapezoidal with 4-vertices. Thus the ith sub-polygon as shown in Fig.3, has an area equals Ai and its centroid $\overline{x}_i$ is computed as follows:

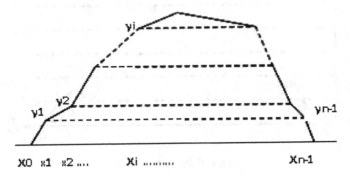

Fig. 2. Polygon characterized

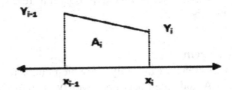

Fig. 3. Sub- polygon area

$$\overline{x}_i = \frac{\int_{x_{i-1}}^{x_i} x f_i(x) dx}{\int_{x_{i-1}}^{x_i} f_i(x) dx}$$

$$= \frac{I_1}{I_2} = \frac{f_i(x) - y_{i-1}}{x - x_{i-1}} = \frac{y_i - y_{i-1}}{x_i - x_{i-1}}$$

$$f_i(x) = \frac{y_i - y_{i-1}}{x_i - x_{i-1}} (x - x_{i-1}) + y_{i-1}$$

$$= \frac{y_i - y_{i-1}}{x_i - x_{i-1}} x - \frac{y_i - y_{i-1}}{x_i - x_{i-1}} x_{i-1} + y_{i-1}$$

$$= \frac{\Delta y_i}{\Delta x_i} x + y_{i-1} - \frac{\Delta y_i}{\Delta x_i} x_{i-1}$$

Therefore,

$$f_i(x) = \alpha_i x + \beta_i$$

Where

$$\alpha_i = \frac{\Delta y_i}{\Delta x_i} \quad , \quad \beta_i = y_{i-1} - \frac{\Delta y_i}{\Delta x_i} x_{i-1}$$

Then, performing the integrations for both I_1 and I_2, we have:

$$\overline{x}_i = \frac{1}{3}[(x_i + x_{i-1}) + (\frac{x_i y_i + x_{i-1} y_{i-1}}{y_i + y_{i-1}})]$$

Also, it is clear that the area of the ith sub-polygon shown in fig. 3 is:

$$A_i = \frac{(x_i - x_{i-1})(y_i + y_{i-1})}{2}$$

3.3   Arithmetic Operations

The arithmetic operations of two polygon fuzzy numbers should satisfy the following two rules:

i - Both fuzzy numbers should have the same number of vertices.
ii- Both fuzzy numbers should have the same level set.

Thus, if we have to add two different polygon fuzzy numbers, e.g triangle $T(a_0, a_1, a_2)$ and hexagon $H(b_0, b_1, b_2, b_3, b_4, b_5)$. If the level set of the triangle is: $\{T(a_0)=T(a_2)=0, (a_1)=1\}$, and the level set of the hexagon is: $\{H(b_0)=H(b_5)=0, H(b_1)=H(b_4)=0.6, H(b_2)=H(b_3)=1\}$.

Then additional level set value at 0.6 should be added to the triangle fuzzy number and consequently more vertices appears for that triangle to be in the form: $T(a_0, a_1, a_2, a_3, a_4, a_5)$ so that its level set becomes: $\{T(a_0)=T(a_5)=0, T(a_1)=T(a_4)=0.6, T(a_2)=T(a_3)=1\}$.

Thus, keeping the above two rules in mind, assume that there two PFNs $A(a_0, a_1, a_2, \ldots, a_{n-1})$ and $B(b_0, b_1, b_2, \ldots, b_{n-1})$, then the arithmetic operations can be defined as follows:

PFNs Addition

$$A \oplus B = (a_0+b_0, a_1+b_1, a_2+b_2, \ldots, a_{n-1}+b_{n-1})$$

PFNs Subtraction

$$A \ominus B = (a_0-b_{n-1}, a_1-b_{n-2}, a_2-b_{n-3}, \ldots, a_{n-1}-b_0)$$

PFNs Multiplication

$$A \otimes B = (a_0 \times b_0, a_1 \times b_1, a_2 \times b_2, \ldots, a_{n-1} \times b_{n-1})$$

PFNs Divisions

$$A \Phi B = (a_0/b_{n-1}, a_1/b_{n-2}, a_2/b_{n-3}, \ldots, a_{n-1}/b_0)$$

$$\text{Where } b_0 \neq 0, b_1 \neq 0, \ldots \text{ and } b_{n-1} \neq 0.$$

4   Illustrative Example

Assume three alternatives and A_1, A_2 A_3 are evaluated with respect to five criteria C_1, C_2, C_3, C_4 and C_5 as shown in table1.

Table 1. Criteria-Alternatives Evaluation Matrix

Alternatives	Criteria				
	C_1	C_2	C_3	C_4	C_5
A_1	A_{11}	A_{12}	A_{13}	A_{14}	A_{15}
A_2	A_{21}	A_{22}	A_{23}	A_{24}	A_{25}
A_3	A_{31}	A_{32}	A_{33}	A_{34}	A_{35}

Let the decision maker put his evaluation values in the form of polygon fuzzy numbers as shown in table2.

According to table 2, the whole level set is $\{0, 0.4, 0.5, 0.6, 1\}$.

Therefore, all the above PFNs should be rewritten according to the whole level set. This of course will add more vertices as shown in Table 3. It is clear, that the obtained unified PFNs are all having ten vertices at the above five values of the level sets (five Left, five Right).

Table 2. The Evaluation PFNs

Fuzzy No.	Level Sets	Type
A_{11}	A(1)=A(4)=0, A(2)=A(3)=1	Quadrilateral
A_{12}	A(1)=A(3)=0, A(2)=1	Triangle
A_{13}	A(1)=A(6)=0, A(2)=0.4, A(3)=A(4)=1, A(5)=0.6	Hexagon
A_{14}	A(1)=A(3)=0, A(2)=1	Triangle
A_{15}	A(1)=A(6)=0, A(2)=0.4, A(3)=A(4)=1, A(5)=0.6	Hexagon
A_{21}	A(1)=A(4)=0, A(2)=A(3)=1	Quadrilateral
A_{22}	A(1)=A(5)=0, A(2)=0.5, A(3)=1, A(4)=0.6	Pentagon
A_{23}	A(1)=A(3)=0, A(2)=1	Triangle
A_{24}	A(1)=A(4)=0, A(2)=0.5, A(3)=1	Quadrilateral
A_{25}	A(1)=A(5)=0, A(2)=0.4, A(4)=4.6 ,A(3)=1	Pentagon
A_{31}	A(1)=A(3)=0, A(2)=1	Triangle
A_{32}	A(1)=A(5)=0, A(2)=0.6, A(3)=1, A(4)=0.4	Pentagon
A_{33}	A(1)=A(4)=0, A(2)=0.4, A(3)=1	Quadrilateral
A_{34}	A(1)=A(5)=0, A(2)= A(4)=0.5, A(3)=1,	Pentagon
A_{35}	A(1)=A(5)=0, A(2)=0.6, A(3)=1, A(4)=0.6	Pentagon

Table 3. The new forms of Evaluation PFNs

level set	The fuzzy numbers corresponding to the level sets														
	A^{11}	A^{12}	A^{13}	A^{14}	A^{15}	A^{21}	A^{22}	A^{23}	A^{24}	A^{25}	A^{31}	A^{32}	A^{33}	A^{34}	A^{35}
0L	1	1	1	1	1	1	1	1	1	1	1	1	1	1	1
0.4L	1.4	1.4	2	1.4	2	1.4	1.8	1.4	1.8	2	1.4	1.7	2	1.8	1.7
0.5L	1.5	1.5	2.2	1.5	2.2	1.5	2	1.5	2	2.2	1.5	1.8	2.2	2	1.8
0.6L	1.6	1.6	2.3	1.6	2.3	1.6	2.2	1.6	2.2	2.3	1.6	2	2.3	2.2	2
1L	2	2	3	2	3	2	3	2	3	3	2	3	3	3	3
1R	3	2	4	2	4	3	3	2	3	3	2	3	3	3	3
0.6R	3.4	2.4	4.8	2.4	5	3.4	4	2.4	3.4	4	2.4	3.7	3.4	3.8	3.8
0.5R	3.5	2.5	5	2.5	5.2	3.5	4.2	2.5	3.5	4.2	2.5	3.8	3.5	4	4
0.4R	3.6	2.6	5.2	2.6	5.3	3.6	4.3	2.6	3.6	4.3	2.6	4	3.6	4.2	4.2
0R	4	3	6	3	6	4	5	3	4	5	3	5	4	5	5

Then, get the normalized ranked PFN for each alternative as in (5) as show in Table 4:

$$R(A_i) = \sum A_{ik} \Big/ \sum\sum A_{ik} \qquad (5)$$

Where k=1, 2,..,5 and i=1,2,3

Table 4. The Normalized Ranking PFNs

level set	$R(A_1)$	$R(A_2)$	$R(A_3)$
0L	0.077	0.077	0.077
0.4L	0.145	0.149	0.151
0.5L	0.163	0.169	0.172
0.6L	0.181	0.190	0.194
1L	0.279	0.302	0.326
1R	0.385	0.359	0.359
0.6R	0.610	0.583	0.578
0.5R	0.683	0.653	0.652
0.4R	0.769	0.734	0.740
0R	1.467	1.400	1.467

Finally, applying the above centroid theorem for PFN to get the final crisp ranking value for each alternative as follows:

$COA(R(A_1)) = 0.476$
$COA(R(A_2)) = 0.462$
$COA(R(A_3)) = 0.472$

Thus it is clear that alternative A_1 should be selected as the best choice, and we can arrange them as the highest priority as $A_1 > A_3 > A_2$.

5   Conclusions

This paper introduced a way for adopting PFNs in FMCDM problems. The proposed form of PFN ensures its generality over other popular forms of fuzzy numbers.

A general formula for ranking PFNs is given using the centroid method. The arithmetic operations for PFNs are presented and an explanation example shows how to adopt such generalized fuzzy numbers for solving FMCDM problems.

References

1. Hong, D.H.: Strong laws of large numbers for t-norm-based addition of fuzzy set-valued random variables. Fuzzy Sets and Systems 223, 449–728 (2013)
2. Chen, S.-M., Wang, C.-Y.: Fuzzy decision making systems based on interval type-2 fuzzy sets. Applied Mathematical Modeling Science Direct 424, 1–21 (2013)
3. Herrera, F., Herrera-Viedma, E.: Spain Linguistic decision analysis: steps for solving decision problems under linguistic information. Computer Science and Artificial Intelligence 115, 67–82 (2000)
4. Chen, S.-J., Chen, S.-M.: Fuzzy Risk Analysis Based on Similarity Measures of Generalized Fuzzy Numbers. IEEE Transaction on Fuzzy Systems 11(1), 45–56 (2003)
5. Yong, D., Wenkang, S., Feng, D., Qi, L.: A new similarity measure of generalized fuzzy numbers and its application to pattern recognition. Pattern Recognition Letters 25, 875–883 (2004)

6. Chen, S.-M., Chen, J.-H.: Fuzzy risk analysis based on ranking generalized fuzzy numbers with different heights and different spreads. Expert Systems with Applications 36, 6833–6842 (2009)
7. Sun, C.-C.: A performance evaluation model by integrating fuzzy AHP and fuzzy TOPSIS methods. Expert Systems with Applications 37, 7745–7754 (2010)
8. Anisseh, M., Piri, F., Shahraki, M.R., Agamohamadi, F.: Fuzzy extension of TOPSIS model for group decision making under multiple criteria 38(4), 325–338 (2011)
9. Roghanian, E., Rahimi, J., Ansari, A.: Comparison of first aggregation and last aggregation in fuzzy group TOPSIS. Applied Mathematical Modelling 34, 3754–3766 (2010)
10. Chang, Y.-C., Chen, S.-M., Liau, C.-J.: Fuzzy Interpolative Reasoning for Sparse Fuzzy-Rule-Based Systems Based on the Areas of Fuzzy Sets. IEEE Transaction on Fuzzy Systems 16(5), 1285–1301 (2008)
11. Cheng, C.H.: A new approach for ranking fuzzy numbers by distance method. Fuzzy Sets and Systems 95, 307–317 (1998)
12. Wang, Y.M., Yang, J.B., Xu, D.L., Chin, K.S.: On the centroids of fuzzy numbers. Fuzzy Sets and Systems 157, 919–926 (2006)
13. Hassanien, A.E., Suraj, Z., Slezak, D., Lingras, P.: Rough computing: Theories, technologies and applications. IGI Publishing Hershey, PA (2008)

Improving the Performance of TAntNet-2
Using Scout Behavior

Ayman M. Ghazy and Hesham A. Hefny

Department of Computer & Information Sciences,
Institute of Statistical Studies and Research, Cairo University, Giza, Egypt
aymghazy@yahoo.com, hehefny@ieee.org

Abstract. Dynamic routing algorithms play an important role in road traffic routing to avoid congestion and to direct vehicles to better routes. TAntNet-2 algorithm presented a modified version of AntNet algorithm to dynamic traffic routing of road network. TAntNet-2 uses the pre-known information about the expected good travel time between sources and destinations for road traffic networks. Good travel time is used as a threshold value to fast direct the algorithm to good route, conserve on the discovered good route and remove unneeded computations. This paper presents a modified version of the TAntNet-2 routing algorithm that employs a behavior inspired from bee behavior when foraging for nectar. The new algorithm tries to avoid the effects of ants that take long route during searching for a good route. The modified algorithm introduces a new technique for launching ants according the quality of the discovered solution. The presented algorithm uses forward scout instead of forward ant and uses two forward scouts for each backward ant, in case of failing the first scout in finding accepted good route. The experimental results show high performance for the modified TAntNet-2 compared with TAntNet and TAntNet-2.

Keywords: Swarm Intelligence, Road networks, Dynamic traffic routing, AntNet, TAntNet-2, Forward ant, Forward scout, Backward ant, Check ant, bee behavior, bad route.

1 Introduction

Ant routing algorithms is one of the most promising swarm intelligence (SI) methodologies that are capable of finding near optimal solutions at low computational cost. Ant routing algorithms have been studied in many researches [1-7]. AntNet is a distributed agent based routing algorithm inspired by the behavior of natural ants [8]. Since its first appearance in 1998, AntNet algorithm has attracted many researchers to adopt it in both of data communication networks and road traffic networks.

On data networks, it has been shown that under varying traffic loads, AntNet algorithm is amenable to the associated changes and it shows better performance than that of Dijkstra's shortest path algorithm [9]. Several enhancements have been made to the AntNet algorithm. Baran and Sosa [10] proposed to initialize the routing table at each node in the network. The proposed initialization reflects previous knowledge about

A.E. Hassanien et al. (Eds.): AMLTA 2014, CCIS 488, pp. 424–435, 2014.

network topology rather than the presumption of uniform probabilities distribution given in original AntNet algorithm. Tekiner et al. [11] produced a version of the AntNet algorithm that improved the throughput and the average delay. In addition, their algorithm utilized the ant/packet ratio to limit the number of used ants. A new type of helping ants has been introduced in [12] to increase cooperation among neighboring nodes, thereby reducing AntNet algorithm's convergence time. A study for a computation of the pheromone values in AntNet has been given in [13]. Radwan et al. [14] proposed an adapted AntNet protocol with blocking–expanding ring search and local retransmission technique for routing of Mobile ad hoc network (MANET). Sharma et al. [15] showed that load balancing is successfully fulfilled for ant based techniques [15].

On road traffic networks, An Ant Based Control (ABC) algorithm has been applied in [2] for routing of road traffic through a city. In [3] a modification of Ant Based Control (ABC) and AntNet has been presented for routing vehicle drivers using historically-based traffic information. Claes and Holvoet [4] proposed a cooperative ACO algorithm for finding routes based on a cooperative pheromone among ants. Yousefi and Zamani in [6] proposed an optimal routing method for car navigation system based on a combination between Divide and Conquer method and Ant Colony algorithm. According to their proposed method, road network is divided into small areas. Then the learning operation is done in these small areas. Then different learnt paths are combined together to make the complete paths. This method causes traffic load balance over the road network. A version of the AntNet algorithm has been applied in [16] to improve traveling time over a road traffic network with the ability to divert traffic from congested routes. In [17] a city based parking routing system (CBPRS) that used Ant based routing has been proposed. Kammoun et al. in [18] introduced an adaptive vehicle guidance system instigated from the ant behavior. Their system allows adjusting the route choice according to the real-time changes in the road network, such as new congestions and jams. In [19] an Ant Colony Optimization combined with link travel time prediction has been applied to find routes. The proposed algorithm takes into account link travel time prediction, which can reduce the travel time. Ghazy et al. [20] proposed a threshold based AntNet algorithm (called TAntNet) for dynamic traffic routing of road networks, which used the pre-known information about good travel times among different nodes as a threshold value.

In the last decade, many researches were directed their efforts to produce hybrid algorithms that combine features from ants and bees behavior [21, 22]. Rahmatizadeh et al. [23] proposed an Ant-Bee Routing algorithm, which inspired from the behavior of both ant and bee to solve the routing problem. The algorithm is based on the AntNet algorithm and enhanced via using bee agents, it use forward agent inspired from ant and backward agent inspired from bee [23]. Pankajavalli et al. [24] presented and implemented an algorithm based on ant and bee behavior called BADSR for Routing in mobile ad-hoc network. The algorithm aimed to integrate the best of ant colony optimization (ACO) and bee colony optimization (BCO), the algorithm uses forward ant agents to collect data and backward bee agents to update the links state, the bee agent update data based on checking a threshold. Simulation results represented better result for the BADSR algorithm in terms of reliability and energy consumption [24]. Kanimozhi Suguna et al. [25] showed an algorithm for on demand ad-hoc routing algorithm, which is based on the foraging behavior of Ant colony optimization and bee colony optimization. The proposed algorithm uses bee agents to collect data about the neighborhood of the node, and uses forward ant agents to update the pheromone state of the links. The results showed that the proposed algorithm has the potential to become an appropriate routing strategy for mobile ad-hoc networks [25].

In this paper, a new modified version of the TAntNet-2 algorithm is proposed for dynamic routing of road traffic networks, where a performance of the algorithm is enhanced by avoiding the bad effect of forward ants that take a bad route. The new modified algorithm uses a threshold to measure the quality of the solution that found by forward ant. When the result of measuring represents bad solution, the algorithm ignores the solution of first forward ant and retransmits another forward ant.

For the purpose of this paper, the standard TAntnet algorithm is presented in Section 2. While, the proposed modified version of the algorithm is introduced in Section 3. The simulation experiment is given in section 4. Section 5 concludes the paper.

2   Threshold Based AntNet-2 algorithm

TAntNet algorithm was proposed by Ghazy et.al. [20]. TAntNet is a modified version of AntNet algorithm for traffic routing of road network. The main idea of TAntNet algorithm is to get benefit of the pre known information about the good travel time between a source and a destination. And use this good travel times as threshold values. TAntNet used a new type of ants called "check ants". Check ants are responsible of periodically checking the discovered good route whether it is still good or not.

When running TAntNet, it was noticed that the good route between a source and a destination may disappear after some amount of time of running ants over the network. The reason was the bad effect of the sub path update on the discovered good route. To overcome this problem, TAntNet-2 was suggested in ([26], [27]) to prevent the sub path updates for the already discovered good routes. Figure 1 illustrates the pseudo code of The TAntNet-2 algorithm ([26], [27]):

Algorithm. Threshold-based AntNet (TAntNet-2)

```
/* Main loop */
FOR each (Node s)          /*Concurrent activity*/
t=current time
WHILE t ≤ T      /* T is the total experiment time */
    Set d := Select destination node;
    Set T_sd = 0     /* T_sd travel time from s to d */
    IF (G_d = yes)
            Launch Check Ant (s, d);      /* From s to d*/
    ELSE
            Launch Forward Ant (s, d); /* From s to d*/
            IF (T_sd<=T_Good_sd)
                Set G_d = yes
            END IF
    END IF
    END IF
END WHILE
END FOR
```

Fig. 1. The TAntNet-2 Algorithm

```
CHECK ANT ( source node: s , destination node: d)
T_sd = 0
WHILE (current_node ≠ destination_node)
      Select next node using routing table
      (node with highest probability)
      Set travel_time= travel time from current node to
                       next_node
      Set T_sd = T_sd + travel_time;
      Set current_node = next_node;
END WHILE
          IF   (T_sd>T_GoodSd)
Set G_d = No
END IF
END CHECK ANT

Forward Ant ( source node: s , destination node: d)
WHILE (current_node ≠ destination_node)
      Select next node using routing table
      Push on stack(next_node, travel_time);
      Set current_node = next_node;
END WHILE
Launch backward ant
Die
END Forward Ant

Backward Ant ( source node: s , destination node: d)
WHILE (current node ≠ source node) do
    Choose next node by popping the stack
    Update the traffic model
    Update the routing table as follows:
    IF (T_sd<=T_Good_sd)
       P_hd' ← 1
       P_nd' ← 0 ,   ∀n ≠ h,n ∈ N_k
       /* where: h is the node "come from", k is the
       current node, NK is the set of neighbors nodes,
          d' is the destination or sub path destination */
    ELSE if (G_sd' = No)
       P_hd' ← P_hd' + r(1 − P_hd')
       /* where r is the reinforcement value*/
    END IF
END WHILE
END Backward Ant
```

Fig. 1. (*continued*)

3   The Improved TAntNet-2 Algorithm

The Behavior of Bee during collecting the nectar is an attractive behavior. The employed forager bee memorizes the location of food source to exploiting it. After the foraging bee loads a portion of nectar from the food source, it returns to the hive and save the nectar in the food area. After that, the bee enters to the decision making process which includes the decision if the nectar amount decreased to a low level or exhausted, in this case it abandons the food source ([28], [29])

This paper uses the previous idea to enhance the performance of TAntNet algorithm. In TAntNet algorithm, forward ant explores a path between a source and destination. Because of the probabilistic selection of route, forward ant can take a bad path. The new modified algorithm tries to treat this bad effect by using an idea inspired from the bee foraging behavior, when bee takes a decision of completing foraging depend on the quantity level of nectar. In the modified TAntNet-2, we will use forward scout instead of forward ant. Forward scout does not launch backward ant immediately after it finishes its trip, but after forward scout finishes its trip it will enter in a decision step depend on the quality of the discovered route, to determine whether to launch backward ant or abandons the forward scout and retransmit another forward scout to search for another solution. The second forward scout will acts the same as forward ant, it will launch backward ant after finishing of its trip.

After forward scout finished its trip and before launching the corresponding backward ant, the modified algorithm checks the quality of the discovered route. Quality is checked compared by the mean value in the local traffic statistics table of the source node. Formula (1) represents the formula that determined the accepted forward ant.

$$F_{sd} \leq \propto \mu_d \qquad (1)$$

Where:

F_{sd}: is the total travel time of the discovered route by the forward ant that launched from s to d.

α : weighs the threshold level.

μ_d : mean of the trip times of ants that launch from node s to node d

The first forward scout with at most total travel time less than or equal $\alpha\mu$ will be accepted, otherwise the algorithm will ignore this first forward scout and second forward scout will be launched. Second scout will be accepted whatever its travel time. Accepting of second forward scout is return to avoid the stuck of the algorithm when critical changes occur in the traffic situation. The pseudo code for the Modified TAntNet-2 algorithm is illustrated in Figure 2. The lines of codes appear in bold font represent the new modifications compared with the TAntNet-2 algorithm. Figure 2 shows the main loop and the forward scout procedure of the modified algorithm while the procedures of Check Ant and Backward Ant will be the same as them of the TAntNet-2 algorithm shown in Figure1.

The new enhancement in the algorithm can be seen as adding a scouting process before launching of backward ant. The scouting process includes the following tasks:

— Sending first scout to search for route between specific source and specific destination.

— Test the quality of the discovered route by the first scout.

- If the route is accepted [i.e. total travel time of the route is less or equal than the threshold ($\alpha\mu$)], the algorithm will launch the backward ant.
- Otherwise if the route is not accepted [i.e. total travel time of the route is higher than the threshold ($\alpha\mu$)], the algorithm will launch second scout to search for another route, and then launch backward ant.

The Proposed Modified TAntNet-2 Algorithm

```
/* Main loop */
FOR each (Node s)                   /*Concurrent activity*/
t=current time
WHILE t ≤ T              /* T is the total experiment time */
   Set d := Select destination node;
   Set T_sd = 0          /* T_sd travel time from s to d */
   IF (G_d = yes)
       Launch Check Ant (s, d);      /* From s to d*/
   ELSE
       Launch Forward Scout (s, d); /* From s to d*/
           IF ( F_sd > ∝ μ_d )
               Die (Forward Scout);
               /* Die of First Forward Scout From s to d*/
               Launch Forward Scout(s,d);
               /*second Forward Scout From s to d*/
           END IF
           IF (T_sd<=T_Good_sd)
               Set G_d = yes
           END IF
       Launch Backward Ant (d, s)
       Die (Forward Scout);  /* Die of Second Forward Scout*/
   END IF
END WHILE
END FOR

Forward Scout (source node: s, destination node: d)
WHILE (current_node ≠ destination_node)
     Select next node using routing table
     Push on stack(next_node, travel_time);
     Set current_node = next_node;
END WHILE
END Forward Scout
```

Fig. 2. The Modified TAntNet-2 Algorithm

The appearance of μ in AntNet and TAntNet-2 algorithms arises in three main places (In the procedure of backward ant only) as follows:

— The first appearance is for computing μ itself.
— The second appearance of μ is during computing the pheromone values of the routing tables.
— The third appearance of μ is arising as a threshold to determine the degree of goodness of the sub path update to determine if the algorithm performs sub path update or not.

But in the new enhanced algorithm the appearance of μ is increased to be four times, the fourth one arises as a part of the decision making during the scouting process, this using of μ here represents as a part of the threshold formula; which determines the acceptance of the solution returned by the forward scout or re launches a new forward scout to search for a different solution.

The increasing of using μ reflects an enhancement in the learning process with the new proposed algorithm. Which consequently increases the intelligence of the algorithm. Table 1 represents a comparison between the three types of algorithm AntNet, TAntNet-2 and the new enhanced algorithm.

Table 1. Comparison between the three types of algorithm

	AntNet	TAntNet-2	The Modified TAntNet-2
Number of Using of μ	(Three Times)	(Three Times)	(Four Times)
Threshold using	* Sub path update is performed for only a solution with a specified degree of goodness.	* Sub path update is performed for only a solution with a specified degree of goodness. * In case of found good solutions (Launching check ant instead of forward-backward ants)	* Sub path update is performed for only a solution with a specified degree of goodness. * In case of found good solution (Launching check ant instead of forward-backward ants) * In case of Bad solution (when forward ant returned with a bad solution retransmit another forward ant)
The expected sequence of launching different type of ants	* Forward-Backward	* Forward-Backward * Check	* Forward-Backward * Check *Forward-Forward-Backward

4   Experiment

A simulation is used to test and compare the performance of the modified TAntnet-2, TAntNet-2 and the original AntNet algorithms. The used network has 16 nodes with the topology shown in Figure 3. The objective is to get best routes between the source node 1 and any other node in the network over a certain period of time.

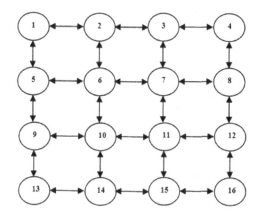

Fig. 3. The topology used for a network with 16 nodes

The simulation runs to test the original AntNet, TAntNet-2 and the modified TAntNet-2 algorithms. The modified TAntNet-2 is tested for different α parameter (α = 0.5; α = 1; α = 1.5; α = 2). The simulation experiment starts by continuously launching forward (or check) ants from the source node 1 to any arbitrary node. The time of each simulated experiment is set to 20 minutes. The experiment is repeated 40 times for the original AntNet, TAntNet-2 and the modified TAntNet-2 (with α =0.5; α = 1; α =1.5; α =2) algorithms on the same processing unit with completely new generated data at each run.

The simulation experiments show the following results:

- The modified TAntNet-2 (with α=2) allows increase in the number of launched ants compared with the original AntNet and TAntNet-2 algorithms. This increase is, even, accelerated further with modified TAntNet-2 (with α=1; 1.5). The modified TAntNet-2 (with α=0.5; 1.5; 2) allows a reduction in average travel time. The reduction is increased further under the modified TAntNet-2 (with α=1) as shown in Table 2.

Table 2. Number of launched ants and the average ants travel time over the simulation period

Algorithm Name	Average No. of ants		Average travel time	
	Value	Percentage of Increase comparing with AntNet Alg.	Value	Percentage of decrease comparing with AntNet Alg.
AntNet	2058.35±71.31		33.84±2.02	
TAntNet-2	2683.92±424.76	23.31 %	32.11±2	5.11 %
The Modified TAntNet-2, Min-threshold= 0.5 μ	2177.22±156.92	5.46 %	31.64±2.1	6.50 %
The Modified TAntNet-2, Min-threshold= μ	3040.8±168.37	32.31 %	27.39±1.93	19.06 %
The Modified TAntNet-2, Min-threshold= 1.5 μ	3063.98±178.78	32.82 %	28.27±2.02	16.46 %
The Modified TAntNet-2, Min-threshold= 2 μ	2895.78±336.74	28.92 %	29.61±2.1	12.5 %

Average ±Standard deviation

The increasing in the number of ants reflects a decreasing in computational complexity, which return to avoiding the ants the takes bad route and in most cases these ants passes many nodes and the corresponding Backward ant takes a lot of computations.

Related t-test is used to show the significance of the new enhancement. A one-tailed t-test in the positive direction is used with degrees of freedom equal to 39, the tabulated α is set to 0.05, so the value t_{crit} equal to +1.69.

Related t-test is applied to the experimental results of AntNet against TAntNet-2 algorithm, AntNet against the Modified TAntNet-2 (α=1) algorithm and TAntNet-2 against the Modified TAntNet-2 (with α=1).

The related t-test analysis applied on the performance index of average travel time over the simulation period, indicates significant decrease in the three cases as (T(experimental results) > 1.69) as illustrated in Table 3.

Table 3. Related t-test between Average Travel Time over the Simulation Period

AntNet with TAntNet-2	AntNet with The Modified TAntNet-2, Min-threshold= μ	TAntNet-2 with The Modified TAntNet-2, Min-threshold= μ
17.32*[1]	61.54.11 *	44.38 *

At each simulation minute, the average travelling time to all network nodes for The Modified TAntNet-2 (with α=0.5; 1.5) were less than that of the original AntNet and TAntNet-2. The reduction is increased further under the Modified TAntNet-2 (with α=1) as shown in Table 4 and Figure 4.

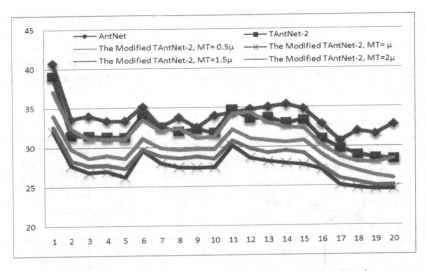

Fig. 4. The average travel time at each minute for all network nodes

---

* Means significant at $\alpha = 0.05$

Table 4. Average Travel Time at Each Minute

Minute	AntNet	TAntNet-2	The Modified TAntNet-2, Min-threshold= 0.5μ	The Modified TAntNet-2, Min-threshold= μ	The Modified TAntNet-2, Min-threshold= 1.5μ	The Modified TAntNet-2, Min-threshold= 2 μ
1	40.68±4.88	39.05±6.27	37.08±6	32.08±4.45	32.62±4.88	34.02±5.2
2	33.6±4.46	31.51±5.22	32.41±6.21	27.69±4.87	28.32±4.85	29.74±5.36
3	33.98±4.73	31.46±6.16	31.13±5.6	26.8±4.44	27.67±4.47	28.63±4.9
4	33.36±3.94	31.34±5.59	31.06±6.46	26.94±4.91	27.74±4.63	28.95±5.33
5	33.35±4.28	31.29±5.85	31.03±6.35	26.18±4.57	27.25±4.85	28.53±5.33
6	35.14±4.4	34.13±6.13	33.47±5.11	29.58±4.97	29.71±4.75	31.08±4.8
7	32.61±4.29	32.54±5.75	32.04±5.43	27.97±4.77	28.85±4.8	29.91±4.76
8	33.74±4.17	32.02±6.12	32.42±6.13	27.42±4.9	28.57±4.93	29.64±5.22
9	32.47±4.67	32.26±5.73	31.1±5.78	27.37±4.67	28.8±5	29.82±5.17
10	33.95±4.19	31.81±5.64	31.34±5.75	27.38±4.65	28.51±4.87	29.68±4.93
11	34.37±3.32	34.69±4.26	34.21±4.47	30.19±4.23	30.79±4.01	32.21±4.07
12	34.78±3.52	33.55±4.42	34.36±4.32	28.56±3.89	29.74±3.9	31.08±4.54
13	35.02±4.35	33.68±4.63	33.09±4.91	28.14±3.7	29.26±4.44	30.79±4.6
14	35.37±3.91	33.01±4.46	32.48±4.42	27.9±4	29.53±4.34	30.59±4.53
15	34.77±3.73	33.38±4.56	32.32±4.22	27.69±3.82	29.18±3.79	30.79±4.57
16	32.83±3.97	31.02±4.08	30.15±3.94	27.07±4.34	27.38±3.6	29±3.57
17	30.69±3.79	29.76±4.51	28.73±4.75	25.09±3.64	25.95±3.92	27.59±3.87
18	31.94±3.65	28.95±4.5	28.04±4.77	24.73±3.81	25.44±3.73	26.89±3.97
19	31.56±3.52	28.53±4.68	27.98±4.53	24.53±3.85	24.98±3.91	26.27±4
20	32.74±3.87	28.36±4.88	28.42±5.1	24.58±3.84	25.05±3.93	25.94±4.06

5   Conclusion and Future Works

In this paper, a modified version of the TAntNet-2 algorithm is presented to be applied to dynamic traffic routing of road networks. The new algorithm inspires a new feature from bee foraging behavior, to enhance the performance of TAntNet-2 algorithm. The new algorithm performs a scouting process before launching of the backward ants. The scouting process uses a threshold to determine the accepted solution. The threshold uses the historical data saved in the local traffic statistics table. Scouting process use retransmits of new scout, in case of rejected first scout. The new algorithm works on preventing the bad effect of bad forward ant. Also the new enhancement decreases the processing time that used by backward ant which corresponds to forward ant that takes bad route and passes many nodes. Experimental results show high performance for the modified TAntNet-2 compared with TAntNet and TAntNet-2. Among different values of α for threshold of the modified TAntNet-2, α =1 represents the best value.

We will work in the future on extend the simulated experiments that compared the modified TAntNet-2 with AntNet and TAntNet-2 algorithms and using the statistics to test and analyze the performance of modified TAntNet-2. Also we will work on test the modified algorithm on a larger network.

References

1. Kassabalidis, I., El-Sharkawi, M.A., Marks, R.J., Arabshahi, P., Gray, A.: Adaptive-SDR: adaptive swarm-based distributed routing. In: Proceedings of the International Joint Conference on Neural Networks, Honolulu, HI, vol. 1, pp. 351–355 (2002)
2. Kroon, R., Rothkrantz, L.: Dynamic vehicle routing using an ABC-algorithm. In: Transportation and Telecommunication in the 3rd Millennium, Prague, pp. 26–33 (2003)
3. Suson, A.: Dynamic routing using ant-based control. Master thesis, Faculty of Electrical Engineering, Mathematics and Computer Science, Delft University of Technology (2010)
4. Claes, R., Holvoet, T.: Cooperative ant colony optimization in traffic route calculations. In: Demazeau, Y., Müller, J.P., Rodríguez, J.M.C., Pérez, J.B. (eds.) Advances on PAAMS. AISC, vol. 155, pp. 23–34. Springer, Heidelberg (2012)
5. Shah, S., Bhaya, A., Kothari, R., Chandra, S.: Ants find the shortest path: a mathematical proof. Swarm Intelligence 7(1), 43–62 (2013)
6. Yousefi, P., Zamani, R.: The Optimal Routing of Cars in the Car Navigation System by Taking the Combination of Divide and Conquer Method and Ant Colony Algorithm into Consideration. International Journal of Machine Learning and Computing 3, 44–48 (2013)
7. Jabbarpour, M.R., Malakooti, H., Noor, R.M., Anuar, N.B., Khamis, N.: Ant colony optimisation for vehicle traffic systems: applications and challenges. International Journal of Bio-Inspired Computation 6(1), 32–56 (2014)
8. Di Caro, G., Dorigo, M.: AntNet: distributed stigmergetic control for communications networks. J. Articial. Intell. Res. (JAIR) 9, 317–365 (1998)
9. Dhillon, S.S., Van Mieghem, P.: Performance analysis of the AntNet algorithm. Computer Networks 51, 2104–2125 (2007)
10. Baran, B., Sosa, R.: AntNet routing algorithm for data networks based on mobile agents. Inteligencia Artificial, Revista Iberoamericana de Inteligencia Artificial 12, 75–84 (2001)
11. Tekiner, F., Ghassemlooy, F.Z., Al-khayatt, S.: The AntNet Routing Algorithm - Improved Version. In: Proceedings of the International Symposium on Communication Systems Networks and Digital Signal Processing (CSNDSP), Newcastle, UK, pp. 22–28 (July 2004)
12. Soltani, A., Akbarzadeh, T.M.-R., Naghibzadeh, M.: Helping ants for adaptive network routing. Journal of the Franklin Institute 343(4), 389–403 (2006)
13. Gupta, A.K., Sadawarti, H., Verma, A.K.: Computation of Pheromone Values in AntNet Algorithm. International Journal of Computer Network & Information Security 4(9), 47–54 (2012)
14. Radwan, A., Mahmoud, T., Houssein, E.: AntNet-RSLR: a proposed ant routing protocol for MANETs. In: Proceedings of the First Saudi International Electronics, Communications and Electronics Conference (SIECPC 2011), April 23-26, pp. 1–6 (2011)
15. Sharma, A.K.: Simulation of Route Optimization with load balancing Using AntNet System. IOSR Journal of Computer Engineering (IOSR-JCE) 11(1), 1–7 (2013)
16. Tatomir, B., Rothkrantz, L.: Dynamic traffic routing using Ant based control. In: IEEE International Conference on Systems, Man and Cybernetics (SMC 2004) on Impacts of Emerging Cybernetics and Human-machine Systems, vol. 4, pp. 3970–3975 (October 2004)
17. Boehlé, J., Rothkrantz, L., van Wezel, M.: CBPRS: a city based parking and routing system. Technical report ERS-2008-029-LIS, Erasmus Research Institute of Management, ERIM, University Rotterdam (2008)

18. Kammoun, H.M., Kallel, I., Adel, M.A.: An adaptive vehicle guidance system instigated from ant colony behavior. In: 2010 IEEE International Conference on Systems Man and Cybernetics (SMC), pp. 2948–2955. IEEE (2010)
19. Claes, R., Holvoet, T.: Ant colony optimization applied to route planning using link travel time predictions. In: 2011 IEEE International Symposium on Parallel and Distributed Processing Workshops and Phd Forum (IPDPSW), pp. 358–365. IEEE (2011)
20. Ghazy, A.M., El-Licy, F., Hefny, H.A.: Threshold based AntNet algorithm for dynamic traffic routing of road networks. Egyptian Informatics Journal 13(2), 111–121 (2012)
21. Kashefikia, M., Nematbakhsh, N., Moghadam, R.A.: Multiple Ant-Bee colony optimization for load balancing in packet switched networks. International Journal of Computer Networks & Communications 3(5), 107–117 (2011)
22. Raghavendran, C.V., Satish, G.N., Varma, P.S.: Intelligent Routing Techniques for Mobile Ad hoc Networks using Swarm Intelligence. International Journal of Intelligent Systems and Applications (IJISA) 5(1), 81–89 (2013)
23. Rahmatizadeh, S., Shah-Hosseini, H., Torkaman, H.: The Ant-Bee Routing Algorithm: A New Agent Based Nature-Inspired Routing Algorithm. Journal of Applied Sciences 9(5), 983–987 (2009)
24. Pankajavalli, P.B., Arumugam, N.: BADSR: An Enhanced Dynamic Source Routing Algorithm for MANETs Based on Ant and Bee Colony Optimization. European Journal of Scientific Research 53(4), 576–581 (2011)
25. Kanimozhi Suguna, S., Uma Maheswari, S.: Bee - Ant Colony Optimized Routing for Manets. European Journal of Scientific Research 74(3), 364–369 (2012)
26. Ghazy, A.: Enhancement of dynamic routing using ant based control algorithm. Master thesis, Institute of Statistical Studies and Research, Cairo University (2011)
27. Ghazy, A.: Ants Guide You to Good Route: Dynamic Traffic Routing of Road Network using Threshold Based AntNet. Lap Lambert Academic Publishing (2012)
28. Baykasoglu, A., Ozbakir, L., Tapkan, P.: Artificial bee colony algorithm and its application to generalized assignment problem. In: Chan, F.T.S., Tiwari, M.K. (eds.) Swarm Intelligence. Focus on Ant and Particle Swarm Optimization, pp. 113–144. ITech Education and Publishing, Vienna (2007)
29. Akbari, R., Mohammadi, A., Ziarati, K.: A novel bee swarm optimization algorithm for numerical function optimization. Communications in Nonlinear Science and Numerical Simulation 15(10), 3142–3155 (2010)

Fuzzy Query Approach for Crops Planting Dates Optimization Based on Climate Data

Ahmed M. Gadallah, Assem H. Mohamed, and Hesham A. Hefny

Institute of Statistical Studies and Research, Cairo University
{ahmgad10,eng_asemhm}@yahoo.com, hehefny@ieee.com

Abstract. Climate change is considered one of the most environmental phenomena of interest in the world nowadays. It affects many aspects of our life. One of the most affected aspects by climate change is agriculture. It is obvious that the ongoing changes in climate variables like temperature affects the suitability of crops plantation. That is, it make some crops became not suitable to plant in its traditional places at its traditional dates while it became more suitable to plant in other new places and/or dates. Based on the available historical spatial agro-climatic database, this paper presents a fuzzy query approach for discovering the new more suitable planting dates of crops in a given governorate of Egypt. The proposed approach consists of three phases, one phase for fuzzy clustering of the year days according to climate data, the second phase for defining crop suitability fuzzy membership functions and the last phase for fuzzy selection and optimization of suitable periods to plant a given crop like squash in a given governorate like Alexandria. The proposed approach proved that most of traditional plantation dates of squash in Alexandria become not suitable compared with the new discovered more suitable periods with a suitability measure for each period.

Keywords: Fuzzy query, Relational database, SQL, Spatial Agro-Climatic Database, Fuzzy set theory, Prediction.

1 Introduction

A spatial database is a collection of data concerning objects located in some reference space that attempts to model some enterprise aspects in the real world. Spatial agro-climatic database is a spatial database that contains the data of the climate variables for some places during specific periods of time. Almost, the data stored in such databases are needed to be searched in a more flexible human-like manner. For example, there is a need for a query approach that allows queries like "retrieve each period suitable to plant squash in Alexandria with matching degree around 75%".

Structured query language (SQL) for relational databases was initially presented by Chamberlin and Boyce for data retrieval and manipulation [2].SQL uses the two-value logic (crisp logic) in querying process. This limitation of SQL can be avoided by fuzzy logic [14]. Commonly, real world abounds in uncertainty, and any attempt to model aspects of the world should include some mechanism for handling uncertainty such as fuzzy logic [1], [12]. Fuzzy set theory was initiated by Zadeh [2]. Since then, many researches and applications in many fields have been achieved. Fuzzy queries have

A.E. Hassanien et al. (Eds.): AMLTA 2014, CCIS 488, pp. 436–445, 2014.

appeared in the last 30 years to cope with the necessity to soften the Boolean logic in database queries. A fuzzy query system is an interface for users to retrieve information from a database using human linguistic words which are qualitative by nature [3]. This area of research is still interesting as there are needs for more improvements of existing approaches. The goal of this paper is to propose a fuzzy query approach for querying a spatial agro-climatic database depending on human like queries. Such query approach helps in determining the more suitable planting date of crops in specific places.

The rest of this paper is organized as follows. The second section introduces the related works. The proposed approach is presented in the third section. The fourth section shows a case study. The conclusion is presented in the fifth section.

2   Related Works

Many approaches have been proposed for the problem of the affection of climate changes on agriculture. Some approaches aimed mainly to show the impact of climate change on crop production like [15] and [16]. They proved that agricultural productivity had been affected reflecting the climate changes. They also advise to make some adaptation on planting dates or by planting crops that are less sensitive to climate changes to get over such effects. Also, climate changes greatly affect the water resources in regions in which agriculture depends on rain. As water resources are one of the most important parameter in plantation process, a new approach has been developed to make adaptation to crop planting dates with climate changes like in [4] and [5]. Another approach for crop yield forecasting was presented in [6] to map the relations between climate data and crop yield. This technique based on time series data of 27 years for yield and weather data. Other approaches have been developed such as Fuzzy-based Decision Support Systems for evaluating land suitability and selecting the more suitable crops to be planted is provided as in [7] and [8]. In these words, fuzzy rule based systems were developed for evaluating land suitability and selecting the appropriate crops to be planted considering the decision maker's requirements in crops selection with the efficient use of the powerful reasoning and explanation capabilities of DSS.

Unfortunately, all of the above algorithms do not provide weight or matching measures for selected period's suitability of plantation for the underlying crop after making adaptation on climate changes. Also, in all of them no clustering for the climate data is made. This mean that the algorithm calculate the suitability of the period day by day every time of searching suitability for planting any crop so that it take long time. Some algorithms depend on the average values of climate data like in [5], [6] and [7] and this is not true as most of crops that have minimum and maximum suitable values of climate variables.

Generally, temperature degree represents one of the most important climate variables affecting crops plantation in Egypt. This paper presents a new approach that handles the effects of the change in temperature on the dates of squash crop plantation in the governorate of Alexandria. It makes an automatic fuzzy clustering on the predicted climate data for the next year. Each cluster (period) consists of some continuous days with length more than or equal to the age of the crop under study. Consequently, it is more convenient to generate a fuzzy query statement to retrieve the matching clusters of days with the underling crop suitable conditions. Also, the selection for suitable clusters of days takes into account the maximum and minimum values of climate variables suitable for the plantation of the underlying crop. Finally, the retrieved clusters of days will be optimized to enhance their suitability degrees. Finally, the resulted optimized clusters are ranked according to their suitability degree for the underling crop plantation.

3   The Proposed Approach

The architecture of the proposed fuzzy query approach for crop planting date optimization based on climate Data (temperature degree) is shown in fig. 1. It consists of three main phases:

1- Automatic fuzzy clustering phase: in this phase the expected data of the next year is clustered to continuous periods. Each period size is equal to or more than the period required in the crop requirements.

2- Define a set of fuzzy membership functions phase: this phase is to define a set of fuzzy membership functions describing the crop suitable climate variables values. These functions are used to evaluate the suitability of each period of days for planting the crop of interest.

3- Fuzzy query and optimization phase: this phase is to perform a fuzzy query selection from the clustered periods of days based on the required climate data of the crop defined in phase 2. After that, an optimization operation to each resulted cluster takes place aiming to increase the suitability degree as possible as it can.

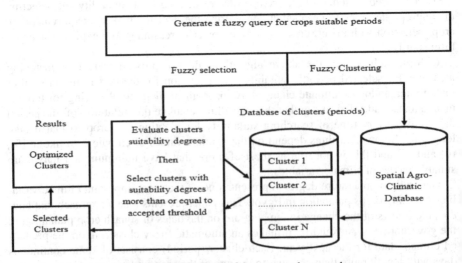

Fig. 1. The architecture of the proposed approach

3.1   Automatic Fuzzy Clustering Phase

In classical hard clustering, data is divided into distinct clusters, where each data element belongs to exactly one cluster. On the other hand, in fuzzy clustering, data elements can belong to more than one cluster with a matching degree for each cluster. Accordingly, the matching degree indicates how strongly an element belongs to a particular cluster [9]. In the proposed approach, the clustering operation for a year days is temperature-based. The inputs of clustering operation are the predicted temperature degrees of the days of the incoming year, the crop plantation age and the maximum accepted variance of temperature in each cluster. The clustering operation takes place according to Algorithm 1. The output of the clustering operation is a set of clusters each represent a period of days with variance in temperature not greater than the maximum accepted variance.

Algorithm 1. Automatic Fuzzy Clustering Algorithm:

Input: climate data for the under study governorate for a year, permissible standard deviation, accepted threshold for standard deviation membership function.
Output: clustered periods.
Cluster items = 0.
For each day in year_ data_table
selected_day =day.
If Cluster items = 0
   Add selected_day to Cluster items.
Else
   Temp cluster items =Cluster items + selected_day.
   New cluster items variance = variance (Temp cluster items).
   Membership_value = cluster_mem_function (accepted variance, Max accepted
                                 variance , New cluster items variance).
   If Membership_value >= accepted_threshold
      Add selected_day to Cluster items.
   Else
      Save new cluster (Cluster items).
      Cluster items = 0.
   End if
End for each

3.2   Defining Crop Suitable Climate Fuzzy Membership Functions Phase

This phase allows defining a set of fuzzy membership functions describing the relationship between a crop and the suitable values of a specific climate variable. For example, assuming that the suitability of temperature degrees for a specific crop is as follow:

- Temperature degrees in [b, c] are the most suitable with full matching degree of 1,
- Temperature degrees in [a,b[and [c,d[are partially suitable with a matching degree in [0,1[and
- Temperature degrees greater than d or less than a are not suitable at all with 0 matching degree.

Consequently, the above description of the suitability of temperature degrees for a specific crop can be easily defined as the trapezoidal fuzzy membership function depicted in (1).

$$\mu_A \ (temp\_suitability, \ x) = \begin{cases} 0 & x < a \\ \frac{x-a}{b-a} & a \le x < b \\ 1 & b \le x < c \\ \frac{d-x}{d-c} & c \le x < d \\ 0 & x \ge d \end{cases} \quad (1)$$

where a=0, b=16, c=25 and d=32.

3.3 Fuzzy Selection and Optimization Phase

This phase in the proposed approach takes the resulted clusters (periods) from the fuzzy clustering phase, crop plantation temperature data and accepted threshold for suitability membership function as inputs. Consequently, it selects the suitable periods from the clusters with threshold equal to or more than the accepted threshold. After that the approach tries to optimize the selected clusters by shifting to left or to right then test the suitability degree for the new cluster. In other words, the optimization can easily achieved by removing days, for example 5 days, from the beginning of a cluster and adding same no of days to the end of the cluster and test the modified cluster suitability. At the end, the set of suitable clusters includes all suitable periods that have the highest suitability degree reached by the optimization operation. The following example shows selection query form:

Select period from clusters of periods where temp_suitability (cluster_max_ temp, cluster_min_temp)>= suitability_threshold.

Algorithm 2. Fuzzy selection and optimization algorithm:

Input: Fuzzy clusters resulted from the clustering phase, crop plantation temperature data, accepted threshold for temperature suitability membership function.
Output: suitable periods.
Suitable periods =0.
For each Period in clustered Periods
  Max__temp_Md = max_temp_suitability (Period).
  Min_temp_Md = min_temp_suitability (Period).
  If min (Max__temp_Md, Min_temp_Md)> = accepted threshold
    Optimized period=Period optimization (period).
    Add Optimized period to Suitable periods.
  End if
End for each

4   An Illustrative Case Study

The proposed approach considers the agro climatic spatial database for Alexandria, the temperature as climate change attribute and the squash requirements data. The traditional periods of squash plantation are given in table1. Commonly, the climate requirements for squash plantation are as follows: [10], [11]

a- The average age of the plant is 100 days.
b- The maximum temperature degree less than 32
c- The minimum temperature degree more than Frost point
d- The Best temperature degree from 16 to 25

The proposed approach is applied for testing the effects of changes in temperature degrees on squash plantation in the governorate of Alexandria. The fuzzy clustering phase is performed according to the flowing requirements:

a- The average age of the plant is 100 days.
b- Standard deviation (S.D) less than or equal to 0.1.
c- Accepted threshold 0.8.

While table1shows the traditional plantation dates for squash in Alexandria, The results of the proposed approach are shown in table2. Where in Temperature column Min means the average of the minimum Temperature of the period, Max indicates the average of the maximum Temperature of the period and the Avg denotes the average of the average Temperature of the period.

Table 1. Squash plantation traditional dates in Alexandria

Traditional plantation periods	Period start and end dates	Temperature			Suitability degree
		Min	Avg	Max	
Winter buttonhole (fig. 2)	from 1 Dec to 10 mar	14	17	20	81.91%
Summer buttonhole (fig. 3)	from 1 Feb to 10 may	15	18	22	87.85%
Nile buttonhole (fig. 4)	from 1 Jul to 10 Oct	24	28	31	11.6%

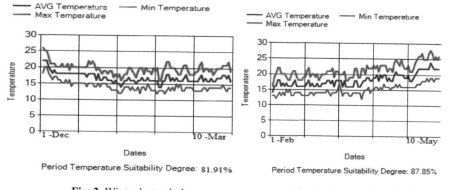

Fig. 2. Winter buttonhole   **Fig. 3.** Summer buttonhole

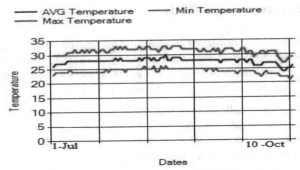

Period Temperature Suitability Degree: 11.6%

Fig. 4. Nile buttonhole

Table 2. Squash plantation suitable dates in Alexandria resulted from applying the proposed approach

Plantation periods	Periods start and end dates	Temperature			Suitability degree
		Min	Avg	Max	
First period (fig. 5)	from 1 Jan to 10 Apr	14	17	20	80.81%
Second period (fig. 6)	from 20 Feb to 30 May	16	20	23	91.77%
Third period (fig. 7)	from 1 Oct to 20 Jan	81	21	23	85.46%
Fourth period (fig. 8)	from 1 Nov to 8 Feb	16	19	21	88.06%

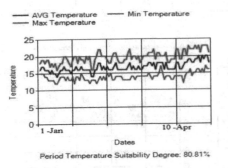

Period Temperature Suitability Degree: 80.81%

Fig. 5. The first resulted period

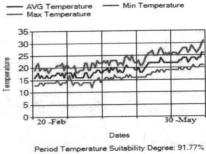

Period Temperature Suitability Degree: 91.77%

Fig. 6. The second resulted period

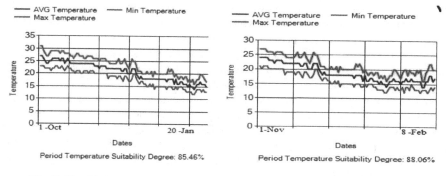

Fig. 7. The third resulted period   **Fig. 8.** The fourth resulted period

Considering both of table 1 and table 2, it is obvious that climate changes namely temperature degree affected the squash plantation dates. As shown in table1, it is clear that the traditional period "Nile buttonhole" became not suitable for squash plantation since its suitability became 11.6%. Hence, planting of squash at this period will damage the crop. On the other hand, some other periods became more suitable for squash plantation like periods suggested by the proposed approach as shown in table 2.The proposed approach discovered new suitable periods for squash plantation like Second period, an ideal period, with around 92% suitability degree that starts from 20 Feb to 30 May, as shown in fig. 6. Also, the fourth period starts from 1 Nov to 8 Feb that has around 88% suitability degree as shown in fig. 8. Both of first and third periods represent also suitable periods for squash plantation with suitability degrees around 81% and around 85% as shown in fig.5 and fig.7 respectively. As noted, all periods resulted from the proposed approach are wholly discovered or at least adjusted by shifting at least 20 days. Also, there is no traditional period near to the discovered fourth period.

According to the explained sample of results, the squash plantation suitable dates must take considerable attention. Such plantation dates should be changed reflecting climate changes in order to increase the productivity of the crop and reducing the cost of production. Consequently, climate change makes some crops not suitable to plant in its old places at the same traditional period. Yet, it can be planted in its traditional places but at different periods of time. On the other hand, such crops may be plant at the same at the same traditional dates but in other deferent more suitable places.

By showing the results of the proposed approach to some experts in Agriculture Research Center in Egypt they accept and admire the results. And they explained that they are doing this adaptation by human observation. Also, they wish to apply he proposed approach to optimize the plantation dates for some other crops.

5   Conclusions

This paper presented a proposed fuzzy query approach for crops plantation dates selection and optimization. The approach proved that the ongoing changes of climate variables like temperature degrees caused that some crops become not convenient for planting in its traditional places at its traditional periods. Based on the available historical spatial agro-climatic database for Alexandria governorate in Egypt, applying the proposed approach leads to discover new periods of time that are more suitable for squash planting than old periods. Also, it discovers that some traditional periods become not suitable for planting squash in Alexandria at all. Accordingly, the proposed approach guides and helps agriculture investors in a flexible manner to adjust the plantation plans for any crop at any location given that the historical spatial agro-climatic data for such location are available. In consequent, such approach greatly helps in agriculture strategical planning to enhance the plantation process of any crop. On the other hand, it directly increases the profit and decreases the cost of any crop plantation. Also, selecting the more suitable dates for a specific crop plantation strongly reduces the chances of crop diseases to appear in a catastrophic fashion.

References

1. Zhang, J., Goodchild, M.: Uncertainty in Geographical Information. Taylor & Francis, London (2002)
2. Zadeh, L.: Fuzzy sets. Information and Control 8, 338–353 (1965)
3. Branco, A., Evsukoff, A., Ebecken, N.: Generating fuzzy queries from weighted fuzzy classifier rules. In: ICDM workshop on Computational Intelligence in Data Mining, pp. 21–28 (2005)
4. Moussa, W., Patrick, L., Traore, S.B., Moussa, S., Harald, K.: A Crop Model and Fuzzy Rule Based Approach for Optimizing Maize Planting Dates in Burkina Faso, West Africa. Journal of Applied Meteorology and Climatology 53 (2014)
5. Mohaddes, S.A., Mohayidin, M.G.: Application of the Fuzzy Approach for Agricultural Production Planning in a Watershed, a Case Study of the Atrak Watershed, Iran. American-Eurasian J. Agric. & Environ. Sci. 3(4), 636–648 (2008)
6. Kumar, P.: Crop Yield Forecasting by Adaptive Neuro Fuzzy Inference System. Mathematical Theory and Modeling 1, 3 (2011)
7. Hartati, S., Sitanggang, I.S.: A Fuzzy Based Decision Support System for Evaluating Land Suitability & Selecting Crops. Journal of Computer Science 6(4), 417–424 (2010)
8. Joshi, R.G., Bhalchandra, P., Khmaitkar, S.D.: Predicting Suitability of Crop by Developing Fuzzy Decision Support System. IJETAE 3(2) (2013)
9. Yang, J., Watada, J.: Fuzzy Clustering Analysis of Data Mining. International Journal of Innovative Computing, Information and Control 8, 8 (2012)
10. http://kenanaonline.com/users/zidangroup/posts/95467
11. http://www.caae-eg.com/new/index.php/2012-12-25-10-49-19/2010-09-18-17-00-51/2011-01-15-19-27-42/234-2011-05-25-17-20-36.html
12. Beaubouef, T., Petry, F.E.: Fuzzy and Rough Set Approaches for Uncertainty in Spatial Data. In: Jeansoulin, R., Papini, O., Prade, H., Schockaert, S. (eds.) Methods for Handling Imperfect Spatial Information. STUDFUZZ, vol. 256, pp. 103–129. Springer, Heidelberg (2010)
13. Miroslav, H.: Fuzzy Improvement of the SQL. Yugoslav Journal of Operations Research 12(2), 239–251 (2011)

14. Galindo, J., Urrutia, A., Piattini, M.: Fuzzy Databases, Modeling, Design and Implementation. Idea Group Publishing, London (2006)
15. Defang, N., Manu, I., Bime, M., Tabi, O., Defang, H.: Impact of climate change on crop production and development of Muyuka subdivision – Cameroon. International Journal of Agriculture, Forestry and Fisheries 2(2), 40–45 (2014)
16. Hamid, R., Mohammad, A., Mohammad, H.: Consideration of Climate Conditions in Reservoir Operation Using Fuzzy Inference System (FIS). British Journal of Environment & Climate Change 3(3), 444–463 (2013)

Flexible Querying of Relational Databases: Fuzzy Set Based Approach

Adel A. Sabour, Ahmed M. Gadallah, and Hesham A. Hefny

Institute of Statistical Studies and Research, Cairo University
1adelsabour@gmail.com, ahmgad10@yahoo.com, hehefny@ieee.com

Abstract. This paper presents a flexible fuzzy-based approach for querying relational databases. Although, many fuzzy query approaches have been proposed, there is a need for a more flexible, simple and human-like query approach. Most of previously proposed fuzzy query approaches have a disadvantage that they interpret any fuzzy query statement into a crisp query statement then evaluate the resulted tuples to compute their matching degrees to the fuzzy query. The main objective of the proposed fuzzy query approach of this paper is to overcome the above disadvantage by evaluating each tuple directly through the use of stored database objects namely packages, procedures and functions. Consequently, the response time of executing a fuzzy query statement will be reduced. This proposed approach makes it easy to use fuzzy linguistic values in all clauses of a select statement. The added value of this proposed approach is to accelerate the execution of fuzzy query statements.

Keywords: Fuzzy Query, Fuzzy Logic, InformationRetrieval, Fuzzy SQL.

1 Introduction

Structured Query Language (SQL) is a very essential language for querying relational databases. It manipulates and retrieves data which is crisp and precise by nature. In contrary, it is unable to respond to human-like queries which are uncertain, imprecise and vague in nature. Almost, human queries have a lot of vagueness and ambiguity due to using his/her subjective linguistic words. For example, excellent students have different definitions that depend on each person searching for them. However, while applying one's thoughts as a query in terms of linguistic words into the database, a lot of problems are experienced due to the inefficiency of DBMS to handle such queries. Consider the query "retrieve the names and addresses of the university students who have height around the ideal height for a handball player". This query cannot be expressed and manipulated directly by a traditional SQL statement. In contrary, it can be expressed and manipulated easily through a fuzzy query statement in a very flexible and human-like manner based on the ideas of fuzzy set theory. Although classical SQL has great querying capabilities, it lacks the flexibility to support human-like queries. Human-like queries depend essentially on manipulation of linguistic values rather than numeric ones. Linguistic values such as: short, tall, hot and calls human

A.E. Hassanien et al. (Eds.): AMLTA 2014, CCIS 488, pp. 446–455, 2014.

concepts that cannot be handled as authorized attributes values in standard SQL statements. Moreover, standard SQL adopts classical Boolean expressions with crisp logical connectors which are too rigid and very limited in manipulation of linguistic values or linguistic modifiers such as: about, nearly, fairly, etc. which can be represented easily based on the concept of fuzzy sets [1].

The rest of this paper is organized as follows: section 2 presents previous works of fuzzy querying in databases. The proposed fuzzy query approach is introduced in Section three. Section 4 addresses a developed tool based on the proposed approach with an illustrative case study. The conclusion is addressed in section 5.

2   Previous Works of Fuzzy Querying in Databases

This section presents several architectures and applications that have been proposed and developed for allowing fuzzy queries of databases. Unfortunately, most of previously developed fuzzy SQL architectures have set of drawbacks. Both of previously proposed approaches in [1] and [2] have the inability for dealing with multi subjective views when interacting with multiusers. Instead, preferences are used to help in reducing the amount of information returned in response to user queries [4]. Also, the proposed approachesin [11], [2] and [1] lack of a standardized format in writing statements that is each tool has its own syntax.On the other hand, most of the proposed approaches depend mainly on using a time consuming parser/translator to check and convert fuzzy queries to crisp SQL queries like in [1],[2], [10] and [11]. Also, such approaches define a Meta base named Fuzzy Meta-knowledge Base (FMB) that includes a set of tables to store all necessary information to describe and manipulate fuzzy attributes and terms [2].Such FMB must be accessed each time a fuzzy query statement is generated. This operation is essential in order to obtain the definitions of each used fuzzy term in order to complete the processing of the generated fuzzy query statement which is a time consuming[5], [10] and [11].This slows down the process of querying because each generated fuzzy query must be analyzed and translated into a standard SQL statement respecting the definitions of all used fuzzy terms or operators stored in FMB. [7]. Also, the approaches proposed in[2],[5] and [10] process nested and correlated fuzzy queries inefficiently [3].

3   The Proposed Fuzzy Query Approach

This section presents the proposed fuzzy query approach that has the architecture shown in Fig. 1. It aims mainly to allow the generation and manipulation of more flexible human-like queries. This approach aims mainly to overcome the drawbacks mentioned in previous works section. It deals with multi subjective views with multiusers and it fully depends on PL/SQL statements. Accordingly, all generated fuzzy queries will be Standard SQL compatible without any need for an interpreter/parser. The proposed approach allows generating fuzzy query statements using the standard SQL language. Accordingly, a fuzzy term can be used easily by write its package name followed by a dot and its function or procedure name with its defined parameters if required. On the other hand, using a procedural language, as PL/SQL, has better performance because data and queries are directly managed and manipulated by RDBMS without the need for intermediate software programs [10]. Also, the proposed approach will benefit from the advantages of using PL/SQL [12] which include: Tight Integration with SQL, High Performance by reducing traffic between the application and the database, High Productivity, Portability on any operating system, Scalability, Manageability and Support for Object-Oriented. Also, using database programming language approach does not suffer from the impedance mismatch that includes [13]:

- Existence of differences between programming language and database models.
- The need to have a binding for each programming language that determines for each attributes type the compatible programing language type.
- The need for a binding that maps the results multi-set of tuples into a corresponding data structure like a record set or a cursor.

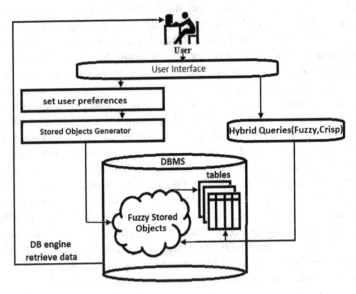

Fig. 1. Thearchitecture of the proposed Fuzzy query approach

The proposed approach depends mainly on the idea of fuzzy set. Commonly, a fuzzy set is a set with smooth boundaries. Accordingly, fuzzy set theory generalizes classical set theory to allow partial membership of its elements to the set, and it was developed to cover areas that cannot be covered by classical set theory. A fuzzy set A in the universe of discourse U is characterized by the membership function μ_A given by $\mu_A : U \rightarrow [0, 1]$ and A is defined as the set of ordered pairs A = {(x, μ_A(x)): x $\in$ U}, where μ_A (x) is a membership function that determines the membership degree of element x in the set A. Commonly, a fuzzy set of continuous universe of discourse is defined as a membership function. A membership function maps an element in the universe of discourse U to some value in the interval [0, 1] that is $\mu_A : X \rightarrow [0,1]$.

A full membership in a fuzzy set has a value of 1 while a membership of 0 indicates excluding the element from the fuzzy set at all. A membership between 0 and 1 indicates a partial membership to the set. As traditional crisp set theory, fuzzy set theory includes a set of operations like complement, intersection and union. Yet, fuzzy set operations are not uniquely defined i.e. as membership functions but they are also context and user dependent.

3.1   The Processing Scenario of the Proposed Approach

- The proposed fuzzy query approach aims mainly to support human-like queries which almost contain linguistic terms and fuzzy connectors. Such linguistic terms include linguistic variables, linguistic values, fuzzy hedges and fuzzy numbers. The proposed approach enhances the definition and manipulation of such linguistic terms. A user can easily define his own linguistic terms which may be stationary or non-stationary using suitable fuzzy membership functions like triangular, trapezoid, S-shape, .etc. Stationary linguistic terms have fixed definition. On the other hand, non-stationary linguistic terms have dynamic definitions which may change from time to time like "excellent students" in a very simple course and in a very complex course. Each defined linguistic term will be stored as a database object namely stored function within a related stored package. In consequence, a user can use any of his defined linguistic terms in any clause in a select statement as using a user defined stored function. Accordingly, the defined linguistic terms can be used easily in even nested, complex and correlated query statements without the need for complex procedures to execute the query statement. A user can interact with a developed tool based on the proposed approach as follows:
- A user interacts with the tool via a graphical user interface to define the user own linguistic terms describing the user preferences like linguistic variables, linguistic values, fuzzy connectors, fuzzy numbers and hedges. All of user linguistic terms will be stored as database objects resembling the user profile. Hence, there is no need for additional database.
- The tool generates PL-SQL statements that create database stored objects (such as packages, procedures, functions, views, etc.) for the defined linguistic terms. These objects contain the description of its linguistic term instead of using Fuzzy Meta Knowledge Base (FMB). Accordingly, the execution of a fuzzy query statement will not need more access for additional database to obtain the definition of the used linguistic terms. Hence, the proposed approach saves the processing time and reduces the response time.
- Finally, the user can write a fuzzy query statement or generate it using the tool query builder. It is obvious that, the fuzzy query syntax is identical to the standard SQL language due to the use of stored database objects for representing the fuzzy terms. In other words, any generated fuzzy query statement agrees with the standard SQL language. So, there is no need for an analyzer or interpreter to map a fuzzy query statement to a standard one then evaluate the resulted tuples with additional algorithms for manipulating the used linguistic terms. The resulted tuples are associated with matching degrees to the generated fuzzy query.

4   A Fuzzy Query Tool Based on the Proposed Approach

This section describes a developed fuzzy query tool (FQ) based on the proposed approach.

4.1   Login to FQ Service

At the first time the user aims to interact with the tool, he/she must create an account with a user name and a password. After that, the user can define his/her own linguistic terms via the tool GUI. Consequently, each created stored database object within the user session becomes part of the user fuzzy profile. Also, the user can generate a fuzzy query statement using his/her predefined linguistic terms and execute it.

4.2   Linguistic Variables Service

This service allows creating, modifying or deleting a linguistic variable that has a set of linguistic values that will be defined over it. The user enters just the name of the linguistic variable and a description for it. For example, a linguistic variable may be a computer grades, temperature degree, height or salary as shown in Fig.2.

Fig. 2. Linguistic Variables Screen

4.3   Linguistic Values Services

This service is responsible for defining a set of linguistic values over a specific linguistic variable. A set of linguistic values can be defined easily for each predefined linguistic variable to be used when constructing a fuzzy SQL statement. For example, the user can define the linguistic values "Tall", "Short" and "Medium" for the "Height" linguistic variable. Each linguistic value can be defined in a very flexible representation, that the user can select the more suitable membership function and modulate its control point's values to be more closed to the meaning of the linguistic value. Through a graphical user-friendly representation, the user can check the linguistic values definitions and its overlapping for a specific linguistic variable. The graphical representation helps the user to make the effective modifications in the selected membership function to satisfy the meaning of the linguistic value, as shown in Fig.3.

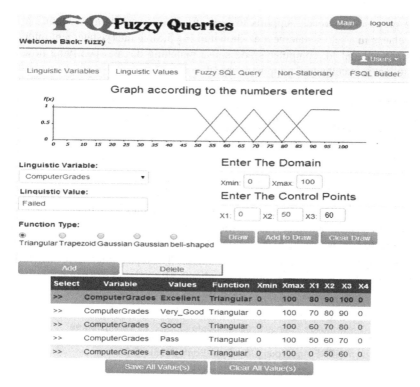

Fig. 3. Defining a set of linguistic values over a Linguistic Variable

4.4   Fuzzy SQL Query Service

This service enables the user to generate fuzzy query statements using the syntax of traditional SQL language. After the execution of the generated fuzzy query statement, the result it displayed as shown in Fig. 4. This query statement is generated to obtain each employee first name, salary and a matching degree specifying how much such salary is high.

SELECT	first_name, salary, round(SAL.HIGH(salary) ,2) as HighSal_MD
FROM	employees

FIRST_NAME	SALARY	HIGHSAL_MD
Steven	24000.00	1
Daniel	9000.00	0.5
John	8200.00	0.37
Ismael	7700.00	0.28
Jose Manuel	7800.00	0.3
...		

Fig. 4. A fuzzy query statement and the result of its execution

The proposed approach supports to write fuzzy expressions as a condition in the *where* clause in a select statement as shown in Fig.5. Such query statement intended to show each student Id, computer grade and a matching degree specifying how much such student is excellent in computer subject.

```
SELECT  SID, Computer,
        ComputerGrades.Excellent( Computer )
FROM    StudentScores
WHERE   ComputerGrades.Excellent( Computer ) > 0.5
```

Fig. 5. Using a fuzzy expression in where clause

Also, it allows passing fuzzy expressions as arguments to a function or a stored procedure in a select statement, as shown in Fig. 6. The generated query statements in Fig. 6 aims to retrieve each student name, age and a matching degree specifying how much such age represents a Young one. The resulted tuples must have matching degrees greater than the specified threshold value 0.20.

```
SELECT sname, round((SysDate-Bdate)/365,1) as age,
round(Age.Young((SysDate-Bdate)/365),2) as YoungDegree
FROM student
WHERE Age.Young((SysDate-Bdate)/365) > 0.20
```

SNAME	AGE	YOUNGDEGREE
Iman Mouhamad	21.1	0.39
Basmala	1.2	1
Sozy Aly	22.8	0.22
Mohimen Adel	3.5	1

Fig. 6. Passinga fuzzy expression as an argument to a function

On the other hand, complex queries including joining of two or more tables are supported. The resulted tuples can be sorted using their matching degree to the specified fuzzy criteria in the generated fuzzy query statement, as shown in Fig. 7. This query statement aims to retrieve each student Id, student name, physics grade and a matching degree specifies how much such student if failed in physics.

```
SELECT S.SID, S.SName, Physics,
ROUND( PhysicsGrades.Failed(Physics),3 ) as Degree
FROM Student S, StudentScores SS
WHERE S.SID = SS.SID
AND PhysicsGrades.Failed(Physics) > 0.30
Order By Degree
```

SID	SNAME	PHYSICS	DEGREE
20	Said Frag	53.00	0.4
19	Nagwa Aly	53.00	0.4
11	Sozy Aly	45.50	1
13	Sameer Abd Allah	20.00	1

• • •

Fig. 7. The result of a complex fuzzy query statement sorted ascendingly

Also, it is available to creatingdatabase views based on fuzzy expressions, as shown in Fig. 8.

```
Create Or replace View AgeDegree_View
as
select  round((SysDate-Bdate)/365) age,
round(AgeDegree.OLD((SysDate-Bdate)/365),2)AgeDegree
from student
where AgeDegree.OLD((SysDate-Bdate)/365) >0
```

Fig. 8. Creating a database view based of fuzzy expressions

Correlated subqueries known as synchronized subqueries are also allowed by the tool. Fig. 9 shows a correlated query statement that retrieves each employee name and salary for each employee has salary around the average salary in his/her department respecting a threshold value of 0.7.

```
Select First_Name, Salary
FROM   employees emp1
where aroundSalary( emp1.salary,
        (select avg(salary)
        from employees
        where emp1.department_id = department_id ) ) >0.7
```

Fig. 9. An example of a correlated Fuzzy query statement

In order to enhance the previous query statement to show each employee name, salary, the average salary of his/her department and a matching degree representing how much the employee salary is closed to such average, the fuzzy query statement shown in Fig. 10 is used. The result of such a query statement shows that the more closed salary to the average, the higher the matching degree.

```
Select first_name, salary,
       round((select avg(emp2.salary)
       from employees emp2
       where emp1.department_id = emp2.department_id ),2) as deptavg,

       round(aroundSalary( salary,
       (select avg(emp3.salary)
       from employees emp3
       where emp1.department_id = emp3.department_id ) ),2 ) as MatchDegree

FROM   employees emp1
where aroundSalary( emp1.salary,
           (select avg(emp4.salary)
           from employees emp4
           where emp1.department_id = emp4.department_id ) ) >0.7
```

Run Query

FIRST_NAME	SALARY	DEPTAVG	MATCHDEGREE
Renske	3600.00	3475.56	0.75
Trenna	3500.00	3475.56	0.95
Peter	9000.00	8955.88	0.91
Allan	9000.00	8955.88	0.91
Julia	3400.00	3475.56	0.85
Jennifer	3600.00	3475.56	0.75
		...	

Fig. 10. Another example of a more complex correlated fuzzy query statement

5   Conclusion

This paper presents a more simple fuzzy set based approach for flexible querying of relational databases. Commonly, many approaches have been proposed allowing fuzzy querying of relational databases. Yet, such approaches present complicated solutions which are two-fold. Firstly, they depend, mainly, on defining a fuzzy meta database (FMD) that is used for storing the definitions of all fuzzy terms that can be used. Accordingly, each generated fuzzy query statement needs to be analyzed to fetch each fuzzy term and brings its definition from FMD which is time consuming. If such approach depends on converting the generated fuzzy query to an equivalent traditional SQL query statement, it reads the definition and then passes it to its counterpart stored database object then process the query. Secondly, some other approaches needs to convert any generated fuzzy query statement to classical SQL statement by eliminating all used fuzzy terms. In consequence, the definition of such fuzzy terms, stored in FMD, is used to fuzzily evaluating the resulted tuples and assigning a matching degree for each resulted tuples. In contrary, the proposed approach depends mainly on storing all used fuzzy terms as database objects namely stored procedures and functions within packages in an organized fashion. Such stored database objects resembles the user fuzzy profile. Consequently, a fuzzy query statement can be generated directly as a traditional SQL statement using the fuzzy terms predefined as database objects. In consequent, there is no need for an analyzer or a translator to covert a generated fuzzy query into an executable query statement. Also, there is no need for a Fuzzy Meta-Knowledgebase that existed in most of previously proposed approaches for fuzzy queries. Hence, not only the complexity of allowing a fuzzy query is reduced but also the response time of executing such query. The proposed approach allows simple, complex, nested and correlated fuzzy query statements in a human-like fashion which almost contain linguistic terms and fuzzy connectors. Also, because linguistic terms are represented as a set of user defined stored functions; the proposed approach allows writing any simple, nested, correlated and complex fuzzy query statements as traditional select statements in a very flexible manner.

References

1. Mishra, J.: Fuzzy Query Processing. International Journal of Research and Reviews in Next Generation Networks 1(1) (March 2011)
2. Grissa, A., Ben Hassine, M.: New Architecture of Fuzzy Database Management Systems. The International Arab Journal of Information Technology 6(3) (July 2009)
3. Qi, Y., et al.: Efficient Processing of Nested Fuzzy SQL Queries in a Fuzzy Database. IEEE Transactions on Knowledge and Data Engineering 13(6) (November/December 2001)
4. Abbaci, K., Lemos, F., Hadjali, A., Grigori, D., Liétard, L., Rocacher, D., Bouzeghoub, M.: Selecting and Ranking Business Processes with Preferences: An Approach Based on Fuzzy Sets. In: Meersman, R., Dillon, T., Herrero, P., Kumar, A., Reichert, M., Qing, L., Ooi, B.-C., Damiani, E., Schmidt, D.C., White, J., Hauswirth, M., Hitzler, P., Mohania, M., et al. (eds.) OTM 2011, Part I. LNCS, vol. 7044, pp. 38–55. Springer, Heidelberg (2011)
5. Singh, K., et al.: Study of Imperfect Information Representation and FSQL processing. International Journal of Scientific & Engineering Research 3(5) (May 2012)
6. Garg, A., Rishi, R.: Querying Capability Enhancement in Database Using Fuzzy Logic. Global Journal of Computer Science and Technology 12(6) (Version 1.0 March 2012)
7. Wahidin, I.: Fuzzy Control (2007)
8. Gadallah, A., Hefny, H.: An Efficient Database Query Processing Tool Based on Fuzzy Logic. In: The 37th Annual Conference on Statistics and Computer Science (2002)
9. Shawky, A., et al.: FRDBM: A Tool For Building Fuzzy Relational Databases. The Egyptian Computer Journal (2006)
10. Galindo, J., et al.: Handbook of Research on Fuzzy Information Processing in Databases. Chapter XI FSQL and SQLf: Towards a Standard in Fuzzy Databases (2008)
11. Bosc, P., Pivert, O.: SQLf Query Functionality on Top of a Regular Relational Database Management. In: Pons, O., Vila, M.A., Kacprzyk, J. (eds.) Proceedings of Knowledge Management in Fuzzy Databases. STUDFUZZ, vol. 39, pp. 171–190. Springer, Heidelberg (2000)
12. Moore, S.: Oracle Database PL/SQL Language Reference, 12c, p. 1 (2014)
13. Elmasri, R., Navathe, S.B.: Fundamental of Database Systems (2011)

Part VII
Web-Based Application and Case Based Reasoning Construction

A Survey on Conducting Vulnerability Assessment in Web-Based Application

Nor Fatimah Awang[1], Azizah Abdul Manaf[1], and Wan Shafiuddin Zainudin[2]

[1] Advanced Informatics School (UTM AIS),
UTM International Campus,
Kuala Lumpur, Malaysia
[2] MySEF Lab, Cyber Security Malaysia
The Mines Resort City, Seri Kembangan
Selangor, Malaysia
norfatimah@upnm.edu.my, azizah07@ic.utm.my,
wanshafi@cybersecurity.my

Abstract. Many organizations have changed their traditional systems to web-based applications to make more profit and at the same time to increase the efficiency of their activities such as customer support services and data transactions. However web-based applications have become a major target for attackers due to some common vulnerability exists in the application. Assessing the level of information security in a web-based application is a serious challenge for many organizations. One of the important steps to ensure the security of web application is conducting vulnerability assessment periodically. Vulnerability assessment is a process to search for any potential loopholes or vulnerability contain in a system. Most of the current efforts in assessments are involve searching for known vulnerabilities that commonly exist in web-based application. The process of conducting vulnerability assessment can be improved by understanding the functionality of the application and characteristics of the nature vulnerabilities. In this paper, we perform an empirical study on how to do vulnerability assessment with the aim of understanding how the functionality, vulnerabilities and activities that would benefit for the assessment processes from the perspective of application security.

1   Introduction

Many organizations are rushing to get online systems implemented and deployed. The development and deployment of web-based applications are becoming easier due to the simplicity of its usage and its high accessibility. Organizations have a tremendous opportunity to use web-based applications technologies to increase their productivity. Web applications have been applied into many environments such as e-commerce, tax payments, human resource systems and student registration portal. As the number of organizations that conduct their activities electronically grows continuously, information security becomes one of major concerns because confidentiality, integrity and availability of data, tools and transactions are critical requirements for organizations to stay functional, legal and competitive [1]. Web-based applications

A.E. Hassanien et al. (Eds.): AMLTA 2014, CCIS 488, pp. 459–471, 2014.
© Springer International Publishing Switzerland 2014

are opened to all web users, both legitimate users and malicious users, where all the information that will be accessed from anywhere will be exposed to cyber-attacks and malicious hackers. According to [2], web-based application attacks have become more sophisticated nowadays. A survey conducted by Ernst and Yong [3], showed that almost half of the companies in the United States have experienced on web applications attacks. In Global Internet Security Threat Report, Symantec Security reported that almost 247,350 web-based application attacks in 2012 and this is increase almost 30 percent from the previous year [3]. Web-based application attacks are become popular due to some reasons such as web applications often represent the most effective entry point to a computer network, because, by design, they are almost always reachable through firewalls and a significant part of their functionality is often available to anonymous users [4]. Web applications often interact with back-end database. In order to successfully perform online services web applications often interact with back-end components and have access to sensitive user information, such as passwords, credit card numbers, user personal information, etc. Thus, they have become an attractive target for attackers who are becoming more and more interested in gaining financial profit [4].

With the explosion of web-based application attacks, the vulnerabilities of web-based application vulnerabilities have also increased. Web applications vulnerabilities and web servers are prime target to criminal hackers. Security flaws in the Web application layer can allow attackers to steal data, plant malicious code or break into other internal systems. Some of the most common vulnerabilities include SQL injection, cross-site scripting flaws, authorizations and authentication errors. Several recent studies discussing the vulnerability of the web applications, proves that the security problem in web applications is an issue far from being solved and exposed to hackers to compromise the applications. As a result, a high percentage of web applications deployed on the Internet are exposed to security vulnerabilities. With the majority of vulnerability exists in web applications today, it is important to evaluate and detect the vulnerability of website or web-based application before it is sent to production. Today, to minimize the probability of vulnerabilities exist in web applications, organizations need some methodologies or approaches to increase efforts to protect against web-based application attack or data breaches. In order to protect organization's confidential data, there are several methodologies or standards that have commonly used by industry to identify and detect vulnerability exist in web applications. One of the important steps to ensure the security of web application is conducting vulnerability assessment periodically [5]. The goal of a vulnerability assessment, also known as a security audit or security review, is to search any potential loopholes contain in a system that lead to compromise the systems. It is important to do assessment on the system to make sure that it will be safely release and not offer any illegitimate access that can affect availability, confidentiality and integrity of the system [5]. Assessment, in general, can be described as an examination or review of compliance against a set of guidelines or standards [6]. Assessment or auditing usually takes place when something has already gone wrong [7, 8]. Goan [9] as well as Ramim and Levy [10] suggested implementing information security assessment as a possible solution to detect vulnerabilities or intentional wrong doings before they can be exploited. Similarly, Chuvakin and Peterson [11] acknowledged that security assessment is particularly important for systems that run over the Web because of the security and design challenges in protocols such as Hypertext Transfer Protocol (HTTP).

The main objective of this paper is to discuss and understand the process of conducting vulnerability assessment in order to identify and report on security vulnerabilities to allow the organizations to close the issues in a planned manner.

The structure of this paper is as follows. Section 2 briefly describes the related works in conduction assessment and testing process, section 3 describes assessment process. Meanwhile, section 4 and 5 discusses more detail on the understanding of functionality and potential vulnerability in web application. Section 6 discusses the attack scenario commonly used for web application, section 7 presents sample of test cases and finally, section 8 presents the conclusion.

2   Related Works

Existing methods focus on various aspects to assess web-based application such as design of threat analysis, attacker behavior using tools and attacker goal that can cause system exploitation. Huang et al. [30] propose an assessment framework for web-based application and they develop one tool called WAVES to detect vulnerabilities. However, Huang's framework discovered two types of vulnerabilities only, SQL Injection and Cross Site Scripting. Meanwhile, Jensen et al. did passive mode testing which are used search engine and port scanning techniques to discover potential vulnerabilities in website [34]. Khaled et al. proposed an assessment framework for web server by using two scanning tools, Nikto and Nessus and this model focuses on detecting vulnerability in web server [35]. Thompson [36] in his paper proposes four main steps in order to do security assessment process for web-based application such as; create a threat model, build a test plan, execute test cases and create the problem report and execute postmortem evaluation. A threat model is designed in order to get a detail potential risks that threatens the application. Through threat modeling, organization can assess the risk and cost and understand the impact to organization. However, threat modeling works better if involve a security expert in the project. Several techniques help software testers, developers, and security auditors create meaningful threat models. According to author, one of the best and most widely used is the STRIDE method. The second step is to build a test plan. The test plan acts as a road map for the total security testing effort. It is created to get a high-level overview of the security test cases, an overview of how exploratory testing will be conducted. An overview of the security test cases, identify the tools that is needed to help tester to conduct testing is important in this phase. In this work, Thompson did not provide detail out all the processes on how to create test cases and which tools are used to conduct a testing. Moa [37], proposed security testing framework for analyzing all information of error messages appear in web application. Unlike with Huang's work, the proposed framework used their tool called WASViewer to collect and analyze all error information and then author will check manually to identify that the tool has detected the expected error message or not. Test cases will be created based on information error given by application. However the framework discovered only a number of web application vulnerabilities such as SQL Injection and cross site scripting and input validation.

3   Vulnerability Assessment Process Framework

Fig. 1 below summarizes the vulnerability assessment process. The process consists of five steps. The processes of each step are:

1. Identify and understand web application function. This process will be described in understanding web application functionality section in section 4. Refer Fig. 2 and 3 for more detail.
2. Identify Vulnerability. The potential vulnerability will be identified based on some functionality described in process 1. For example login.html page. Based on login page, several possible vulnerabilities will be listed. Refer Fig. 4.
3. Develop test cases as shown in section 7.
4. Run scanning and exploit. In this activity, selected scanning tools will be used to do scanning. In our scanning phase, we always set the scanner to run in automated mode to maximize vulnerability detection capability. In this activity, results produced by different tools will be compiled for further analysis purposes.
5. Analyze vulnerability and result. Analysis of scanning result and analysis of exploit result.

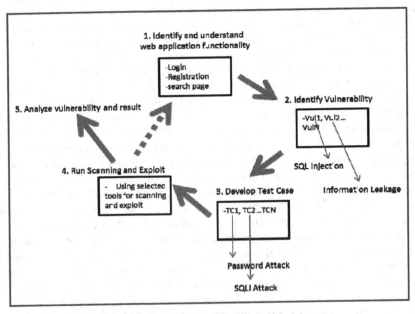

Fig. 1. Vulnerability Assessment Model

4   Understanding Web Application Functionality

Finding security vulnerabilities in web-based application typically requires detailed knowledge of it functionality. With the support of a large number of web technologies, web applications can implement rich functionalities to support complex business logic. The content presented by a web application to users is generated dynamically on the fly, and is often tailored to each specific user. Most of web-based applications provide a set of common functionalities to make the business logic available to end users. These include Login and Logout, Session Tracking, User Permissions Enforcement, Role Level Enforcement, Data Access such as HTML form-based data collection and Search, and Application Logic [12]. Most of web application must have at least login and password for their functionality for minimum authentication to access their web systems. For example in Figure 2, there are some commons functionalities used in digital library website available for student and staff such as login and logout, search portal, user comment and help function. In order to borrow all library materials online, student or staff must be as a library member and use a valid username and password to access the application. Login functionality as shown in Figure 3 allows the application to authenticate a specific user by allowing a valid username and password. The implementation of these functionalities can cut across all the different components of a web application architecture including browser, web server, application, and database. Any flaws happened during the development of a functionality may cause a security vulnerability. Hackers often try to exploit these vulnerabilities against a web application by analyzing the application's functionalities to identify potential vulnerabilities and then launch attacks to attempt to compromise them. By using the login function, hackers will try several conditions to compromise the system such as SQL Injection, manipulate session cookie, brute force and etc.

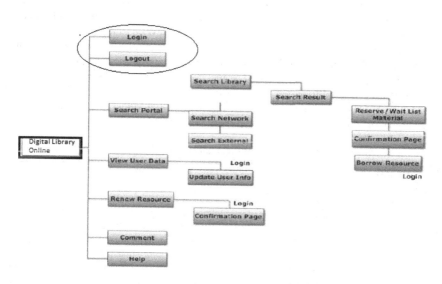

Fig. 2. Sample of functionality in digital library website

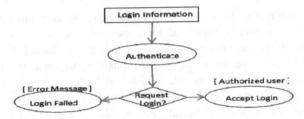

Fig. 3. Scenario for accessing Web Application with Authentication Process

5   Understanding Web Application Vulnerability

Many of web applications attack because of exploitation of vulnerabilities in a system. There are no accepted standard in information security regarding the word vulnerability. We are trying to define the word vulnerability related to web application based on some literature researches. According to Wang et al., vulnerability is a security flaws, defect or mistakes in software that can be directly used by a hacker to gain access to system or network [13]. According to [14], vulnerability is a weakness of an asset or group of assets that can be exploited by one or more threats where an asset is anything that can has value to the organization. Open Web Applications Security Project (OWASP) [15] as an online community that offers free open source security tools and information to assist on developing security in web application, refer the vulnerability as a hole or a weakness in the application, which can be a design flaw or an implementation bug that allows an attacker to cause harm to the stakeholders of an application. Stakeholders include the application owner, application users, and other entities that rely on the application.

A lot of vulnerabilities appear in web applications make system or network administrator more difficult to protect core assets such as personnel information, confidential data and customer credit card numbers. Several studies and report have shown that the security of a web application, in general is unsafe. According to Curphey and Araujo, vulnerability can be classified into a variety of category types [16]. Meanwhile, Popa [17], have identified some common vulnerabilities in information system that may be used to assess web applications. This paper assessed some potential vulnerability in web application. Table 1 lists the potential vulnerabilities studied from literature that may be found in web applications.

Table 1. Lists of Potential Vulnerabilities

Vulnerabilities/Source (Proceeding/Journal)	Description
SQL Injection Paper [17, 18, 19, 20, 21,24]	Vulnerability caused when SQL code is inserted or appended into application or user input parameters that are later passed to a back-end SQL database for parsing and execution. It affects the database without the knowledge of the database administrator. Any procedure that constructs SQL statements could potentially be vulnerable. The primary form of SQL injection consists of direct insertion of code into parameters that are concatenated with SQL commands and executed. The exploitation of a SQL injection vulnerability can lead to the execution of arbitrary queries with the privileges of the vulnerable application and, consequently, to the leakage of sensitive information and/or unauthorized modification of data.

Table 1. (*continued*)

Cross Site Scripting Paper [22,23,24]	Vulnerability happened because vulnerable applications fail to sanitize malicious input at either server side or client side, allowing an attacker to be injected into response page with embedding malicious hypertext markup language (HTML) or script inside of dynamic Web applications
Cross Site Request Forgery Paper [24,25]	Vulnerabilities happened when a malicious web site interferes with a victim user's ongoing session with a trusted website. The malicious web site tricks the web browser into attaching a trusted site's authentication credentials to malicious requests targeting the trusted site. This attack forces a logged-on victim's browser to send a forged HTTP request, including the victim's session cookie and any other automatically included authentication information, to a vulnerable web application.
Broken Authentication and Session Management Paper [24, 25]	This vulnerability allows an attacker to bypass the authentication system or escalate their privileges without using an injection attack. A typical vulnerable application would allow an attacker to access restricted sections without being identified as a valid user.
Buffer Overflows Paper [25, 26]	Vulnerabilities associated with security failures due to exceeding the buffer's storage capacity. Buffer overflow vulnerability is the cause of many cyber- attacks such as worms, zombies, and botnets.
Custom Cookies/Hidden Fields Paper [25, 27]	Vulnerabilities happened when cookies stored on the client side that can be manipulated by malicious users
Denial of Service Attacks Paper [28]	Vulnerability happened when attackers use multiple computers as the source of the attack to target some resources such as web server. In the majority of DoS attacks, the computers used as the source of the attack are not aware that there systems are being compromised.
Application Runtime Configuration Paper [25]	Vulnerabilities caused by the improper configuration of the runtime environment. Improper configuration of the runtime environment can lead to a variety of potential vulnerabilities in the internal or external runtime environment of application.
Backdoor, Trojans, and Remote Controlling Paper [29]	Vulnerabilities associated with concealing unauthorized access. These programs may appear to be legitimate or have justifiable use but they disguise malicious functionality such as downloading unauthorized files or attacking other computers. According to [29], a backdoor is a program which allows programmers to gain access to an information system without going through the established security measures. Backdoors are usually left intentionally and they may permit the original programmer to gain access even after software is sold or licensed.
Information Leakage and Improper Error Handling Paper [30]	Vulnerabilities happened when application reveal some vulnerable information through the error messages. By analyzing some error messages, attackers can perform an attack to target application.
Insecure Communications Paper [25]	Vulnerabilities happened when attackers obtaining sensitive information by sniffing the unencrypted HTTP data. Data is transmitted without encryption process.

Table 1. (*continued*)

Insecure Cryptographic Storage Paper [25]	Vulnerabilities happened when data is stored without cryptographic functions. Many web applications do not properly protect sensitive data, such as credit cards, identification ID and authentication credentials, with appropriate encryption or hashing. Attackers may steal or modify such weakly protected data to conduct identity theft, credit card fraud, or other crimes.
Insecure Direct Object Reference Paper [24, 25]	This vulnerability occurs when a developer exposes a reference to an internal implementation object, such as a file, directory, or database key. Without an access control check or other protection, attackers can manipulate these references to access unauthorized data.
Parameter Tampering Paper [31]	Vulnerabilities happened when attackers performing some modification of parameters that has been sent between the user and the information system. According to [31], parameter tampering is mostly carried out by attempting to alter the parameters of a Web URL, modifying the HTML form fields in the browser or modifying the query string when a user makes a request.
Insecure/Weak Password Paper [32]	Vulnerabilities happened when application have an account to access the system by using weak password characteristics. Some websites allow user to register with weak password. The weak password can be on of dictionary word, user of either lower or upper case only, only alphabets, small length password. According to [32], insecure passwords may allow unauthorized access to information systems or to users' personal information. Most information systems have a form of password security but information systems without passwords or with weak passwords increase the probability that a malicious user or attacker can gain access to the system By using password cracker tools, attackers can guess any password and give them the ability to test millions of passwords per second.
Insufficient Authentication Paper [32]	Vulnerability occurs when a web site permits an attacker to access sensitive content or functionality without having to properly authenticate.
Path Traversal Paper [25, 30]	Vulnerability occurs when attacker forces access to files, directories, and commands that potentially reside outside the web document root directory. The error message has disclosed very important information to the attacker showing the directory structure of web server.

OWASP [33] has recommended the top ten lists of application level vulnerabilities in web application. The top ten classes of vulnerabilities outlined by OWASP are Injection, Broken Authentication and Session Management, Cross-Site Scripting, Insecure Direct Object References, Security Misconfiguration, Sensitive Data Exposure, Missing Function Level Access Control, Cross-Site Request Forgery, Using Components with Known Vulnerabilities and Unvalidated Redirects and Forwards. Another worldwide security organization, the SANS (SysAdmin, Audit, Network, Security) Institute, also lists cross-site scripting, SQL injection and cross-site forgery as major web application vulnerabilities.

6   Vulnerability Scenario for Web Application

Our approach aims to identify the intentions and goals of the possible attackers, identify possible functionality of the system as shown in Figure 2 that commonly used by the attacker to attack the system, identify possible vulnerability based on the functionality, identify test cases to the system and how to exploit these attacks to improve the web application security. By analyzing the intentions of functionality and vulnerability, the testers can obtain information that helps to understand where the function and vulnerability that commonly used by the attacker to exploit the system.

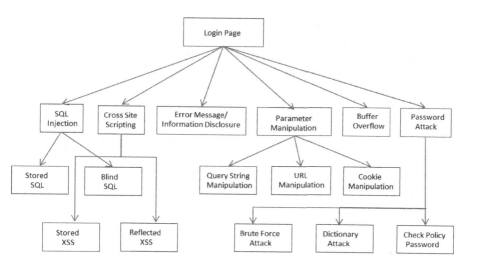

Fig. 4. Attack Scenario for Login Page

Analyzing functionality and vulnerability within web application allows testers to understand how system security can be compromised. Figure 3, helps security tester or assessor to understand possible vulnerabilities in login page that would be used for security testing. These also assist in preparing security test cases. Similarly, it will provide a basis of mitigating planning based on several vulnerabilities in web application. It can be said that almost all attacks on web application make use of Login Page to exploit and access illegally within application. Another common vulnerability is the danger of weak authentication and session management practices. Access control and authentication schemes are seriously hard to secure properly.

7   Example of Security Test Cases

Table 2. Vulnerability Password Attack

AIM	Brute/Dictionary attack for password in login page to obtain valid username and password	
Pre Test Step	**Description**	OK/Not(X)
1	In Attacker Machine, use and open Backtrack software	
2	Create a text file that contains several password such as root, admin, administrator. Save the file as username.txt in /root	
3	Create a text file that contains several password such as root, admin, administrator. Save the file as password.txt in /root	
4	Open Web Browser and go to login page, https://x.x.x.x. Try login using fake account and password. For example user: admin and password:password	
5	Right click at the page and view source. Find the parameter for username(username_form), password(password_form) and failure response message (Login Failed)	
Test Step	**Description**	OK/Not(X)
1	Use Hydra to launch password attack.	
2	Type: Hydra x.x.x.x –s 81 https-post-form "/index.php:username_form=^USER^&password_form=^PASS^:Login Failed." –L username.txt –P password.txt –t 10 –w 30 –o hydra-https-post-attack.txt	
3	Go to /root folder. Find a text file "hydra-https-post-attack.txt" that store the result of the attack.	
4	Result also can be seen in the terminal itself	
5	Result stored in hydra-https-post-attack.txt: #Hydra v7.3 run at XXXX on XXXX http-post-form (hydra –s 81 –L username.txt –P password.txt –t 10 –w 30 –o hydra-https-post-attack.txt x.x.x.x http-post-form /index.php:username_form=^USER^&password_form=^PASS^:Login Failed. [81] …login:admin password:admin Valid username and password is obtained	
Expected Test Results	Dictionary attack could not be launch due to account locking/limit fail attempt for authentication.	
Actual Test Result	Attacker successfully obtained valid username and password. Login Page does not enforce account locking/limit fail attempt for authentication.	

8 Conclusion

In this paper, we have presented vulnerability assessment process and also an attack or vulnerability scenario purposely for web application security testing or assessment. The main idea is to help security tester or assessor to perform security assessment within their organization and identify possible vulnerability and attack that would affect web application system.

Acknowledgements. This work was supported by the Advanced Informatics School, University Technology of Malaysia and Cyber Security Malaysia.

References

1. Andoh-Baidoo, F.K., Osei-Bryson, K.-M.: Exploring the characteristics of Internet security breaches that impact the market value of breached firms. Expert Systems with Applications 32(3), 703–725 (2007) ISSN 0957-4174
2. Anastacio, M., Blanco, J.A., Villalba, L., Dahoud, A.: E-Government: Benefits, Risks and A Proposal to Assessment including Cloud Computing and Critical Infrastructure. In: International Conference on Information Technology (2013)
3. Abusaimah, H., Shkaukani, M.: Survey of Web Application and Internet Security Threats. IJCSNS International Journal of Computer Science and Network Security 12(12) (2012)
4. Felmetsger, V., Cavedon, L., Kruegel, C., Vigna, G.: Toward automated detection of logic vulnerabilities in web applications (2013), http://static.usenix.org/event/sec10/tech/full_papers/Felmetsger.pdf
5. Meier, J.D., Mackman, A., Dunner, M., Vasireddy, S., Escamilla, R., Murukan, A.: Improving Web Application Security: Threats and Countermeasures. Microsoft Corporation (2003), http://msdn.microsoft.com/en-us/library/aa302420.aspx
6. Carlin, A., Gallegos, F.: IT audit: A critical business process. IEEE Computer, 47–49 (2007)
7. Balwin, A., Shiu, S.: Enabling shared audit data. International Journal of Information Security, 263–276 (2005)
8. Peisert, S., Bishop, M., Marzullo, K.: Computer forensics in forensis. ACM SIGOPS Operating Systems Review 42(3), 112–122 (2008)
9. Goan, T.: A cop on the beat: Collecting and appraising intrusion evidence. Communications of the ACM 42(7), 46–52 (1999)
10. Ramim, M., Levy, Y.: Securing e-learning systems: A case of insider cyber attacks and novice it management in a small university. Journal of Cases on Information Technology 8(4), 24–34 (2006)
11. Chuvakin, A., Peterson, G.: Logging in the age of web services. IEEE Security and Privacy 7(3), 82–85 (2009)
12. Cross, M., Kapinos, S., Meer, H.: Web applications vulnerabilities, Detect, Exploit, Prevent. Syngress (2007)
13. Wang, J.A., Guo, M., Wang, H., Xia, M., Zhou, L.: Environmental metrics for software security based on a vulnerability ontology. In: Third IEEE International Conference on Secure Software Integration and Reliability Improvement, pp. 159–168 (2009)

14. Stuttard, D., Pinto, M.: The web application hacker's handbook: discovering and exploiting security flaws. Wiley Publishing, Inc. (2007)
15. Category Vulnerability (2013), https://www.owasp.org/index.php/Category:Vulnerability
16. Curphey, M., Araujo, R.: Web Application Security Assessment Tools. IEEE Security & Privacy (2006)
17. Ezumalai, R., Aghila, G.: Combinatorial Approach for Preventing SQL Injection Attacks. In: IEEE International Advance Computing Conference, IACC 2009 (2009)
18. Claarke, J., et al.: SQL Injection Attacks and Defense. Syngress Publishing (2009) ISBN 13: 978-1-59749-424-3
19. Huang, Y., Yu, F., Hang, C., Tsai, C.H., Lee, D.T., Kuo, S.Y.: Securing Web Application Code by Static Analysis and Runtime Protection. In: Proceedings of the 12th International World Wide Web Conference (WWW 2004) (May 2004)
20. Asagba, P.O., Ogheneovo, E.E.: A Proposed Architecture for Defending Against Command Injection Attacks in a Distributed Network Environment. In: Information Technology for People-Centred Development, pp. 134–142 (2011)
21. Bisht, P., Madhusudan, P., Venkatarishnan, V.N.: CANDID: Dynamic Candidate Evaluations for Automatic Prevention of SQL Injection Attacks. ACM Transactions on Information and System Security 13(2), Article 14 (2010)
22. Shahriar, H., Zulkernine, M.: Taxonomy and classification of automatic monitoring of program security vulnerability exploitations. Journal of Systems and Software 84, 250–269 (2011), doi:10.1016/j.jss.2010.09.020, ISSN 0164-1212
23. Avancini, A.: Security testing of web applications: A research plan. In: 2012 34th International Conference on Software Engineering (ICSE), June 2-9, pp. 1491–1494 (2012)
24. Scambray, J., Shema, M., Sima, C.: Hacking Exposed: Web Applications, 2nd edn. McGraw-Hill, San Francisco (2006)
25. Popa, M.: Detection of the security vulnerabilities in web applications. Informatica Economica 13(1), 127–136 (2009)
26. Wang, W., Pan, C., Liu, P., Zhu, S.: SigFree: A signature-free buffer overflow attack blocker. IEEE Transactions on Dependable and Secure Computing 7(1), 65–79 (2010)
27. Park, J.S., Sandhu, R.: Secure cookies on the web. IEEE Internet Computing 4(4), 36–44 (2000)
28. Badishi, G., Keidar, I., Sasson, A.: Exposing and eliminating vulnerabilities to denial of service attacks in secure gossip-based multicast. IEEE Transactions on Dependable and Secure Computing 3(1), 45–61 (2006)
29. Hahn, R.W., Layne-Farrar, A.: The law and economics of software security. Harvard Journal of Law and Public Policy 30(1), 283–353 (2006)
30. Huang, Y., Tsai, C., Lin, T., Huang, S., Lee, D.T., Kuo, S.: A testing framework for web application security assessment. Computer Networks 48(5), 739–761 (2005)
31. Smith, R.: Information security-a critical business function. Journal of GXP Compliance 13(4), 62–68 (2009)
32. Garrison, C.P.: An evaluation of passwords. The CPA Journal 78(5), 70–71 (2009)
33. OWASP Top 10 2013-Top 10, https://www.owasp.org/index.php/Top_10_2013-Top_10 (retrieved May 5, 2014)
34. Zhao, J.J., Zhao, S.Y., Zhao, S.Y.: Opportunities and threats: A security assessment of state e-government websites. Government Information Quarterly 27(1), 49–56 (2010), doi:10.1016/j.giq.2009.07.004, ISSN 0740-624X

35. Alghathbar, K.S., Mahmud, M., Ullah, H.: Most known vulnerabilities in Saudi Arabian web servers. In: 4th IEEE/IFIP International Conference on Internet, ICI 2008, pp. 1–5 (2008)
36. Thompson, H.H.: Application penetration testing. IEEE Security and Privacy 3(1), 66–69 (2005)
37. Mao, C.: Experiences in Security Testing for Web-based Applications. In: ICIS 2009, Seoul, Korea (2009)

An Adaptive Information Retrieval System for Efficient Web Searching

Safaa I. Hajeer, Rasha M. Ismail, Nagwa L. Badr, and Mohamed Fahmy Tolba

Faculty of Computer & Information Sciences, Ain Shams University, Cairo, Egypt
safaahajeer@yahoo.com, rashaismail@fcis.asu.edu.eg,
nagwabadr@cis.asu.edu.eg, fahmytolba@gmail.com

Abstract. Stemming algorithms (stemmers) are used to convert the words to their root form (stem), this process is used in the pre-processing stage of the Information Retrieval Systems. The Stemmers affect the indexing time by reducing the size of index file and improving the performance of the retrieval process. There are several stemming algorithms, the most widely used is porter stemming algorithm because of its efficiency, simplicity, speed, and also it easily handles exceptions. However there are some drawbacks, although many attempts were made to improve its structure but it was incomplete. This paper provides an efficient information retrieval technique as well as proposes a new stemming algorithm called Enhanced Porter's Stemming Algorithm (EPSA). The objective of this technique is to overcome the drawbacks of the porter algorithm and improve the web searching. The EPSA was applied to two datasets to measure its performance. The result shows improvement of the precision over the original porter algorithm while realizing approximately the same recall percentage.

Keywords: Information Retrieval (IR), Stemming (Stemmer), over-stemming, under-stemming, Porter Stemmer.

1 Introduction

Over the years Information Retrieval Systems play critical roles to obtain relevant information resources based on indexing techniques technology. [1, 2]

The basic idea of information retrieval systems is to find a degree of similarity between the documents collection and the query after eliminating Stop Words. Stop words are useless in the documents collection, it occurs in 80% of the document [2, 3, 4]. Stemming is an aspect supported by indexing and searching technique. It's one of the pre-processing stages used to combat the vocabulary mismatch problem, in which query words don't match exactly document words.-The stemming process will reduce the size of the documents representations by 20-50% compared to full words representations, according to van Rijsbergen [5]. Furthermore, the relevancy of the retrieved documents will be improved and their number will also be increased [2] [5].

However the stemming can cause errors in the form of words; there are mainly two types of errors, over-stemming and under-stemming. These errors decrease the effectiveness of stemming algorithms.

A.E. Hassanien et al. (Eds.): AMLTA 2014, CCIS 488, pp. 472–482, 2014.

- **Over-stemming** is when two words with different stems are stemmed to the same root. This is also known as a **false positive**.
  For example, past and paste →past (same root)
- **Under-stemming** is when two words that should be stemmed to the same root are not. This is also known as a **false negative**.
  For example: Adhere → adher, Adhesion → Adhes

The challenge is to reduce these errors as much as possible, because of that many researches tried to improve the structure of porter algorithm; however it still has several drawbacks. This paper presents a new stemming algorithm (EPSA) to reduce stemming errors and improve information retrieval systems effectiveness.

The rest of the paper is organized as follows. Section 2 presents related Work. Section 3 describes Porter's algorithm. Section 4 discusses The System Architecture. Section 5 presents the evaluation to ESPA and its experimental results. Finally, the conclusion and future work of the paper appears in Section 6.

2   Related Work

Lovins in [6, 7] described the first stemmer in 1968, which was developed specifically for the information-retrieval applications. The stemmer performs a lookup on a table of 294 endings, 29 conditions and 35 transformation rules. The Lovins stemmer removes the longest suffix from a word. Once the ending is removed, the word is recoded using a different table that makes various adjustments to convert these stems into valid words. It always removes a maximum of one suffix from a word. The result shows that this algorithm is very fast but it has some drawback like many suffixes are not available in the table of endings. It is sometimes highly unreliable and frequently fails to form words from the stems [8].

The Paice/Husk stemmer in [9], is an iterative algorithm with one table containing about 120 rules indexed by the last letter of a suffix; on each iteration, it tries to find an applicable rule by the last character of the word. If there is no such rule, it terminates. It also terminates if a word starts with a vowel and there are only two letters left or if a word starts with a consonant and there are only three characters left. Otherwise, the rule is applied and the process repeats in [10]. The disadvantage is it may be a very heavy algorithm and over stemming may occur.

Dawson Stemmer is an extension of the Lovins approach except that it covers a much more comprehensive list of about 1200 suffixes. The suffixes are stored in the reversed order indexed by their length and last letter. The rules define if a suffix found can be removed. The disadvantage is it is very complex and lacks a standard reusable implementation [11, 12].

Porters stemming algorithm is one of the most popular stemming methods proposed in 1980. It is based on the idea of defining five or six successively applied steps of word transformation. Each step consists of a set of rules that is applied until one of them passes the conditions. If a rule is accepted, the suffix is removed accordingly, and the next step is performed. The resultant stem at the end of the sixth step is returned [8] [13].

Porter stemmer produces a less error rate than the Lovins stemmer. However, Lovins's stemmer is a heavier stemmer that produces a better data reduction [13, 14]. The Lovins algorithm is noticeably bigger than the Porter algorithm, because of its very extensive endings list [8]. Many modifications and enhancements have been done and suggested on the basic of Porter stemmer algorithm in order to reduce its errors.

STANS algorithm is a modification of the porter's stemming algorithm. In this algorithm totally 31 modifications are done among that 15 rules were added, 11 rules were modified and 5 rules were deleted from the origin. For example: Modify (*v*) ING→ as (*v*) ING→E, add rule of less→ . STANS still has some drawbacks like the rule LESS→ the meaning of the word is changed to the opposite one (Careless will change to care, which cause errors) [15, 16].

In 2013, Megala et. al. developed an algorithm called TWIG, it is an improvised Porter stemming algorithm [16], it holds 31 modifications done on the STANS algorithm, among that 29 modifications are adapted to the improvised stemming algorithm except (LESS→ , FUL→) and 15 new rules are introduced to improve the performance of the IR System with meaning full stems.

Based on the previous work, many researches tried to improve the structure of porter algorithm; however they still have several drawbacks. Some of them concentrated on plural and singular words only, others concentrated on the semantic of some words without being careful about singular and plural. Also, the previous work didn't discuss the past tense of words ending by "ed" and the verbs ending by "en". So, In this paper we presents the Enhanced Porter Stemming Algorithm (EPSA) to overcome these problems by collecting the important rules from the previous researches and proposed new rules that solve the errors mentioned in the next section.

3 Porter's Stemmer

Porter Stemming Algorithm in [14] operates in six steps. At each step the input word is transformed based on a list of rules. These steps are:

Step 1: Gets rid of plurals and -ed or -ing suffixes, the rules here like:
SSES → SS, SS → SS, IES → I, S →, (*v*) ED→

Step 2: Turns terminal y to i when there is another vowel in the stem, the rules here like: (*v*) Y→I

Step 3: Maps double suffixes to single ones: -ization, -ational, etc. the rules here like: (m>0), ational→ate

Step 4: Deals with suffixes, -full, -ness etc. the rules here like: full→, ness→

Step 5: Takes off -ant, -ence, etc. the rules here like: ence→, al→

Step 6: Remove a final e. the rules here like: (m>1) E →, (m=1 and not *o) E →
Note:
v (Vowel) :The letters A, E, I, O, U are considered vowels, sometimes y.
m: number of vowels in the word.
c (Consonant): Any letter, not a vowel

Languages are not completely regular constructs, and therefore stemmers operating on natural words unavoidably make mistakes. Although Porter stemmer is known to be powerful, it still faces many Errors. The major ones are Over-stemming and Under-stemming errors. These errors like the concept denotes the case where a word is reduced too much for example, "Appendicitis" after stemming may become "Append" Which may lead to the point where completely different words are conjoined under the same stem or stem incorrect in the Language. For instance [3] [17]:

- Past, Paste return to the same root Past
- Appendicitis, Append according to porter algorithm are from Append
- Dying→ dy (impregnate with dye)
- Dyed→ dy (passes away)
- Political, Polite → polit
- Generative, General → gener
- child/children→ child/children
- Words ends by feet, men retrun as it is
- careless→ careless (remove less, will change to care will cause error)

Another instance of errors come from our testing on porter algorithm like written→written, complier→compil, languages→ languag, described→describ

4   The System Architecture

This section we studied the original porter stemming algorithm [14], against the pervious researches done to enhance it. From these studies we proposed a new stemming algorithm called Enhanced Porter Stemming Algorithm (EPSA). This algorithm was applied to the following system architecture that is shown in figure 1 Our system consists of 4 main modules:

- **Module 1**: Tokenization: this stage for breaking a stream of text up into words, and keeping the words in a list called Word's List.
- **Module 2**: Data Cleaning: it removes useless words from the Word's List, These useless words are stored in a stop words database as appear in the figure. The database has 311 English stop words with a size 3KB.
- **Module 3**: the ESPA stemming Algorithm: In this stage, we applied our proposed stemming algorithm to overcome the drawbacks of the porter algorithm. The algorithm is explained in the next section.
- **Module 4**: Indexing & Ranking: Indexing is a process for describing or classifying a document by index terms; index terms are the keywords that have meaning of its own (i.e. which usually has the semantics of the noun). This index terms are grouped in an indexer and stemmer is service this stage by improving the group of these keywords in the indexer. Then the user's query is matched with the index terms to get the relevant documents to the query. Documents are then ranked using ranking algorithms according to the most relevant to the user's query.

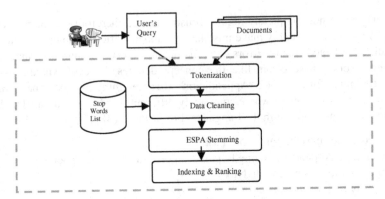

Fig. 1. The System Architecture

4.1   The Enhanced Porter Stemming Algorithm (EPSA)

The ESPA stemming includes the original porter rules and our new rules that are proposed to solve the previous errors (appears in bold font in table 1). Also, we collected the critical rules from other researches [3], [14], [16, 17], [20], and added them to the EPSA stemming. Table 1 shows the details of these rules.

Table 1. The EPSA Rules

If the word:		
ends with "e", function must keep e at the end of the word	ends with "ize" - m=2, keep it - m>1, "ize" removed	**ends with "er"** **after it constant** **then delete "r"**
ends with "ches" or with "shes"… remove "es" only	ends with "ive" - m=1, keep it - m>1, "ive" removed	**If end "es",** **Remove "s", keep** **"e"**
ends with "is", don't delete	ends by "iral" , m=2, start with vowel, keep it	**If end "en", Keep** **"e"**
ends with "ying"→i & "yed"→y	ends "al", m=2, delete "al" and add "e"	**If the word end by** **"y", Replace it** **with "I"**
m=2, consonant, vowel, consonant, vowel, then remove "al"	ends –knives, -knives→ -knife	**ends "ed" or "ing",** **keeping "e" while** **removing "ed" or** **"ing"**
ends by "ative" and m=2, ative→"ate"	ends "ic" ,m=2, delete "ic"	ends –staves, - staves→ -staff
Ends with "ness", m=1, Consonant, vowel & consonant, "ness"→ "ness"	ends "icate", delete "ate"	ends –xes, -xis→ -x
m=2, ends with "ness", ness→	m=1, ends "ical" , "ical"→ "ic"	ends –trixes, - trixes→-trix
ends "ousness", m=1, Consonant, vowel & consonant, ousess→ ous	m> 0, Ends with "ator", Remove "ator" and replace it with "ate"	ends –ei, -ei→ -eus
m> 0, Ends with "less", "less"→ "less"	ends with "ceed", m>0, remove "ceed" and replace it with "cess"	ends –pi, -pi→ -pus

Table 1. (*continued*)

m> 0, Ends with "lessly", "lessly"→"less"	ends –wives, -wives→ -wife	ends –ses, -sis→-s
m> 0, Ends with "fully", delete "ly"	ends –feet, -feet→ -foot	ends with "ence", m=2, delete "ence"
ends with "ous", m>1,Delete "ous"	ending in –men, -men→ -man	ends with "ment", m=2, Keep it
Ends with "ous", m=2, Keep "ous"	ends –ci, ci→ -cus	ends with "ment", m>2, remove "ment"
ends with "eer", m=2 Then remover "er"	end with "eed" -m=0, then Keep "eed" -m>0, remove "d"	ends with "tion", m=2, replaced with "e"
ends with "ible", m=2, Starts with a consonant, not ending with series of consonant vowel consonant vowel Then keep it as it is	m> 0, Ends with "ator", Remove "ator" and replace it with "ate"	ends with "ional" then delete "al"
ends with "nate" or "ate" -m=2, keep it as it is m>2, "ate" or "nate" removed m=1, then "at" is kept - m=0, then left it as it is	m=2, ends "able", delete "able" m>2, remove "ible"	ends with "ance", m=2, Consists of series consonant, vowel, consonant, vowel , Replaced with "e", Else, removed "ance"

5   Performance Studies

In order to study the performance of our technique, we used different evaluation measures. These measures are discussed in section 5.1. Then, the data sets used and the experimental results are shown in section 5.2.

5.1   Evaluation of the EPSA Algorithm

To evaluate our algorithm, the Paice's Evaluation Methods [18] are used. The standard evaluation method of a stemming algorithm is to apply it to an IR test collection and calculate the recall and precision to assess how the algorithm affects those measurements. However, this method does not show the specific causes of errors therefore it does not help the designers to optimize their algorithms. Paice in [19, 20] proposed an evaluation method in which stemming is assessed against predefined groups of semantically related words. He introduced three new measurements: over and under-stemming index and the stemming weight. This test requires a sample of different words partitioned into concept groups containing forms that are morphologically and semantically related to one another. The perfect stemmer should have all words in a group to the same stem and that stem should not occur in any other group.

For each concept group two totals are computed:

- Desired merge total (DMT), which is the number of different possible word form pairs in the particular group, and is given by the formula:

$$DMT_g = 0.5 \ n_g \ (n_g - 1) \tag{1}$$

Where n_g is the number of words in that group.

- Desired Non-merge Total (DNT), which counts the possible word pairs formed by a member and a non-member word and is given by the formula:

$$DNTg = 0.5 \ ng \ (W - ng) \tag{2}$$

Where W is the total number of words. By summing the DMT for all groups we obtain the GDMT (global desired merge total) and, similarly, by summing the DNT for all groups we obtain the GDNT (global desired non-merge total). After applying the stemmer to the sample, some groups still contain two or more distinct stems, incurring in understemming errors. The Unachieved Merge Total (UMT) counts the number of understemming errors for each group and is given by:

$$UMT_g = 0.5 \ \sum_{i=1..s} u_i \ (n_g - u_i) \tag{3}$$

Where s is the number of distinct stems and ui is the number of instances of each stem. By summing the UMT for each group we obtain the Global Unachieved Merge Total (GUMT). The understemming index (UI) is given by: GUMT/GDMT.

After stemming, there can be cases where the same stem occurs in two or more different groups, which means there are over-stemming errors. By partitioning the sample into groups that share the same stem we can calculate the Wrongly Merged Total (WMT), which counts the number of over-stemming errors for each group. The formula is given by:

$$WMT_g = 0.5 \ \sum_{i=1..t} v_i \ (n_s - v_i) \tag{4}$$

Where t is the number of the original groups that share the same stem, ns is the number of instances of that stem and v_i is the number of stems for each group t. By summing the WMT for each group we obtain the Global Wrongly-Merged Total (GWMT). The over-stemming index (OI) is given by: GWMT/GDNT. The Stemming Weight (SW) is given by the ratio OI/UI. SW gives some indication whether a stemmer is weak (low value) or strong (high value).

Also we evaluate our algorithm by measuring the performance of the IR system and compare its result with the result when using the exact match (without stemming) and when using the porter stemmer. The performance measured by the recall and precision measurements, which are represented in the following formulas:

$$Precision = \frac{|\{relevant \ documents\} \cap \{retrieved \ documents\}|}{|\{retrieved \ documents\}|} \tag{5}$$

$$Recall = \frac{|\{relevant\ documents\} \cap \{retrieved\ documents\}|}{|\{relevant\ documents\}|} \quad (6)$$

$$AveP = \frac{\sum_{i=1}^{N_0} P_i(r)}{N_q} \quad (7)$$

Where: AveP: Average precision at recall level r. $P_i(r)$: the precision at recall level r for the ith query. N_q: the number of queries used.

5.2   Experimental Results

EPSA Algorithm was applied to 23,531 words (Porter's collection) [21] and compared to the Original porter algorithm. This Dataset used to measure the Over-stemming, Under-stemming and stemming weight.   In addition, it was applied to a famous collection called CISI; the collection is available in [22]. This collection contains 1460 documents with different sizes, and tested the system with 35 queries in order to evaluate the IR system performance.

Firstly, we applied the Porter Algorithm and ESPA algorithm to the porter's Collection Dataset to measure how many words are stemmed correctly from both algorithms. This measure calculated by counting the correct stem words coming from each stemmer then finding the percentage. Figure 2 shows that ESPA returns a better indication of correct stem of words, it reaches nearly over 80%, on the other hand, porter reaches only 58.60%.

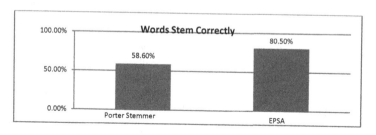

Fig. 2. Percentage of the Words Stem Correctly

Fig. 2 represents the word reduction, which is the process to reduce the words to the collection of words. For example, if the collection contains the words: "fishing", "fished", and "fisher", they are reduced to one word (the root) word, "fish". Although Porter's Stemmer holds a good reduction (31%), EPSA reached a much better percentage (43.40%). The word reduction is calculated by the next equation

$$\frac{\sum unique\ stem\ words}{\sum whole\ number\ of\ words\ in\ thecollection} * 100\% \quad (8)$$

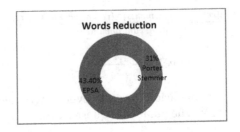

Fig. 3. Words Reduction

Paice's Evaluation Methods were applied to the same collection (Porter's Collection). Table 2 shows the results obtained from the porter and the ESPA algorithms against the Over-stemming, Under-stemming and the stemming weight measures. These measures are calculated by using the equations 1, 2, 3, and 4 that are mentioned in the evaluation section.

The results show that our algorithm succeeds to reach much less over-stemming and under-stemming values than the original porter algorithm. The results also reflect that our stemmer is stronger than the Porter stemmer, this appears clearly on the stemming weight, which is larger than the Porter's one.

Table 2. Paice's Evaluation Methods

	Under-stemming (UI)	Over-Stemming (OI)	Stemming Weight (SW)
Porter Stemmer	0.31	$0.51 * 10^{-5}$	$0.1645 * 10^{-5}$
ESPA	0.183	$0.40 * 10^{-5}$	$2.186 * 10^{-5}$

Finally, EPSA was tested on the CISI collection using IR evaluation to measures, which is mentioned in the evaluation section. Figure 3 shows the precision and recall results for each query without stemming (exact match) and with stemming (using the porter and EPSA algorithms). The average precision for the ESPA algorithm is 75%, which is the highest value in comparison to the one without stemming (nearly 70%) and the porter algorithm (approximately 72%). These results are shown in Table 3. From these results, the EPSA improves the precision over the porter algorithm by about 2.3% while realizing approximately the same recall percentage.

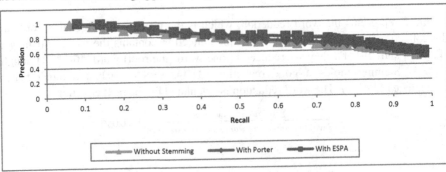

Fig. 4. Precision & Recall against the CISI 35 queries

Table 3. Average Precision and Average Recall

	Precision	Recall
Without Stemming	70.9%	63.6%
With Porter	72.9%	65.7%
With ESPA	75.2%	66%

6   Conclusion and Future Works

Many researches were introduced to improve the structure of the porter algorithm; however it still has several drawbacks. Thus, in this paper we proposed an efficient information retrieval technique as well as a new proposed stemming algorithm called Enhanced Porter's Stemming Algorithm (EPSA). The objective of this new technique is to overcome the drawbacks of the porter algorithm and improve the web searching.

The algorithm was applied to two datasets for testing. The results show that the ESPA algorithm gives fewer errors than the original porter algorithm for both over-stemming and under-stemming measures. ESPA algorithm also holds a good Stemming weight that's mean its stronger in stemming than the porter algorithm. In addition, EPSA improves the performance of the Information Retrieval systems respecting the recall and precision measures. The EPSA improves the precision over the porter algorithm by about 2.3% while realizing approximately the same recall percentage.

Rough set theory is a way of representing and reasoning imprecision and uncertain information in data. Therefore, in our future works, rough sets-based feature extraction, rule generation and classification will provide more challenging and may allow us to refine our learning algorithms and/or approaches to the Web-based support systems and research on information retrieval support systems [23].

References

1. Singhal, A.: Modern Information Retrieval: A Brief Overview. IEEE Data Engineering Bulletin 24(4), 35–43 (2011)
2. Baeza-Yates, R., Ribeiro-Neto, B.: Modern Information Retrieval. ACM Press, New York (1999)
3. Yamout, F., Demachkieh, R., Hamdan, G., Sabra, R.: Further Enhancement to the Porter's Stemming Algorithm. In: Machine Learning and Interaction for Text based Information Retrieval, Germany, pp. 7– 23 (2004)
4. Maurya, V., Pandey, P., Maurya, L.S.: Effective Information Retrieval System. International Journal of Emerging Technology and Advanced Engineering 3(4), 787–792 (2013)
5. Sembok, T., Abu Ata, B., Bakar, Z.: A Rule and Template Based Stemming Algorithm for Arabic Language. International Journal of Mathemtical models and Methods in Applied Sciences 5(5), 974–981 (2011)
6. Lovins, J.: Development of a stemming algorithm. Mechanical Translation and Computational Linguistics 11, 22–31 (1968)

7. Bijal, D., Sanket, S.: Overview of Stemming Algorithms for Indian and Non-Indian Languages. International Journal of Computer Science and Information Technologies (IJCSIT) 5(2), 1144–1146 (2014)
8. Jivani, A.: A Comparative Study of Stemming Algorithms. Int. J. Comp. Tech. Appl. 2(6), 1930–1938 (2011)
9. Paice, C.: Another stemmer. ACM SIGIR Forum 24(3), 56–61 (1990)
10. Sharma, D.: Stemming Algorithms: A Comparative Study and their Analysis. International Journal of Applied Information Systems 4(3), 7–12 (2012)
11. Smirnov, I.: Overview of Stemming Algorithms, http://the-smirnovs.org/info/stemming.pdf
12. Dawson, J.: Suffix removal and word conflation. ALLC Bulletin 2(3), 33–46 (1974)
13. Willett, P.: The Porter stemming algorithm: then and now. Program: Electronic Library and Information Systems 40(3), 219–223 (2006)
14. Porter, M.F.: An algorithm for suffix stripping. Program 14(3), 130–137 (1980)
15. Srinivasan, S., Thambidurai, P.: STANS Algorithm for Root Word Stemming. Information Technology Journal 5(4), 685–688 (2006)
16. Megala, S., Kavitha, A., Marimuthu, A.: Improvised Stemming Algorithm – TWIG. International Journal of Advanced Research in Computer Science and Software Engineering 3(7), 168–171 (2013)
17. Karaa, W.: A New Stemmer To Improve Information Retrieval. International Journal of Network Security & Its Applications (IJNSA) 5(4), 143–154 (2013)
18. Moral, C., Antonio, A., Imbert, R., Rmirez, J.: A survey of stemming algorithms in information retrieval. Information Research 19(1) (2014)
19. Paice, C.D.: An evaluation method for stemming algorithms. In: Proceedings of the 17th Annual International ACM SIGIR Conference on Research and Development in Information Retrieval, pp. 42–50. ACM, Dublin (1994)
20. Karaa, W.B.A., Gribâa, N.: Information Retrieval with Porter Stemmer: A New Version for English. In: Nagamalai, D., Kumar, A., Annamalai, A. (eds.) CCSEIT-2013. AISC, vol. 225, pp. 243–254. Springer, Heidelberg (2013)
21. The Porter Stemming Algorithm, http://tartarus.org/~martin/PorterStemmer/index.html
22. Common IR Test Collection, http://web.eecs.utk.edu/research/lsi/corpa.html
23. Hassanien, A.E., Suraj, Z., Slezak, D., Lingras, P.: Rough computing: Theories, technologies and applications. IGI Publishing Hershey, PA (2008)

EHR Data Preparation for Case Based Reasoning Construction

Shaker El-Sappagh[1,2], Mohammed Elmogy[1], A. M. Riad[1], Hosam Zaghlol[3], and Farid A. Badria[4]

[1] Faculty of Computers & Information, Mansoura University, Egypt
[2] Faculty of Computers & Information, Minia University, Egypt
[3] Faculty of Medicine, Mansoura University, Mansoura, Egypt
[4] Pharmacognosy Department, Faculty of Pharmacy, Mansoura University, Mansoura, Egypt

Abstract. Case Based Reasoning (CBR) is the first choice in experience-based problems as diagnosis. However, building a case base for CBR is a challenging. Electronic Health Record (EHR) data can provide a starting point for building case base, but it needs a set of preprocessing steps. In this paper, we propose a case-base preparation framework for CBR systems. This framework consists of three main phases including data preparation, fuzzification, and coding. This paper will focus only on the data-preprocessing phase to prepare the EHR database as a knowledge source for CBR cases. It will use many machine-learning algorithms for feature selection and weighing, normalization, and others. As a case study, we will apply these algorithms on diabetes diagnosis data set. To check the effect of data preparation steps, a CBR prototype will being designed for diabetes diagnosis and prediction of its complications as kidney failure. The results show an enhancement to the case retrieval process of the implemented CBR system.

Keywords : case based reasoning, data preprocessing, diabetes diagnosis, clinical decision support system, electronic health record, and case-base knowledge.

1 Introduction

Diabetes Mellitus (DM) is a serious disease. If it has not treated on time and properly, it can lead to serious complications including death. This makes diabetes one of the main priorities in medical science research, which in turn generates huge amounts of data. These data are transactional and distributed in the patient's EHR. Early diabetes diagnosis is the most critical step in diabetes management. The diagnosis of diabetes is an ill-formed problem and depends on the physician experience. Case Based Reasoning (CBR) is considered as the most suitable Clinical Decision Support System (CDSS) for dealing with these problems where physicians share their experience [1, 2]. Therefore, case-base creation is a challenging step. On the other hand, CBR is appealing in medical domains because a case-base already exists as storing symptoms, medical history, physical examinations, lab tests, diagnoses, treatments, and outcomes for each patient [3]. However, because clinical data are usually incomplete,

A.E. Hassanien et al. (Eds.): AMLTA 2014, CCIS 488, pp. 483–497, 2014.
© Springer International Publishing Switzerland 2014

inconsistent, and noisy, these data need a set of preparation steps before converted into CDSS knowledge [4]. The first step is the data preprocessing stage that is applied to enhance data quality. The second step is the data fuzzification stage that is used to handle vague knowledge. Finally, the third step is the coding stage that is used to represent the processed data with standard coding systems such as SNOMED CT [5].

Authors in [6, 7] stated that the introduction of health information technology like EHRs has not led to improvements in the quality of the data being recorded, but rather to the recording of a greater quantity of bad data. As a result, Lei [8] has proposed what he called the first law of informatics: "data shall be used only for the purpose for which they were collected." In the same time, EHR contains all the current and history of medical data of the patient. These data can be used as a complete source for building the CBR's case-base [4]. The quality of CBR is based on the quality of case-base content [3]. EHR data quality measurement and improvement must be an essential step before using its data in CDSS's knowledge base [4]. As a result, data preprocessing steps are the first and the foremost to improve the accuracy of CBR systems [9]. By focusing on DM diagnosis, its medical dataset is seldom complete [10]. Moreover, because diabetes is a lifelong disease, even data available for an individual patient may be massive and difficult to interpret. Data preprocessing steps include deleting of low quality rows and columns, feature selection, feature mining, integration, transformation (i.e. normalization and discretization), data cleaning, feature weighting, etc. [11]. An example of a system focusing on feature mining is the dietary counseling system by Wu et al. [12]. Jagannathan and Petrovic [13] have concluded that missing values in the case-base pose a common and serious problem that impairs the performance of the system, and they have provided an imputation method to deal with missing value. However, they have only handled missing values of some attributes, and they have left others. Floyd et al. [14] have concluded that applying preprocessing techniques to a case-base as feature selection can increase the performance of a CBR system.

Case retrieval is the most important phase in CBR system. However, it is mainly depend on the types of case-base knowledge and its quality. All existing case retrieval algorithms depend on one type of knowledge for retrieval [43]. We will combine the most important three techniques for data preparation to enhance the case retrieval process especially for medical domain. In this paper, our proposed framework will result in cleaned, normalized, fuzzified, and encoded knowledge. These types of knowledge will support different queries and different types of similarity algorithms, which improve the retrieval phase of CBR system. Diabetes diagnosis is used as a case study for applying this framework. There are not CBR systems that utilize machine-learning algorithms to prepare case-base knowledge. Moreover, most of CBR systems for diabetes diagnosis have used a diabetes specific data set. However, diabetes as a chronic disease results in many other diseases as nephropathy, retinopathy, neuropathy, heart diseases, stroke, and others [42]. In our data set, patient is described by 70 different features, as shown in Table 1. These features link diabetes with other diseases, such as cancer, kidney disease, and liver diseases. A CBR-based CDSS for diabetes diagnosis will be designed using myCBR 3 protégé plug-in [36]. The diagnosis will be the patient's diabetes status plus his future conditions of having other diseases, such as cancers. The paper is organized as follows. Section 2 provides related works, section 3 provides a description of our diabetes data set, section 4 provides the proposed preparation framework, section 5 provides a case study for CBR system, and finally section 6 provides the conclusion and future works.

2 Related Work

Data quality of EHR should be measured before using it as a knowledge source. Weiskopf and Weng [15] have determined five dimensions to measure the quality of EHR data. Klompas et al. [16] has asserted that EHR data can improve diabetes management even if its raw data quality is low. For achieving data quality, a preprocessing or data preparation step is critical for any knowledge based CDSS system [17]. For example, the case retrieval algorithms, such as nearest neighbor, require data cleaning and normalization steps [13, 18]. Low quality data need handling in CBR. For example, Xie et al. [19] have handled missing values and unmatched features in case retrieval algorithm. Guessouma et al. [20] have proposed five approaches for managing missing data problem in a medical CBR system. The data preprocessing steps include data cleaning [13], data transformation [10], feature mining and selection [21], etc. Especially for CBR, the conversion of database structure to case-base structure is a critical step [4]. Database record is similar to case-base case, but transformation from generic EHR to specialized case-base is not a straightforward mapping of attributes. In order to change one of simple database records into a case, it is required to associate an experience to the record [22]. Abidi et al. [4] have assumed that the structure of case-base is defined in advance. They have mapped the structure then contents of EHR to case-base. However, this assumption is not realistic. The structure of the case-base depends on the structure and contents of EHR, and it must be inferred from it. As EHR contains patient raw chronicle data, a temporal abstraction preparation step is critical to aggregate and provide trends from patient data [23]. Moreover, features weights must be specified for case retrieval algorithms [1]. It can be determined manually by domain expert and CPGs [20] or automatically using machine learning algorithms [24], e.g. neural network [4]. The choice of case features that best distinguish classes of instances has a large impact on the similarity measure [22], and it improves the performance and decrease the complexity [3]. This process can be done automatically [25] (using techniques as information gain [26] and Relief [27]) or manually [28, 29] according to domain expert or CPGs. Kotsiantis [30] has surveyed data preprocessing algorithms for each step. Han et al. [31] have proposed the preprocessing steps for a DM data set using RapidMiner [37]. There is no single sequence of data pre-processing algorithms with the best performance [30]. As a result, data preprocessing is a set of not ordered steps [17]. It can be an inherent component of a CBR system, or it can be performed as a preprocessing step. For example, eXiT*CBR [32] system incorporate basic preprocessing steps as discretization, normalization and feature selection techniques. However, the complete list of needed preprocessing steps is different according to the nature of data and the CBR system purpose [3]. As a result, this paper will provide the preprocessing step as a separate phase before CBR system processes.

3   Data Set Description

The paper uses a data set from diagnostic biochemical lab, AutoLab of Mansoura institution, Mansoura University, Mansoura, Egypt. This data was collected in the period from January 2010 through August 2013. The control subjects were healthy and recruited from the diagnostic biochemical lab and were matched by age, sex and ethnicity to the case subjects. The eligibility criteria for controls were the same as those for patients, except for having a cancer diagnosis. A short structured questionnaire was used to screen for potential controls based on the eligibility criteria. Analysis of the answers received on the short questionnaire indicated that 80% of those questioned agreed to participate in clinical research. A total of 67 eligible subjects were ascertained in the current study. However, 7 control subjects were excluded due to limited blood samples for testing AFP. Blood samples (5 mL) were taken, centrifuged, and the serum separated and stored at 220uC until analyzed. Serum samples were assayed for AFP by enzyme-linked immunosorbent assay with commercial kits (Abbott, North Chicago, IL), transferase (ALT) and aspartate aminotransferase (AST), with an auto-analyzer (Hitachi Model 736, Japan) and commercial kits.

The problem features include: Demographics (Residence, Occupation, Gender, Age, BMI); Lab tests (HbA1C, 2h PG, FPG); Hematological Profile (e.g. Prothrombin INR, Red cell count, Hemoglobin, Haematocrit (PCV), MCV, MCH, MCHC, Platelet count, White cell count, Basophils, Lymphocytes, Monocytes, Eosinophils); Symptoms (Urination frequency, Vision, Thirst, Hunger, Fatigue); Kidney Function Lab tests (Serum Potassium, Serum Urea, Serum Uric acid, Serum Creatinine, Serum Sodium); Lipid Profile (LDL cholesterol, Total cholesterol, Triglycerides, HDL cholesterol); Tumor Markers (FERRITIN, AFP Serum, CA-125); Urine Analysis (Chemical Examination (Protein, Blood, Bilirubin, Glucose, Ketones, Urolibingen), Microscopic Examination (Pus, RBcs, Crystals)); Liver Function Tests (S_Albumin, Total Protein, Total Bilirubin, Direct Bilirubin, SGOT (AST), SGPT (ALT), Alk_Phosphatase, γ GT); Females History (Amenorrhea, Birth, Dysmenorrhea). Solution features include: Diabetes Diagnosis; Nephropathy check; Hypercholestremia check; Tumor Markers; Liver problem; Radiological Diagnosis (Glomerulonephrinitis, Shrunken kidney, HCC, HCV, Ovarian cancer, Fatty Liver, Splenomegaly). Table 1 is a sample of the tested features, and we have selected a unified Unit of Measurement (UoM) for each lab test. All features' values are converted to the selected UoM.

Table 1. Patient features (Data type:N=Numerical, C=Categorical, and O=Ordinal}.

Feature type	Feature name	Data type	Normal Range	UoM	Min-Mean-Max
Demographics	BMI	N	18.5 - 25	kg/m²	20-33.117-45
Diabetes Lab Tests	HbA1C	N	<=5	mmol/L	5-6.373-7.4
Hematological Profile	Hemoglobin	N	12 - 16	g/dL	9.8-12.332-13.4
	White cell count	N	4 - 11	10^3/cmm	6-8.055-9.2
Symptoms	Urination frequency	O	-	-	-
Kidney Function Lab tests	Serum Uric acid	N	3.0 - 7.0	mg/dL	3-4.237-7.9
	Serum Creatinine	N	0.7 - 1.4	mg/dL	0.9-1.35-3.6
Lipid Profile	LDL cholesterol	N	0 - 130	mg/dL	50-94.917-170
Tumor Markers	FERRITIN	N	28 - 397	ng/mL	-
Liver Fun. Tests	S. Albumin	N	3.5 - 5.0	g/dL	1.9-4.082-5.4
Females History	Amenorrhea	O	-	-	-
Diagnosis	Diabetes Diagnosis	C	-	-	-
	Nephropathy check	C	-	-	-
	Hypercholestremia check	C	-	-	-
	Tumor Markers	C	-	-	-
	Liver problem	C	-	-	-

Table 2 show a sample of 11 case with all features describing diabetic patients. These raw data will be prepared as a case base knowledge for diabetes diagnosis CDSS based on CBR technique. Our case base will contain 60 cases.

Table 2. A sample of 11 cases from our raw data set before preprocessing

Case	Case 1	Case 2	Case 3	Case 4	Case 5	Case 6	Case 7	Case 8	Case 9	Case 10	Case 11
BMI	20	23	31	32	40	45	24	31	29	28	29
Gender	M	F	F	M	M	F	F	M	F	F	F
Age	39	47	32	35	53	43	41	32	35	53	60
Vision	blurred	Allergy /redness	Non	Non	Allergy /retinopathy	retinopathy	non	Aller-gy/redness	Non	Non	Allergy/ retinopathy
Fatigue	Non	Non	+	+	+	+	+	+	++	+	+
Hunger	Non	Non	+	+	++	+	+	Non	+	+	+
Thirst	Non	+	+	+	++	++	+	Non	+	+	Non
Urination Freq.	Non	+	Non	Non	++	+++	Non	Non	++	++	Non
Residence	urban	Rural	Rural	urban	urban	urban	urban	urban	urban	urban	Rural
Occupation	Teacher	Merchant	Farmer	Engineer	worker	Pharmacist	Doctor	Non	Worker	Merchant	Teacher
2hPG	180	190	200	210	220	230	165	225	175	185	230
FPG	120	125	130	135	140	150	118	148	117	121	150
HbA1C	6.4	6.5	6.6	6.7	6.9	7.1	6.1	6.8	6.2	6.3	7.1
Prothrombin (INR)	1.3	1.1	1	1.1	1.4	1	1.4	1.1	1	1.3	1.4
Basophils	0	1.1	0	1.2	0.5	1	1.5	1.2	0.5	1	1
Eosinophils	1	2	3.4	1.8	1.4	2.4	1.2	1.8	1.4	2.4	2.4
Monocytes	2	3.8	2.5	3.3	2	4	2.2	3.3	2	4	4
Lymphocytes	25	29	24.7	26.8	23.2	27	21.2	26.8	23.2	27	27
White cell count	7.9	9.2	8.8	8.4	7	8.9	6	8.4	7	8.9	8.9
Platelet count	232	2000	195.4	210	198	210	192	210	198	210	210
MCHC	32.8	37.8	41.7	38.7	30.8	32.8	30	41.7	38.7	37.8	30.8
MCH	23.4	28.4	29.4	28.4	21.4	23.4	21.3	29.4	28.4	28.4	21.4
MCV	71.3	74.5	76.4	74.5	70	71.3	70	76.4	74.5	74.5	70
Haematocrit	34.8	36.8	34.5	36.8	34.3	36.8	31.1	34.5	36.8	36.8	34.3
Hhg	11.4	12.8	13.4	12.8	10.4	12.5	9.8	13.4	12.8	12.8	10.4
Red cell count	4.88	5.4	5.88	5.4	4.1	5.2	3.8	5.88	5.4	5.4	4.1
Serum Potassium	3.8	2.4	4.1	4.3	2.9	4	3.9	3.5	4.3	4.1	3.9
Serum Sodium	135	149	135	134	152	134	135	135	134	135	135
Serum Creatinin	2.6	1.9	1	0.9	2.4	1.1	1	1	0.9	1	1
Serum Uric acid	7.8	5.2	3.4	4.9	7.9	3	3.4	3.4	4.9	3.1	3.4
Serum Urea	56	47	18	18	52	28	17	17	28	19	17
LDL cholesterol	90	85	55	67	165	81	79	89	165	167	55
HDL cholesterol	51	55	60	61	35	65	65	65	38	38	65
Triglycerides	158	141	152	150	180	142	134	101	181	167	111
Total cholesterol	200	200	200	200	240	200	200	200	243	255	200
FERRITIN	Normal	Normal	Normal	Normal	Normal	Normal	Normal	Normal	Normal	Normal	Normal
AFP Serum	Normal	Normal	Normal	Normal	Normal	Normal	Normal	Abnormal	Normal	Normal	Normal
CA-125	Normal	Normal	Normal	Normal	Normal	Abnormal	Normal	Normal	Normal	Normal	Abnormal
Crystals	+	Nil	Nil	Nil	++	Nil	Nil	Nil	Nil	Nil	Nil
RBcs	+	Nil	Nil	Nil	++	Nil	Nil	Nil	Nil	Nil	+
M.Pus	++	Nil	Nil	Nil	+++	Nil	Nil	Nil	Nil	Nil	+
Urolibingen	+	Nil	Nil	Nil	+++	+++	Nil	Nil	Nil	Nil	Nil
Ketones	++	+	Nil	Nil	++	+++	Nil	Nil	Nil	Nil	Nil
Glucose	+++	++	Nil	Nil	+++	+++	Nil	Nil	Nil	Nil	++
Bilirubin			Nil	Nil	++	Nil	Nil	Nil	Nil	Nil	+++
Blood	++	+	Nil	Nil	++	Nil	Nil	Nil	Nil	Nil	Nil
Protein	++	++	Nil	Nil	+++	Nil	Nil	Nil	Nil	Nil	+
Albumin	4.5	4.1	5	5.4	4.5	4.4	2.4	1.9	4.1	5	2.8
Total protein	4.7	3.2	6.4	3.8	3.1	4.1	8.1	8.2	3.2	6.4	7.5
γ GT	22	20	18	27	25	28	64	98	20	18	68
Alk. phosphatase	250	189	300	210	170	184	350	289	189	300	210
SGPT (ALT)	35	45	42	40	35	42	110	82	45	42	123
SGOT(AST)	40	41	39	35	40	40	98	78	39	35	145
Direct bilirubin	0.3	0.4	0.3	0.3	0.5	0.5	1.4	1.3	0.3	0.4	1.6
Total bilirubin	1	1.1	1	1.2	0.9	1	2.8	3	1.1	1	2.2
Birth	Normal	Normal	Normal	Normal	Normal	twins	Normal	Overweight baby	Normal	Normal	Normal
dysmenorrhea	Normal	Normal	++	Normal	Normal	Normal	Normal	Normal	Normal	Normal	Normal
Amenorrhea	Normal	Normal	+	Normal	Normal	Normal	Normal	Normal	Normal	Normal	Normal
Diabetes Diagnosis	Prediabetic	Diabetic '	Diabetic	Diabetic 'gestational'	Diabetic	Diabetic	Prediabetic	Diabetic 'gestational'	Prediabetic	Prediabetic	Diabetic
Nephropathy Diagnosis	Nephropathy	Nephropathy	Normal	Normal	Nephropathy	Normal	Normal	Normal	Normal	Normal	Normal
Hypercholestremia Diagnosis	Normal	Normal	Normal	Normal	Hypercholestremia	Normal	Normal	Normal	Hypercholestremia	Hypercholestremia	Normal
Cancer Type	Normal	Normal	Normal	Normal	Normal	Ovarian cancer	Normal	Hepatocellular carcinoma	Normal	Normal	Hepatocellular carcinoma
Liver Diagnosis	Normal	Normal	Normal	Normal	Normal	Normal	HCV	HCC	Normal	Normal	HCV
Glomerulonephrinitis	No	Yes	No	No	Yes	No	No	No	No	No	No
shrunken kidney	No	Yes	No	No	Yes	No	No	No	No	No	No
HCC	No	No	No	No	No	No	No	Yes	No	No	No
HCV	No	No	No	No	No	No	Yes	No	No	No	Yes
Ovarian cancer	No	No	No	No	No	Yes	No	No	No	No	No
Fatty Liver	No	No	No	No	Yes	No	No	No	No	No	No
Splenomegaly	No	No	No	No	No	No	Yes	No	No	No	No

4   Case-Base Preparation Framework

In this section, we propose a case-base preparation framework for extracting a diabetes diagnosis case-base from EHR databases, as shown in Fig. 1. This framework discusses the conversion of both structure and content of EHR database to a derived case-base structure and content. There are three sequential phases for creating a case-base from EHR including data preprocessing, data fuzzification, and data encoding. We have proposed, in other work, a standard data model based on HL7 RIM [38] to standardize the structure of case base. This data model will be utilized to prepare the case-base structure. Moreover, we have created an OWL 2 ontology for diabetes concepts from SNOMED CT to be used for coding and standardizing the case-base contents. Encoding phase is performed to standardize the textual contents of case-base according to standard medical ontology [5]. Because of space restrictions, the data preparation phase will be discussed in this work, and the other two phases will be considered in future works. Because it is the general case, we will assume that EHRs (the case sources) are in relational database format [33], and do not use any coding mechanisms (e.g. the data are in primitive types only like numerical, textual, etc.)

4.1   Data Extraction, Integration, and Anonymization

EHR contains medical, administration, financial, security, and privacy information. By concentrating on medical data, it contains heterogeneous data about patient, related to his lifelong health states. Moreover, that information structure and representation format of EHR varies across different HISs. Interoperability of EHRs is out of scope. We need to collect only data fields related to DM diagnosis from all of patients EHRs into a coherent data store. According to our domain expert and Clinical Practice Guidelines (CPGs), the list of diagnostic data elements are identified including lab tests, symptoms, and others. These data elements are extracted from multiple sources (i.e. EHRs). These sources require object matching and schema integration strategies. Data integration can help reducing and avoiding redundancies and inconsistencies in the resulting data set. We will follow a manual process for schema integration, i.e. attributes names conflict. Object matching requires attribute tuple redundancy detection and prevention. Anonymization is a privacy preservation step. We have removed all patient personal details, and the paper will depend on an artificial identifier for each case.

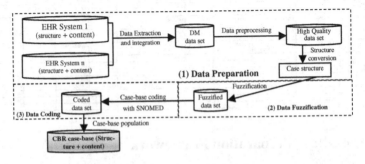

Fig. 1. Case-base preparation phases

4.2   Data Preprocessing Steps

This paper focus on the representation of case-base in attribute-value tables, because it is one of the most common kind of case representation in CBR. Moreover, the attributes types will be nominal and numerical only with no object and concept types. Currently available options in our CBR are: feature weighting, feature selection, outlier detection and removal, handling missing values, categorical feature coding, and others. These steps are performed sequentially on the raw case-base data to produce a new high quality case base. Fig. 2 shows that case-base pre-processing steps are sequential, where P_i is a pre-processing step.

Fig. 2. The case-base pre-processing steps

4.2.1   Handling Missing Values and Categorical Features Coding

The effect of incomplete data on retrieval performance is an important issue because it affects both precision and recall of retrieval algorithm [34]. There are so many methods for handling missing values. In our case, we will remove: (1) all cases with more than or equal to 50% of missing features, (2) all features with more than or equal to 50% of missing values. Missing values will be filled using the k-NN algorithm [41], where the corresponding feature of the most similar case will fill the missed feature. The features data types are numerical, ordinal, and nominal.

Table 3. Categorical and diagnosis features codes

Categorical Feature		Coded Values	Original value				# of cases
Urination Frequency		0	Non (Normal) *i.e. 3-5 times urination per day*				42
		1	+ *i.e. 6-8 times urination per day*				4
		2	++ *i.e. 9-10 times urination per day*				11
		3	+++ *i.e. more than 10 times urination per day*				3
CA-125		0 (i.e. Normal)	Numerical values				57
		1 (i.e. Abnormal)	Numerical values				3
FERRITIN		0 (i.e. Normal)	Numerical values				58
		1 (i.e. Abnormal)	Numerical values				2
AFP Serum		0 (i.e. Normal)	Numerical values				56
		1 (i.e. Abnormal)	Numerical values				4
Urine Analysis		0 (i.e. Nil)	1 (i.e. +)	2 (i.e. ++)	3 (i.e. +++)		
	Protein	50	1	7	2		
	RBcs	49	9	1	1		
	Crystals	55	4	1	0		
Diabetes Diagnosis		0	Normal				8
		1	Prediabetic				19
		2	Prediabetic/Gestational				1
		3	Diabetic				29
		4	Diabetic/Gestational				3
Nephropathy Check		0= Normal	Empty				49
		1= Abnormal	Nephropathy				11
Hypercholestremia Check		0= Normal	Empty				46
		1= Abnormal	Hypercholestremia				14
Cancer Type		0	Empty (i.e. Normal)				52
		1	Ovarian cancer (Preneoplasm)				3
		2	Liver cirrhosis				1
		3	HCC (liver cancer or Hepatocellular carcinoma)				4
Liver Problem		0	Normal				50
		1	Fatty (bright) liver				3
		2	HCV (virus C)				3
		3	HCC (liver cancer)				4

Numerical features (i.e. lab tests) will not be coded. As a special case of numerical features, FERRITIN, AFP Serum, and CA-125 will be coded as Normal and Abnormal. On the other hand, all categorical and ordinal features (e.g. patient symptoms) will be coded. Moreover, the diagnosis features (the case solution) which include diabetes diagnosis, Nephropathy check, Hypercholestremia check, kidney problems, liver problems, and Radiological diagnoses will also be coded. RapidMiner Studio 6.0 has done the coding process, using Discretize Operator. This operator discretizes the selected attributes into user-specified classes using a simple mapping process. Table 3 shows a sample of the coded features.

4.2.2 Feature Selection

A sufficient and appropriate description for the problem is necessary to describe it in a case-base form [35]. The collected features for patients are huge, and they have different levels of importance for determining the diagnosis of diabetic patient. Initial feature vector is collected from domain expert, EHR database schema and diabetes diagnosis CPGs, as discusses in section 3. The large number of features affects the performance of the case retrieval algorithm. If extraneous and misleading features are removed, the similarity measures can retrieve cases that are more useful. As a result, feature selection algorithms must be applied on the collected features to determine the most important ones. There are two types of feature selection methods: Wrapper methods and Filter methods [39]. As the two methods are complementary, we will try algorithms from both approaches. We have applied a set of machine learning algorithms [39], and we have compared their results. For example, in WEKA, we have examined K-NN classifier, Naïve Bayes, C4.5 decision tree, and fuzzy rough feature selection. Moreover, in RapidMiner, we have examined rule induction technique in backward elimination optimization technique, rule induction technique in forward selection optimization technique, k-NN technique in backward elimination optimization technique, and decision tree technique in genetic algorithm optimization. For space restrictions, we will not discuss any details about these algorithms. We can range the most important features repeated in feature selection algorithms. The most important feature is HbA1c, and it will have the highest weight in CBR system. Moreover, it can be noticed that feature selection algorithms have selected overlapped sets of features. As a result, we will combine the collected features from these algorithms to participate in case representation in our CBR system.

4.2.3 Feature Weight Assignment

Feature weighting process is used to improve the performance of the case retrieval algorithm. Weights can be attached to cases, so that cases considered more important for a given application when have higher weights. Weight vectors can also be assigned to description variables: one can either use the same weight vector for all cases, or assign individual weight vectors to each case. As a result, high significant attributes will receive higher weights. This paper calculates, using machine learning algorithms and tools, a single weights vector for features of the entire case-base. Feature weighting will be calculated by using variety of algorithms. In Table 4, we have calculated features weights using the following algorithms [37, 40]: A= Genetic

Algorithm + decision tree, B= Genetic Algorithm + rule induction, C= Particle swarm optimization, D= Information Gain technique, E= Correlation Technique. Because some features have been selected in feature selection, but they have weight = zero in feature weighting algorithms, we will take the maximum of these weights as the feature weight vector in our CBR system. Table 3 shows a sample of features weights that are vales ∈ [0, 1]. All selected features in previous section will have weights.

Table 4. Features weights

	A	B	C	D	E	Maximum
2hPG	0.060	0.001	1.000	0.894	0.038	1
Birth	0.128	0.551	0.396	0.206	0.248	0.551
CA125	0.515	0.011	0.000	0.043	0.298	0.515
Blood	0.124	0.241	0.000	0.110	0.410	0.410
Ketones	0.254	0.129	0.000	0.290	0.736	0.736
D. bilirubin	0.000	0.004	0.000	0.082	0.172	0.172
FERRITIN	0.312	0.204	0.000	0.024	0.305	0.312
Hunger	0.377	0.202	0.929	0.074	0.308	0.929
HbA1C	0.030	0.025	0.370	1.000	0.049	1
Hemoglobin	0.092	0.176	0.619	0.157	0.699	0.699
Platelet count	0.020	0.118	0.000	0.111	0.422	1
Prothrombin INR	0.134	0.057	0.056	0.051	0.689	0.689
Red cell count	0.047	0.089	0.897	0.157	0.745	0.897
Residence	0.063	0.391	1.000	0.031	0.520	1
SGOT_AST	0.150	0.106	0.000	0.070	0.271	0.271
SGPT_ALT	0.061	0.000	0.000	0.059	0.311	0.311
S. Potassium	0.099	0.165	0.047	0.077	0.746	0.746
S. Sodium	0.104	0.144	0.095	0.098	0.363	0.363
S. Uric acid	0.000	0.136	1.000	0.087	0.522	1
Triglycerides	0.014	0.127	0.000	0.101	0.790	0.790
White cell count	0.096	0.020	1.000	0.134	0.221	1

4.2.4 Outlier Detection and Handling

Retrieval performance in CBR is likely to be adversely affected if noise is presented in the case library. While many learning algorithms can be modified to be noise tolerant, it is difficult to make instance-based learning (IBL) algorithms such as k-nearest neighbor (k-NN) algorithm or case-based reasoning (CBR) robust against noise. Moreover, outliers will affect the normalization process in the next section. In our data set, we have checked for outliers using RapidMiner's Detect Outlier (Distances) operator. This operator identifies n outliers in the given dataset based on the distance to their k nearest neighbors. We have found that all outlier cases are tightly affected by extreme values of some specific features, and these values are replaced by our domain experts.

4.2.5 Normalization of Numerical Attributes

Most CBR retrieval algorithms (e.g. k-NN) are depending on distance similarity functions as Euclidean distance. Therefore, all attributes should have the same scale for a fair comparison between them. Normalization of the data is very important when dealing with attributes of different units and scales. Regarding normalization of numerical attributes, a key issue is having all the local similarity measures in the same scale [0, 1] to combine them at the global similarity measure. There are many normalization techniques as Z-score and Min-Max. RapidMiner has normalize operator, which normalize all of the numerical features using Min-Max technique. All features are normalized in specific [C, D] range using Eq.1 where A is the old value, B is the normalized value. The used range in our case is [0.0, 1.0].

$$B = \left(\frac{A - minimum\ value\ of\ A}{maximum\ value\ of\ A - minimum\ value\ of\ A} \right) * (D - C) + C \qquad (1)$$

After the previous preprocessing steps, our normalized data set is now ready for building the diabetes diagnosis case-base. In the following section, we will build a CBR system prototype using myCBR3 protégé plugin [36]. The myCBR is an open-source similarity-based retrieval tool, and it comes in three forms: protégé plugin, workbench, and software development kit (SDK). In this paper, we will use the first form.

5   Case Study

To check the effects of data preprocessing on the quality of CBR results, we will provide a prototype of a CBR system using myCBR framework [36]. The following sub-sections describe the developing phase of CBR.

5.1   Case-Base Formulation

We have a case-base containing 60 cases in the form: *CaseBase= {case$_1$, case$_2$ ... case$_{60}$}*, where *case$_i$=case (F$_i$, S$_i$)*, present the ith case of database. Each case has ID. $F_i = (f_{i1}, f_{i2}...f_{in})$ presents the problem description features of case i, and f_{in} present the nth feature of case C_i. These are 58 features includes patient's *lab tests* as HbA1c and *symptoms* as Thirst. $S_i = (s_{i1}, s_{i2}...s_{in})$ presents the solution sets of case C_i, whereas s_{in} presents the nth solution of case C_i. Each solution has 12 features and determines the diabetes diagnosis of a patient plus his probability of having a kidney problem (e.g. Glomerulonephrinitis, shrunken kidney, etc.), liver problem (e.g. Fatty liver, HCV, HCC, etc.), Hypercholestremia, and Nephropathy. The *myCBR* supports case representation by CSV data file in the form of attribute-value using CSV data import module. Fig. 3 shows the myCBR screen after importing the data set into a new class Patient that will be used as query and case values for retrieval step.

5.2   Modeling Similarity Measures

The k-NN algorithm is used for case retrieval. It follows the local-global approach that divides the similarity definition into a local similarity measures for each attribute, weight for each attribute, and a global similarity measure for calculating the similarity of cases. Different types of features have different similarity comparison methods. For attributes of the data type float, integer, and ordinal, we used distance functions. We used similarity tables for symbolic value ranges. In our case, if each case has n attributes then the similarity between query q and case c is calculated using Eqs. 2, 3, 4, as shown in Fig. 4.

$$Dist(q,c) = \sqrt{\sum_{i=1}^{n} w_i \times D_i(q_i, c_i)} \qquad (2)$$

$$D_i(q_i, c_i) = \begin{cases} (q_i - c_i)^2, & \text{if } q_i \text{ and } c_i \text{ are numerical or ordinal} \\ 1, & \text{if } q_i \text{ and } c_i \text{ are nominal and } q_i \neq c_i \\ 0, & \text{if } q_i \text{ and } c_i \text{ are nominal and } q_i = c_i \end{cases} \qquad (3)$$

$$Sim(q,c) = f\big(Dist(q,c)\big), \quad \text{e.g. } Sim(q,c) = \frac{1}{1 + \alpha\, Dist(q,c)} \qquad (4)$$

Where D_i and w_i denote the local similarity measure and the weight of attribute i, respectively. Sim is the similarity between q and c. $Dist$ represents the weighted Euclidian distance between two cases which is special case of Minkowski Distance (MD). The choice of the parameter p depends on the importance we give to the differences.

$$MD(q,c) = \left(\sum_{i=1}^{n} |q_i - c_i|^p \right)^{\frac{1}{p}} \qquad (5)$$

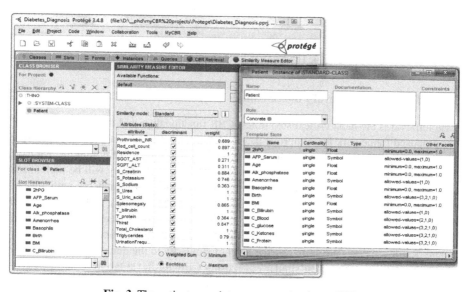

Fig. 3. The patient case data representation in myCBR

The weight values of the case Id and Class features are zero. The myCBR support the modification of the default similarity functions for nominal and numerical feature. Moreover, it supports the definition of similarity between UNKNOWN and UNDEFINED values, and it allows the creation of other special values.

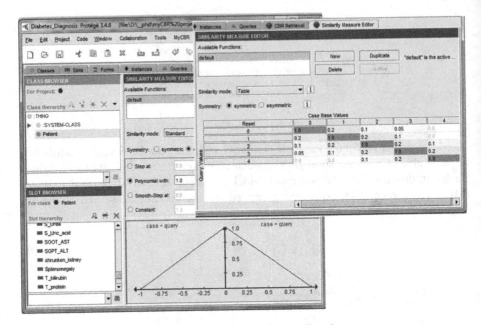

Fig. 4. The features local similarity functions

5.3   Testing of Retrieval Functionality

Case retrieval is the most critical step to test CBR functionality, and the whole purpose of our proposed framework is to improve it. The definition of an optimal similarity measure is often a difficult and tricky task. It requires repeatedly testing and fine-tuning. We have tested the 60 cases that are exist in the case base. The physician query will be normalized and coded before entered to the CBR system. We have tested our framework with a list of new cases. Cases descriptions will be entered to the system as queries, and the output will be the patient diabetes diagnosis including normal, prediabetic, diabetic, etc. Moreover, the systems will provide the patient probability to produce other complications as kidney failure. Retrieved cases are sorted by the level of similarity to the query. Fig. 5 shows one query after retrieving the most similar cases using CBR Retrieval tab in protégé. Moreover, the Queries tab in protégé can be used to build queries. We have measured the retrieval accuracy of the system. The accuracy means the ability of the system to retrieve the right or the similar case. The applied preprocessing steps have increased the accuracy of the CBR system to find the most suitable case. The value of k has not determine because myCBR framework will return all the case with their similarity level ranging from 100 (i.e. exact similarity) to 0 (i.e. not similar). All tested cases are retrieved with 100% accuracy. The retrieved case may need some adaptation to generate the final solution, but this task is out of scope.

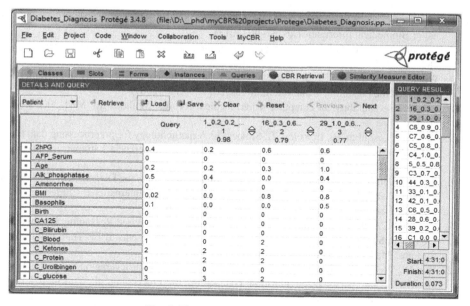

Fig. 5. The case-base query retrieval

6   Conclusion

In this paper, we have proposed a case-base preparation framework. The paper concentrated on the data preprocessing steps including missing values handling, feature selection, feature weighing, outlier detection, and normalization. Multiple machine learning algorithms have been applied to preprocess the case-base data. A CBR system has been implemented as a case study on diabetes diagnosis data set, and the results has been recorded. The preparation of EHR data enhances the accuracy of the retrieval phase of our CBR system. In our future works, we will complete the implementation of our case-base preparation framework including the fuzzification of the case base, implementation of fuzzy similarity technique and codifying of the case-base using a standardized ontology as SNOMED CT, ICD, or UMLS. Moreover, we will increase our case-base size with new cases to enhance the accuracy the CBR system decision.

References

1. Richter, M., Weber, R.: Case-Based Reasoning. Springer, Heidelberg (2013)
2. Blanco, X., Rodríguez, S., Corchado, J.M., Zato, C.: Case-Based Reasoning Applied to Medical Diagnosis and Treatment. In: Omatu, S., Neves, J., Rodriguez, J.M.C., Paz Santana, J.F., Gonzalez, S.R. (eds.) Distrib. Computing & Artificial Intelligence. AISC, vol. 217, pp. 137–146. Springer, Heidelberg (2013)
3. Andritsos, P., Jurisica, I., Glasgow, J.: Case-Based Reasoning for Biomedical Informatics and Medicine. In: Springer Handbook of Bio-/Neuroinformatics, Part C, pp. 207–221. Springer, Heidelberg (2014)
4. Abidi, S., Manickam, S.: Leveraging XML-based electronic medical records to extract experiential clinical knowledge an automated approach to generate cases for medical case-based reasoning systems. International Journal of Medical Informatics 68, 187–203 (2002)

5. Lee, D., Cornet, R., Lau, F., Keizer, N.: A survey of SNOMED CT implementations. Journal of Biomedical Informatics 46, 87–96 (2013)

6. Burnum, J.: The misinformation era: the fall of the medical record. Ann. Intern. Med. 110, 482–484 (1989)

7. Weiner, M., Embi, P.: Toward reuse of clinical data for research and quality improvement: the end of the beginning? Ann. Intern. Med. 151, 359–360 (2009)

8. Lei, J.: Use and abuse of computer-stored medical records. Methods Inf. Med. 30, 79–80 (1991)

9. Borges, K., Aquino, R., Barcelos, T., Simoes, J.: A methodology for preprocessing data for application of case based reasoning. In: 2012 XXXVIII Conferencia Latinoamericana En IEEE Informatica (CLEI), pp. 1–8 (2012)

10. Jayalskshmi, T., Santhakumaran, A.: Impact of Preprocessing for Diagnosis of Diabetes Mellitus Using Artificial Neural Networks. In: IEEE Second International Conference on Machine Learning and Computing, pp. 109–112 (2010)

11. Begum, S., Ahmed, M., Funk, P., Xiong, N., Folke, M.: Case-Based Reasoning Systems in the Health Sciences: A Survey of Recent Trends and Developments. IEEE Transactions on Systems, Man, and Cybernetics, Part C 7(1), 39–59 (2010)

12. Wu, D., Weber, R., Abramson, D.: A case-based framework for leveraging nutrigenomics knowledge and personalized nutrition counseling. In: Proceeding of Workshop CBR Health Sci.s, pp. 71–80 (2004)

13. Jagannathan, R., Petrovic, S.: Dealing with Missing Values in a Clinical Case-Based Reasoning System. In: Second IEEE International Conference on Computer Science and Information Technology (ICCSIT), pp. 120–124 (2009)

14. Floyd, M.W., Davoust, A., Esfandiari, B.: Considerations for Real-Time Spatially-Aware Case-Based Reasoning: A Case Study in Robotic Soccer Imitation. In: Althoff, K.-D., Bergmann, R., Minor, M., Hanft, A. (eds.) ECCBR 2008. LNCS (LNAI), vol. 5239, pp. 195–209. Springer, Heidelberg (2008)

15. Weiskopf, N., Weng, C.: Methods and dimensions of electronic health record data quality assessment: enabling reuse for clinical research. Journal American Medical Informatics Association 20(1), 144–151 (2013)

16. Klompas, M., Eggleston, E., McVetta, J., Lazarus, R., Li, L., Platt, R.: Automated detection and classification of type 1 versus type 2 diabetes using electronic health record data. Diabetes Care 36(4), 914–921 (2013)

17. Esfandiari, N., Babavalian, M., Moghadam, A., Tabar, V.: Knowledge discovery in medicine: Current issue and future trend. Expert Systems with Applications 41, 4434–4463 (2014)

18. Kuhn, M., Johnson, K.: Data Pre-processing. Applied Predictive Modeling, 27–59 (2013)

19. Xie, X., Lin, L., Zhong, S.: Handling missing values and unmatched features in a CBR system for hydro-generator design. Computer-Aided Design 45, 963–976 (2013)

20. Guessouma, S., Laskrib, M., Lieberc, J.: RespiDiag: A Case-Based Reasoning System for the Diagnosis of Chronic Obstructive Pulmonary Disease. Expert Systems with Applications 41(2), 267–273 (2014)

21. Piramuthu, S.: Evaluating feature selection methods for learning in data mining applications. European Journal of Operational Research 156(2), 483–494 (2004)

22. Baig, M.: Case-based reasoning – an effective paradigm for providing diagnostic support for stroke patients. Master Thesis, Queen's University, Kingston, Ontario, Canada (2008)

23. Bottrighi, A., Leonardi, G., Montani, S., Portinale, L., Terenziani, P.: Intelligent Data Interpretation and Case Base Exploration through Temporal Abstractions. In: Bichindaritz, I., Montani, S. (eds.) ICCBR 2010. LNCS (LNAI), vol. 6176, pp. 36–50. Springer, Heidelberg (2010)

24. Gopal, K.: Efficient case-based reasoning through feature weighting, and its application in protein crystallography. Ph.D. Thesis,Texas A & M University (2007)
25. Xiong, N., Funk, P.: Combined feature selection and similarity modelling in case-based reasoning using hierarchical memetic algorithm. In: IEEE Congress on Evolutionary Computation (CEC), pp. 1–6 (2010)
26. Shanga, C., Min, M., Fenga, S., Jianga, Q., Fana, J.: Feature selection via maximizing global in-formation gain for text classification. Knowledge-Based Systems 54, 298–309 (2013)
27. Huang, Y., McCullagh, P., Black, N., Harper, R.: Feature selection and classification model construction on type 2 diabetic patients' data. Artificial Intelligence in Medicine 41, 251–262 (2007)
28. Kwiatkowska, M., Atkins, S.: Case Representation and Retrieval in the Diagnosis and Treatment of Obstructive Sleep Apnea: A Semio-fuzzy Approach. In: Proceedings of 7th European Case Based Reasoning Conference (ECCBR), pp. 25–35 (2004)
29. Balakrishnan, V., Shakouri, M., Hoodeh, H.: Integrating association rules and case-based reason-ing to predict retinopathy. Maejo International Journal of Science and Technology 6(03), 334–343 (2012)
30. Kotsiantis, S., Kanellopoulos, D., Pintelas, P.: Data Preprocessing for Supervised Leaning. International Journal of Computer Science 1(2), 111–117 (2006)
31. Han, J., Rodriguze, J., Beheshti, M.: Diabetes Data Analysis and Prediction Model Discovery Using RapidMiner. In: 2008 Second International Conference on Future Generation Communication and Networking, pp. 96–99 (2008)
32. Pla, A., López, B., Gay, P., Carles, C.: eXiT*CBR.v2: Distributed case-based reasoning tool for medical prognosis. Decision Support Systems 54(3), 1499–1510 (2013)
33. Rea, S., Pathak, J., et al.: Building a robust, scalable and standards-driven infrastructure for secondary use of EHR data: The SHARPn project. Journal of Biomedical Informatics 45, 763–771 (2012)
34. McSherry, D.: Precision and Recall in Interactive Case-Based Reasoning. In: Aha, D.W., Watson, I. (eds.) ICCBR 2001. LNCS (LNAI), vol. 2080, pp. 392–406. Springer, Heidelberg (2001)
35. Wang, X., Dong, J.: Fuzzy Based Similarity Adjustment of Case Retrieval Process in CBR System for BOF Oxygen Volume Control. In: Sixth International Conference on Advanced Computational Intelligence, Hangzhou, China, pp. 19–21 (2013)
36. The myCBR3 Project, http://www.mycbr-project.net/download.html (last accessed on May 11, 2014)
37. RapidMiner, http://rapidminer.com/ (last accessed on May 20, 2014)
38. HL7 Version 3: Reference Information Model (RIM), http://www.hl7.org (last accessed on May 11, 2014)
39. Molina, L.C., Belanche, L., Nebot, A.: Feature selection algorithms: a survey and experimental evaluation. In: IEEE International Conference on Data Mining ICDM, pp. 306–313 (2002)
40. Kar, D., Chakraborti, S., Ravindran, B.: Feature Weighting and Confidence Based Prediction for Case Based Reasoning Systems. In: Agudo, B.D., Watson, I. (eds.) ICCBR 2012. LNCS, vol. 7466, pp. 211–225. Springer, Heidelberg (2012)
41. Jagannathan, R., Petrovic, S.: Dealing with missing values in a clinical case-based reasoning system. In: Second IEEE International Conference on Computer Science and Information Technology (ICCSIT), pp. 120–124 (2009)
42. Michael, F.: Microvascular and Macrovascular Complications of Diabetes. American Diabetes Association, Clinical Diabetes 26(2), 77–82 (2008)
43. Hassanien, A.E., Abdelhafez, M.E., Own, H.S.: Rough sets data analysis in knowledge discovery: a case of kuwaiti diabetic children patients. Advances in Fuzzy Systemsol. 8, 2 (2008)

Solutions for Time Estimation of Tactile 3D Models Creation Process

František Hrozek, Štefan Korečko, and Branislav Sobota

Department of Computers and Informatics,
Faculty of Electrical Engineering and Informatics,
Technical University of Košice,
Letná 9, 042 00 Košice, Slovakia
{frantisek.hrozek,stefan.korecko,branislav.sobota}@tuke.sk

Abstract. Blind people see world through touch. However, objects are sometimes too big or too small to be observed in this way. Tactile scale 3D models are a solution to this problem. In this paper we deal with the process of creation of such models and some problems related to it. We present two solutions that assist in planning the process and estimating the time needed for its realization. The first one is a set of timed Coloured Petri net simulation models for an estimation of a total time of the process. The second one is a software tool for planning a 3D scanning job of an exterior area, which also computes the time needed for an actual scanning. An example of a tactile 3D model, created by the authors, is presented, too.

Keywords: visually impaired people, tactile 3D models, virtual reality technologies, time estimation.

1 Introduction

Visually impaired people do not see, but through other senses they perceive information about the world around them. There are many aids, which help them substitute their vision. According to Hersh and Johnson [4], these aids can be divided to: haptic low-tech aids (e.g. the long cane, the guide dog or braille), matrices of point stimuli (e.g. aids for reading text and pictures), computer based aids for graphic information (e.g. aids for graphic user interface or tactile computer mouses) and haptic displays. But which aid can be used to transfer information about objects, whose size prevents them to be observable by touch?. Tactile scale 3D models present a suitable solution and there is a lot of ongoing research and development work in this area. Celani et al. use tactile models for teaching architecture. In [2] they present four models of buildings, designed by Oscar Niemeyer, which they modeled in CAD software and 3D printed in ZPrinter 310 Plus. Models were 3D printed as solids, because hollow models could be easily broken with hands [2].

Tactile 3D models can be also used to help with the spatial orientation of blind people in large areas such as a university campus [11]. However, creation

A.E. Hassanien et al. (Eds.): AMLTA 2014, CCIS 488, pp. 498–505, 2014.
© Springer International Publishing Switzerland 2014

of such complex models using 3D modeling or CAD applications can be very time consuming. Therefore in [8] Moustakas et al. present a method, where 3D models are created from video, deriving structure from the motion. The method offers an automated way to generate models of these areas.

Other examples of objects *too big to be observed by touch* are architectural elements, statues and archaeological findings. 3D modeling is not the only way to create their virtual representations, 3D scanners can be used instead. For example, in [10] Tucci and Bonora scanned these kinds of objects for various purposes, including creation of tactile models for museums. The objects can be sometimes too small to be observed by touch and printed 3D models can solve this problem. In the work of Teshima et al. [9], 3D models of Radiolaria and Foraminifera (plankton) were created using X-ray computer tomography, enlarged and 3D printed.

The works mentioned focused on the process of tactile 3D models creation and printing, but didn't deal with some significant issues related to the process. One of the most interesting questions is how to estimate the time needed to create a model and, eventually, the price of its creation. To be able to estimate the price is very important with respect to the use of the models for the benefit of visually impaired people, because organizations, which support them often operate with limited funding.

In this paper two solutions for the time estimation, developed at the home institution of the authors, are presented. The first one is a set of configurable timed Coloured Petri nets models, which allows to estimate the time of the whole process for a large number of objects by means of simulation based analysis. The second solution focus on a specific subtask of the process, namely on 3D scanning. It is a software tool where scanning time is determined on the basis of a virtual scene consisting of a building to be scanned and scanners to be used.

2   3D Model Creation Process

Before a 3D model of an object can be printed, its virtual representation has to be created. The first step of the creation process is data gathering and analysis (preparation phase). When the data are prepared a 3D model creation begins (modeling phase), followed by checking for errors (verification phase).

A 3D model of an object can be created, in the modeling phase, using 3D modeling applications (e.g. SketchUp [16]), a 3D scanning (e.g. by Leica ScanStation 2 [15]) or their combination [3]. Which of these modeling methods will be used has to be decided at the beginning of the preparation phase according to the availability of required software and hardware, intended use of the model to be created (e.g. simulation, visualization or 3D printing) and also to the object properties (e.g. size, position or its existence in the real world). Chosen modeling method also affects the next step in the preparation phase - data gathering. For example, if the model will be created from scratch in a 3D modeling application, detailed measurements are necessary and data for textures have to be obtained. On the other hand, a 3D scanner is capable to obtain most of these data automatically. However, positions from which the object will be scanned and 3D scanning parameters need to be determined. When all the necessary data are gathered, the analysis step begins. Here the data are checked if they are adequate for the 3D model creation (e.g quality of pictures for textures is adequate or there is enough scanning position to scan the whole surface of the object). After successful analysis, creation of a 3D model (the modeling phase) can begin utilizing the chosen modeling method and gathered data. The created model is subsequently verified in the verification phase, where a modeler checks whether the model satisfies requirements for its further use. For example, if the model is about to be 3D printed, it is needed to check that it is "water-tight", i.e. there are no holes in its exterior surface.

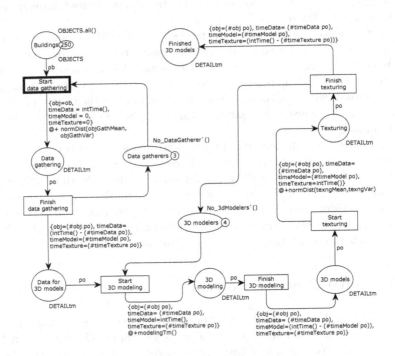

Fig. 1. A timed CPN simulation model of the creation process

3 Simulation Models for Time Estimation

To estimate a time needed for creation of a large number of relatively complex 3D models, such as buildings, is not an easy task. This is because of a number of parameters that need to be taken into consideration. The most important ones are related to diversity of individual buildings (e.g. size or number of details). Other ones are associated with modelers – their skills (e.g. modeling or texturing speed), their quantity (i.e. how many modelers will be modeling) and distribution of tasks among them (i.e. do we have individual modelers for each modeling phase or one modeler is doing everything). The last set of parameters reflects limitations given by VR system utilized (e.g. maximum of polygons per model or maximum texture resolution). In many cases these parameters can be characterized by existing random distributions, so 3D model creation processes can be modeled as stochastic discrete event systems and simulation-based performance analysis can be used for the time estimation.

There are several languages available for modeling of discrete event systems and one of them is (timed) Coloured Petri nets (CPN) [5], [6]. CPN allows to create models, which can be both relatively easy to understand (thanks to a graphical representation) and comprehensive (thanks to the use of a functional language for data representation and manipulation). In addition, there is an editor, simulator and analyzer of CPN, called CPN Tools [14], which provides facilities for data extraction during simulations. Because of these properties, we used CPN to create a set of simulation models that allow to estimate the 3D models creation time. The simplest of them is shown in Fig. 1.

As it is common in CPN, the simulation model (Fig. 1) has a form of bipartite digraph, where round vertices are places and rectangular ones are transitions. Places hold values, called tokens and these tokens define a state or marking of the net. Each place can hold only one type of tokens. Marking of the net can be changed by firing of its transitions. A firing of a transition t can occur when there are enough tokens in its pre-places (i.e. places from which there is an arc to t). The firing removes tokens from pre-places of t and adds tokens to its post-places (i.e. places to which there is an arc from t). Number and values of removed and added tokens are defined by expressions written next to arcs. Detailed information about CPN behavior can be found in [5] or at [14].

The simulation model in Fig. 1 represents a process where we have to create 250 models of buildings (250 tokens in the place Buildings) using three employees for data gathering (3 tokens in Data gatherers) and four employees for 3D modeling and texturing (4 tokens in 3D modelers). Creation of each 3D model is divided into three steps. In the first one, which corresponds to the preparation phase as defined in section 2, data necessary for 3D model creation are collected. This step starts by a firing of Start data gathering and ends by a firing of Finish data gathering. The next step is a creation of the 3D model (transitions Start 3D modeling and Finish 3D modeling) and the final step is its texturing (Start texturing and Finish texturing). These two steps correspond to the modeling phase. The verification phase is not covered by this model but can be added in the form similar to the previous ones. Times needed to accomplish the steps are drawn from normal distributions of chosen mean and variance. For example, the data gathering time is computed by the expression normDist(objGathMean,objGathVar), where normDist is a function that returns a random number, objGathMean is a constant that defines mean and objGathVar define variance of the gathering time. The normal distribution and values of its parameters have been chosen on the basis of experience of the authors with 3D modeling using contemporary software and hardware, including the pieces mentioned in the previous section. The time in the simulation model is measured in seconds and data about total time of 3D model creation and time of its steps are collected during simulation.

4   3D Scanning Time Estimation Tool

The simulation models, as the one presented in the previous section, are able to provide a rough estimate of costs of planned 3D model creation. This is useful when a decision to create the model or not is to be made. However, after the decision is made (and is positive), more precise estimates are necessary to plan actual realization of the model creation. As we mentioned above, one of the methods of 3D model creation is 3D scanning. In the case of large objects, namely buildings, the critical parameters to estimate a 3D scanning time are a ground plan and segmentation of the building, number and technical parameters of scanner(s) to be used and how the surroundings of the building limit possible scanning positions.

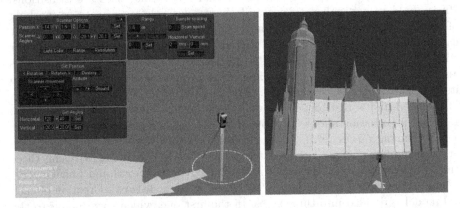

Fig. 2. Interface for parameters setting (left) and preview of a scanned area (right)

To assist in planning a 3D scanning job we developed a software tool (Fig. 2), which computes the time estimate on the basis of a virtual representation of a scanning site. The virtual representation has to be prepared by a user of the tool and consists of a 3D model of the building and scanners positioned around it. For the purpose of the estimation a simplified 3D model is enough. It just needs to contain the basic shape and features of the building (e.g. windows) and overhanging parts (e.g. balconies) that can overlap other components when looking at the building from some angle. Still, one may find the task to create the simplified model too laborious. But usually only historically or architectonically significant buildings are scanned and 3D models of many of them already exist and are available for free. If not, the simplified model can be quickly created in an easy to learn and use 3D modeling application, such as SketchUp. And in many cases the data needed to create it may be obtained from the Internet for free: a ground plan for the building and its surroundings can be created from satellite maps, its dimensions can be found in online encyclopedias and its shape and features can be derived from images, available via services like Google Street View. Thanks to this the tool allows a detailed planning of a scanning job without a preliminary visit of the scanning site. For precise planning the virtual representation should also include buildings, trees and other obstacles in the neighborhood of the building. As they will affect only positions of scanners, they can have a form of primitive box models.

After the building and its surroundings are prepared the user can add 3D scanners into the virtual representation, choose their position and set scanning parameters for individual positions. For each scanner the tool highlights the area to be scanned by selected color as it can be seen in Fig. 2. In Fig. 2 the highlighting colors are green and red (or darker and lighter shades of gray in the printed version). The tool also indicates overlapping areas of individual scanners by mixing the colors selected for them. This is very important, because adjacent scanned areas have to overlap in order to merge them into a single 3D model (point cloud).

When the user is satisfied with the settings and position of 3D scanners (i.e. all important parts of the building are highlighted), he sets up a scanning resolution and scanning speed for individual scanners. After this, the total time of 3D scanning is estimated by the tool. If the time is bigger than expected, the user can change scanning resolution or the whole layout of the scanning site. The layout can then be used to set up the actual scanning job.

5   Related Work and Discussion

Our solutions can be seen as an adjunct to the above mentioned works as they deal with time and price estimation of the 3D model creation process, which is not covered by [2], [11], [8],[10] and [9]. There are also other works, which deal in detail with 3D modeling pipeline (e.g. [1], [7]). A lot of research effort is focused on process analysis using languages as CPN [5], [6] or BPMN [13]. However, the works about the modeling pipeline describes processes, but don't deal with the time estimation. On the other hand, the research on process analysis focus on the time estimation, but not on 3D modeling. Our results try to interconnect both approaches, so 3D modelers can easily estimate the 3D modeling time and price. Of course, the solutions presented are not without drawbacks. The random distributions and parameters of the simulation models have been selected on the basis of the practical experience of the authors and can benefit from inclusion of findings of others. An online survey can be used to collect these findings. The current version of the second solution only supports the Leica ScanStation 2 scanner and the scanning speed has to be entered manually. In the future we would like to support other scanners, too and, if possible, implement an algorithm to compute the scanning speed from the basic parameters of the scanner.

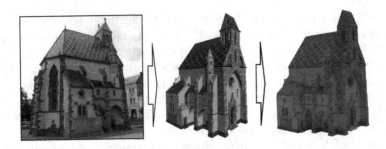

Fig. 3. Creation process of a tactile 3D model of St. Michael Chaple in Košice: real building (left), virtual 3D model created by SketchUp (middle), tactile 3D model printed by ZPrinter 450 (right)

Both solutions have been successfully used at the home institution of the authors. One of the created tactile models, printed at 3D Systems ZPrinter 450 [12], can be seen in Fig. 3. The model is suitable for an extensive use by blind persons as it doesn't include any small detail that could be easily broken by manipulation. One of the features we would like to add to the tactile models is a certain level of interactivity. Namely, we would like to implement an embedded system that will play informational recordings when certain parts of the model are touched.

6   Conclusions

In this paper we presented two solutions that can assist in tactile 3D model creation for visually impaired people. The solutions focus on planning a creation of models of buildings. Buildings are a typical example of objects that cannot be satisfactory observed by blind people in their real form, so their scaled 3D models can help significantly. The first solution, the set of timed CPN simulation models, can be used for a rough estimation of a total time of 3D model creation. It can be utilized to determine overall feasibility of the 3D model creation job and amount of resources needed for it. The simulation models are fully configurable, so they can be adjusted to specific parameters of given 3D modeling jobs. The second one is a software tool for planning a 3D scanning of a building or an arbitrary exterior area. Thanks to the provided functionality the tool not only estimates the time needed for a 3D scanning job, but also helps to prepare a detailed plan of the corresponding scanning site. It highlights areas covered by the scanners, so the user can easily see problematic places and reposition the scanners.

Acknowledgments. Paper is the result of the Project implementation: University Science Park TECHNICOM for Innovation Applications Supported by Knowledge Technology, ITMS: 26220220182, supported by the Research & Development Operational Programme funded by the ERDF.

References

1. Bernardini, F., Rushmeier, H.: The 3D model acquisition pipeline. Computer Graphics Forum 21(2), 149–172 (2002)
2. Celani, G., Zattera, V., de Oliveira, M.F., da Silva, J.V.L.: Seeing with the Hands: Teaching Architecture for the Visually-Impaired with Digitally-Fabricated Scale Models. In: Zhang, J., Sun, C. (eds.) CAAD Futures 2013. CCIS, vol. 369, pp. 159–166. Springer, Heidelberg (2013)
3. Hrozek, F.: 3D Interfaces of Systems. Information Sciences and Technologies Bulletin of the ACM Slovakia 5(2), 17–24 (2013)
4. Hersh, M., Johnson, M.A.: Assistive Technology for Visually Impaired and Blind People. Springer, Heidelberg (2010)
5. Jensen, K., Kristensen, L.M.: Coloured Petri Nets: Modelling and Validation of Concurrent Systems. Springer, Heidelberg (2009)
6. Jensen, K., Kristensen, L.M., Wells, L.: Coloured Petri Nets and CPN Tools for Modelling and Validation of Concurrent Systems. International Journal on Software Tools for Technology Transfer 9(3-4), 213–254 (2007)
7. Liu, J., Guo-hua, G.: Research on 3D reality-based modeling and virtual exhibition for cultural sites Taking the Small Wild Goose Pagoda in Tang-Dynasty as the case. In: Proc. of 2010International Conference on Computer Application and System Modeling (ICCASM), vol. 15, pp. V15-307–V15-313 (2010)
8. Moustakas, K., Nikolakis, G., Kostopoulos, K., Tzovaras, D., Strintzis, M.G.: Haptic Rendering of Visual Data for the Visually Impaired. IEEE MultiMedia 14(1), 62–72 (2007)
9. Teshima, Y., Matsuoka, A., Fujiyoshi, M., Ikegami, Y., Kaneko, T., Oouchi, S., Watanabe, Y., Yamazawa, K.: Enlarged Skeleton Models of Plankton for Tactile Teaching. In: Miesenberger, K., Klaus, J., Zagler, W., Karshmer, A. (eds.) ICCHP 2010, Part II. LNCS, vol. 6180, pp. 523–526. Springer, Heidelberg (2010)
10. Tucci, G., Bonora, V.: From Real to "Real". A Review of Geomatic and Rapid Prototyping Techniques for Solid Modeling in Cultural Heritage Field. International Archives of the Photogrammetry, Remote Sensing and Spatial Information Sciences 38 (2011)
11. Voigt, A., Martens, B.: Development of 3D Tactile Models for the Partially Sighted to Facilitate Spatial Orientation. In: Proc. 24th eCAADe Conference (Communicating Space(s)), Volos, pp. 366–370 (2006)
12. 3D Systems ZPrinter 450 homepage, http://www.zcorp.com/en/Z-Corp/450-Video-and-Snapshot-Kit-Deliverables/spage.aspx
13. Business Process Model and Notation homepage, http://www.bpmn.org/
14. CPN Tools homepage, http://cpntools.org/
15. Leica ScanStation2 homepage, http://hds.leica-geosystems.com/en/Leica-ScanStation-2_62189.htm
16. SketchUp homepage, http://www.sketchup.com/

Part VIII
Social Networks and Big Data Sets

An Observational Study to Identify the Role of Online Communication in Offline Social Networks

Mona A.S. Ali[1,3] and Aboul Ella Hassanien[2,3]

[1] Faculty of Computers and Information, Minia University, Egypt
[2] Faculty of Computers and Information, Cairo University, Egypt
[3] Scientific Research Group in Egypt (SRGE)
Mona.sedeek@mu.edu.eg
http://www.egyptscience.net

Abstract. Social networks are frequently categorized as online (supported electronically without necessitating participants meeting) or offline (defined though physical interactions). Often these social networks are treated as mutually exclusive but in many situations they coexist across the same community. In these circumstances online social networks provide communication to enhance information flow and support different types of relationship. In this paper we undertake an observational study of a social network of 98 undergraduate students to determine the role of on-line communication in offline social networks. We exploit social network analysis to examine the structure of the underlying communication. We examine the role of electronic social networks (web-based, email, telephone and SMS) to establish how different methods of communication and communication frequency support different types of relationship. Interesting trends emerge on how different technologies are used. The results reaffirm the importance of on-line social networks in facilitating structurally important weak links and reinforcing strong links.

1 Introduction

Social networks have become increasingly important for understanding and explaining complex behaviour within groups and communities. The importance of social and technological networks is now widely acknowledged [5]. The Internet and widespread availability of electronic communication have added much greater opportunity for social relationships [6] to be established and maintained. As such the electronic social networking tools have emerged as a popular and effective mechanism for communication. Popular approaches to such communication now include Facebook, Twitter, SMS as well as email and mobile telephone. It is often the case that online and offline social networks are studied independently. However in many cases an online social communication coexists with an offline social network across the same community. In these circumstances it is possible to study the effect that online communication has on the offline social network.

Electronic communication makes it much easier to connect people who may not have a strong offline relationship. This makes online communication technology a powerful phenomenon with possibly significant changes to information flow in a social network.

A.E. Hassanien et al. (Eds.): AMLTA 2014, CCIS 488, pp. 509–522, 2014.

We undertake an observational study of a social network of 98 undergraduate students over which online and offline social networks coexist. We seek to determine the role of online communication in the offline social network, in particular the role that online communication tools plays in augmenting face-to-face interactions. Through a participant questionnaire we are able to determine the communication structures and the patterns of communication supporting the underlying relationships. These are examined using extensive social network analysis.

Currently relatively few studies examine both online and offline social communication structures over a single community. The *hybrid* online-offline social network has been defined as a social network in which social links are maintained using both online and offline methods of communication [3]. Some of the few studies tackling the hybrid network include [1] where the authors consider the embedding of social networks in different technologies. Further work from the psychology viewpoint [10] establishes how networks of "friends" of young adults relate to their online social networks. These studies examine the social network structure using social network analysis approaches including an ego-centric viewpoint. Haythornthwaite et al [4] studies the impact of social media on existing relationships and the influence on strength of relationship between parties.

A focus of a number of studies concerning hybrid social networks concerns the issues of trust and identity. This affects the sharing of information and in some cases it is claimed that a mixture of virtual and physical social networks may overcome these difficulties. For purposes of knowledge sharing, [7] [3] investigate communities where physical interactions are extended in to the virtual world. Subrahmanyam et al [10] attempt to answer the question as to whether having physical interactions combined with virtual interactions reduces problems concerning trust and online knowledge sharing. In a reverse study Xie [11] found that conversely, as well as enabling the creation of online social relationships, the Internet can affect offline relationship formation.

Social networks provide a structure for information flow. Consequently much of the literature in this field is relevant and we highlight some of the most interesting ways in which social network analysis has been adopted and exploited. This paper presents a comprehensive survey on hybrid online and offline social networks and the effect of online social network on the connectivity of offline social network.

The rest of this paper is organized as follows. Section 2 discusses our contributions by exploring issues of a students social network online and offline and presents the effect of online social network on the offline one. Section 3 addresses the proposed study to understand the role of online communication on offline social networks. Section 4 several issues that arise from the founded results. Finally, we summarize and conclude the outlined issues and the future research directions are presented in Section 5.

2   Exploring Issues of Students Social Network

In this paper we explore the role of communication technologies in a social network of students in higher education. This is a well-defined and self-contained social network with hybrid online and offline properties. We explore the role of the online communication on the social network and we are seeking to address a number of issues, namely:

(1) communication technologies and network structure . In particular we are able to determine the dependency on and role of different communication structures (Section 4.1); (2) relationship strength and network structure. In particular we determine the role of weak links and the modes of communication relative to relationship strength (Section 4.2); (3) frequency of interaction and network structure. In particular we able to determine the critical structures that are facilitating most of the communication traffic and the frequency with which different technologies are used. We also assess the frequency of interaction across different strengths of relationship (Section 4.3); (4) community detection and clustering. In particular we assess the extent to which clustering occurs within the network and how this occurs (Section 4.4).

The results give a measure of the potential *communication gain* that participants have over communication a solely offline network. Our analysis also reveals the clustering characteristics that are due to online social networks and the role of key players within the hybrid network for global information spreading. From these questions we are able to significantly increase our understanding of the interplay between online communication and offline social networks and the augmentation effect of online communication technologies.

3   Proposal and Experimental Study

In order to measure the potential of communication gain, a questionnaire was designed to identify the relationships maintained with others for each participant. Then we investigated how these relationships were maintained. Specifically considered was the intensity of the relationship (relationship strength in three categories), the frequency of interaction and and the mode of interaction (offline, online and by which communication mode). Relationship strength was categorized as a *strong friendship* (someone with whom you have significant level of trust and interaction), *friendship* (someone with whom you have empathy or common views with and may socialize with them) or *course-mate* (someone you know and would acknowledge but with whom you have little other contact).

Frequency of interaction was categorized as the most frequent option from daily, at least a few times per week, at least every month, or at least once per semester. Communication was categorized as either face-to-face, mobile phone text messaging, telephone, email, micro-blogging, chatting on the Internet (VoIP), or Facebook, which was a-priori known to be the dominant social networking service used by this group. Thus each possible relationship could be maintained in $3 \times 4 \times 7 = 78$ different combinations of relationship intensity, frequency of interaction and mode of interaction. The maximum reach by any participant was 18 incidences.

Each incidence of mode of interaction, relationship strength and frequency of interaction represents an edge (a,b) in a directed graph from the participant a who has completed the questionnaire to b with whom she communicates. Similarly other graphs can be created by considering a subset of the experimental variables (mode of interaction, relationship strength and frequency of interaction). We use this approach to create a range of graphs that allow us to explore the relationship between the experimental variables.

This study has been conducted using a cohort of mainly 18-19 year old undergraduate students in the first year (freshman) of bachelor degree programmes in Computer Science and Information Systems. This is a well-defined community that exists as an offline social network. A feature of this age group is their disposition toward online interaction which makes it easy to observe the impact of online communication. An online questionnaire was distributed to 137 subjects with a response rate of 76% (104 students). Of these 6 responses were incomplete and were discounted from analysis, leaving a sample of 98 subjects, with a 15%-85% female-male gender balance.

The questionnaire was designed to establish the characteristics of the social network from each participant's personal view point (i.e., an ego-centric perspective). This required participants to express their perceived communication activity with other subjects. This approach allows us to ask questions about a variety of online communication technologies but a disadvantage is the potential for mis-perception of a subject's own activity (in certain circumstances this has been measurable by considering the different perceptions of two parties engaged in a relationship). Anonymity has been preserved by recoding personal identifiers in the data prior to analysis.

4   Experimental Results and Discussion

Firstly we consider the social network structure. In Section 4.1 we look at the social network's dependency on different types of communication (electronic or physical). We are able to determine the marginal effects of additional online communication for particular types of technology. In Section 4.2 we consider the effect of relationship strength on network structure. In Section 4.2 characteristics of networks formed from different intensity of relationship is assessed. Different structures support different frequencies of interaction. In Section 4.3 we consider the way in which different technologies are used with different frequency. We also consider the relationship strength and frequency of interaction in Section 4.3. In Section 4.4 we determine the how social network is affected in terms of clustering by the combination of the physical and social networks. Finally in Section 4.5 we determine the characteristics of the social network from the perspective of key player analysis with particular interest in the effect on key players in the virtual network.

4.1   Communication Technologies and Network Structure

We consider the different modes of communication through a directed network $G_C = (V, E)$ where V is the set of participants and an edge $(i, j) \in E$ exists if and only if node i communicates with node j via at least one mode of communication, where $C \subseteq \{c_1, \ldots, c_5\}$, the set of possible modes of communication. While keeping the population V fixed, we vary the set C and examine the effect on shortest path lengths between all pairs of nodes. We label the communication modes in C in order of popularity, the greatest first, where popularity is measured by the number of participants who indicate they use a particular mode of communication. This is displayed in Table 1. We examine the effect of additional types of communication, starting with the most popular and progressively observing the effect of additional (less popular) modes of communication. This results in a progressively more dense network structure. We examine the

Table 1. Popularity of Different Modes of Communication

Identifier	Communication Mode	% Participants
c_1	Face to Face	98.2%
c_2	Facebook	42.6%
c_3	SMS	28.7%
c_4	email	25.2%
c_5	Phone call	14.7%

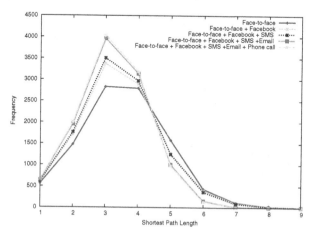

Fig. 1. The distribution of shortest path lengths for different modes of communication

effect of additional connectivity by considering the profile of all shortest path lengths from within the network. Table 2 shows the mean path progressively decreasing. From Figure 1 we can see that the mode shortest path length of three dominates for all combinations of communication - however the addition of electronic communication significantly strengthens this. In particular Facebook and email have the biggest individual effect.However from Table 1 we can see that while Facebook is well adopted (42.6%), email is a much less popular choice for communication (in terms of number of adopters) than all other forms of communication with the exception of the phone call.

From Table 2 the addition of email communication to the network only marginally increases participant inclusion, with the percentage of participants having no path between them decreasing from 19.4% to 17.0%. The small number of participants who do use email (25.2%) make key connections across the network and they provide an important effect - Table 2 shows the largest decrease in the mean shortest path length (from 3.37 to 3.22) as compared to when other technologies were added. This is due to a small number of email users who have a relatively high out-degree. In this regard email communication has a powerful structural effect.

Table 2. Statistics on Average Path Lengths in Graph G_C

Communication set C	Mean Shortest Path Length	Standard Deviation on Path Length	% with no path
$\{c_1\}$	3.53	1.287	25.3%
$\{c_1,c_2\}$	3.4	1.239	21.1%
$\{c_1,c_2,c_3\}$	3.37	1.20	19.4%
$\{c_1,c_2,c_3,c_4\}$	3.22	1.08	17.0%
$\{c_1,c_2,c_3,c_4,c_5\}$	3.2	1.06	16.0%

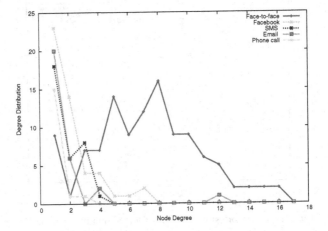

Fig. 2. Degree Distribution for each communication topology

A further observation concerns the degree distribution for each of the communication methodologies (Figure 2). The face-to-face network's degree distribution is approximately Gaussian as compared to each online network's degree distribution which are closer to a power law distribution and indicates that online networks are approximately scale free. We consider that the face-to-face network is is much more imprecisely defined and there is likely to be spurious edges included particularly from participants subjectively making a decision on whether someone is a course-mate.

4.2 Relationship Strength and Network Structure

We examine different levels of relationship strength and assess their effect on the shortest path lengths. To do this we consider a directed network $G_R = (V, E)$, where set V is the set of participants and there exists an edge $(i, j) \in E$ if and only if node i perceives it has a relationship with node j with strength from set $R \subseteq \{R_1, R_2, R_3\}$. Here R_1 is the strongest relationship strength, denoted as "strong friendship" (someone with whom you have significant level of trust and interaction), R_2 is the medium level of relationship strength, denoted as "friendship" (someone with whom you have empathy or common views with and may socialize with them) and R_3 is the weakest relationship strength, denoted as "course- mate" (someone you know and would acknowledge but with whom you have little other contact) - see Table 3. We observe the effect of relationship strength by combining different relationship types together, starting with the strongest strength (which is least popular) and progressively observing the effect of adding weaker links. The effect of this additional connectivity is analysed using shortest paths.

Role of Weak Links in Network Connectivity. Figure 3 shows the different density and structure of relationship types. In isolation the strong friendships provide little overall connectivity (Figure 3a) with improvements for friendship (Figure 3b) and course-mate relationships (Figure 3c).

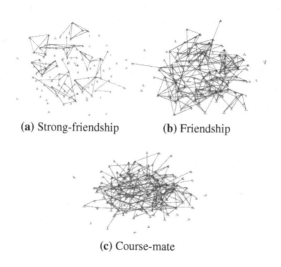

(a) Strong-friendship   (b) Friendship

(c) Course-mate

Fig. 3. Different relationships social network

Table 3 shows statistics on the different types of relationship strength. For strong friendships, 61.7% of participants have at least one strong-friendship while 86.9% of participants have at least one friendship and 92.1% of participants have at least one course-mate. Interestingly there is also often a mismatch between reciprocation of friendship - for example if node i has a strong friendship with node j, the inverse relationship is not always a strong friendship. 43.5% of bidirectional relationship are mis-matched in this way. To count the different types of links we consider a graph of undirected links. An undirected link $\{i, j\}$ is defined if and only if either the links (i, j) or (j, i) exist. Consequently an undirected link indicates the existence of some relationships (in either direction or both ways) between two nodes. The number of undirected links for strong friendship and friendship nearly equals half of the number of directed links but this is not the case for course-mate. The course-mates, as we can see in Table 3, the number of undirected links equals 205 however the number of directed links equals 240. This means that most course-mate links are uni-directional. This result reaffirms the significance of the course-mate relationship in adding more connectivity to the network rather than the other relationships (strong friendship and friendship).

Table 3. Popularity of Relationship Strength

Relationship Strength	% Included	# of Directed Links	# of Undirected Links
Strong Friendship (R_1)	61.7%	111	51
Friendship (R_2)	86.9%	273	166
Course-mate (R_3)	92.1%	240	205

The role of the weak relationships (i.e., course-mate) is notable. In Table 4 we look at the effect of combining different types of relationships, starting with the strongest relationships and progressively adding the weaker ones. The weakest relationships (i.e., course-mate) play a significant role in reducing the mean path length and improving connectivity. The addition of the course-mate relationships to friendships and strong friendships leads to a reduction in mean shortest path length from 5.55 to 3.2. The path length reported for strong friendship (1.633) is low due to the graph having a small connected component.

Table 4. Statistics on Average Path Lengths within connected components

Communication set R	Mean Shortest Path Length	Standard Deviation on Path Length	% with no path
$\{R_1\}$	1.633	0.85	98.1%
$\{R_1, R_2\}$	5.55	2.79	70%
$\{R_1, R_2, R_3\}$	3.2	1.06	16%

4.3 Frequency of Interaction in the Social Network

Different interaction frequencies between the participants result in different communication topologies. These can be characterized from a directed network using four types of frequencies to select edges, namely: daily, few times a week, once a month and once a semester. Figure 4 displays the resulting network structures For daily interaction, the participants interact via a connection of clusters - see Figure 4a. For few-times-a-week interactions, the communication between participants is more regularly distributed -see figure 4b. For the other two frequencies that represent weak-links in this context, the networks are disconnected and sparse. Arguably the daily network is most significant in terms of regular influence on the population.

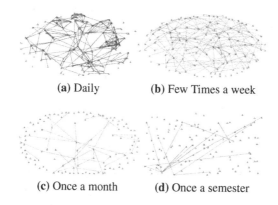

(a) Daily   **(b)** Few Times a week

(c) Once a month   **(d)** Once a semester

Fig. 4. Figures of Different communication Frequencies

Frequency of Interaction and Relationship Strength. In Figure 5 we can see a correlation between friendship strength and frequency of interaction. Strong relationships are sustained by frequent interactions with the strong friendships and friendships seeing daily interactions.

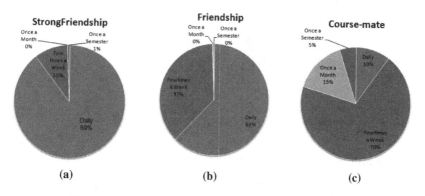

(a)   **(b)**   **(c)**

Fig. 5. The frequency of interaction and relationship strength

In contrast the weaker relationship (i.e., course-mate) are dominated by less frequent interaction albeit still quite frequent (i.e., few times a week).

Frequency of Interaction and Mode of Communication. In Figure 6 we address the types of communication and the frequency with which they are used. A number of findings are notable. Firstly highly frequent communication (e.g., daily) are sustained primarily by face to face interactions. As the frequency of communication for a relationship is reduced, face to face interactions reduce and email interactions are increasingly used, especially in the case highly infrequent interactions (i.e., once a semester) which are primarily sustained by email interactions. At the same time, it seems that when excluding highly infrequent interactions, Facebook becomes a substitute for face to face communication.

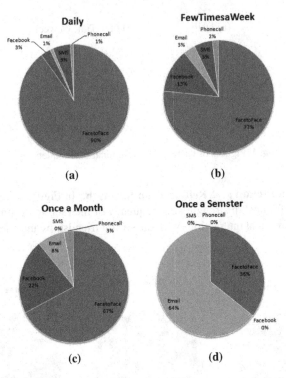

Fig. 6. Frequency of usage for different communication technologies

4.4   Community Sub-structures

An interesting feature of the hybrid online and offline social network is the extent to which different technologies support dense sub-structures and the role played by individuals in the network. To explore this we have analysed the clustering characteristics induced by different technologies. We apply the Grivan-Newman clustering method [9]. Using betweenness centrality Grivan and Newman focus on constructing a measure to indicate the edges which are least central to the cluster and they remove them. This divisive technique is repeatedly applied as described in [9]. To assess the strength of different clustering levels we use modularity as an external measure [8, 9]. Modularity compares the number of edges inside a cluster with the expected number of edges that one would find in the cluster if the network were a random network with the same number of nodes and where each node keeps its degree but edges are randomly connected. It does not provide a guide to how many clusters a network should ideally be split into but is a useful measure on the quality of a division of a network into clusters, with higher modularity measures indicating increased density within the clustering.

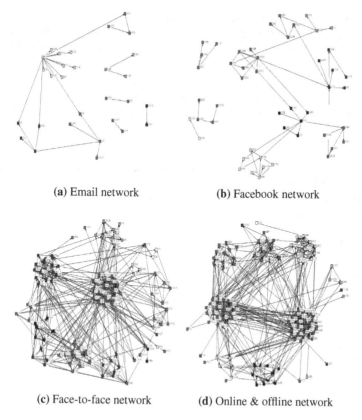

(a) Email network   (b) Facebook network

(c) Face-to-face network   (d) Online & offline network

Fig. 7. Clustered sub-networks at the highest modularity value

Interestingly the strongest clustering effect, as measured by modularity, is seen through Facebook closely followed by email. The aggregated online/offline network and face to face network exhibit similar levels of clustering strength that are significantly lower. It is likely that the strong clustering occurs for email and Facebook because of the "opt-in" nature of these technologies as compared to casual face-to-face interactions. Figures 7a, 7b, 7c and 7d show the clustered sub-networks that occur at maximum modularity.

4.5 Key Players

Key players represent a minimal subset of nodes from which all others can be reached within a particular maximum path length. Introduced in [2], we employ key player analysis to determine the trade-off between a minimum number of informed participants and extent of possible dissemination across the population within a given path length. This allows us to explore the susceptibility of a network to possible information spreading effects. We achieve this by searching for generalized dominating sets (i.e., key players) that maximize the proportion of the participants that are reachable from at least one key player within a given path length. The heuristic technique that we adopt for this search is presented in Algorithm 1 and we modify the fitness function to consider a given path length on a directed (rather than undirected) graph. This achieves local optima that forms an upper bound on the global solution. Figure 8 shows the results for the aggregated networks and modest increases in path length provide a significant decrease in the minimum number of key players required. It is likely that a longer path length is applicable to this scenario.

Algorithm 1 The key player greedy optimization algorithm

Require: *Graph* of the social network (adjacency matrix)
    Select *k* nodes at random to populate set S
    Set F = fit using appropriate key player metric
    for each node *u* in *S* and each node *v* not in *S* **do**
        DELTAF = improvement in fit if *u* and *v* were swapped
        Select pair with largest DELTAF;
        if DELTAF $<=$ **then**
            terminate
        else
            swap pair with greatest improvement in fit
            Set $F = F + DELTAF$
        end if
    end for

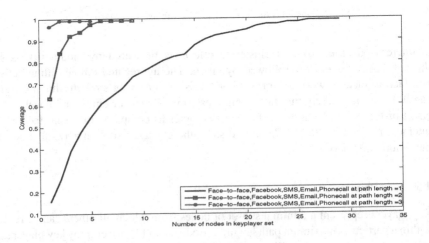

Fig. 8. Key players and coverage for offline and online communication technologies

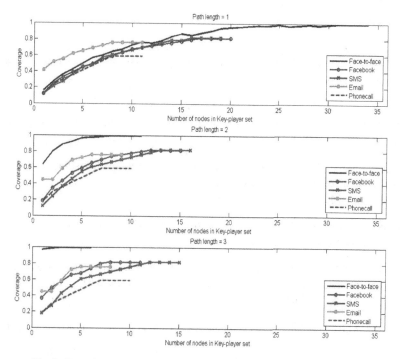

Fig. 9. Key players and path length for each communication technology

5 Conclusion

Through the obtained results we can see that facebook and email have a great effect on the participant social networks. Surprisingly, we found that although weak ties supporting the connectivity of the network structure but the participants with weak ties have a low frequency of interactions. The state-of-the-art examining both online and offline social network are limited where they study the relationship strengths and some communication technologies. In this paper we focuses on most common communication technologies and the frequency of communication with the relationship strengths. We close our discussion with what we find to be an interesting analysis. This study results can be used for many applications such as information flow by the optimum way where we extract the keplayers or the most popular participant whom able to disseminate information fast. Additionally, they can be used in identifying the similarity between the extracted social network and real life opportunistic network. There are many studies can utilize the resulted information. All in all these are some issues as a future work for our study.

Acknowledgement. The first author would like to thank prof. Roger Marcus Whitaker whose encourage and support her to enable her to develop her research. Also would like to thank Dr Matthew J. W. Morgan whose helped her a lot in filling the captured survey from students at School of Computer Science and Informatics, Cardiff University, United Kingdom.

522   M.A.S. Ali and A.E. Hassanien

References

1. Acquisti, A., Gross, R.: Imagined communities: Awareness, information sharing, and privacy on the Facebook. In: Danezis, G., Golle, P. (eds.) PET 2006. LNCS, vol. 4258, pp. 36–58. Springer, Heidelberg (2006)
2. Borgatti, S.P.: Identifying sets of key players in a social network. Computational & Mathematical Organization Theory 12(1), 21–34 (2006)
3. Gaved, M., Mulholland, P.: Grassroots initiated networked communities: A study of hybrid physical/virtual communities. In: Proceedings of the 38th Annual Hawaii International Conference on System Sciences, HICSS 2005, p. 191c (2005)
4. Haythornthwaite, C.: Social networks and internet connectivity effects. Information, Communication & Society 8(2), 125–147 (2005)
5. Kleinberg, J.: The convergence of social and technological networks. Communications of the ACM 51(11), 66–72 (2008)
6. Licoppe, C., Smoreda, Z.: Are social networks technologically embedded?: How networks are changing today with changes in communication technology. Social Networks 27(4), 317–335 (2005)
7. Matzat, U.: Reducing problems of sociability in online communities: Integrating online communication with offline interaction. American Behavioral Scientist 53(8), 1170 (2010)
8. Newman, M.E.J.: Finding community structure in networks using the eigenvectors of matrices. Physical Review E 74(3), 036104 (2006)
9. Newman, M.E.J.: Modularity and community structure in networks. Proceedings of the National Academy of Sciences 103(23), 8577 (2006)
10. Subrahmanyam, K., Reich, S.M., Waechter, N., Espinoza, G.: Online and offline social networks: Use of social networking sites by emerging adults. Journal of Applied Developmental Psychology 29(6), 420–433 (2008)
11. Xie, B.: Using the internet for offline relationship formation. Social Science Computer Review 25(3), 396 (2007)

Prioritization of Public Expenditure for a Better Return on Social Development: A Data Mining Approach

Hisham M. Abdelsalam[1,*], Abdoulrahman Al-shaar[1],
Areej M. Zaki[1], Nahla El-Sebai[2], Mohamed Saleh[1], and Miral H. khodeir[1]

[1] Faculty of Computers and Information, Cairo University
Cairo, Egypt
[2] Faculty of Economics and Political Science, Cairo University, Cairo, Egypt
h.abdelsalam@fci-cu.edu.eg

Abstract. Public expenditure affects people both directly, through subsidies and transfers, and indirectly through affecting consumption and production activities. The effects of public expenditure depend not only on its absolute values but also on both its composition and the efficiency of this spending. This paper uses data mining techniques to reach a model that maximizes social develoment through efficient allocation of public expenditure and assesses the current state of Egypt with respect to the model reached. Out of five tested models, decision tree was the one found more appropriate given this research focus and data available.

1 Introduction

Public expenditure policies, as a key component of fiscal policies, play an important role in the economy in terms of their ability to allocate resources among various economic sectors. Public expenditure plays an important role in pursuing economic growth objectives while ensuring that gains are widely distributed to promote broad-based increases in living standards. Governments' relative fiscal positions, how much they spend, and the composition of that spending are likely to make a difference in achieving these objectives [1].

Governments that want to improve their citizens' well-being can spend their financial resources in different ways. The effect of each type of expenditure differs from the other; on the one hand spending on areas such as research and development, education, and infrastructure may facilitate the achievement of economic growth in the long term but at the same time it is possible to ignore those who do not reach the fruits of growth in the short term. On the other hand, spending on health and cash transfers to the poor will meet the immediate needs of the poor but may neglect productive investments. Hence, policymakers should consider different types of government spending and the impact of each type on development, and the time range in which the yield of each type of expenditure achieved when determining the priorities

---

* Corresponding author.

A.E. Hassanien et al. (Eds.): AMLTA 2014, CCIS 488, pp. 523–530, 2014.

of this spending [2]. As such, policy recommendations regarding the impact of each type of government spending must be built depending on the circumstances of each country and must be based on applied studies [3].

Reviewing published economic research we came across many empirical studies that linked government expenditure to long term economic growth, e.g. [4][5] [6]. Literature analyzing public expenditure effects on economic development is much scarcer, and to the best of our knowledge there is no previous research that attempted to prioritize different types of government expenditure according to their effects on human development. Hence, this study is trying to fill an important gap in the available literature.

Given the inadequacy of public economics theory in providing the necessary guidance on expenditure allocation to policy-makers and development practitioners it is important to think about how a government should allocate public expenditure across various sectors to maximize prospects for achievement of its development objectives [3].

In the current research we use data mining tools for building a quantitative model that helps determining the best possible composition of public expenditure in order to maximize its benefits for all the society. The human development index (HDI) is used as an indicator for those benefits and the effect of five good governance indicators will be tested.

2   Methodology

This objective of this study is to determine the best allocation of public expenditure that would lead to higher HDI. To do so, data mining will be used to reach the model based on the use of data variables relevant to a large number of countries (all countries or years of data is available) without prior hypotheses about the nature of the relationship between them. This study follows a genetic data mining process found in literature (e.g. [7]) that consists of the following nine phases: (1) understanding the domain and goal of the application also collecting prior knowledge about the study; (2) determining a target data set, starting data gathering and selection; (3) data cleansing and preparation; (4) finding useful variables and reduction of data; (5) selecting suitable functions for data mining; (6) selecting a data mining algorithm(s); (7) data mining process, searching for useful and meaningful patterns; (8) evaluating and understanding the patterns and presenting them by an understandable way; and (9) using the discovered hidden patterns and knowledge.

The issue under investigation was initiated by our fellow economists. As such, for the first phase, they provided needed knowledge and explanations regarding the problem formulation and worked with the data mining technical team throughout the following phases schematically illustrated by Figure 1.

For the purpose of the study, needed data was collected from several data sources and, then, was aggregated into Microsoft Excel® sheets [8] in which data cleansing and preparation took place (phases 2 and 3) using Visual Basic for Applications

(VBA®) and Macros [9] to ensure the accuracy of the data and avoid any duplication. Then, Toad® software (Tool for Oracle Application Developers) [10] was used to read and extract cleansed data into a developed database and then into SAS environment. SAS® Enterprise Guide® was then used [11] to replace any missing data with dashes so that the data mining software can deal with it and converted the excel tables into SAS tables. SAS tables were then introduced into SAS® Enterprise Miner [11] for the modeling phases (6 and 7) can take place. The two final phases included reviewing the results and coming out with the suitable recommendations.

Fig. 1. Implementation Phases – Schematic Illustration

2.1   Data Gathering

Data items of this study were divided into three main categories: public expenditure allocation of various countries, governance factors of these countries, and their Human Development Index (HDI). Table 1 lists these items – referred to hereinafter as 'factors'. For detailed definition, kindly refer to [12] [13] [14][15]. These data items were gathered from various different sources to ensure that they include all available countries and also to enhance accuracy, these sources include: World Bank Data [12], Human Development Report (Data set) [13], World Governance Indicators[14], and Ministry of Finance of Egypt. Cross Sectional (all countries) and time series data (from 1990 to 2010) were compiled.

Table 1. Factors Definition

No.	ITEM (Factor)
1	Public health expenditure as % of total Public expenditure.
2	Public health expenditure as % of GDP
3	Public Expenditure on education as % of total public expenditure.
4	Public Expenditure on education as % of GDP
5	Military Public expenditure as % of Total Public Expenditure.
6	Military expenditure as % of GDP
7	Public Expenditures on research and development as % of total public expenditure.
8	Public Expenditures on research and development as % of GDP
9	Public expenditure on subsidies and other transfers.
10	A statistical index used to measure a country's overall achievement in its social and economic dimensions.

2.2   Data Cleansing and Preparation

A database was designed and constructed to include all the data gathered and to provide the base for the checking and preparation process. The database for the study was designed to include the following tables: Country; Factors; User Modification; Users; and Country Factor Facts. A star schema infrastructure was deployed to help relating tables with each other. The schema was designed in order to minimize the number of tables in the database and so ease the process on the user [16].

An excel sheet was created containing all different country-related data and VBA® and Macros® on excel were used to specify the factors and countries with primary keys to be uniquely identified; thus avoiding any duplication in the countries' names. The same steps were repeated for the all factors' names with different and unique keys.

Several checkups have been made on the data to avoid inconsistency, conflicts and missing data. Data cleansing was conducted on continuous basis which is the process of detecting and correcting (or removing) corrupted or inaccurate records from a record set, table, or database. It is used to identify incomplete, incorrect, inaccurate and irrelevant parts of the data and then replacing, modifying, or deleting this data. The inconsistencies detected or removed may have been originally caused by user entry errors; or by corruption in transmission or storage but after data cleansing all this inconsistency was removed.

The next step included using Toad® to start migrating the modified excel sheets into the database developed. Toad® is a software application from Quest Software used for developing and managing different relational databases using SQL®, as SQL® toad is used to conduct some quires on the data. It was used as a simple application to support inserting the data from excel sheets into the database rapidly and efficiently. Microsoft Excel® sheets were imported into Toad® software; then the processes of classification and insertion in the database were conducted.

For the replacement of the missing data algorithms will be used with different tested models, and SAS will automatically select the algorithm leading to the best results. The two algorithms are: Most Correlated Branch algorithm, and the Largest Branch algorithm.

3   Implementation

SAS® Enterprise Miner was deployed for the modeling part to streamline the data mining process to create highly accurate predictive and descriptive models based on large volumes of data. SAS is a recognized industry leader in business analytics software (including data mining). Given its availability for the researchers and their past experience working with it, it was selected to be used in the current study.

Figure 2 shows the model diagram. An input Data node was added including the data source in which in this input data the property of each variable will be identified.

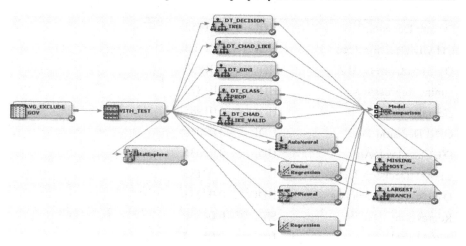

Fig. 2. Model Diagram - snapshot

A Data Partitioning node was added to enable partitioning data sets into training, test, and validation data sets. The training data set is used for preliminary model fitting, 60% of the data was considered as training data. The validation data set is used to tune and monitor the model weights during the running estimation and assures that the built model fits a real and valid data set, 20% of the data was considered as validation data. The test data set is used for model assessment and making the final comparisons to the data, 20% of the data was considered as testing data.

Five different Models were developed each representing a specific technique or algorithm in data mining, theses were: (1) Decision Tree; (2) Neural Network; (3) Auto Neural; (4) Regression; and (5) D-mine Regression. Different configurations (parameters' settings) for the Decision Tree algorithm were tested with different characteristics. The aim was to check whether there is a better representation in the characteristics of the Decision Tree algorithm; or the default characteristics representation is the best. These were: Default Decision Tree; CHAID like Decision Tree; GINI Decision

Tree; CART like Class Probability Decision Tree; and CHAID LIKE and Valid Decision Tree. Detailed configurations are provided in Appendix B.

4   Results

Different models were introduced to the Model Comparison node. The Comparison node compared all the models according to their accuracy. The results showed that the default Decision Tree Model was the most accurate and efficient model in showing and representing the data in the most meaningful way. Table 2 shows the comparison between the different models according to their selection criteria, which is based on the test average squared root error value. The most efficient model is the model with the least selection criteria value which is the Default Decision Tree.

Table 2. Selection Criteria for all Models

Model Description	Target	Test: Average Squared Error
Default Decision Tree	HDI	0.01752
CHAID Decision Tree	HDI	0.01763
DT Most Correlated Alg.	HDI	0.01857
Dmine Regression	HDI	0.01878
Neural	HDI	0.01985
GINI Decision Tree	HDI	0.02036
DT Class Prop Alg.	HDI	0.02036
CHAD Decision Tree	HDI	0.02183
DT Largest Branch Alg.	HDI	0.02231
Regression	HDI	0.04840
AutoNeural	HDI	0.05217

As an example, figure 3 will be used to represent the main points of the tree characteristics. The first node (Health Expenditure GDP) is the parent node that shows that the trained records were about 231 records and will apply the prioritized factors on them; it is also given that the average HDI is about 0.86. The tree have been split into two branches. The first branch on the left hand side introduces the records that had the Health Expenditure as % of GDP greater than or equal 5.3588 % for 221 records with 0.86 average HDI units. The second branch indicates that only 10 records have Health Expenditure as % of GDP less than 5.3588 with 0.78 average HDI. The difference in the colored boxes indicates that the dark node is most preferable than the lighter one. To conclude, this figure indicates that it is preferable for the government to spend more than or equal 5.36% of GDP on Health.

Table 3 shows the importance of each factor, the importance rating of the factors starting from the factor with the highest importance which in this case will be the Subsidi Expenditure EXP factor to the least importance factor which will be the Education GDP factor. The best path, shown in Figure 4, was identified showing the best preferable path (indicated by stars) leading to the country with the best HDI values. The figure indicates that it is preferable to spend greater than or equal 29.5654 % on the Subsidies, then spending greater than or equal 5.1234 % of GDP on Health, then spending over than or equal 1.0601% of GDP on Research and Development.

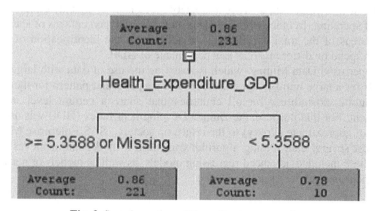

Fig. 3. Sample section of the entire tree - snapshot

Table 3. Importance of factors

ITEM (Factor)	Importance
Public expenditure on subsidies and other transfers.	1.000
Public health expenditure as % of GDP	0.758
Public Expenditures on research and development as % of GDP	0.463
Military expenditure as % of GDP	0.249
Public Expenditure on education as % of GDP	0.201

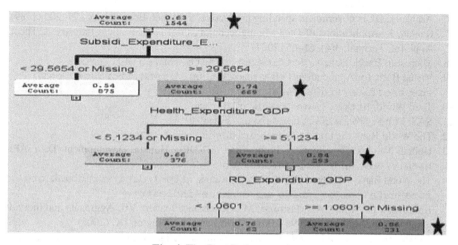

Fig. 4. The Best Path - snapshot

5   Conclusions

The main objective of the current study was to identify the best distribution of public expenditure on different areas of education, health, research and development and other, which maximizes the benefit to society (social development) under the same volume of spending. In other words, to determine the best percentages of spending on different areas of the total expenditure, in addition to the identification of the best amount to spend on different areas as a percentage of GDP.

The paper used Data Mining, which is based on the use of data with large number of records for a large number of countries to draw a particular pattern for the distribution of public expenditure for all countries that share a certain level of human development. For that purpose, the Human Development Index (HDI) was used as an indicator of approximate (Proxy) to the return on society. SAS Enterprise Miner was used to test several data mining algorithms to reach a model that achieves the best results. These included advanced regression models, as well as models of networks of artificial neurons and decision tree models.

References

1. Dewan, S., Ettlinger, M.: Comparing public spending and priorities across OECD countries. Cent. Am. Prog (2009),
   http://www.BoellOrgdownloadsewanEittinglerComparingPublicSpending.Pdf
2. Ali, A.G.A., Fan, S.: Public policy and poverty reduction in the Arab region. Arab Planning Institute (2007)
3. Paternostro, S., Rajaram, A., Tiongson, E.R.: How does the composition of public spending matter? Oxf. Dev. Stud. 35(1), 47–82 (2007)
4. Barro, R.J.: Government spending in a simple model of endogenous growth. National Bureau of Economic Research Cambridge, Mass (1991)
5. Landau, D.: Government expenditure and economic growth: a cross-country study. South. Econ. J., 783–792 (1983)
6. Aschauer, D.: Is government spending productive? J. Monet. Econ. 23(2), 177–200 (1989)
7. Ranjan, J.: Applications of Data Mining techniques in Pharmaceutical Industry. J. Theor. Appl. Inf. Technol. 3(4), 61–65 (2007)
8. Microsoft: Excel, http://office.microsoft.com/en-001/excel
9. Visual Basic for Applications (VBA) macros, http://msdn.microsoft.com/en-us/office/ff688774.aspx
10. Toad World, https://www.toadworld.com
11. SAS, http://www.sas.com/en_us/home.html
12. The World Bank, http://data.worldbank.org
13. United Nations Development Programme (UNDP): Human Development Data API, http://hdr.undp.org/en/data/ap
14. The World Bank: Worldwide Governance Indicators, http://data.worldbank.org/data-catalog/worldwide-governance-indicators
15. Kaufmann, D., Kraay, A., Mastruzzi, M.: Governance matters VII: Aggregate and individual governance indicators 1996–2007. World Bank, Washington DC (2008)
16. Giovinazzo, W.A.: Object-oriented data warehouse design: Building a star schema. Prentice Hall PTR (2000)

Visual Browsing of Large Image Databases

Gerald Schaefer

Department of Computer Science, Loughborough University, Loughborough, U.K.

Abstract. The amount of user-generated and -contributed data, online and offline, is growing at a rapid rate. This is particularly true for visual information in form of images. On the other hand, efficient and effective tools for managing these growing repositories are relatively scarce. In this paper, we present approaches that, rather than being directly retrieval-based, allow visual interactive exploration of large image collections. We introduce the underlying methods that are being employed to effectively visualise image databases as well as the browsing operations that enable interaction. We then present the Hue Sphere Image Browser, an efficient and intuitive hierarchical image browser, including its recent ports to large multi-touch screens and to mobile devices.

Keywords: image databases, content-based image retrieval, image database browsing, image database navigation, visualisation, Hue Sphere Image Browser.

1   Introduction

Visual information, in particular in form of images and videos, is becoming increasingly important. Consequently, effective tools to manage these vast media repositories are highly sought after. In this paper, we focus on image databases and on how to effectively and efficiently access these. Unfortunately, only a small minority of images are annotated [22] which in turn has significant implications for search systems as there is no textual information to base searches on. Content-based retrieval methods [33,7,29,28], which extract various image features (describing e.g. colour, texture or shape properties) as descriptors and allow retrieval of images based on a derived visual similarity, seem necessary but have only shown limited usefulness so far.

Image browsing approaches present a visual overview of a whole image collection, coupled with various browsing operators to allow for an intuitive and effective exploration of image repositories, and thus present an interesting alternative to retrieval-based approaches [10,20]. In this paper, we introduce the underlying methods that are being employed to effectively visualise image databases and the browsing operations that enable interactive exploration. We then present the Hue Sphere Image Browser, an efficient and intuitive hierarchical image browser, and also cover its recent ports to large multi-touch screens and mobile devices.

A.E. Hassanien et al. (Eds.): AMLTA 2014, CCIS 488, pp. 531–539, 2014.
© Springer International Publishing Switzerland 2014

2   Image Browsing Approaches

Common image browsing software tools display images in a one-dimensional linear format where only a limited number of thumbnail images are visible on screen at any one time, thus requiring the user to scroll back and forth through thumbnail pages. Obviously, this constitutes a time consuming, impractical and exhaustive way of searching images, especially in larger collections. Also, the order in which the pictures are displayed is often based on attributes such as filenames that do not reflect the actual image contents and hence cannot be used to speed up the search.

In recent years, various approaches have been introduced which provide a more intuitive interface to browsing and navigating through image collections [10,20]. In general, we can divide these into mapping-based, clustering-based and graph-based approaches [19]. Once a database has been visualised using one of these, it should then be possible to explore the collection in an interactive, intuitive and efficient manner [18].

The basic idea behind mapping-based techniques is similarity-based visualisation which places images which are visually similar, as established through the calculation of image similarity metrics based on features derived from image content [33], also close to each other in the visualisation space, a principle that has been shown to decrease the time it takes to localise images [23]. Various techniques for establishing this mapping have been proposed. For example, [15] uses principal component analysis (PCA) in order to visualise image collections, while [12] employs a PCA visualisation to present images in a 3D interface based on texture features.

In contrast to PCA, multi-dimensional scaling (MDS) [14] attempts to preserve the original relationships (i.e., distances) in the high-dimensional feature space, as best possible in the low-dimensional projection. MDS was employed in [25] where, based on colour signatures of images, image thumbnails are placed at the co-ordinates derived by the algorithm. All images in a database are (initially) shown simultaneously with their locations having been derived based on their visual similarity compared to all other images in the database. The user can browse the database easily from a top-down point of view in an intuitive way. One disadvantage of the MDS approach however is its computational complexity which makes interactive visualisation of a large number of images difficult. Another drawback, which MDS displays share with PCA-based visualisations, is that many images are occluded while others overlap each other, leading to a less intuitive browsing experience [23].

Apart from PCA and MDS, other dimensionality reduction techniques have also been employed, for example ISOMAP (isometric mapping), SNE (stochastic neighbour embedding), LLE (local linear embedding) as well as some combinations of these techniques [17].

Dimensionality reduction techniques applied to image database visualisation are limited by the number of images that can be displayed simultaneously on screen. This can be addressed by clustering groups of similar images together and showing representative images of clusters. Content-based clustering uses extracted image features in order to group perceptually similar images together. For example, [13] uses local colour histograms (extracted from image sub-regions) to cluster images and each cluster is visualised by a representative image.

A hierarchical tree structure for browsing image databases was suggested in [5], where images are grouped in a quadtree similarity pyramid derived through clustering. Images at the top of the pyramid are rather dissimilar but become more similar as the user navigates through the different levels of the data structure. Unfortunately, this method is not very well suited for providing an overview of the image collection as little can be inferred from the images at the top levels leaving the user unsure which branch of the tree to select for further investigation. In terms of computational complexity, even though an efficient sparse clustering method is used, the associated overheads are significant.

A hierarchical browsing strategy was also pursued in [2], however in this approach the underlying tree structure can also be interactively modified by the user. This is particularly useful, as all similarity-based browsing approaches rely on imperfect features to establish similarity [33]. Changing the tree hence provides a possibility of 'correcting' the visualisation. Nodes corresponding to groups of images or single images can be adaptively moved to a different part of the tree by the user which in turn leads to an improved navigation sytem. Other systems that cluster images in a hierarchical manner for image database browsing include [21,6,9].

Graph-based visualisations utilise links between images to construct a graph where the nodes of the graph are the images and the edges the links between similar images. [8] and [34] use a mass spring model to generate a visualisation based on associated keywords between images. In order to visualise this high-dimensional data in two dimensions, connected images are placed closer together, while unconnected images are moved further apart. However, a display of the whole graph can appear rather confusing to the user, and navigation hence not as intuitive.

Image browsing based on Pathfinder networks (the Pathfinder algorithm removes all but the shortest links by testing for triangle inequality [3]) were introduced in [4], while [11] proposed NN^k networks to browse through an image database. The basic principle here is that a directed graph is formed between every image and its nearest neighbours (based on certain features) if there exists at least one possible combination of features for which the image is the top ranked of the other. However, an overview of the complete database can only be obtained through the application of other visualisation techniques.

3   Hue Sphere Image Browser

In our Hue Sphere Image Browser [31,32,27], we arrange images by visual similarity so that similar images are located close to each other in a spherical visualisation. Large datasets are efficiently accessed through a hierarchical browsing structure, while utilisation of visualisation space is maximised through application of image spreading techniques.

G. Schaefer

Among features used for image retrieval, those describing the colour content are certainly the most popular [33]. We follow this and describe each image by its median colour. However rather than employing the standard RGB colour space we use the HSV space [26] which humans find more intuitive. Of this, we take only the hue H and value V attributes and calculate the median image colour as descriptor. The advantage of such simple features is that they greatly reduce the overall computational complexity of our approach, as on the one hand the features themselves can be calculated extremely fast, while on the other hand no computationally intensive dimensionality reduction technique is necessary. Rather, the location of each image is derived directly from the hue and value (brightness) co-ordinates. At the same time, we do not compromise the achieved image database visualisation, since, as has been shown [24], the average image colour is at least as effective a descriptor for images browsing as high-dimensional feature vectors (e.g. based on colour distributions).

For visualisation, a spherical visualisation space is employed. The derived median hue and value attributes are used directly to define longitude and latitude of the position of the image on a sphere. Since most users will be familiar with the concept of the earth globe, this provides an immediately intuitive visualisation and browsing interface as users are already aware of how to locate and find something on its surface.

The visualisation space is divided using a regular lattice. This means that images cannot overlap nor occlude each other, which in turn has been shown to lead to an improved browsing experience [23]. Each image in the database will fall into exactly one cell on the lattice, and we can hence make use of the advantages of clustering-based methods without actually having to employ a computationally expensive clustering technique, which again makes our approach decisively less demanding in terms of computational load.

Large databases are handled by employing a hierarchical approach to visualising and browsing images. If multiple images fall into a specific cell, a representative image (that closest to the centre of the cell) will be shown in the browser, while the user has the possibility to open that image cluster and hence navigate to the next level of the browsing hierarchy. Here, the colour space is again divided into (now smaller) cells and the same principles as on the root layer are employed.

Browsing operations at the user's disposal are panning, where the user can rotate the globe to focus on images of a different colour respectively tilt it to bring up darker or brighter images, and optical zooming which allows to narrow down on a specific set of images. In addition, exploiting the hierarchical database structure, users can select an image and expand the corresponding image cluster, hence delving deeper into the browsing structure and showing images that were not visible before.

While constraining images to a grid lattice prevents overlapping effects, in essence it provides a 'quantised' form of the visualisation space. Thus, it still suffers from the relatively unbalanced view that is usually generated where certain areas are not filled. To address this problem and to provide a more balanced browsing screen, local search strategies are employed which move images across grid boundaries to previously unoccupied cells. First the positions of all empty cells are retrieved and then immediate neighbouring cells inspected to move a relative percentage of their images (the images closest to the borders) to fill the previously empty cell. In the tree nodes of the cells it will commonly occur that only a few images are present, most of which will be visually fairly similar. To avoid them from being mapped to the same cell and hence to trigger another tree level, a spreading algorithm is applied which displays them on the same screen once only a certain percentage of cells are filled for a cluster.

Fig. 1 shows the initial view of our image database sphere based on the MPEG-7 common colour dataset [16]. As can be seen, a sphere as visualisation space provides an intuitive interface for navigation where it is clear to the user in which part to look for certain image classes.

In Figs. 2 and 3 we show the results of some user interactions where the user first rotated the sphere to focus on images with a different hue followed by a tilt operation to bring up darker images resulting in the view given in Fig. 2.

Fig. 3 then shows the result of a zoom operation where the user chose one of the images to bring up those photos that are contained in that selected part of the tree. To aid navigation, the previous hierarchy level is also displayed and the current position within those grid marked by the red dot.

A variant of our Hue Sphere Image browser that we have implemented can be used on a large multi-touch screen, a PQ Labs G^3 screen with a resolution of 1600x1200 pixels, for which we have defined relevant multi-touch gestures to operate the browsers [30]. Users now literally have image collections 'at their fingertips' and are able to effectively and efficiently navigate through image databases directly on the screen. For this, the browser captures and reacts to gestures for panning (2-finger drag), zooming (2-finger pinch), opening image

Fig. 1. Hue Sphere Image Browser visualisation of the MPEG-7 dataset

Fig. 2. View after rotation and tilt operations

clusters (double tap) and other browsing operations. Fig. 4 shows the multitouch application 'in action'.

Since nowadays a significant proportion of images are taken with mobile devices rather that bespoke cameras, we have also ported the Hue Sphere Browser to run on smartphones and tablets and hence allow management of the image collections contained there [1]. Fig. 5 demonstrates this mobile version running on a tablet.

Fig. 3. View after zoom operation

Fig. 4. Multi-touch Hue Sphere Image Browser

Fig. 5. Mobile Hue Sphere Image Browser on a tablet

4 Conclusions

In this paper, we have given an overview of the main approaches to image database browsing, in particular mapping-based, clustering-based and graph-based techniques and the various browsing operations. We have then presented the Hue Sphere Image Browser which organises images, based on colour, on a spherical visualisation space and makes large image collections accessible through a hierachical data structure, thus enabling intuitive and effective browsing of image databases. As was shown, this can be performed on standard computers, on large multi-touch screens, or on mobile devices.

References

1. Ahlstroem, D., Schoeffmann, K., Hudelist, M., Schaefer, G.: A user study on image browsing on small screens. In: 20th ACM Int. Conference on Multimedia (2012)
2. Bartolini, I., Ciaccia, P., Patella, M.: Adaptively browsing image databases with PIBE. Multimedia Tools and Applications 31(3), 269–286 (2006)
3. Chen, C.: Information Visualization, 2nd edn. Springer (2004)
4. Chen, C., Gagaudakis, G., Rosin, P.: Similarity-based image browsing. In: Int. Conference on Intelligent Information Processing, pp. 206–213 (2000)
5. Chen, J.Y., Bouman, C.A., Dalton, J.C.: Hierarchical browsing and search of large image databases. IEEE Trans. Image Processing 9(3), 442–455 (2000)
6. Chen, Y.-X., Butz, A.: PhotoSim: Tightly integrating image analysis into a photo browsing UI. In: Butz, A., Fisher, B., Krüger, A., Olivier, P., Christie, M. (eds.) SG 2008. LNCS, vol. 5166, pp. 224–231. Springer, Heidelberg (2008)
7. Datta, R., Joshi, D., Li, J., Wang, J.Z.: Image retrieval: Ideas, influences, and trends of the new age. ACM Computing Surveys 40(2), 1–60 (2008)
8. Dontcheva, M., Agrawala, M., Cohen, M.: Metadata visualization for image browsing. In: 18th Annual ACM Symposium on User Interface Software and Technology (2005)
9. Gomi, A., Miyazaki, R., Itoh, T., Li, J.: CAT: A hierarchical image browser using a rectangle packing technique. In: 12th Int. Conference on Information Visualization, pp. 82–87 (2008)
10. Heesch, D.: A survey of browsing models for content based image retrieval. Multimedia Tools and Applications 40(2), 261–284 (2008)
11. Heesch, D., Rüger, S.: NNk networks for content-based image retrieval. In: European Conference on Information Retrieval, pp. 253–266 (2004)
12. Keller, I., Meiers, T., Ellerbrock, T., Sikora, T.: Image browsing with PCA-assisted user-interaction. In: IEEE Workshop on Content-Based Access of Image and Video Libraries, pp. 102–108 (2001)
13. Krischnamachari, S., Abdel-Mottaleb, M.: Image browsing using hierarchical clustering. In: IEEE Symposium on Computers and Communications, pp. 301–307 (1999)
14. Kruskal, J.B., Wish, M.: Multidimensional scaling. Sage Publications (1978)
15. Moghaddam, B., Tian, Q., Lesh, N., Shen, C., Huang, T.S.: Visualization and user-modeling for browsing personal photo libraries. Int. Journal of Computer Vision 56(1-2), 109–130 (2004)
16. Moving Picture Experts Group: Description of core experiments for MPEG-7 color/texture descriptors. Tech. Rep. ISO/IEC JTC1/SC29/WG11/ N2929 (1999)
17. Nguyen, G.P., Worring, M.: Interactive access to large image collections using similarity-based visualization. Journal of Visual Languages and Computing 19(2), 203–224 (2008)
18. Plant, W., Schaefer, G.: Navigation and browsing of image databases. In: Int. Conference on Soft Computing and Pattern Recognition, pp. 750–755 (2009)
19. Plant, W., Schaefer, G.: Visualising image databases. In: IEEE Int. Workshop on Multimedia Signal Processing, pp. 1–6 (2009)

20. Plant, W., Schaefer, G.: Visualisation and browsing of image databases. In: Lin, W., Tao, D., Kacprzyk, J., Li, Z., Izquierdo, E., Wang, H. (eds.) Multimedia Analysis, Processing and Communications. SCI, vol. 346, pp. 3–57. Springer, Heidelberg (2011)
21. Platt, J., Czerwinski, M., Field, B.: PhotoTOC: automatic clustering for browsing personal photographs. Tech. rep., Microsoft Research (2002)
22. Rodden, K.: Evaluating Similarity-Based Visualisations as Interfaces for Image Browsing. Ph.D. thesis, University of Cambridge Computer Laboratory (2001)
23. Rodden, K., Basalaj, W., Sinclair, D., Wood, K.: Evaluating a visualisation of image similarity as a tool for image browsing. In: IEEE Symposium on Information Visualization, pp. 36–43 (1999)
24. Rodden, K., Basalaj, W., Sinclair, D., Wood, K.: A comparison of measures for visualising image similarity. In: The Challenge of Image Retrieval (2000)
25. Rubner, Y., Guibas, L., Tomasi, C.: The earth mover's distance, multi-dimensional scaling, and color-based image retrieval. In: Image Understanding Workshop, pp. 661–668 (1997)
26. Sangwine, J., Horne, R.E.N.: The Colour Image Processing Handbook. Chapman & Hall (1998)
27. Schaefer, G.: A next generation browsing environment for large image repositories. Multimedia Tools and Applications 47(1), 105–120 (2010)
28. Schaefer, G.: Content-based image retrieval: Advanced topics. In: Czachórski, T., Kozielski, S., Stańczyk, U. (eds.) Man-Machine Interactions 2. AISC, vol. 103, pp. 31–37. Springer, Heidelberg (2011)
29. Schaefer, G.: Content-based image retrieval: Some basics. In: Czachórski, T., Kozielski, S., Stańczyk, U. (eds.) Man-Machine Interactions 2. AISC, vol. 103, pp. 21–29. Springer, Heidelberg (2011)
30. Schaefer, G., Fox, M., Plant, W., Stuttard, M.: Exploring image databases at the tip of your fingers. Information 17(5), 1951–1960 (2014)
31. Schaefer, G., Ruszala, S.: Image database navigation: A globe-al approach. In: Bebis, G., Boyle, R., Koracin, D., Parvin, B. (eds.) ISVC 2005. LNCS, vol. 3804, pp. 279–286. Springer, Heidelberg (2005)
32. Schaefer, G., Ruszala, S.: Hierarchical image database navigation on a hue sphere. In: Bebis, G., Boyle, R., Parvin, B., Koracin, D., Remagnino, P., Nefian, A., Meenakshisundaram, G., Pascucci, V., Zara, J., Molineros, J., Theisel, H., Malzbender, T. (eds.) ISVC 2006. LNCS, vol. 4292, pp. 814–823. Springer, Heidelberg (2006)
33. Smeulders, A.W.M., Worring, M., Santini, S., Gupta, A., Jain, R.: Content-based image retrieval at the end of the early years. IEEE Trans. Pattern Analysis and Machine Intelligence 22(12), 1249–1380 (2000)
34. Worring, M., de Rooij, O., van Rijn, T.: Browsing visual collections using graphs. In: Int. Workshop on Workshop on Multimedia Information Retrieval, pp. 307–312 (2007)

Author Index